Fundamentals of Friction and Wear of Materials

Other ASM Materials Science Seminar Books

Grain-Boundary Structure and Kinetics
Fatigue and Microstructure
Interfacial Segregation
Metallic Glasses
Radiation Damage in Metals
New Trends in Materials Processing
Diffusion
The Inhomogeneity of Plastic Deformation

Fundamentals of Friction and Wear of Materials

Papers presented at the
1980 ASM Materials Science Seminar
4-5 October 1980
Pittsburgh, Pennsylvania

Sponsored by the Seminar Committee of the
Materials Science Division of the
American Society for Metals
in cooperation with
The Metallurgical Society of AIME

Edited by
David A. Rigney
Metallurgical Engineering
The Ohio State University

AMERICAN SOCIETY FOR METALS
Metals Park, Ohio 44073

Library of Congress Cataloging in Publication Data

ASM Materials Science Seminar (1980 : Pittsburgh, Pa.)
Fundamentals of friction and wear of materials.

Includes bibliographical references.
1. Materials—Congresses. 2. Friction—Congresses.
3. Mechanical wear—Congresses. I. Rigney, David A.
II. American Society for Metals. Materials Science
Division. III. Metallurgical Society of AIME. IV. Title.
TA401.2.A77 1980 621.8′9 81-10789
ISBN 0-87170-115-4 AACR2

PRINTED IN THE UNITED STATES OF AMERICA

Contents

Introduction

D. A. Rigney
Metallurgical Engineering
The Ohio State University
Columbus, Ohio 43210

Friction and wear are subjects of obvious practical significance. Therefore it is not surprising to find that man has tried for many centuries to control and understand friction and wear processes.[1] In modern times, these subjects have attracted investigators from many different scientific and technical backgrounds. However, if one surveys the contents of journals which concentrate on friction and wear, it becomes clear that metallurgists and materials scientists comprise a small minority among researchers in these fields. Many authors writing on friction and wear have been trained as mechanical or chemical engineers. They have made valuable contributions to our understanding of those aspects of friction and wear which depend on machine design, surface topography, solid mechanics, fluid dynamics, and lubrication chemistry. These topics tend to be emphasized at most conferences devoted to friction and wear.

In recent years, interest in the materials aspects of friction and wear has grown dramatically,[2] and this has been accompanied by many new ideas and observations. The 1980 ASM Materials Science Seminar on the Fundamentals of Friction and Wear of Materials was scheduled in recognition of this recent progress.

The ASM seminars have a healthy tradition of offering some tutorial value as well as presenting some of the latest advances in a given field. Therefore the conference and the book are intended to provide a convenient overview of the friction and wear of materials, but they are also intended to stimulate further progress.

Some of the speakers and authors are well known for their research on friction and wear. Others are well known for aspects of materials science which are closely related to these subjects. This mix was intentional. It assured a wide range of ideas and participation by persons from many disciplines.

The terminology used by those interested in friction and wear is very rich. However, the terms used are not always well understood or clearly defined. In some cases, a given term has quite different meanings to persons in different disciplines. An example is the word *texture*, which a metallurgist associates with the degree of preferred orientation of the crystalline grains in a material, but which others may associate with the nature of surface topography. Such differences in terminology can create real barriers to interdisciplinary cooperation. Fortunately, the American Society for Testing and Materials (ASTM) offers its Standard Terminology Relating to Corrosion and Wear, which is regularly updated as new findings lead to new and modified terminology. In recognition of the importance of clear definitions, particularly for multidisciplinary subjects like friction and wear, a few terms are selected for detailed discussion in this introductory chapter.

A good place to begin would be with the word *wear* itself. A general definition would refer to the (usually) gradual damage which may result when two materials in contact move relative to each other. This broad class of phenomena can immediately be divided into two categories. In the first, material may be displaced so that a given object is changed in shape and/or properties, but little or no material is actually lost. In the second, there is a net change of mass or volume from at least one of the interacting components, and loose wear debris may be observed. These distinctions are important. If one measures wear by measuring only a change in volume or mass, one might report that no wear has taken place, whereas damage due to material displacement may be pronounced.

An example will help to illustrate both types of wear. Figure 1 shows a portion of the exit side of a wear scar which developed on a sample block of OFHC (oxygen-free, high-conductivity) copper when a ring of 440C martensitic steel was rotated against it. The sliding speed was relatively low (5 cm/sec), and the atmosphere was dry argon. For several similar tests, the wear scars were formed by ductile smearing, resulting in long finger-like extrusions like those shown in Fig. 1. If these were still present at the end of the test, then the wear, defined in terms of mass or volume change, was very low. Therefore, the wear rate could be reported as close to zero, despite the development of an obvious wear scar and the presence of large plastic strains near the sliding interface. Throughout this book, the definition of wear based on loss of material has been emphasized, but one should remember this other category of wear damage, since it commonly precedes actual loss of material, and it may also occur independently.

Ductile copper samples like the one shown in Fig. 1 are also useful to illustrate several additional points. When a similar sample was tested for a somewhat longer time under the same conditions, some of the finger-like extrusions broke off to form loose wear debris. A cluster of such debris particles is visible at the top of Fig. 1. When this occurred, there was a dramatic rise in the wear rate. Thus, if the chosen test time was less than that needed to generate this type of debris, one would report negligible wear, but one might observe a much higher wear rate at slightly longer test times. This situation could result in excessive

Fig. 1. Secondary electron image of part of the exit region of a wear scar on OFHC copper. Sliding direction is to the right. The marker represents 100 μm.

scatter of reported data if the basic reasons for these variations were not understood. Or, two different investigators might report no wear and high wear, respectively, depending on the testing times used.

The definition of wear rate is a related problem. First, it should be recognized that various definitions are in common use, and the number reported may have different dimensions. For example, wear rate could be defined as the mass or volume of material removed per unit time, or, since sliding speed is frequently held constant, the wear rate is commonly expressed as mass or volume per unit sliding distance. Other forms could be dimensionless, such as the depth of material removed per unit sliding distance, or the volume removed per apparent contact area and per unit sliding distance.

In addition, as the example of ductile copper has shown, it is important to know if an average wear rate is being reported for the entire test or for some shorter time interval, since the reported values could be quite different. Wear rate may start low and later rise, as in the ductile copper case, or the reverse may be true, with high initial wear rate followed by lower values. Wear rate can also fluctuate widely, particularly if test conditions are not carefully controlled. When a transition from one wear mechanism to another occurs during a wear test, then changes in wear rate are expected. In general, wear rate is a complex function of time.

Numerous simple models of wear have been proposed. Unfortunately, the authors of many papers in the literature on wear give the impression that a limited number of simple wear models correlate well with observed wear behavior. However, there seem to be many possible mechanisms for wear, dependent on many variables. A few simple models, depending on only a few variables, are unlikely to be sufficient. One should not expect such simplicity when the materials/design/environment systems can be so varied. The fracture of finger-like extrusions at the exit side of a wear scar (Fig. 1) seems to be an operative wear mechanism under certain special conditions. But many other ways of generating wear debris have been reported.

It is quite common to classify types of wear by using such terms as adhesive wear or rolling contact fatigue. The former is widely used for wear involving sliding of one piece of metal on another, while the latter is applied to most failures involving rotating bearings. Terms like these should be used with caution, because each implies that the dominant mechanism is known.

An alternate classification scheme would be phenomenological. Simple descriptive terms like sliding, rolling, impact, wet/dry, and lubricated/unlubricated wear are based on simple observations about the wear system, and they are more appropriate whenever the wear mechanisms are not well characterized.

It is commonly reported that sliding behavior during an initial break-in (running-in) period is different from that observed at longer times, when a dynamic steady state may be achieved. During steady-state sliding, the average friction coefficient, wear rate, type of wear debris, surface topography, depth of deformation, coating or film characteristics, and near-surface temperature, microstructure, hardness, and crystallographic texture (and their gradients) will all be relatively constant with time. Local changes will contribute to the statistical fluctuations such as the "noise" observed in friction traces. In some systems, these may be large enough so that it is questionable to use the term steady-state. However, it is generally useful to separate sliding behavior into two stages, break-in and steady-state. With some systems, a third stage leading to catastrophic failure may appear. This last stage is not addressed in this book.

Behavior during break-in will depend strongly on the initial material structure and properties and on surface conditions such as the surface finish and the nature of any films present. During this transition period, films may be destroyed or modified, the material may work-harden, phase transformations may occur, temperature will rise as a result of frictional heating, energy will be stored, and the microstructure, texture, and surface topography will evolve toward those consistent with the development of steady-state conditions. These and other changes during break-in can occur with or without the generation of loose wear debris.

Because of these changes, steady-state behavior may be independent of one or more of the initial conditions. For example, surface finish has deservedly received much attention from engineers concerned with sliding systems of practical importance. However, if one waits long enough, and if the design allows the achievement of steady-state conditions, *any* initial surface finish should evolve to the surface finish appropriate to steady-state for the given system. The choice of initial surface finish may affect the time needed to achieve steady-state, but the same steady-state should be reached eventually.

Anyone investigating friction and wear is faced with a large number of variables to control. One of the first choices to make is the selection of a suitable wear test machine. It has become axiomatic that the best wear test is one that closely approximates the actual conditions encountered in service. However, most wear tests are done using simplified arrangements such as those shown in Fig. 2 and 3. Others are described in a convenient collection published by the American Society of Lubrication Engineers.[3]

Despite the apparent simplicity of the test geometries shown in Fig. 2 and 3, there are still several important design features to be determined for each. For example, in Fig. 2, the tendency of loose debris to be retained and reprocessed in the interface region is different for the three arrangements, and this can affect both friction and wear.

One also has the choice of using a conforming or non-conforming geometry (Fig. 4). In the latter case, the size and/or shape of the wear scar changes with time, making stress analysis more difficult and delaying or preventing the achievement of steady-state conditions.

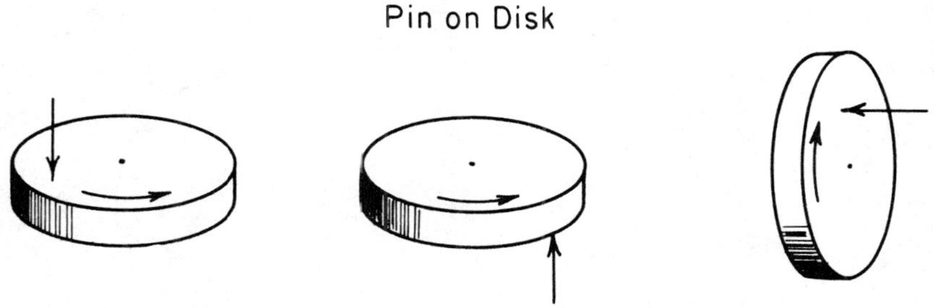

Fig. 2. Three different arrangements for a simple pin-on-disk wear test machine.

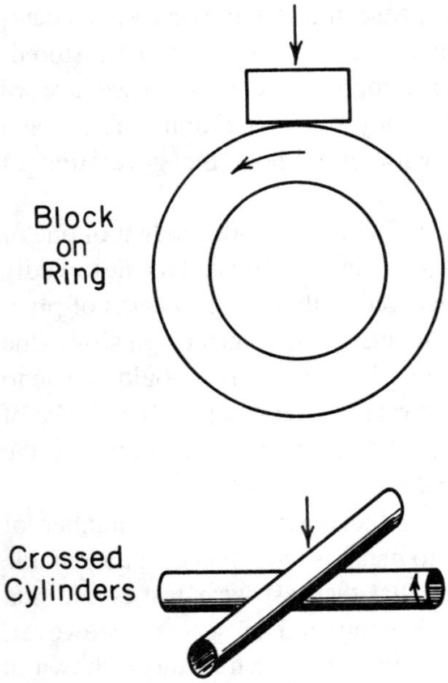

Fig. 3. Basic arrangements for simple block-on-ring and crossed-cylinders wear test machines.

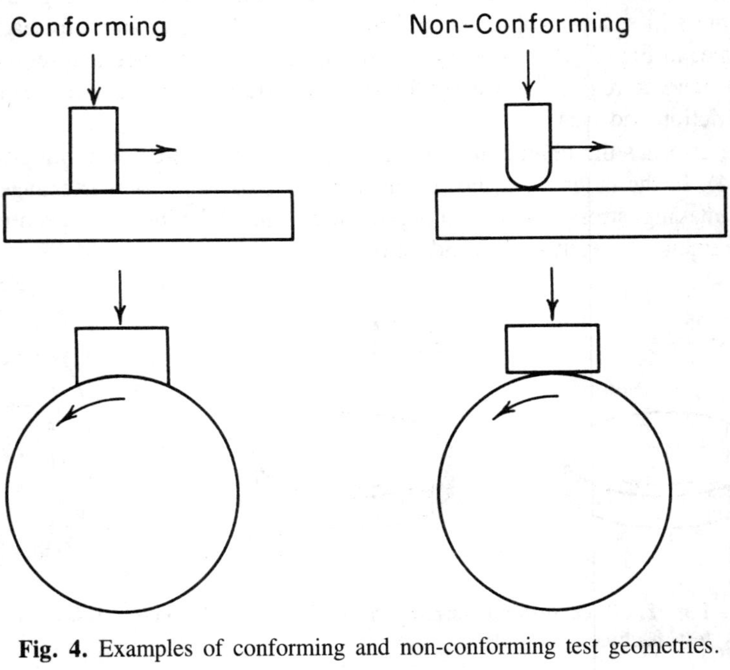

Fig. 4. Examples of conforming and non-conforming test geometries.

Most wear test machines are asymmetric systems. For example, in Fig. 2, if the materials of the pin and the disk are different, and if their positions are interchanged, the results will generally be quite different. Symmetric systems do exist, but they are relatively few. Three examples are (1) disks or rings of the same diameter, face to face, (2) disks or rings of the same diameter, parallel axes, counter-rotating at the same speed, and (3) crossed cylinders of the same diameter, both rotating at the same speed. The first of these has the disadvantage of tending to trap debris particles.

Other variables are numerous. The mode of sliding can be unidirectional as in simple rotating systems, or it can be oscillating. In the latter case, the amplitude can be large or small (as in fretting), and the frequency can be selected from a wide range. Ambient temperature, applied load, and sliding speed are obviously important. The number of passes can be one, a few, or many on a single track. The components of materials pairs can be unlike, similar, or the same (self-mated tests). The removal of debris can be encouraged with the aid of brushes or fluid flow, or the debris can be allowed to accumulate. Lubricants, coolants, and coatings (thick or thin) of various types can be used, and the chemical environment (gas, liquid, solid) can be varied. And, of course, the materials themselves bring in additional variables such as composition, grain size, texture, microstructure, and magnetic properties. A thoughtful article by Peterson contains further discussion of the factors involved in designing friction and wear tests.[4]

The preceding discussion of experimental variables is intended to emphasize the choices which can affect the outcome of friction and wear tests. Certainly such freedom of choice can cause confusion. But realizing the nature of these choices is important if test data are to be meaningful or useful, particularly when comparisons are made between results from different tests. Authors reporting test data should state clearly all details of their tests which may be relevant. This should include information on materials parameters. It is far too common to read that one member of a sliding pair was "steel," without any clue to the type of steel, composition, heat treatment, structure, etc. Such lack of care in reporting critical details raises questions about other aspects of such research and makes the results of questionable value. This topic has been emphasized here in the hope that the yield or efficiency of friction and wear research can be increased.

With most sliding test geometries it is not difficult to monitor the friction force.[3,5] Therefore, even if the primary goal is to obtain wear data, it is generally advisable to measure friction as well. The information obtained can aid interpretation of the wear results at different stages. However, friction and wear data often correlate poorly, and one can have friction without wear, or between wear events. Therefore, one should be careful not to tie friction too closely to wear.

It may also be helpful to monitor temperature near the sliding interface. This is more complicated than it may at first seem. Introducing a temperature probe, e.g., a thermocouple, into the interface region would modify the sliding behavior. A common alternative is to insert a fine thermocouple about 1 mm below the interface and then to estimate the interface temperature by heat flow calculations using appropriate boundary conditions. Results may vary by hundreds of degrees Celsius, and the subject is one of considerable controversy, since the assumed boundary conditions depend on the model chosen to describe the dissipative process.[6,7]

The output from a single thermocouple allows continuous monitoring of sliding conditions, much as monitoring of the friction coefficient can do. In fact, the traces for these two parameters will generally be quite similar, as expected. However, one must remember that a single thermocouple below the surface detects local conditions and is insensitive to rapid fluctuations from surface hot spots. These local temperature fluctuations can be detected by using the sliding members themselves as components of a thermocouple or by selecting a hard transparent material for one component of a sliding pair.[8,9] A single thermocouple is also insensitive to conditions at different parts of the interface. For example, with the block-on-ring geometry, the exit side of the wear scar will be hotter than the entrance side. This can lead to dramatic differences in processes occurring in the two regions. In one interesting test on nickel, recrystallization was observed below the exit side of the scar, but not below the cooler entrance side.[10]

If one is interested in the mechanisms of friction and wear, the completion of a sliding test is just the beginning of the investigation. Both the worn specimens and the debris particles can be characterized by a wide range of experimental techniques. These can provide information about the surface or about the modified material below the surface. Some of these techniques are based on familiar metallographic procedures, perhaps modified to reduce possible mechanical or chemical damage to the critical near-surface material.[11,12]

Since the structures below worn surfaces are usually anisotropic, it is desirable to obtain a full three-dimensional picture of near-surface features. This can be done with the aid of sections parallel and normal to the surface. Useful cross sections are those parallel to the sliding direction (longitudinal section) and perpendicular to the sliding direction (transverse section), as indicated in Fig. 5. The sliding direction itself is defined in this book as the direction of motion of the opposing surface as it slides on a sample material (Fig. 6). In a longitudinal section, the sliding direction is usually evident as the direction in which various structural features tend to bend. With light loads and lubricated sliding, the deformed region may be too thin to show up well in a normal section. In such cases, taper sections can in effect magnify selected features in one direction.[11]

Examination of sections obtained from samples *before* sliding tests has shown that different methods of preparation create "damaged" regions of very different extent. If one wishes to examine a region affected by sliding to a depth of, e.g., 20 μm, then the methods chosen for preparation of the test sample should not

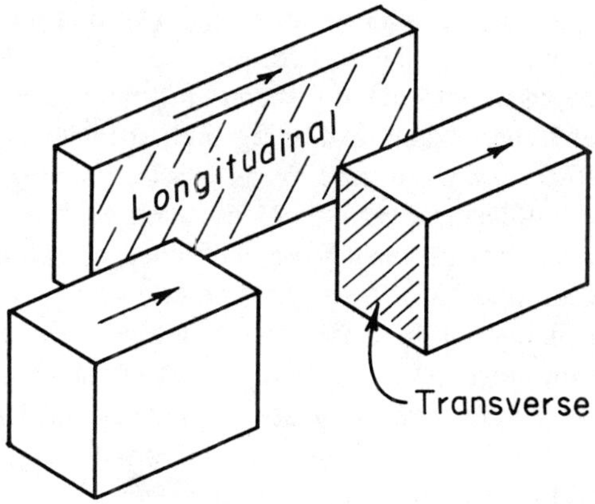

Fig. 5. Convention used in this book to describe cross sections: Figure shows longitudinal section, with sliding direction parallel to the section, and transverse section, with sliding direction normal to the section. Both sections are normal to the wear surface.

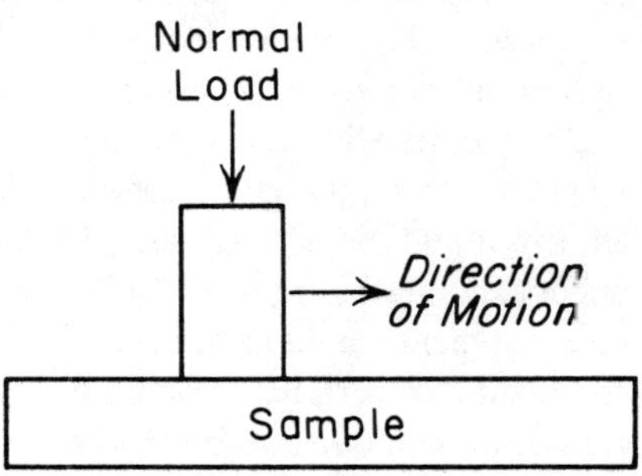

Fig. 6. Convention used in this book to define sliding direction: As shown, it is the direction of motion of the opposing surface as it slides on a sample material.

affect the material that deeply. Blau has reported that fly-cutting creates less sub-surface damage than other commonly used machining processes.[13] To select subsequent grinding and polishing procedures, it is helpful to refer to the book by Samuels.[11]

One of the main reasons for the renewed interest in basic research on friction and wear is the recent availability of sophisticated instruments for local structural and chemical analysis. Of these, the scanning electron microscope (SEM) is probably the most important, particularly if it is equipped to provide information on local chemical composition. A modern SEM can be operated in various modes which provide complementary information. As researchers become more familiar with these modes, the SEM will become even more useful.

It has become clear that sliding can develop very large shear strains near the sliding interface. The resulting microstructures can be quite different from those present at the start of testing (see the paper by Ruff, Ives, and Glaeser in this book). Transmission electron microscopy has shown that these microstructures are typically very much like those reported when high net shear strains are achieved by other types of deformation, such as rolling, extrusion, wire-drawing, creep, and fatigue.[14] The regions exhibiting high shear strains may tend to be localized, as in fatigue, or more general, as in wire-drawing. Such differences tend to obscure the similarities, which include being well up on the stress-strain curve where the shear strength tends to saturate because of appreciable amounts of recovery. Since the study of each of these other deformation processes is well advanced, it seems likely that increased attention to any similarities with sliding processes will aid progress in understanding friction and wear.

Lubrication is not featured in this book, although certain aspects of lubrication are mentioned in several papers. This is an important subject which receives much attention at other conferences on friction and wear. Processes with and without intentional lubrication are often treated quite differently. Yet, experiments show (see, for example, the paper in this book by Ruff, Ives, and Glaeser) that similar high shear strains and fine microstructures are achieved with and without lubrication. The principal difference is that these features are restricted to a smaller volume of material adjacent to the sliding interface when a lubricant is present. Thus, many of the ideas discussed in this book should be applicable to both lubricated and unlubricated systems.

Despite the wide range of topics discussed in this book, it is easy to think of others which might have been included. For example, erosion by solid particles and by liquids is a broad area which is not covered here. Recent advances in understanding the friction of elastomers have been impressive,[15-18] and a chapter on elastomers would have been timely and appropriate. The friction and wear of composite materials, which range from glass-filled polymers to cemented carbides, is another important subject which could have been included. Still another is the subject of material transfer.[19-22] Transfer processes can lead to films that are patchy or continuous, hard or soft, thick or thin. Since transfer can be responsible for dramatic changes in friction and wear, it is important that it be better understood. Fortunately, there is evidence of increasing interest in the mechanisms of transfer.[23] With the aid of modern probes to determine local structure and composition, it may soon be possible to obtain enough information on transfer to design control strategies based on principles of materials science.

Obviously, this book is not a comprehensive review of the friction and wear of materials. This is a vast field, combining as it does surface and sub-surface effects and the interactions of deformation, fracture, and environmental factors. It is a challenging field for materials scientists and engineers to apply their knowledge and skills.

Acknowledgments

The National Science Foundation provided partial support of this conference under Grant No. DMR 8019439.

Parts of this Introduction have been based on research supported by the National Science Foundation under Grant No. DMR 7805719 and by the Army Research Office and Office of Naval Research under Grant No. DAAG29-80-C-0145.

Members of the ASM Seminar Committee and the 1980 Materials Science Seminar Committee helped to guide the Seminar planning, and Dr. Ed Myers and Ms. Patti Gaeta, both with the American Society for Metals, have helped with many important details.

The four seminar sessions were ably chaired by D. H. Buckley, Lewis Research Center; E. F. Finkin, Allegheny Ludlum Industries, Inc.; J. P. Hirth, The Ohio State University; and R. P. Steijn, E. I. DuPont de Nemours and Co.

The help of Dr. Peter Heilmann during the editing process is gratefully acknowledged.

References

1. D. Dowson, History of Tribology, Longman, London and New York, 1978.
2. Proceedings, Wear of Materials—I, II, III, St. Louis (1977), Dearborn (1979), San Francisco (1981), publ. ASME, N.Y.
3. Friction and Wear Devices, Second Edition, American Society of Lubrication Engineers, 1976.
4. M. B. Peterson, Wear Testing Objectives and Approaches, in Selection and Use of Wear Tests for Materials, ed. R. G. Bayer, STP 615, ASTM, 1976 (based on ASTM Symposium, New Orleans, Nov. 1975).
5. F. P. Bowden and D. Tabor, Friction: An Introduction to Tribology, Anchor Press, N.Y., 1973.
6. D. A. Rigney and J. P. Hirth, Wear *53*, 345-370 (1979).
7. S. Malkin and A. Marmur, Wear *42*, 333-340 (1977).
8. K. Takazawa, Bull. Japan Soc. Precision Eng. *2* (1), 14 (1966).
9. G. Boothroyd, Brit. J. Appl. Physics *12*, 238-242 (1961).
10. J. Schell, M. S. Thesis, Metallurgical Engineering, The Ohio State University, 1979.
11. L. E. Samuels, Metallographic Polishing by Mechanical Methods, 2d ed., Pitman and Sons Ltd., Melbourne and London, 1971.
12. T. M. Ahn, P. J. Blau, K. L. Hsu, D. A. Rigney and J. D. Schell, Wear *56*, 409-413 (1979).
13. P. Blau, Ph.D. Thesis, Metallurgical Engineering, The Ohio State University, 1979.
14. J. P. Hirth and D. A. Rigney, The Application of Dislocation Concepts in Friction and Wear, to appear in Dislocations in Solids, ed. F. R. N. Nabarro.
15. A. Schallamach, Wear *17*, 301-312 (1971).
16. A. D. Roberts and A. G. Thomas, Wear *33*, 45-64 (1975).
17. G. A. D. Briggs and B. J. Briscoe, Phil. Mag. *A38*, 387-399 (1978).
18. B. J. Briscoe, The Friction of Polymers: A Short Review, Proceedings Fundamentals of Friction, Leeds, Sept. 9-12, 1980.
19. A. E. Roach, C. L. Goodzeit and R. P. Hunnicutt, Trans. ASME, Nov. 1956, pp. 1659-1667.
20. M. Kerridge and J. K. Lancaster, Proc. Roy. Soc. *A236*, 250-264 (1956).
21. I. Sasada, H. Ohmura and S. Norose, The Wear and Mutual Transfer in Cu/Fe Rubbing, Proc. 15th Japan Congress on Materials Research, Tokyo, Sept. 1971, publ. Soc. of Mat. Sci., Kyoto, 1972.

22. J. Gerkema, Fall Meeting, Electrochemical Society, Los Angeles, Oct. 14-19, 1979.
23. Proceedings, Wear of Materials, III, San Francisco, March-April, 1981, various papers on transfer, publ. ASME, N.Y., 1981.

Sliding Wear Mechanisms

L. E. SAMUELS, E. D. DOYLE AND D. M. TURLEY
Materials Research Laboratories
Defence Science and Technology Organization
Department of Defence
Australia

Abstract

It is now becoming widely accepted that further development of mathematical models of wear systems must await improvements in the understanding of fundamental wear mechanisms and their interactions. Current knowledge of these fundamental mechanisms is reviewed. The approach is based on insights obtained from the study of debris recovered from wear systems, as being indicative of individual material removal events. The structure of worn surfaces is also reviewed briefly, because a knowledge of this structure and the deformation and fracture events that can be expected to occur in the surface layers is also likely to be a prerequisite to the modelling of wear systems.

Introduction

It is now generally accepted that any research program that resulted in even a small percentage reduction in the wear bill of an industrialized nation would probably be very cost effective. As a measure of progress in the field, Blok[1] reached the same conclusion nearly thirty years ago. It could be said, of course, that engineers have learned to live with wear even if at a high cost. Even accepting this complacency, there are still instances where engineers have not learned to live with wear — that is, where developments are wear limited. It is not as though the field has been neglected. Indeed, the literature contains a plethora of reports of wear tests and investigations. The problem is that adequate guiding principles have not evolved that enable reliable predictive design to be made of low-wear systems, particularly under unfamiliar service conditions.

The current situation is perhaps best summarized in a report of a panel discussion on wear research and development held during an international conference on wear of materials in St. Louis in 1977. Pertinent extracts are:[2]

> The published work on wear is still largely descriptive and not yet analytical —many papers are difficult to assess because they provide data on material tested under arbitrary conditions in machines of dubious value. Many authors fail to analyse their results—or study the phenomena and mechanisms involved.

and

> A general consensus appeared to prevail amongst those present that there is a dire need for the understanding of fundamental wear mechanisms and their interactions. Unless a thorough understanding is developed, the likelihood of success in mathematical modelling of a wear system is very small.

Mechanisms of wear were barely considered at all before the classic work of Bowden and his colleagues, which was commenced in the late 1930's. This work, which has been summarized by Bowden and Tabor,[3] gave rise to a general adhesion theory of wear, but it can fairly be said that this theory was largely a corollary of their theory of friction. There have been improvements in the details of the adhesion theory over succeeding years, but little in the way of substantial developments. Moreover, until quite recently, no proposals for alternative or additional mechanisms of wear have emerged.

The basic phenomenon of wear is one of removal of material from a surface, but wear systems are such that with few exceptions the processes of material removal have not been observed directly. Moreover, there is no expectation of this being able to be done in the immediate future. The problem is further compounded by the fact that wear usually is quite slow in real engineering systems; that is, the events which remove material from a surface are generally rare statistically. An indication of the magnitude of the dilemma can be obtained from the wear equation developed by Archard[4] which is based on an asperity-contact model:

$$V = k \cdot \frac{S}{3} \cdot \frac{L}{H} \qquad [1]$$

where V is the volume worn away, S is the sliding distance, L is the normal load, H is the indentation hardness value of the softer of the wearing pair, and k is a constant which effectively is required to make the formula fit reality within an order of magnitude. Values of k of between 10^{-3} and 10^{-8} are found to be needed. An interpretation of these figures is that either only one in 10^3 to 10^8 of the contacting events produces a wear particle or that 10^3 to 10^8 repetitions of a particular contacting event are required for it to produce a wear particle. In either case, the probability of observing the removal event is very small.

The only experimental approach available, to use Tabor's[5] words, is to "investigate the resultants of the wear system at various stages and to attempt to reconstruct a picture of what is going on." This, as we shall see, leaves much room for conjecture. The two resultants of a wear system are the worn surface and a number of particles that constitute the wear debris. The problem with examining a worn surface is that evidence of the critical wear events is likely to have been obscured by subsequent events. The particles of wear debris, however, are less likely to have been damaged by subsequent events, although this possibility cannot be ignored. Thus they are more likely to be representative of single material-removal events, although their origin must always be somewhat conjectural unless they can be correlated with some feature in a worn surface. In any event, the approach based on the examination of wear debris particles by a wide range of investigational techniques is currently the most powerful one available. Although only recently developed, it has already made substantial contributions to the development of the understanding of wear mechanisms. The main purpose of the present review will be to explore what contribution this new approach has made in attempts "to reconstruct a picture of what is going on."

Types of Particles Present in Wear Debris

Sampling of the Wear Particles

To date, most of the work has been carried out on particles that can be recovered from oil-lubricated systems, although there is no reason in principle why particles should not be recovered from dry or essentially dry systems. For oil-lubricated systems, the particles present in a sample of lubricant can be recovered by filtration, by centrifuging, or magnetically, for which purpose a special device has been developed.[6] The particles themselves can be examined in a number of ways, some of which are discussed elsewhere in this book, but the present discussion will concentrate on examination by the following microscopical methods: optical microscopy, scanning electron microscopy (SEM), and transmission electron microscopy (TEM).

First, however, a word of warning. Very little work has yet been done to correlate positively the wear debris to be described with changes in the wearing surfaces, and one such investigation has produced a disturbing result.[7] An experimental gear box was run under controlled conditions in such a way that the debris recovered from the lubricating oil could be established positively as coming from the wearing gear teeth alone. Only about 10% of the loss in mass of the gears could be accounted for by the debris recovered from the oil. This result raises questions about the degree to which debris recovered from an oil system represents the material removed from the wearing surfaces. For the moment, this question will have to be ignored. Even so, the recoverable debris can be taken to be indicative of the mechanisms of some, if not all, of the material removal processes occurring at the wearing surfaces.

A classification system has to be adopted for sensible analysis of wear systems, and there are two possibilities. First, a system can be based on an engineering description of wear, examples being scuffing, rubbing, and fretting. These terms presuppose particular but undefined wear mechanisms. The second possibility is a system based on mechanisms of material removal. The second is considered to be the more desirable of the two, but cannot be adopted yet because of the many doubts about some of the possible mechanisms. A step in this direction, however, is a system based on the morphology of the wear particles, and this is the one that will be adopted here. An attempt will be made to use it as a guide to basic mechanisms of wear.

Plate-Shaped Particles

Laminar Particles

Thin, plate-shaped wear particles are commonly found in wear debris recovered from oil-lubricated systems, the lateral dimensions of the particles ranging from tens to tenths of a micrometer with the thickness usually about one-tenth of the lateral dimensions (Fig. 1). The existence of this type of particle, which was recognized by Seifert and Westcott,[6,8] has been the basis for the important concepts known as the delamination theory of wear, developed by Suh[9-11] and his co-workers. This can be said to have constituted the first major development in wear theory since the Bowden school proposed their adhesion theory.

Suh established that plate-shaped particles do detach from wearing surfaces, and he originally proposed that:[9]

(i) Material nearest the surface has a low dislocation density because of image forces at the surface.
(ii) Pile-up of dislocations occurs below the surface, and this leads to the formation of voids in a layer beneath the surface, particularly if second-phase particles are present.
(iii) The voids coalesce to form cracks parallel to the surface.
(iv) The crack propagates if it exceeds a critical length, and eventually turns towards the surface.

The basic premise of this theory is not in accord with the real structure of worn surfaces as presently known (see below). In any event, it appears that Suh[10] subsequently abandoned this theory and replaced it by the following one, which is illustrated schematically in Fig. 2.

(i) Normal and tangential forces are first transmitted through contacting asperities, which deform to smooth the surface. Plane contact with cyclical loading can then be supposed.
(ii) Plastic strain (shear) accumulates in the surface layers.
(iii) Cracks are nucleated below the surface layers. Triaxial compressive stresses are present in a zone close to the surface and this forces crack nucleation to a lower level.
(iv) The cracks grow and join.
(v) The cracks shear to the surface at weak positions.

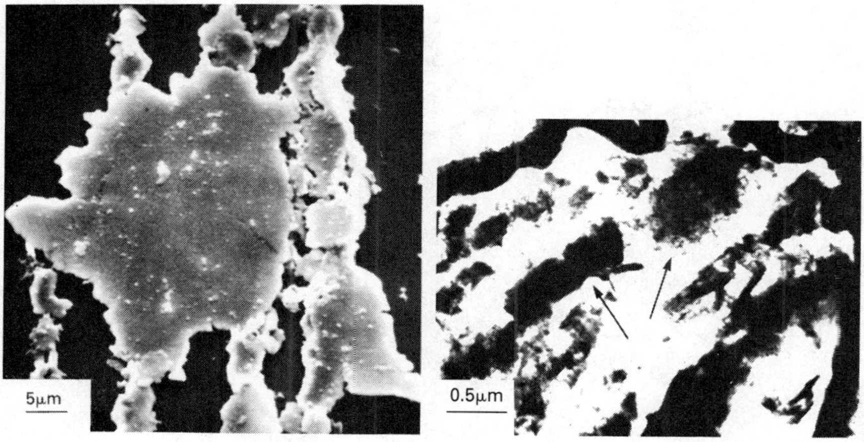

Fig. 1. Laminar wear particles of steel recovered from an oil-lubricated gear box, showing the morphology of the particles and the range of sizes encountered. At left is a scanning electron micrograph; at right, a transmission electron micrograph in which typical laminar particles are arrowed.

A collection of papers reviewing the status and development of the theory has been published recently.[11]

However, other workers have demonstrated by means of *in situ* experiments in the scanning electron microscope that plate-like wear particles may be generated by entirely different mechanisms. For example, Bates et al.[12] and Kayaba and Kato[13] found that plate-like debris is ejected at the leading edge of a hard tool that is made to slide across a surface of a softer material (Fig. 3). A mechanism proposed by Kayaba and Kato for this process is illustrated schematically in Fig. 4. The suggestion is that a plastic zone, ABC, develops early in the interaction between two asperities and that shear displacement occurs along AC, generating a shear tongue `CC. The plastic zone is then transferred to a lower layer, and the process is repeated to generate a number of layers of shear tongues. When a crack eventually develops at the sharp cusp at the root of the tongues, a fragment consisting of a series of plates detaches from the surfaces and the individual plates subsequently separate.

The development of a plastic field such as ABC in Fig. 4 is predicted by a slip-line field analysis developed by Challen and Oxley.[14] This analysis is concerned strictly with the deformation of a soft asperity by a much harder one, but it does predict that a plastic zone of similar shape to ABC in Fig. 4 should form, and that shear could occur along AC under certain circumstances. It cannot say anything, of course, about the repeated shearing proposed by Kayaba and Kato. The circumstances predicated by Challen and Oxley for shear to occur are that, for a given included angle in the soft asperity, the ratio of the strength of the interfacial film along AB to the flow stress of the asperity material should exceed

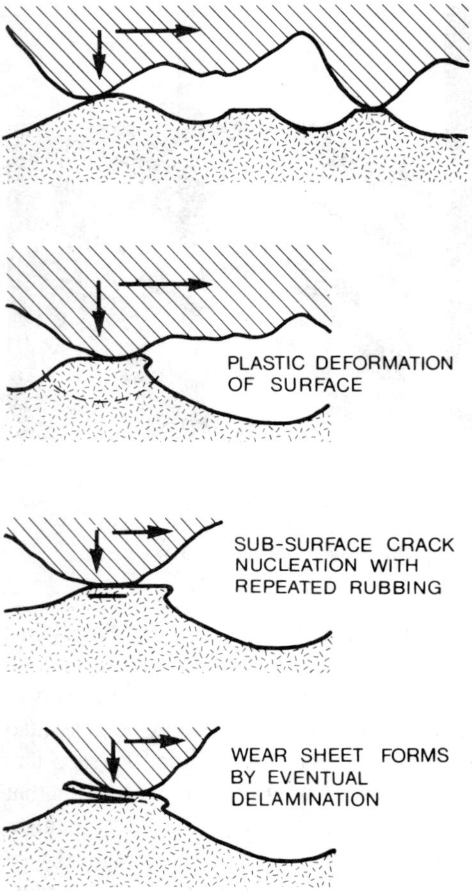

Fig. 2. Schematic representation of the various stages involved in the formation of delamination wear sheets. *(After Suh, Ref 10)*

a certain value, the larger the angle the smaller the value of the ratio. Thus, for an engineering surface, the probability of material removal would depend on both the geometry and the surface condition of a particular asperity; the rougher the surface the better the lubrication would need to be to obviate material removal. This model thus meets an essential requirement in that it predicts that only a fraction of the contacting asperities should produce a wear particle.

Another mechanism for the detachment of plate-like particles has been advanced by Doyle and Turley,[15] who found that delamination of the general type observed by Suh can occur during a single pass of an asperity (Fig. 5, top). They established, moreover, that the delamination occurred by fracture within a surface layer that has undergone severe plastic deformation (Fig. 5, bottom). The structure of this layer will be described later. Doyle and Turley suggest that shear separation occurs within the small subgrains present in the plastically deformed

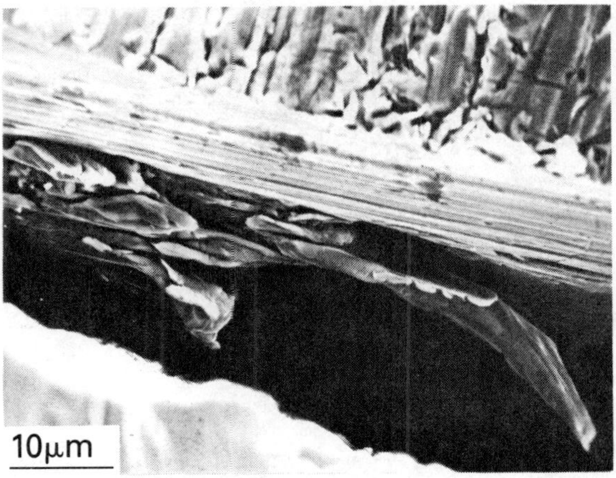

Fig. 3. Flake-like tongues separated from the lower model asperity by a sliding model asperity *(upper)* of the same material. *(Ref 13)*

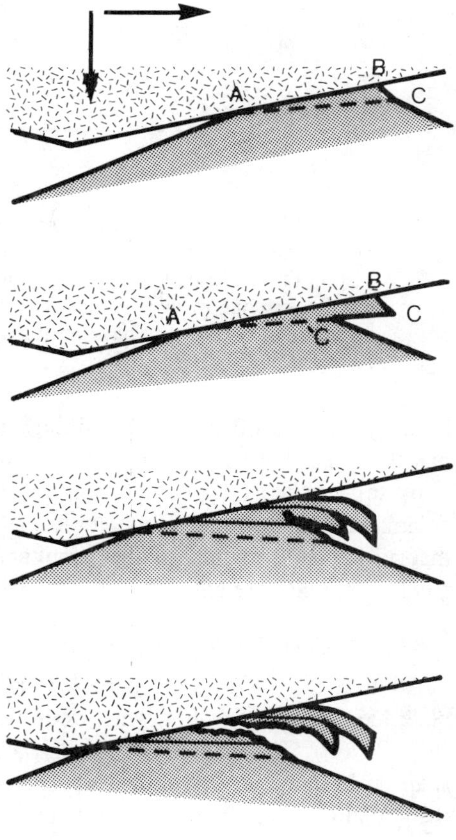

Fig. 4. Schematic representation of the various stages involved in separation of flake-like tongues of material between sliding asperities. *(After Kayaba and Kato, Ref 13)*

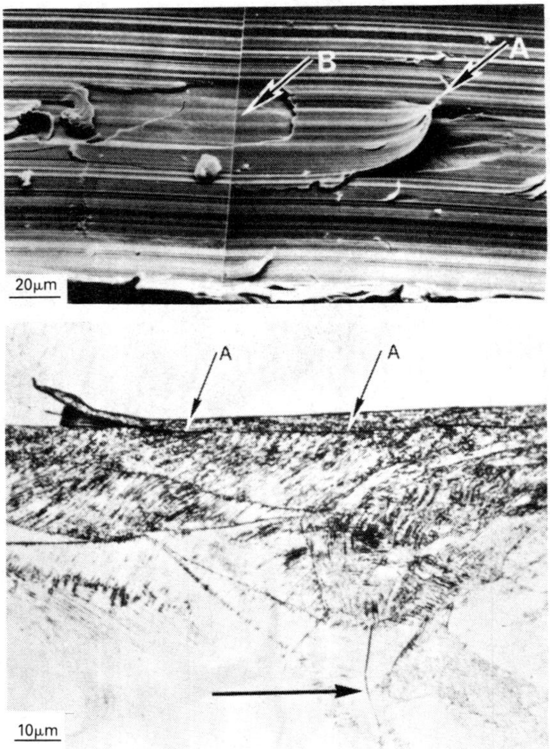

Fig. 5. (Top) Laminar sheet separating from a groove produced in brass by a single pass of an abrasive particle. **(Bottom)** Etched longitudinal section through a partly detached laminar sheet similar to that shown at top. Separation (arrowed A) has occurred within the dark-etching surface layer which is known to have been severely deformed. *(After Doyle and Turley, Ref 15)*

layer by slipping off during the sliding process itself (Fig. 6); Hsu et al.[16] have made a somewhat similar suggestion.

In summary, as Hirth and Rigney[17] have pointed out, the existence of delamination and its phenomenological aspects are now well established, but the theory of the formation of the lamellar particles remains open to alternative mechanistic interpretations.

Wedge-Shaped Particles

Atkin et al.[7] found wear particles originating from gear teeth which had a plate-like morphology but were wedge-shaped rather than lamellar in section (Fig. 7, left). They were able to correlate these particles with correspondingly shaped pits in the surface of the teeth (Fig. 7, right). These pits were much deeper than delamination pits that formed simultaneously and were asymmetric with a

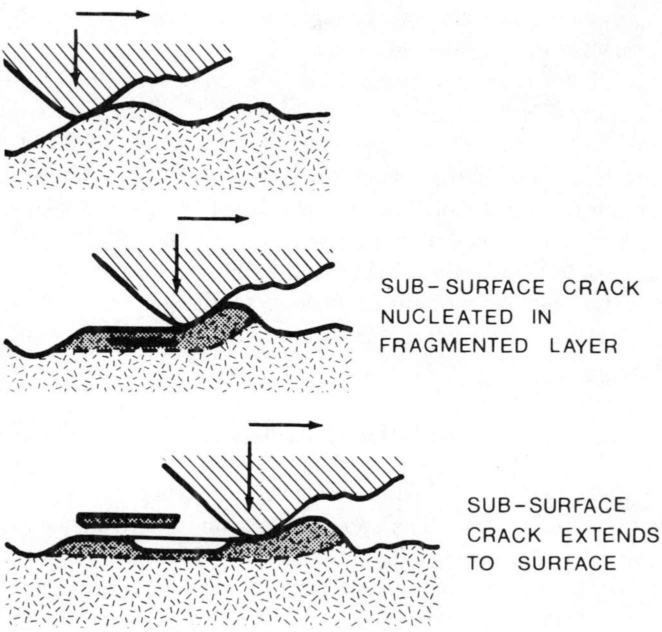

Fig. 6. Schematic representation of a mechanism of delamination occurring during a single pass of a hard asperity. The proposal is that a fracture develops in a surface layer of severely deformed material produced by the passage of the asperity. *(After Doyle and Turley, Ref 15)*

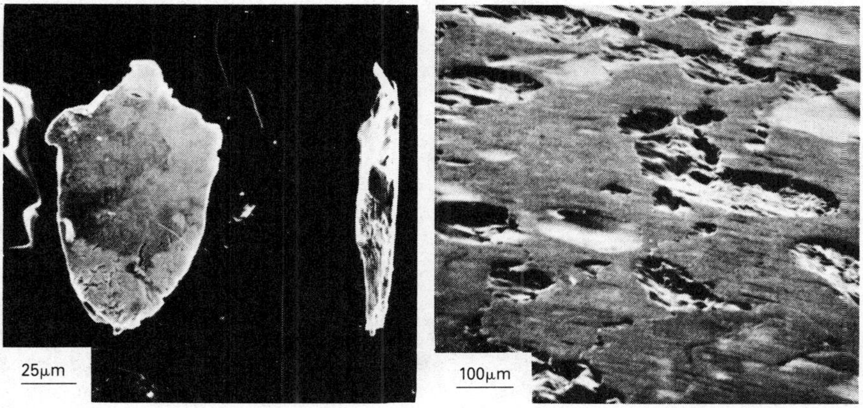

Fig. 7. (Left) Front and side views of a wedge-shaped particle of steel recovered from an oil-lubricated gear box. **(Right)** The worn surface of the gear teeth from which the particle was generated. The pits appear to result from the detachment of wedge-shaped particles. *(After Atkin et al., Ref 7)*

shallow end and an edge with a deep step. An number of repeated stress cycles were necessary for the particles to detach.

Atkin et al. found that the particles develop when the directions of rolling and sliding are opposed, with the shallow ends of the pits pointing in the direction of sliding (see Fig. 7, right). They suggest that they are caused by the development of low-cycle high-stress fatigue cracks which are initiated at the surface and which propagate at an angle of about 35° to the surface; at a certain stage a shear fracture connects the tip of the fatigue crack to the surface (Fig. 8). This mechanism is not to be confused with rolling contact fatigue, which develops macroscopic pits and is thought to be due to the initiation of sub-surface high-cycle low-stress fatigue cracks at the level at which Hertzian elastic stresses are a maximum.

Spherical Particles

Virtually perfect spheres 1-5 μm in diameter are perhaps the most unexpected wear particles found to date. The spheres are found rather frequently in small numbers (Fig. 9) and occasionally in profusion (Fig. 10). They seem to be associated mainly with rolling contact fatigue,[8] and it has been suggested that there is a marked increase in the incidence of the spherical particles just prior to fatigue failure.[18] If so, they would constitute a very useful indicator of impending catastrophic failure. It is less likely, however, that a study of these particles will provide an insight into mechanisms of actual material removal because all of the proposals made to date to explain their spherical shape involve events that occur after a wear particle has been detached from a surface. The main proposals are:

(i) The original particle is rounded by rolling between the rubbing surfaces.[18] A detailed model suggests that wear particles trapped in cavities are burnished under conditions that arise only in uniaxial sliding in fretting, or within cracks propagating by fatigue.[19] Smith[20] has demonstrated that spherical particles are produced on the surfaces of fatigue cracks that are

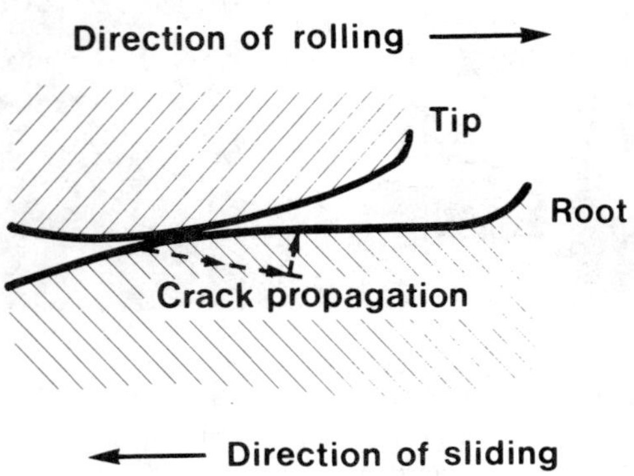

Fig. 8. Schematic representation of the mechanism proposed for the development of the wedge-shaped wear particle illustrated in Fig. 7. *(After Atkin et al., Ref 7)*

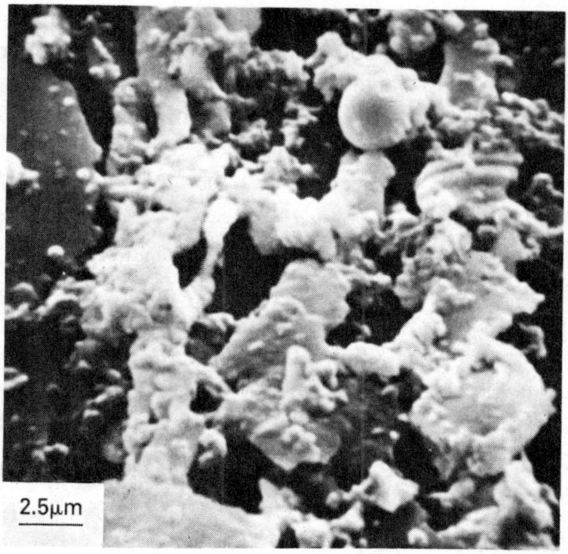

Fig. 9. A spherical particle among much other debris recovered from an oil-lubricated gear box. *(After Atkin et al., Ref 7)*

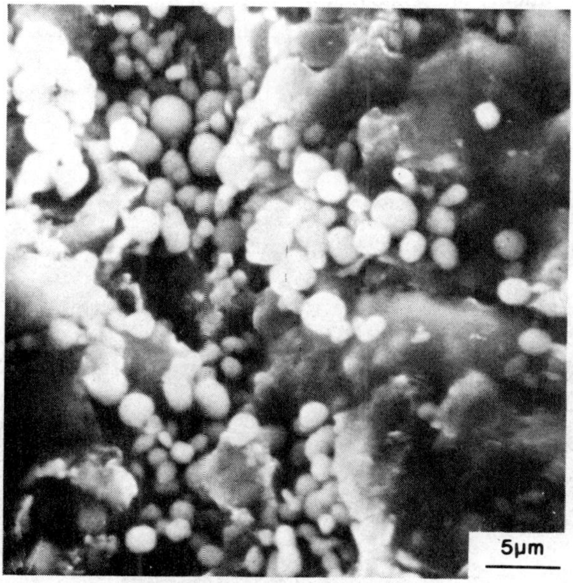

Fig. 10. Profusion of spherical particles present on the surface of a crack produced by a loading cycle that includes a compressive component. *(After Smith, Ref 20)*

propagating with a compressive component to the loading cycle (Fig. 10).

(ii) The original particle is spheroidized by heating.[21] This can occur at temperatures somewhat below the melting point, presumably because the self-diffusion rate in the wear particles is high because of their high defect structure. The driving force would be a reduction in surface energy.

(iii) A group of smaller particles is agglomerated by polymers present in the lubricating oil.[22]

There is also always a possibility that spherical particles are contaminants originating from extraneous operations such as grinding or welding. A case has even been reported where they appeared to be non-metallic inclusions released from the parent steel during wear.[22] Clearly, it is always wise to establish by chemical analysis the origin of any spherical particle found in wear debris.

Ribbon-Shaped Particles

Short Splintery Particles

Particles that tend to be small and straight with aspect ratios of the order of ten (Fig. 11) are characteristically produced during the early running-in period of ground surfaces.[23] It is assumed that they result from detachment of fin-like ridges that commonly are present at the edges of the abrasion grooves in freshly ground surfaces. It is certainly true that regularly shaped fins do form at the edges of such grooves, and it is not difficult to imagine that these fins would detach easily in a wear situation. If so, these particles are not indicative of a major continuing wear event.

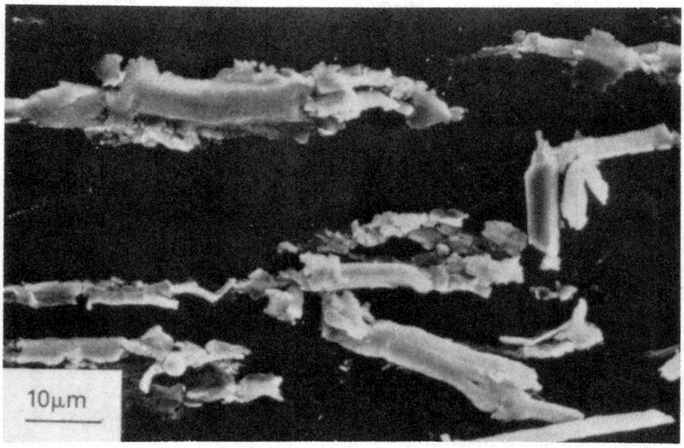

Fig. 11. Splintery particles recovered during the running-in period of a ground steel component. *(After Bowen and Westcott, Ref 23)*

Long Irregular Particles

Ribbon-shaped particles are frequently found which tend to have large aspect ratios, usually are curved and even curly, and characteristically are smooth on one surface and serrated on the opposite surface. They have all the characteristics of machining chips produced by a vee-point tool. In Fig. 12, for example, a chip among the wear debris from an engineering system is compared with those produced in a grinding operation. A much smaller particle from the same engineering system which has the same morphological characteristics but is so small that it can be resolved only in the transmission electron microscope is shown in Fig. 13. Thus the scale of this *abrasion* mechanism of material removal varies over many orders of magnitude.

This is the wear mechanism which probably is understood best of all because the extensive studies that have been made of abrasive machining and polishing processes can reasonably be extrapolated to wearing systems. On this basis, it can be concluded that the following criteria must all be met before an abrasion chip will be removed:

(i) The asperity or particle responsible for the removal of a chip must be plastically rigid compared to the workpiece. Theoretically, the hardness of the cutting point should be several times that of the workpiece surface but, in practice, it seems that the point needs to be harder than the workpiece surface but not much harder.[24,25] The hardness that must be considered in this context is that of the surface layer of the workpiece. This layer is in a severely work-hardened condition soon after wear has

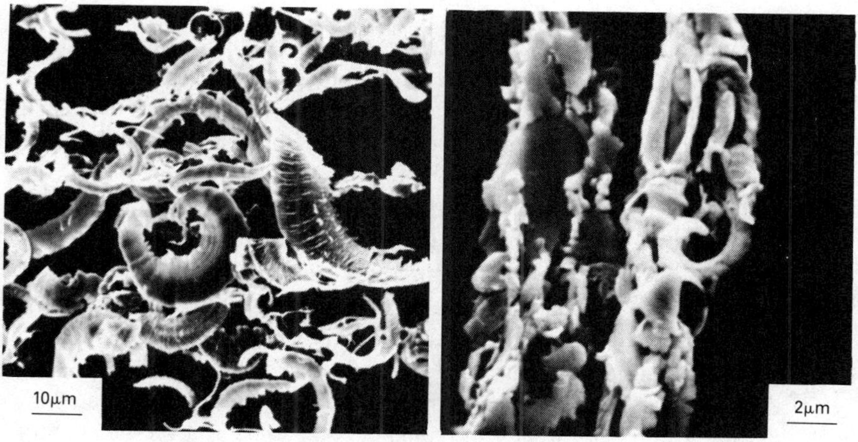

Fig. 12. (Left) Debris produced during grinding of steel. **(Right)** Chip-like particles, similar to those illustrated at left, among the debris recovered from an oil-lubricated gear box.

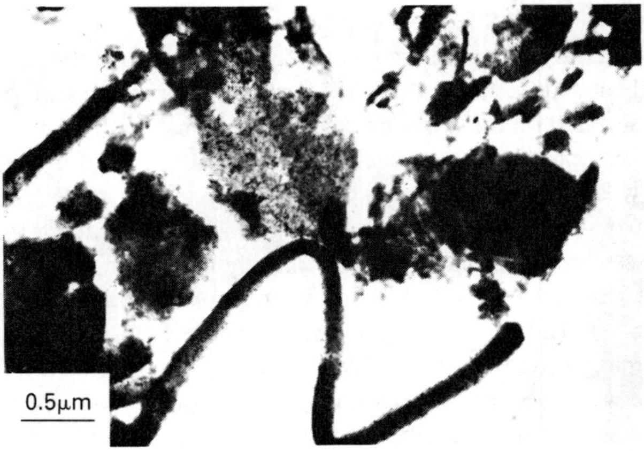

Fig. 13. Minute chip-like particles found by transmission electron microscopy among the debris shown in Fig. 12, right.

commenced (see below), and its hardness is likely to be about twice that of the bulk hardness. Moreover, it is the hardness of the local region from which the chip is removed that is of concern, a factor which may have to be taken into account in multiphase alloys. In general, therefore, it is non-metallic contaminants that are most likely to produce abrasion chips in metals. Foreign contaminants are the ones most commonly thought of in this context, but one source of contamination that should not be overlooked is fragments of abrasive embedded into a ground surface during abrasive machining. Extensive contamination of this nature occurs even with a reasonably hard workpiece.

(ii) The cutting point must be fixed in space for the required length of contact. Rolling particles produce only a series of indentations in the surface; they may then result in a low-order rate of material removal,[24] but they cannot produce a groove in the wearing surface and they cannot remove a chip. Loose contaminant particles will therefore cut a chip only when they embed in one of the wearing surfaces, and usually they will embed in the softer of the two surfaces, when the chips will be removed from the harder of the two. When the contaminant is embedded during preparation of the surface (e.g., during machining by grinding) the chips will be removed from the uncontaminated surface. The situation becomes complicated when both surfaces are heavily contaminated.[26]

(iii) A point must be of suitable geometry before it can separate material by cutting a chip; specifically, if the point is regarded as a vee-point cutting tool it must have a rake angle less negative than a certain critical value (α_c) characteristic of the specimen material (Fig. 14).[27,28] Points with rake

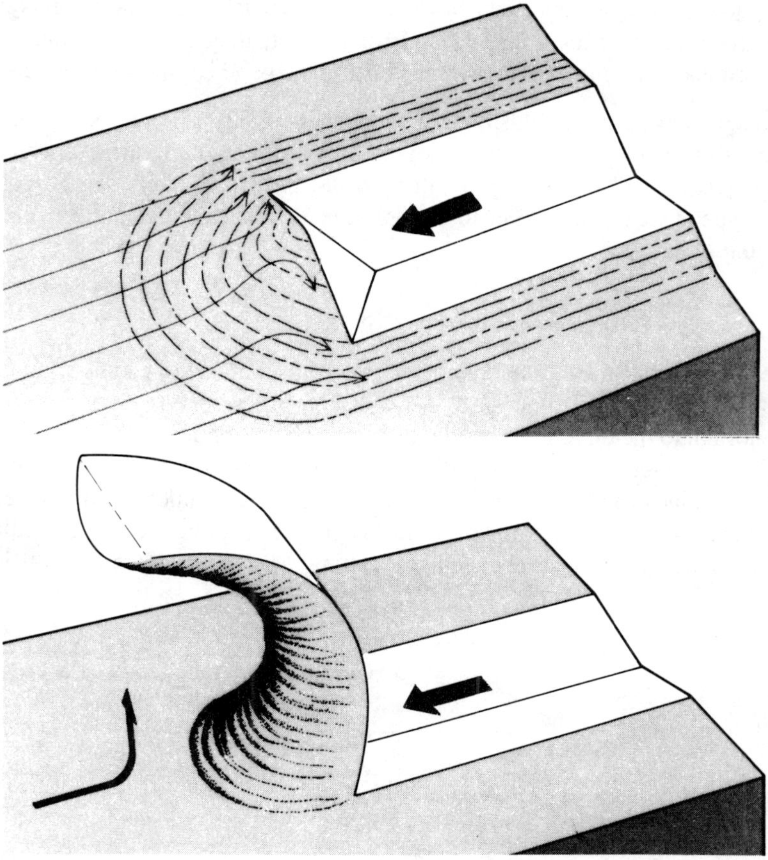

Fig. 14. Sketches illustrating the different effects of vee-point tools which plow and cut, respectively. A standing wave forms in front of a point which has a rake angle more negative than a certain value **(top)**; material moves upward ahead of the rake face and then moves around it into side ridges. In cutting **(bottom)**, a ribbon of material is separated from the surface of the workpiece and moves upward past the rake face continuously; this occurs only when the rake angle of the point is more positive than a certain value which is a characteristic of the workpiece material.

angles more negative than the critical value plow a groove, forming a ridge on each side and a standing-wave prow in front. The prow and portions of the ridge may become detached and so contribute to the wear process, but a chip-like wear particle will not be produced.

(iv) The point must remain in contact with the wearing surface for a reasonable length of traverse. Abrasives are very friable and their points may fracture off after a short length of contact; they may then either lose contact with the wearing surface or lose the geometry needed for cutting.

Thus a number of criteria, illustrated schematically in Fig. 15, must be met before a hard point in one surface can remove a chip from a softer surface, which is perhaps just as well because chip cutting could give rise to very rapid wear.

This type of wear is amenable to an approximate mathematical model along the following lines:[27]

$$V = \frac{f}{\varphi} \cdot S \cdot \frac{L}{H} \qquad [2]$$

where f is related to the number of contacting points that have a geometry suitable for cutting a chip, φ is a measure of the shape of these points projected onto a plane normal to the direction of travel (i.e., the cross-sectional area of the groove that they produce), the remaining symbols being the same as for Archard's equation. Values can be ascribed to both f and φ if the geometry and number of the contacting points are known, which is simple in principle but difficult in practice, and if the critical rake angle for cutting (α_c) is known, which is difficult

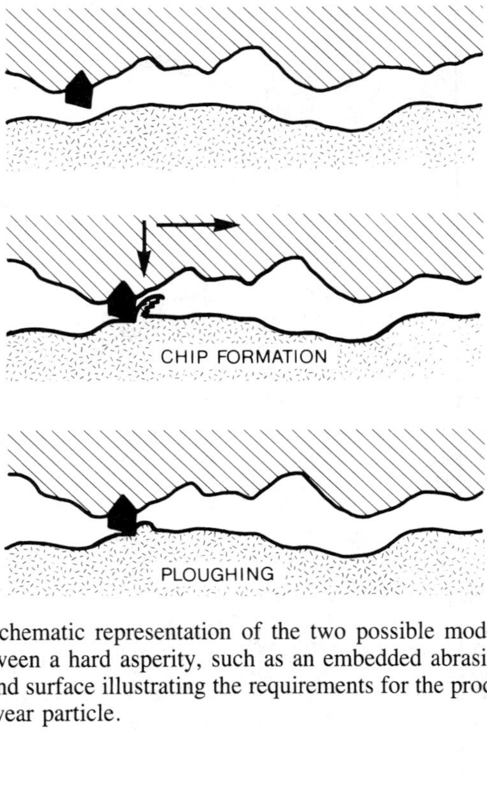

Fig. 15. Schematic representation of the two possible modes of interaction between a hard asperity, such as an embedded abrasive particle, and a second surface illustrating the requirements for the production of a chip-like wear particle.

in principle.[14,28,29] It is known from empirical experiments that α_c is a material-dependent parameter not apparently related to hardness,[28,29] but those material or systems characteristics that determine α_c have not been fully identified. An approximately representative value can be determined for a given material, however, by experiments with model points. The more negative the value of α_c the better the resistance to abrasive wear that can be expected.

Since the probability of abrasive wear can depend critically on the geometry of the asperities that produce it, and since the dependence is likely to vary with different materials, it is apparent that there is a risk in extrapolating laboratory trials to practice unless account is taken of likely difference in the geometries in the two cases. Like risks exist in extrapolating from one set of conditions of use to another.

Particles of Irregular Morphology

The morphology of many wear particles is too irregular to be classified. With the present state of knowledge, such particles could be formed by any one of a number of mechanisms, many of which cannot be said to have advanced beyond the stage of hypotheses. All that can be done here is to discuss a number of these as possibilities.

Adhesion

The adhesion theory of wear has been widely accepted as a basis for understanding sliding wear phenomena, principally because of its apparent success in explaining the classical laws of friction. In essence, the approach points out that the real area of contact of two solids brought into close proximity is very small, the points of contact being only at the summits of the highest irregularities in the surfaces. Consequently, the real area of contact is independent of the apparent area of the surfaces but is determined by the applied load. Under the high pressure at the localized points of contact, elastic and plastic deformation occur until the areas of the contact junctions are large enough to support the load. The suggestion is that the frictional force required to cause sliding is in large measure the force required to shear these junctions. In the discussion of the Archard wear equation based on this model of asperity contact it was pointed out that one of the implications of the equation is that actual removal of material can occur during only a very small fraction of the asperity contacts.

Thus, in its application to wear, the adhesion theory involves the following sequence of events (Fig. 16):

(i) Flattening of asperities by deformation and the development of an interface between the two surfaces, an interface which has a high shear strength.
(ii) Fracture in one of the materials at a position remote from the interface, removing a fragment from one material and transferring it to the other.

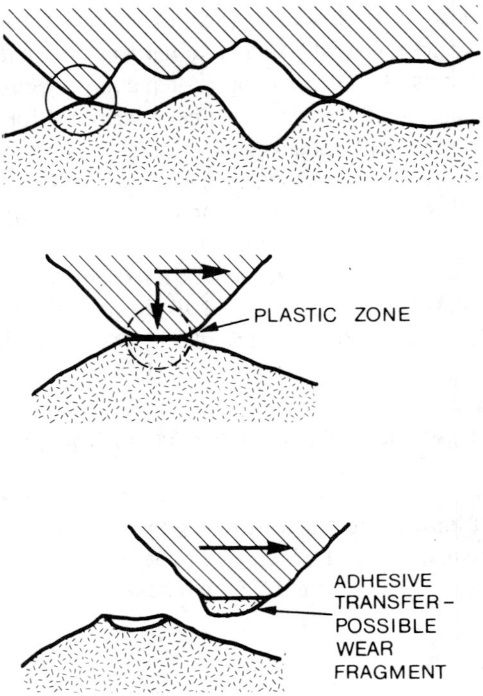

Fig. 16. Schematic representation of the classical adhesion mechanism of generating an irregular wear particle.

The formation of a wear fragment involves a third step, namely:

(iii) The detachment of the transferred fragment.

The theory encounters difficulties at each of the three stages. First, as Bikerman[30] has discussed, contact between solids in air is actually made between adsorbed layers covering oxide layers on the surfaces. Atomic contact, and hence true adhesion, is not achieved unless these layers are dislodged during rubbing. The theory requires that a very strong adhesive bond should form, and at the very least the strength of this bond must be expected to be greatly influenced by the environment that is surrounding it while it forms; little is known about this. There has even been dispute about whether bonding is really necessary, the proposal being that mechanical interlocking between irregularities, as envisaged by Coulomb,[31] formed during deformation of the asperities would suffice.

However, the greatest difficulty with the theory perhaps is that it makes no concrete predictions on how the fracture occurs in one of the surfaces. It has been suggested in general terms that fracture occurs in a region remote from the surface where work hardening during contact has been less severe than at the interface, but this begs the question of how fracture occurs under these special circumstances. Some type of fatigue mechanism can also be envisaged but this is conjectural. Kayaba et al.[13] have proposed a mechanism based on observations of the behavior of model asperities that involves tearing of the rear end of one asperity by the other (Fig. 17). Only mechanical interlocking need be assumed in this model, but it still cannot predict the primary shape of the wear particles that are to be expected.

Conjecture also surrounds the events that might occur after a transferred particle detaches and before it appears as a wear particle. For example, Sasada et al.[32] suggest that an adhered fragment grows with repeated contacts and elongates in the direction of sliding. This, they suggest, causes a gap to develop progressively between the sliding surfaces such that the adhered fragments

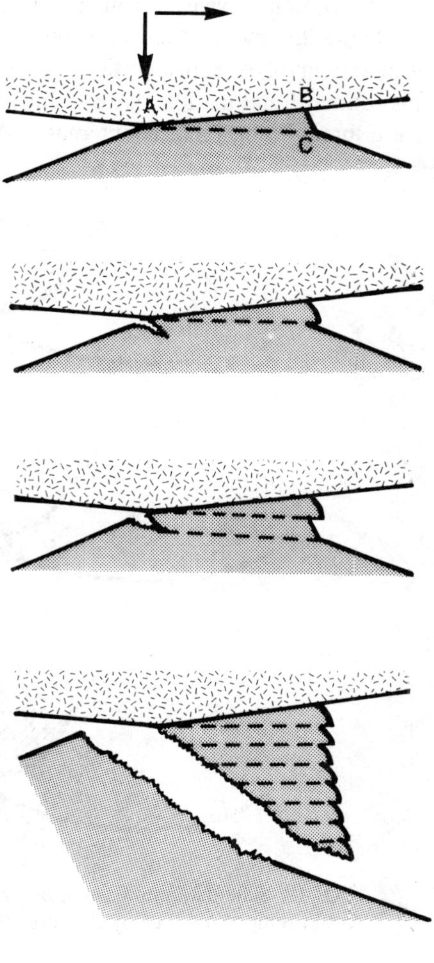

Fig. 17. Schematic representation of a mechanism of producing an irregular wear particle. This is a possible alternative interaction between sliding asperities to that illustrated in Fig. 4. *(After Kayaba and Kato, Ref 13)*

become unstable and finally break away. Again the implication is that at present the adhesion theory cannot predict the shape of the final wear debris particle.

Fracture of Interacting Asperities

The Challen and Oxley[14] slip-line field analysis of the deformation of a soft asperity by a comparatively rigid one indicates that the deformed part of the softer asperity may shear off under certain circumstances (Fig. 18). Whether or not one asperity shears instead of sliding past the contacting asperity depends, according to this analysis, on the angle of inclination of the hard asperity (α in Fig. 18) and the ratio (r) of the resolved shear stress along the interface between the two asperities and the shear flow stress of the softer asperity, the larger the value of α the smaller the value of r need be. Surface contaminant films can be expected to reduce r; consequently, contaminants that are more effective lubricants are required to reduce wear on rougher surfaces.

Note that the model does not suppose any adhesion between the two surfaces but merely involves considerations of the transfer of stress between the two. It is a two-dimensional model, like all slip-line field analyses which are assumed to have a three-dimensional counterpart. Moreover, although the theoretical anal-

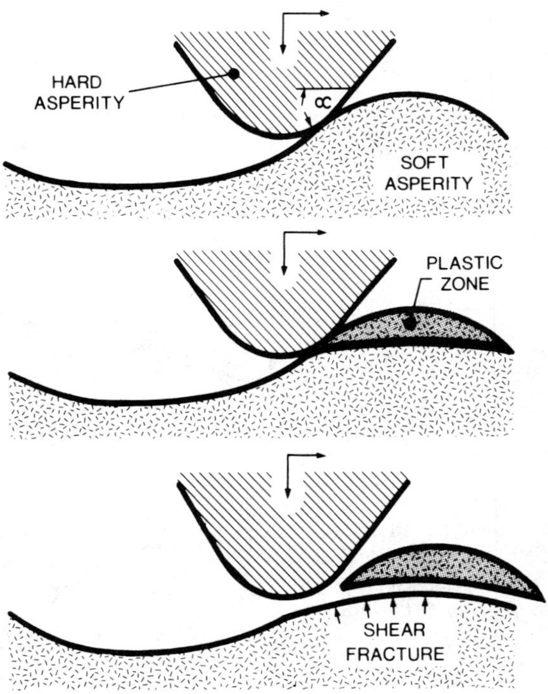

Fig. 18. Schematic representation of a mechanism by which one asperity could remove another by fracture. *(After Challen and Oxley, Ref 14)*

ysis only applies strictly when one asperity is much harder than the other, the concept can be expected to apply in principle to asperities of similar hardness.

Engineering surfaces have a distribution of asperities of varying geometry, and it would be reasonable to suppose on the basis of the Challen-Oxley model that the conditions for most asperity contacts would be appropriate for rubbing; these asperities would carry most of the normal load. Wear would then be the result of the interaction between the relatively small number of asperities that are sufficiently sharp, considering the local interfacial film strength, to detach a wear particle. This gives a possible explanation for the small value of the constant k in Archard's equation.

Note also that hard second-phase particles (e.g., carbides in steels) and abrasive particles embedded in one surface might remove wear particles by the Challen-Oxley mechanism even when their geometry is not suitable for removing a machining chip. The geometry and rigidity of such particles generally would be more appropriate for removing material by this mechanism than the topographical asperities normally present in engineering surfaces.

Fracture in a Work-Hardened Surface Layer

Hsu et al.[16] observed that flat and elongated, but chunky, particles of debris detached from several nickel-chromium austenitic steels during repeated sliding, the particles seeming to result from multiple fracturing in a brittle layer produced on the surface during sliding (Fig. 19). Deformation of alloys of this type causes transformation of the austenite to a martensitic phase and it appears that this is the phase, perhaps contaminated by transferred material, that fractures. One might also expect brittle layers to be produced when appropriate environmental conditions are present. Less marked multiple fracturing of the severely deformed surface layers perhaps might also be expected more generally, particularly in multiphase alloys. It is well established that microfracture occurs in large

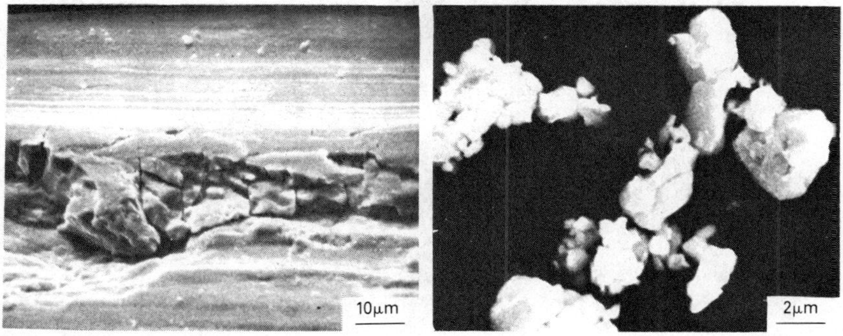

Fig. 19. (Left) Wear surface of an austenitic steel showing irregular fragmentation of the surface. **(Right)** Typical debris particles generated during this wear process.

semi-metallic phases present in otherwise ductile alloys during abrasive machining, but the behavior of smaller and less obviously brittle phases has not been investigated.

Oxide Particles

Limited observations have indicated that a small volume fraction of plate-shaped oxide particles frequently is present in wear debris from oil-lubricated systems (Fig. 20). These particles generally are so small that they are likely to be observed only by transmission electron microscopy; wear debris is only rarely examined by this technique. The plates of oxide are very thin, usually being transparent to electrons in a TEM, and have a characteristic mottled appearance in a micrograph. Their nature as an oxide can be confirmed by electron diffraction (in a TEM).

The origin and significance of these oxide particles is not clear at present. Quinn[33] has developed an oxidational theory of mild wear for dry rubbing of steels which proposes, as a first stage, the formation of a film of oxide on a rubbing asperity and, as a second stage, the removal of the oxide when it reaches a critical thickness and becomes mechanically unstable, the sequence being repeated (Fig. 21). This may well in part explain the formation of the oxide particles visible, for example, in Fig. 20. However, the oxide usually is mixed with metallic debris, in which event oxidation can be only one of the wear mechanisms operating. Quinn developed his theory for the case where oxidation is the dominant mechanism.

Nevertheless, a mechanism of this general type is a distinct possibility whenever chemical interaction between a fresh surface and its environment is likely,

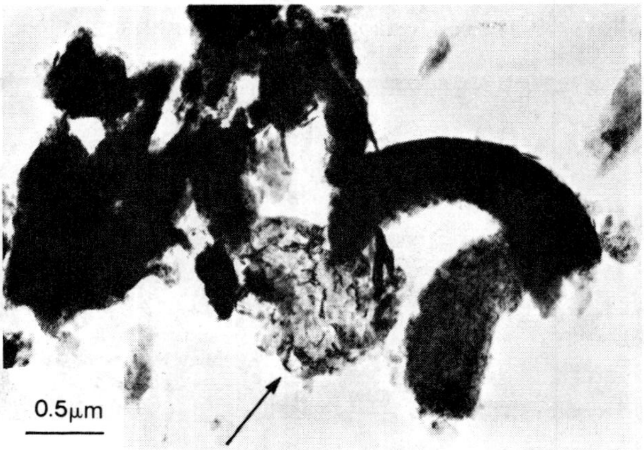

Fig. 20. Transmission electron micrograph of debris recovered from an oil-lubricated gear box. An iron oxide particle is arrowed.

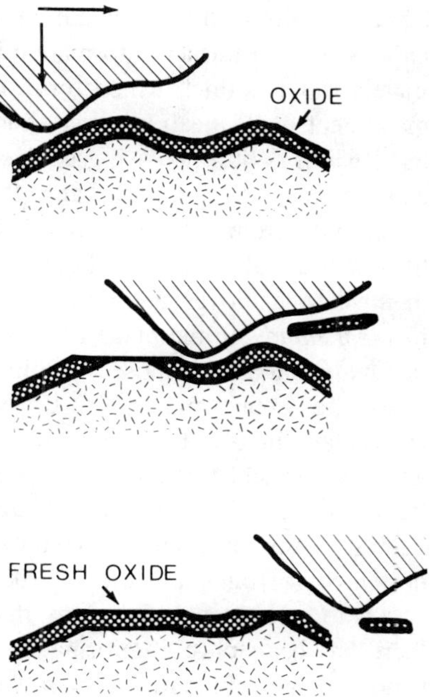

Fig. 21. Schematic representation of a possible mechanism of mild wear resulting from the repeated formation and removal of a surface oxide film. *(After Quinn, Ref 33)*

although the wear rate is likely to be low. It might be instructive if wear debris were examined more frequently by TEM for the presence of oxides or other surface reaction products.

The Structure of Worn Surfaces

Once the general characteristics of a particular wear mechanism have been established, further theoretical development of a mathematical model of wear usually is dependent on considerations of the deformation and fracture mechanisms that are likely to occur in the outermost layers of the wearing surface. These layers almost certainly are severely cold worked before any material-removal mechanism is initiated. Consequently, it is essential that the development of a wear model be based on deformation mechanisms that are known to occur in polycrystals at large strains. It is only recently that information of this nature has started to become available, but enough is now available to indicate that the extrapolation of the classical concepts relevant to deformation close to yield will almost certainly be misleading.

Starting first with a simple single-phase material of high stacking fault energy (e.g., copper),[34] deformation by the classical movement of dislocations on octahedral planes occurs only up to natural strains of about 0.1 (engineering strain of ~10% compression), relaxation after deformation causing a pattern of equiaxed cells with small-angle boundaries to develop. The slip process is then exhausted, and further deformation is concentrated in thin *microbands* approximately 0.3 μm thick which form on {111} planes and involve the cooperative movement of dislocations over short distances that are much smaller than this band width. With increasing strains, the bands rotate to become normal to the compression direction, new bands form, their boundaries sharpen, and the bands cluster into groups. Eventually this process also becomes exhausted because all the bands are aligned normal to the compression axis. A new deformation mode then has to take over, and this takes the form of deformation being concentrated in *shear bands* on the plane of maximum shear stress. Shear bands are 0.5-1.0 μm wide and are composed of thin sheets of elongated narrow cells with sharp boundaries approximately 0.2 μm wide. It is not clear yet what atomic mechanisms are involved in the deformation that occurs within the shear bands, although possibly cooperative dislocation movements constrained within the limits of the individual cells are again involved. Shear bands start to form at natural strains of about 1.0 (~65% reduction engineering strain) and become dominant at strains above 2.3 (~90% reduction).

It is to be expected, therefore, that the outermost layers of a surface in which there has been plastic deformation during wear would be composed of shear bands, perhaps mixed with microbands, aligned parallel to the surface. This has been confirmed by a detailed investigation by transmission electron microscopy, supported by electron diffraction, of the structure of surfaces of copper that had been abraded on a fine grade of silicon carbide paper;[35] such a surface can be taken to be representative of surfaces produced largely by rubbing asperities. The structures, which are summarized in Fig. 22, are complex in detail and are variable, but the essential feature is the presence at the surface of small subgrains aligned approximately parallel to the surface which are either shear band or microband structures. A generally similar structural arrangement was found by Ohmae et al.[36] and by Ives[37] beneath wear tracks on copper, although some of the detail described in Fig. 21 apparently was not found. Work by Scott and Wilman[38] and by Wheeler and Buckley,[39] using reflection electron diffraction techniques, has indicated that the surface subgrains have a characteristic orientation texture which is independent of the orientation of the parent grain and which is basically a tilted rolling texture (Fig. 23).

Much of the shear band structure in abraded surfaces of copper is recrystallized. Some investigators have found recrystallized grains in as-deformed material of this type whereas others have not; in the latter case, however, annealing for as little as 5 min at 100°C caused some recrystallization. The differences possibly are due to difference in the purity of the materials investigated. Ohmae et al.[36] did not find any evidence of recrystallized grains in the wear track that they investigated. In any event, it is apparent that very modest temperature rises (<100°C) in the surface layers during abrasion or wear are likely to introduce small recrystallized grains into the surface layers of a wear

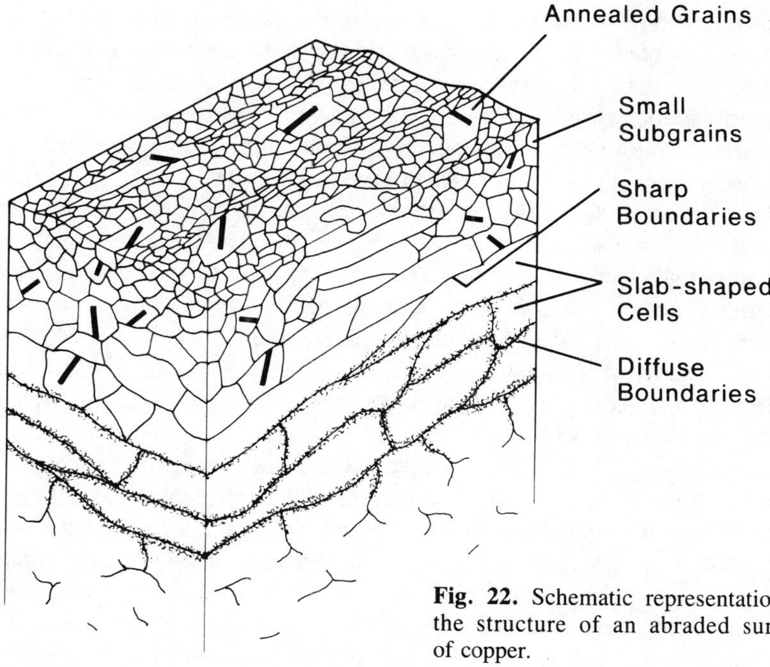

Fig. 22. Schematic representation of the structure of an abraded surface of copper.

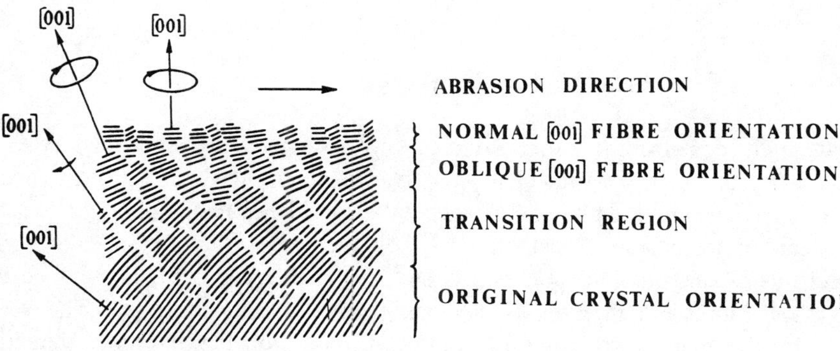

Fig. 23. Schematic illustration of the orientation texture developed on an abraded surface. *(After Scott and Wilman, Ref 37)*

track in copper, the size and distribution of these grains being sensitive to the purity of the base material and the severity of the thermal cycle to which the surface layers were subjected. This is obviously also possible in other metals, to a degree dependent on their melting points.

The deformation structures developed in cartridge brass, a representative material of low stacking fault energy, are basically similar to those just described, but somewhat modified.[40-42] Relaxation to form cells does not occur after small strains in this type of material, and dislocations, stacking faults, and twin faults produced by octahedral slip processes are the dominant features of the deformation structures. However, at natural strains of 0.5-1.0 (~40%-65% compression engineering strain) the deformation twins progressively rotate to become normal to the compression axis, which inhibits further octahedral slip. Shear instabilities in the form of shear bands then develop at strains of 0.7-1.0 (~50-65% compression), and the structure of these bands is similar to that of the shear bands in copper. The shear bands rotate with increasing strain to become aligned normal to the compression axis, and new bands form at the expense of regions containing deformation twins. The bands dominate by strains of 2.5 (~90% compression). The structures found in a finely abraded surface of 70-30 brass are summarized diagramatically in Fig. 24.[43] The essential features again are that the surface layers are composed of small subgrains and the structure changes significantly with depth.

The sketches in Fig. 22 to 24 describe the basic types of structure to be expected in worn surfaces. Many variations will occur with metastable alloys, multiphase alloys, and alloys in which phase changes can occur if the surface temperature is raised significantly. This is not the place to pursue these complexities. The main points are that a surface cannot be thought of in terms of the structure of the bulk crystals of the matrix structure, and that crystal defect interactions cannot be thought of in terms of those occurring in single crystals at small strains.

Remarks

It might seem from this discussion that we have moved from a scarcity to a superabundance of wear mechanisms (by no means have all of them been mentioned here) without having increased the confidence in any of them. This is to a certain extent true, but at least it is now recognized that a number of wear mechanisms are possible in practice and that a mixture of them may occur together. Moreover, thinking is advancing on a number of fronts. The very complexities that are emerging make it even more probable that the systematic design of wear-resistant systems will have to await the unravelling of more of these complexities.

A major inhibiting factor currently is that the pragmatic engineering approach to classifying wear, based as it is on the investigations of failures and maintenance problems, has endowed the engineering literature with a terminology that is based more on the appearance of worn surfaces than on the mechanisms that gave rise to the damage.[44] This terminology is so well embedded in the literature that perforce it must be acknowledged, although it is to be hoped that there will be some rationalization of the proliferation of terms that describe fundamentally

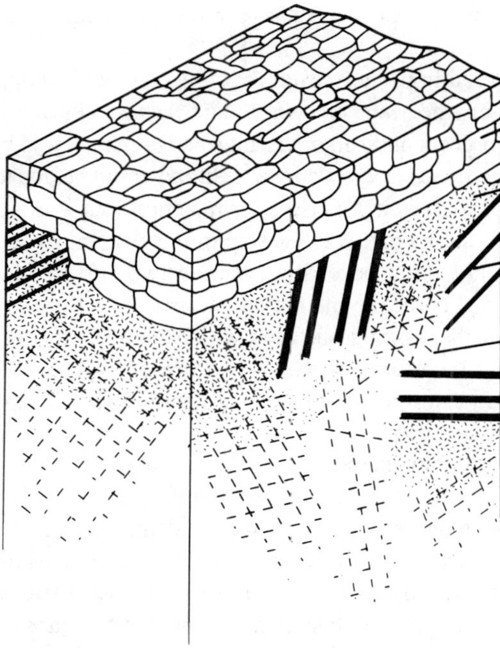

Fig. 24. Schematic representation of the structure of an abraded surface of 70-30 brass.

similar processes. Some attempt at this has already been made,[44] but there is a long way to go.

For example, the term *abrasive wear*, which is regarded by many investigators to be of the greatest industrial significance, appears to be covered by such diverse terms as *scratching*, *scoring*, *grooving*, *polishing*, *grinding*, *gouging*, and *erosion*. All of these processes probably involve a good measure of the basic abrasive cutting mechanism which has been described here, but it is possible in some cases that other basic mechanisms are also involved. Detailed investigation of both the worn surfaces and the wear debris would be necessary to establish this point.

Next, terms such as *scuffing*, *galling*, *smearing*, and *seizure* all imply that adhesion is the dominant wear mechanism. We have seen how difficult it is to attempt to quantify wear in this regime; indeed, no satisfactory empirical criteria exist for predicting the various sub-types of adhesive wear. For example, there is a consensus that the chemical environment is vitally important in scuffing, but just how this can be quantified presently is not known. Moreover, some of these terms imply plastic flow of the surface, but plastic flow alone would result in changes in topography and shape although not necessarily in loss of material. Changes in topography and shape might, of course, initiate other mechanisms of wear.

Other types of wear considered important, even if not of such general application in the field of industrial engineering, are *pitting* and *fretting*. Pitting is the commonest form of damage in gears and roller bearings and is manifested by the appearance of small cavities in the surfaces of contact. Pitting could be due to any number of mechanisms and of itself is not a cause for rejection, but it is a cause for concern because of the vibrations that it can set up in the system. *Fretting* is a term applied to a wear phenomenon occurring between two surfaces having oscillatory motion of small amplitude, the main characteristic being the appearance of oxide debris between the fretting surfaces which induces an abrasion mechanism. The process may therefore involve a mixture of oxidational and abrasion mechanisms. Detailed examination of fretting debris by a full range of modern techniques might be informative.

Although wear is emerging as an interactive multifaceted problem of great complexity, as are most problems of great engineering significance, there is room for optimism. At least the facets are being revealed and identified, and we are approaching the critical stage of the scientific process—namely, that of asking the right questions. It would seem, however, even taking an optimistic view, that there is a fair way to go before reliable mathematical models of wear systems can be expected.

References

1. H. Blok, *Engineering*, 1952 *173*, 594.
2. S. Bahadur, *J. Lubrication Tech.*, 1978, *100*, 449.
3. F. P. Bowden and D. Tabor, *The Friction and Lubrication of Solids*, Vol. 1: 1950; Vol. 2: 1966, Oxford (Clarendon Press).
4. J. F. Archard, *J. Appl. Phys.*, 1953, *24*, 981.
5. D. Tabor, in *Wear of Materials 1977*, p. 1, 1977, New York (ASME).
6. W. W. Seifert and V. C. Westcott, *Wear*, 1972, *21*, 27.
7. M. L. Atkin, R. A. Cummins, E. D. Doyle and G. R. Sharp, *Trans. Inst. Eng. Aust.* 1979, *ME4*, 40.
8. D. Scott, W. W. Seifert and V. C. Westcott, *Scientific American*, 1974, *230*(5), 88.
9. N. P. Suh, *Wear*, 1973, *25*, 111.
10. N. P. Suh, *Wear*, 1977, *44*, 1.
11. N. P. Suh and colleagues, *Wear*, 1977, *44*(1), 1-161.
12. T. R. Bates, K. C. Ludema and W. A. Brainard, *Wear*, 1974, *30*, 365.
13. T. Kayaba and K. Kato, in *Wear of Materials 1979*, p. 45, 1979, New York (ASME).
14. J. M. Challen and P. L. B. Oxley, *Wear*, 1979, *53*, 220.
15. E. D. Doyle and D. Turley, *Wear*, 1978, *51*, 269.
16. K. L. Hsu, T. H. Ahn and D. A. Rigney, *Wear*, 1980, *60*, 13.
17. J. P. Hirth and D. A. Rigney, *Wear*, 1976, *39*, 133.
18. D. Scott and M. H. Mills, *Nature*, 1973, *241*, 115.
19. E. Rabinowicz, *Wear*, 1977, *42*, 149.

20. R. D. Smith, private communication; also, M.I.T. Conference on Fundamentals of Tribology, June, 1978, publ. 1980, ed. N. Saka and N. P. Suh.
21. E. D. Doyle, *J. Aust. Inst. Metals*, 1974, *19*, 276.
22. A. W. Ruff, *Wear*, 1977, *42*, 49.
23. E. R. Bowen and V. C. Westcott, *Wear Particle Atlas*, 1976, Naval Air Engineering Center, Lakehurst, N.J. 08733.
24. E. C. Rabinowicz, *Friction and Wear of Materials*, New York, 1965 (Wiley).
25. R. C. D. Richardson, *Wear*, 1968, *11*, 245.
26. F. Hirano and A. Ura in *Tribology in Steel Works*, Iron & Steel Inst. Publication No. 125, 1970, 163.
27. T. O. Mulhearn and L. E. Samuels, *Wear*, 1962, *5*, 478.
28. A. J. Sedriks and T. O. Mulhearn, *Wear*, 1963, *6*, 457.
29. A. J. Sedriks and T. O. Mulhearn, *Wear*, 1964, *7*, 451.
30. J. J. Bikerman, *Wear*, 1976, *39*, 1.
31. C. A. Coulomb, *Mem. Math. Phys. Acad. Sci.*, 1785, *10*, 161.
32. T. Sasada, S. Norose and N. Mishina, in *Wear of Materials 1979*, pp. 72-80, New York (ASME).
33. T. F. J. Quinn, *Brit. J. Appl. Phys.*, 1962, *13*, 33.
34. A. S. Malin and M. Hatherly, *Met. Sci.*, 1979, *13*, 463.
35. D. Turley and L. E. Samuels, to be published.
36. N. Ohmae, T. Tsukizoe and F. Akiyama, *Phil. Mag.*, 1979, *404*, 803.
37. L. K. Ives, in *Wear of Materials 1979*, p. 246, 1979, New York (ASME).
38. V. D. Scott and H. Wilman, Proc. Roy. Soc., 1958, *247A*, 353.
39. R. Wheeler and D. M. Buckley, *Wear*, 1975, *33*, 65.
40. D. M. Turley, *J. Inst. Metals*, 1969, *97*, 237.
41. D. M. Turley, *Met. Trans.*, 1971, *2*, 3233.
42. M. Hatherly and A. S. Malin, *Met. Tech.*, 1979, *6*, 308.
43. D. M. Turley and L. E. Samuels, *J. Aust. Inst. Metals*, 1972, *17*, 11.
44. Friction, Wear and Lubrication, Terms and Definitions, OECD, Paris, 1969.

Update on the Delamination Theory of Wear

NAM P. SUH
Massachusetts Institute of Technology
Cambridge, Massachusetts 02139

Abstract

The recent work on the delamination theory of wear is reviewed. Particular attention is given to three aspects of the theory: the genesis of friction which affects the wear process; the details of crack nucleation and propagation mechanisms; and the wear of polymeric materials under sliding conditions. In the first part of the paper, it is shown, contrary to the postulates of the adhesion theory of friction, that the frictional force arises from three different sources (i.e., asperity deformation, plowing by wear particles and hard asperities, and adhesion), which depend on the sliding history and environment. At any given loading contact only one of these mechanisms operates. As a consequence, the local surface traction can vary from contact to contact, affecting the crack nucleation and propagation phenomena. The second part of the paper deals with model experiments which show that the crack nucleation around a hard inclusion is dependent on the inclusion size. This result is consistent with theoretical predictions. The effect of various crack configurations on crack propagation is also discussed. Finally, the delamination theory of wear is applied to pure polymers to develop practical criteria for the onset of wear. These criteria have been used to minimize the sliding wear of polymeric materials.

Introduction

The history of tribology is both very old and new. It is old in the sense that mankind has always been concerned with frictional phenomena and has used them to its advantage. Scientific inquiry on the subject dates back to the era of Leonardo da Vinci.[1] On the other hand, it is new in the sense that a great deal of basic knowledge is still being advanced which is drastically changing the nature of scientific inquiry on tribology. Yet, as in most scientific disciplines, the new and the old are intimately interwoven, since new theories can only be generated based on the available facts and through questioning of old theories.

The field of tribology is complex. Different mechanisms are responsible for friction and wear of a given material depending on the operating conditions, and there can be more than one mechanism governing a given tribological phenomenon. This paper is concerned with only the wear of materials under sliding conditions where the interfacial temperature rise is not high enough to induce significant chemical interactions between the sliding surfaces. Therefore, it excludes such chemical behavior dominated phenomena as the wear of carbide tools at high cutting speeds[2,3] and such mechanical behavior dominated wear phenomena as abrasive and erosive wear.[4,5]

The sliding wear of metals with typical machined surfaces is dependent on the history of sliding, the environment, the applied load and the surface topography as well as on the microstructure and the basic properties of metals. To a first approximation, the wear volume of a typical metal may be schematically plotted as a logarithmic function of the sliding distance as shown in Fig. 1. Initially very high wear rates are observed as the original surface asperities are removed (represented by Region I). When the applied normal load is very low or when the surface is lubricated, the removal of these asperities is followed by a period of no observable wear and the appearance of a highly polished surface (Region II). While the surface appears very smooth, the subsurface is subjected to changes which lead to the delamination process. When wear sheets delaminate, the wear rate increases rapidly again (Region III). It should be noted that Region II may be absent when the applied load is very high. When the surface is lubricated, the material will be in Region II for a long time in the absence of pre-existing subsurface cracks.

The foregoing sliding wear process has been investigated extensively by the author and his co-workers.[6-10] It was done to verify the postulates and predictions

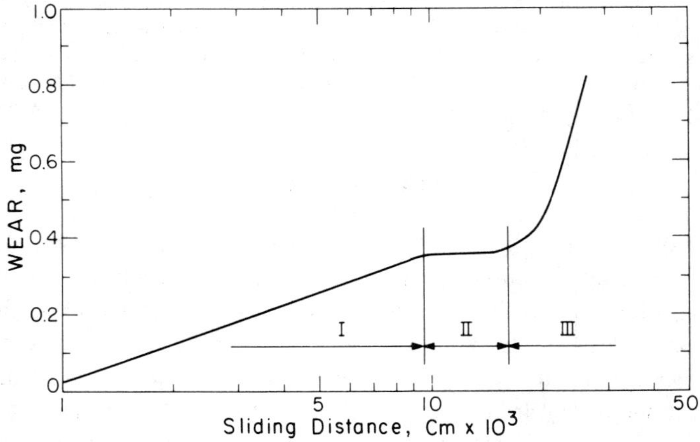

Fig. 1. Wear volume of typical metal vs. sliding distance.

of the delamination theory of wear which were advanced in 1973 to explain the sliding wear of materials,[11] since they were made through theoretical reasoning in the absence of any experimental evidence. Since the advance of the theory, many papers have appeared in the literature supporting the theory.[12-18]

The purpose of this paper is to review the most recent work on the delamination theory of wear. The emphasis will be given to the results obtained since the publication of the early M.I.T. work, which resulted in a special publication of *Wear*.[6] It first reviews the essence of the theory and then examines the details of selected aspects based on recently obtained experimental and theoretical results. It specifically discusses the question of surface plasticity, the genesis of friction between sliding surfaces, and the mechanics of crack nucleation and propagation. Then the theory is extended to polymeric materials, leading to the development of criteria for wear of polymeric materials without any inclusions. Experimental results in support of these criteria are also presented.

Review of the Delamination Theory of Wear

The delamination theory of wear describes the wear mechanisms responsible for the most destructive failure of sliding surfaces. According to the theory, wear takes place by the deformation of the surface layer, nucleation of cracks at the subsurface, and the propagation of these cracks nearly parallel to the surface. Eventually, loose sheet-like wear particles are generated. The wear particles generated by delamination wear differ in shape from those generated by the deformation and fracture of asperities and by plowing. The wear particles generated by the removal of the surface asperities are small, whereas those associated with plowing tend to be long in one dimension. The ultimate wear rate and the failure of many sliding surfaces, however, is controlled by the delamination mechanisms.

The delamination theory of wear describes the following sequential (or independent, if there are pre-existing subsurface cracks) events which lead to loose wear sheet formation:

(1) When two surfaces come into contact, normal and tangential loads are transmitted through the contact points. Asperities of the softer surface are easily deformed and fractured by the repeated loading action. Hard asperities are also removed but at slower rates. A relatively smooth surface is initially generated, either when these asperities are deformed or when they are removed.

(2) The surface traction exerted at the contact points induces incremental plastic deformation per cycle of loading which accumulates with repeated loading. (The increment of permanent deformation remaining after given cyclic loading is small compared with the total plastic deformation that occurs in that cycle as a result of the reversal of shear deformation.[6] Most of the energy is dissipated by this mechanism.)

(3) As the subsurface deformation continues, cracks are nucleated below the surface. Crack nucleation cannot occur very near the surface because of the triaxial state of compressive loading which exists just below the contact region.

(4) Once cracks are nucleated (pre-existing voids and cracks will play a similar role), further loading and deformation causes the cracks to extend and propagate, eventually joining with neighboring cracks. The cracks tend to propagate parallel to the surface at a depth governed by material properties and the state of loading. When the cracks cannot propagate because of the small stress concentration at the crack tip due to an extremely small surface traction at the contact area, crack nucleation is the rate-controlling mechanism. In this case, the wear particles may not be sheet-like, but large thick chunks such as those observed in gears and splines.

(5) When the cracks finally shear to the surface, long and thin wear sheets delaminate. The thickness of the wear sheet is controlled by the location of subsurface crack growth, which is controlled by the normal and tangential loads at the surface.

In the original paper on the delamination theory of wear,[11] it was postulated that the surface layer may be softer than its subsurface because of dislocation instability very near the surface. It was argued that some of the dislocations generated during plastic deformation must escape through the newly generated clean surface because of the image force acting on the dislocations. Obviously, if there is a contaminated surface layer (such as an oxide layer), the dislocations cannot escape and thus accumulate near the surface to make it harder.

Each of these aspects of the delamination theory will be examined in greater detail in light of the knowledge gained since the publication of the original paper.[11]

Flow Stress of the Surface Layer

The possibility that the dislocation density at the *very* surface may be less than that of the subsurface has generated more intense discussions than perhaps any other aspects of the theory.[12,19-21] This is in spite of the fact that the question of whether or not the very surface layer is softer than the subsurface region does not significantly affect the subsequent mechanisms of the delamination wear process. The experimental work done by others will be reviewed here to offer the author's perspective on this matter.

One of the difficulties in verifying the existence of the soft layer (with a lower dislocation density than the substrate) has been experimental. Although the hardness of various metals can be measured by conventional indentation methods, the flow stress of the very surface layer cannot be conclusively determined by these methods. By using indentation techniques, some found the surface to be softer,[22] while others obtained the opposite results (see Ref 23 for a general review).

Transmission electron microscopy was also used by Doyle and Turley[12] and Van Dijck[24] to determine the microstructure of the surface. They reported that there was no dislocation-depleted zone near the surface. Ruff[21] used an electron channeling technique to determine the residual stress distribution near the surface. Concluding from his results that the residual stress is highest at the surface, he thus inferred that the dislocation density must be the highest at the surface. The residual stress is an elastic state of stress, however, and cannot be used as an indication of the dislocation density and/or hardness.

Mitchell and Laufer recently published a most thorough experimental paper on this subject.[20] Using X-ray and electron diffraction and transmission electron microscopy (TEM), they examined the surface of a hexagonal close-packed Ti-6%Al–4%V alloy which had been abraded by extrusion honing.

Based on analysis of X-ray diffraction patterns and the TEM and electron diffraction results, they concluded that the extrusion-honed Ti-6%Al–4%V alloy consisted of thin subgrains, derived from the original grains, which are several tens of nanometers in diameter and of the order of 12 nm thick. From the TEM and electron diffraction analysis they found that dislocations within the subgrain are not observed in general and their absence is consistent with the absence of microstrain line broadening in the X-ray diffraction results. They also determined that the surface layer was under a high hydrostatic compressive stress, indicating that considerable plastic flow has occurred in the surface layer. This finding on compressive stress is consistent with that of Ruff.[21] They also concluded that there appears to be a region of high dislocation density at a depth of a few micrometers. They found the zone of low dislocation density to extend to 2-3 μm, which is an order of magnitude larger than the depth at which the image force can exert a sufficient force to attract the dislocations to the surface.

The work of Mitchell and Laufer supports the original postulate that in metals with no coherent oxide or other similar contaminant layer a dislocation-depleted zone could exist. The significance of such a zone is that the surface layer may undergo large plastic deformation without work hardening and crack nucleation. Additional implications are that the surface can easily be plowed and that the flow stress of the surface layer may be less than that of the subsurface layer. If this is true, the size of the plastically deformed zone at the crack tip will be the same order of magnitude as the depth of the crack. Such a crack cannot propagate.

It should be noted that the results of Mitchell and Laufer are consistent with the results of Doyle and Turley[12] in that both confirm the formation of a fine subgrain structure. However, they differ on dislocation density. The difference might be that the latter relied on only TEM measurements, whereas the former used X-ray and electron diffraction techniques as well as TEM. The formation of subgrains is also supported by Van Dijck[24] and Dautzenberg.[25] Dautzenberg claims that the subgrain formation is a result of recrystallization. However, since recrystallization involves annihilation of dislocations because of their enhanced mobility at higher temperatures, it may be possible to form a similar structure even at low temperatures if dislocations have sufficient mobility. It is interesting to note that none of these papers has found any evidence of the formation of dislocation cells.

The Genesis of Friction and Its Effect on Delamination Wear

It is well known that the coefficient of friction has a significant effect on the wear rate.[26] According to the delamination theory of wear this is due to the effect of the surface traction on the plastic deformation, crack nucleation and crack propagation rate. Therefore, it is imperative that we properly understand the genesis of friction, because the magnitude of the surface traction at each contact point, which will in turn affect the crack nucleation and propagation rates, cannot be determined without this understanding. In the past, since the real genesis of friction was not known,[6,7] some gross assumptions had to be made on the surface traction in analyzing crack nucleation.

The view currently accepted by many tribologists is that the friction is caused by the *adhesion* of asperities.[1,27] The essence of the adhesion theory is that the asperities of the sliding surfaces come into contact with opposing asperities and form welded junctions which must be sheared to satisfy the kinematic requirements of the sliding surfaces. According to this view, surface contaminants do not play a significant role, since clean surfaces eventually become exposed through the sliding action. The frictional force, according to the adhesion theory, depends directly on the actual area of contact which is a function of the applied normal and tangential load. Much of the past research was concerned with the actual area of contact to explain why the measured friction coefficient is larger than predicted values. The surface energy argument[27] and the possibility that some junctions may in fact be supporting tensile loads rather than compressive loads[28] were advanced to explain this discrepancy between the experiment and the predictions of the adhesion theory. In order to explain the compatibility between two sliding surfaces based on the adhesion concept, a mutual solubility criterion has been proposed by Rabinowicz[29]—i.e., the higher the solubility the higher the friction coefficient. However, there are many exceptions to this rule.

There are other difficulties in accepting the adhesion theory of friction. The environment has a very significant effect on the steady state of friction; for example, the coefficient of friction is usually much higher in vacuum than in air. Although the experimental conditions in vacuum satisfy the conditions assumed in plasticity analysis,[28] the discrepancy between adhesion theory and experiment is largest when the experiment is done in a clean environment. Furthermore, the presence of wear particles between the sliding surfaces affects the friction coefficient significantly. Figure 2 illustrates the dramatic decrease in friction

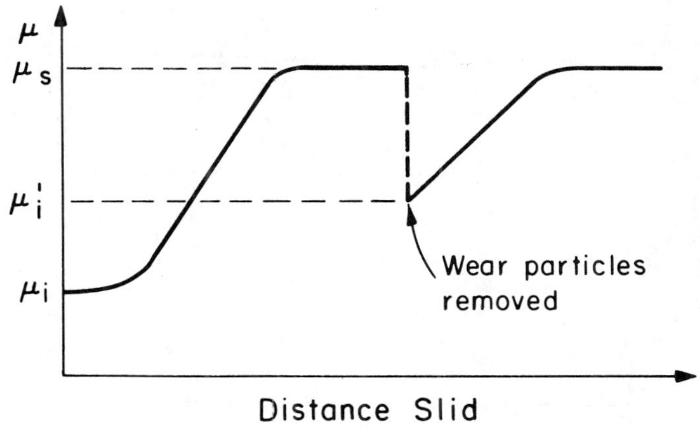

Fig. 2. Effect of removing wear particles. (For Armco iron/Armco iron specimen $\mu_i = 0.13$, $\mu_s = 0.71$, $\mu_i' = 0.4$.)

coefficient when the wear particles are removed. The coefficient of friction eventually reaches again the steady-state friction coefficient when the sliding is continued after the removal of wear particles. Another difficulty with the adhesion theory is that the friction coefficient is dependent on the sliding distance. As shown in Fig. 3, the friction coefficient at the early stages of sliding is very low, being in the neighborhood of 0.1 to 0.2, and then increases to a steady-state value. The initial value of the friction coefficient is not greatly affected even when lubricants are applied to the sliding interface. The most damaging evidence against the adhesion theory is that the steady-state coefficient of friction for the same pair of steels can differ by as much as 0.2, depending on which is the stationary slider and which is the rotating specimen.[30]

In order to explain the foregoing experimentally observed frictional behavior, which was investigated extensively at M.I.T. by the author and his co-workers,[30,31] the author proposed a new theory for the genesis of friction based on the following postulate:[32]

> The coefficient of friction between the sliding surfaces is due to the various combined effects of asperity deformation, μ_d, plowing by wear particles and hard surface asperities, μ_p, and adhesion between the flat surfaces, μ_a. The relative contributions of these components depend on the condition of the sliding interface, which is affected by the history of sliding, the specific materials used, the surface topography and the environment.

In order to clarify the above postulate, the time-dependent behavior of friction was subdivided into the following stages based on experimental observations (see Fig. 4):

Stage I—In this early stage of dynamic friction the coefficient of friction is largely dictated by plowing of the surfaces by hard particles and asperities.

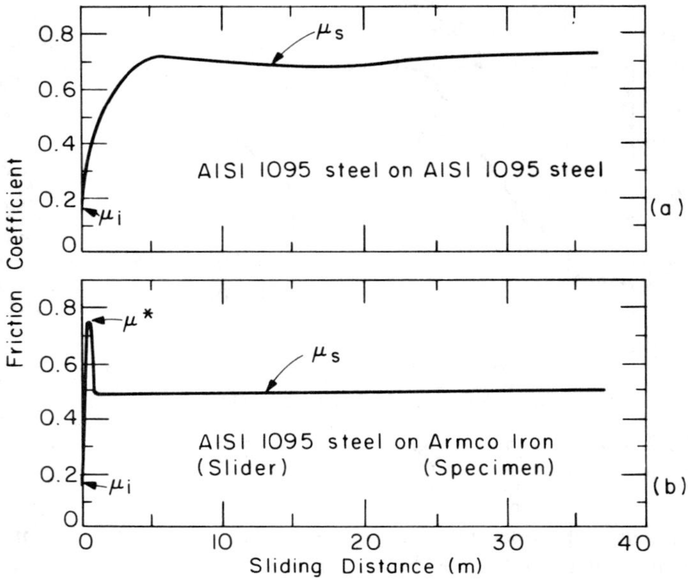

Fig. 3. Coefficient of friction vs. sliding distance.

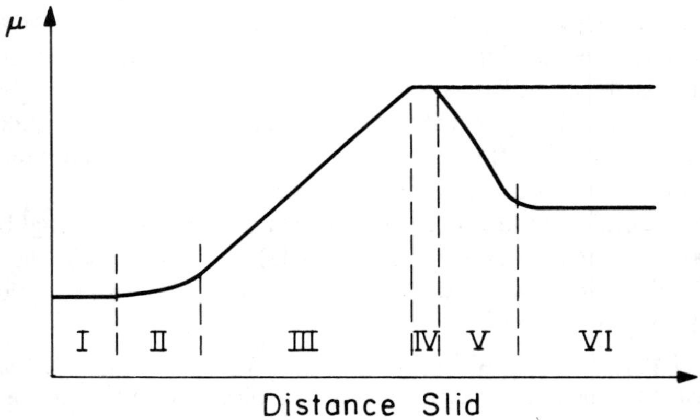

Fig. 4. Six stages of frictional force vs. distance slid.

Because of the contaminated nature of the surface, adhesion does not play a significant role in this stage. Consequently, the coefficient of friction in this stage, μ_i, is largely independent of material combinations, the environmental conditions, and the surface finish. The major effect of initial surface finish is to affect the static coefficient of friction. However, the asperities deform very readily (perhaps in one pass of asperity contact) and consequently do not contribute to the dynamic coefficient of friction in Stage I.

Stage II—In this second stage, the frictional force begins to rise slowly as a result of increase in adhesion between two nearly flat surfaces. When the interface is lubricated, Stage I persists for a long time because of slow delamination and the absence of wear particles created by delamination, and Stage II may not be present. The slope in Stage II can be steeper if the wear particles (generated by the asperity deformation and fracture) are entrapped between the sliding surfaces and plow the surfaces.

Stage III—This stage is characterized by a steep increase in slope because of the rapid increase in the number of wear particles entrapped between the sliding surfaces as a consequence of delaminated wear particle generation. The slope is also affected by the increase in adhesion arising from the increase in clean interfacial areas. The force required to deform the asperities will continue to contribute to the frictional force in this stage if new surface asperities are generated. The wear particles are generated when the process of wear particle formation by subsurface deformation, crack nucleation and crack propagation postulated by the delamination theory of wear[12] is completed. Some wear particles will also be generated as a consequence of asperity deformation and removal. Some of the wear particles get entrapped between the surfaces, causing plowing. The friction will be greater when the wear particles are entrapped between metals of nearly equal hardness, because they will penetrate into both surfaces, hindering slippage between the particle and the surface.

Stage IV—This stage is reached when the number of wear particles entrapped within the interface remains constant. This occurs when the number of the newly entrapped particles equals the number of entrapped particles leaving the interface. The adhesion contribution to friction also remains constant in Stage IV. The asperity deformation also contributes to the frictional force, since the wear by delamination creates new rough surfaces with asperities. When two like metals are slid against each other or when the mechanisms responsible for Stage V do not play a significant role, the coefficient of friction in Stage IV is the steady-state frictional coefficient between the two metals.

Stage V—When a very hard stationary slider is slid against a soft specimen, the asperities of the hard surface are gradually removed, creating a mirror finish as shown in Fig. 5. In this case the frictional force decreases because of the decrease in plowing of the surfaces by the wear particles and the reduction of surface asperities. Wear particles cannot plow since they cannot penetrate and anchor into the hard surface.

Stage VI—Eventually, when the hard surface becomes mirror smooth to a maximum extent, the softer surface also acquires the same mirror finish and the frictional force levels off. The surfaces are never completely smooth since there are always craters caused by the creation of delamination wear particles. These craters provide anchoring points for wear particles. When the hard material is the rotating specimen sliding against the stationary softer surface, in a cylinder-on-cylinder test the hard surface remains rough, probably because polishing cannot

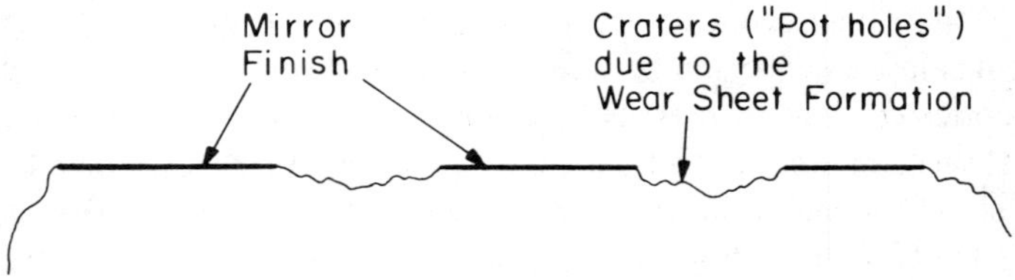

Fig. 5. A hard stationary surface polished by a soft surface.

take place for geometric reasons. In this case, Stages V and VI are not present.

Based on this physical model for the genesis of friction, Suh and Sin[32] investigated the three friction components analytically. The main results of the analyses will be briefly reviewed here.

In order to determine the friction coefficient contributed by the asperity deformation, the interaction of two representative asperities approaching each other was considered. These asperities have to deform in such a manner that the resulting displacement field is compatible with the sliding direction and the sum of the vertical components of the surface traction at the contacting asperities is equal to the applied normal load. A possible slip-line field that satisfies the kinematic condition is given in Fig. 6. Although one or both of the asperities may deform depending on their relative hardness and the geometric shapes, the figure shows the case of one of the asperities deforming plastically. The solution requires that the shear stress along OA be whatever is necessary to constrain the resulting deformation in the sliding direction. This would be possible even under the lubricated conditions if the interface OA is not perfectly smooth but rough enough to allow mechanical interlocking.

The coefficient of friction due to asperity deformation (μ_d) can be analytically determined, assuming that all the externally applied normal load is carried by the asperities. This would be the case when two machined surfaces are statically in contact. Plasticity analysis using the slip-line field shown in Fig. 6 yields the following expression for μ_d:[32]

$$\mu_d = \frac{F}{L} = \frac{\left(1 + \frac{\pi}{2} + 2\alpha - 2\theta - 2\theta'\right) \sin\theta + \cos(2\alpha - \theta)}{\left(1 + \frac{\pi}{2} + 2\alpha - 2\theta - 2\theta'\right) \cos\theta + \sin(2\alpha - \theta)} \qquad [1]$$

Equation 1 states that μ_d is independent of material properties but depends on the geometry of asperities. The solution is plotted in Fig. 7, which shows that μ_d varies from 0.42 to 1 as the slope of the asperities changes from 2 to 45°. (The slope of typical asperities is much less than 45°.) These values are close to the static coefficients of friction between various materials.[33]

Once the sliding commences, however, the value of μ_d given by Eq 1 is expected to be only an upper bound. When the existing asperities deform, there will not be any more asperity interactions unless new asperities are generated by delamination of wear sheets. However, the asperities deform readily (perhaps within a few encounters with the asperities of the counterface), while the generation of new asperities may take a long period of time since the delamination of wear sheets often requires a large number of loading cycles.[6] Therefore, the contribution of asperity deformation to the dynamic coefficient of friction is expected to be smaller than that predicted by the plasticity analysis under the assumed conditions. Experimental results indicate that μ_d is about 0.3 when AISI 1045 steel is slid against AISI 1090 steel.[32]

It should be noted that the slip-line analysis done to determine μ_d is not very different from that of Green,[28] who presented similar solutions by considering the deformation of asperities under the influence of the load exerted by welded asperity junctions of various strengths. It should be emphasized that these solutions are only approximate since they ignore the elasticity of the material, which can affect the nature of deformation at these contact points. However, the predicted values should be reasonable estimates of the forces involved when there is large plastic deformation.

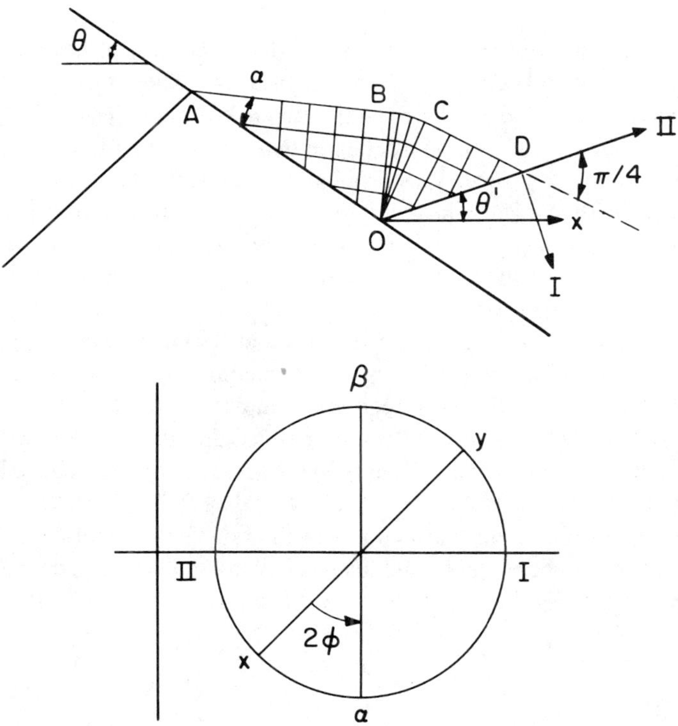

Fig. 6. A slip-line field for asperity contact. *(Ref 32)*

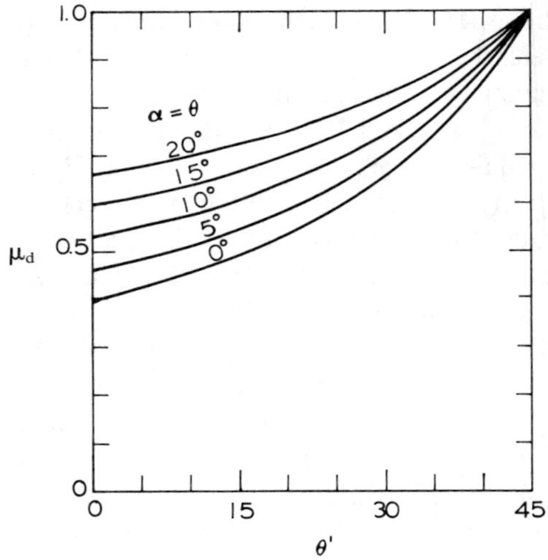

Fig. 7. Slip-line field solution for friction as a function of the slope of asperities. *(Ref 32)*

A frictional force can also arise from the adhesion of two nearly flat surfaces. Unlike the contribution from the deformation of asperities, this frictional force is a function of the adhesion between the two opposing surfaces. The adhesion force arises either because of the welding of two nearly flat portions of the surface or because atoms are brought close enough together for interatomic interactions (but without welding). The adhesion at the slopes of two interacting asperities may be neglected in this analysis since it has already been considered in deriving μ_d. Also, in that case, the deformation is kinematically constrained and is independent of the quality of adhesion.

The adhesion term, μ_a, can be analyzed by considering two nearly flat surfaces coming into contact as shown in Fig. 8. Depending on the nature of adhesion along the interface ED, the force required to move the interacting surfaces with respect to each other will vary. When there is no adhesion the force will be zero and when there is complete adhesion it will reach a maximum. The solutions to this problem can be obtained again using the slip-line fields. The exact geometric shape of the slip-line field will depend on the boundary condition at ED. The solution sought can be adapted from the recent work of Challen and Oxley,[34] who derived an expression for the friction coefficient as

$$\mu_a = \frac{f}{A + \sin(\cos^{-1} f)} \qquad [2]$$

$$A = 1 + \frac{\pi}{2} + \cos^{-1} f \qquad [3]$$

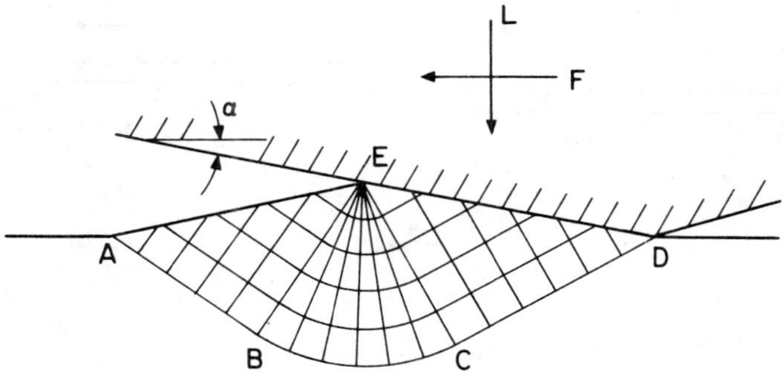

Fig. 8. A slip-line field for rubbing contact. *(Ref 32)*

f equals the strength of the adhesion at ED, expressed as a fraction of the shear flow strength of the softer material; μ_a varies from 0 to 0.39 as f changes from 0 to 1.

The friction coefficient contributions determined by Eq 2 and 3 are based on the assumption that all the applied normal load is carried by the flat interfaces. However, since part of the normal load is also carried by other contacts (such as purely elastic contacts, the interacting asperity junctions discussed in the preceding section, and the entrapped wear particles), μ_a under typical sliding conditions should be less than 0.4. The experimental results obtained with the hard AISI 1095 steel slider and the soft Armco iron specimen showed that the steady-state coefficient of friction reached a value of 0.51 when both surfaces were polished smooth and thus the friction was caused primarily by adhesion, although there must have been some contributions by asperity deformation and wear particles.

The plowing component of the frictional force can be due to the penetration of hard asperities or to the penetration of wear particles. The plowing due to wear particles is schematically illustrated in Fig. 9. When two surfaces are of equal hardness, the particle can penetrate into both surfaces. As the surfaces move with respect to each other, grooves will be formed in one or both of the surfaces. When one of the surfaces is very hard and smooth, the wear particle will simply slide along the hard surface and no plowing can occur. However, when the hard surface is very rough, wear particles can anchor in the hard surface and plow the soft surface.

Investigation by Sin et al.[5] of the friction due to plowing showed that the contribution of plowing to the friction coefficient is very sensitive to the ratio of the radius of curvature of the particle tip to the depth of penetration. The friction coefficient by plowing, μ_p, is given by:

$$\mu_p = \frac{2}{\pi}\left\{ \left(\frac{2r}{w}\right)^2 \sin^{-1}\frac{w}{2r} - \left[\left(\frac{2R}{w}\right)^2 - 1 \right]^{1/2} \right\} \qquad [4]$$

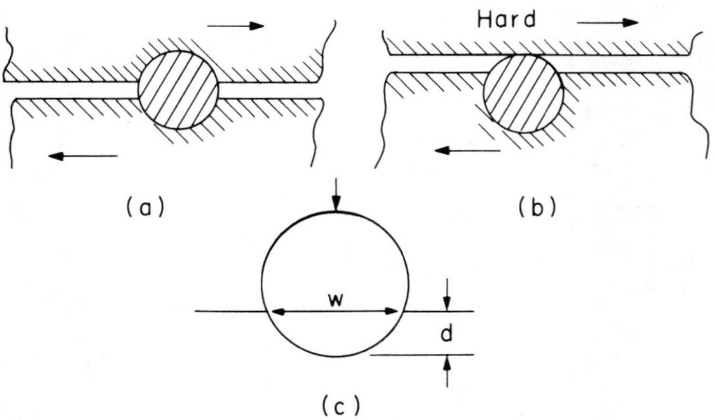

Fig. 9. Idealized model of wear particle interaction between two sliding surfaces.

where w is the width of the penetration and r is the radius of curvature of the tip of the abrasive particle. The ratio of w/r measured by sectioning some of the worn steel specimens is in the neighborhood of 0.8. Substituting this value into Eq 4, the plowing coefficient of friction is found to be 0.2. This value is in the same range as the reduction in the friction coefficient observed when the wear particles are removed from the Armco iron/Armco iron and Armco iron/AISI 1095 steel interfaces, which were 0.31 and 0.16 respectively. The range of possible values of μ_p as a function of the ratio w/2r is 0 to 1 as shown in Fig. 10.

Plowing not only increases the total frictional force and delamination wear but also creates small wear particles which in turn affect the subsequent wear of sliding surfaces. Plowing action forms ridges along the sides of plowed grooves. When these ridges are deformed flat and subjected to repeated loading, some become loose wear particles with continued sliding.

Based on the experimental observations and the analytical results, the components of frictional force can be summarized as follows:

μ_d—From about 0.42 to 1 as the slopes of asperity surfaces change from 2° to 45°, regardless of the nature of metals, if the asperities are the only load-carrying contacts at the interface. In actual interfaces, μ_d is expected to be considerably smaller than these values because the asperities deform readily, while the generation of new asperities takes a long time. This component is present whenever there are surface asperities. The friction due to asperity deformation is the primary friction component at the transition between static and dynamic friction.

μ_a—From about 0 to 0.4, depending on the nature of adhesion between the flat part of the interacting surfaces. The low value is for a well-lubricated surface with light lubricant; the high value is for identical metals sliding

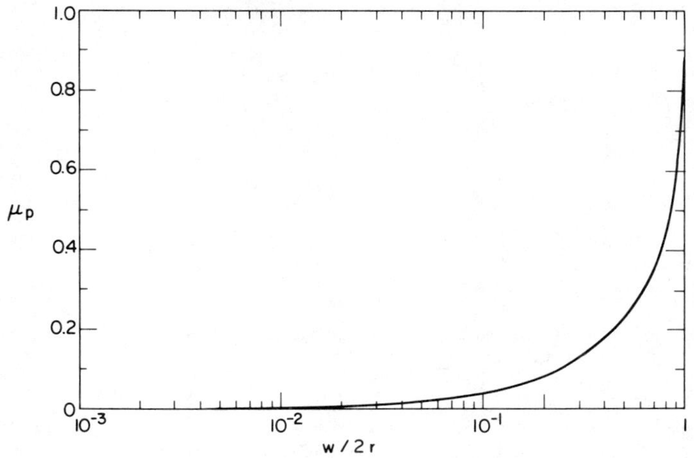

Fig. 10. Plowing component of friction coefficient as a function of the ratio of width to diameter of the entrapped wear particle.

against each other without any surface contaminants and oxide layers.

μ_p—From nearly 0 to 1.0, depending on the depth of penetration, but normally less than 0.2 in a typical situation. The high values are associated with two identical metals sliding against each other with entrapped wear particles penetrating into the surfaces, while the low value is obtained either when wear particles are totally absent from the interface or when a soft surface is slid against a hard surface with a mirror finish.

It is difficult to determine the relative contributions of μ_d, μ_a and μ_p to the total frictional coefficient, because analyses for μ_d and μ_a assume that the total normal load is carried either by asperities or by flat contact areas. In real situations the normal load will be apportioned among the asperity contacts, flat adhesion junctions, and the entrapped particles. However, it is plausible in many real situations that these mechanisms operate sequentially rather than concurrently. Consider, for example, a flat junction and an asperity in contact as shown in Fig. 11. The flat areas may come into contact first and form an adhesion junction. With further sliding the asperities will come into contact and must shear along the dotted line if the materials are identical or along the crossed line if the top asperity is much harder than the bottom asperity. In this case, only a tangential force is required to deform the asperity plastically along the direction of sliding.

The coefficient of friction is plotted in Fig. 12 as a function of environment (i.e., adhesion), plowing and roughness, which is defined as the "friction space." The adhesion is expressed in terms of the non-dimensional interfacial shear strength, f, of flat contacts, which is defined in Eq 2. The roughness is plotted in terms of the slope of the surface, while the plowing is given in terms of the

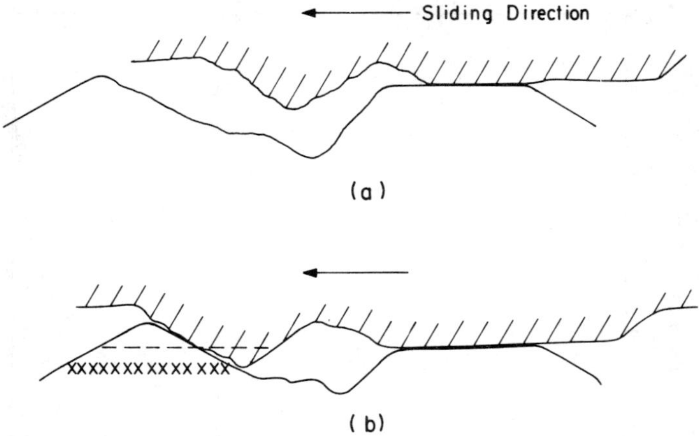

Fig. 11. Two sliding surfaces in contact.

ratio of the width of penetration to the wear particle radius. The friction space is bound by the lower friction surface that represents the case of no surface asperities, i.e., $\theta = 0$. The friction surface $\theta = \theta_i$ represents the case of initial surface roughness, while the $\theta = \theta^*$ friction surface represents the steady-state case. The coefficient of friction under steady-state conditions is bound by the friction surfaces of $\theta = 0$ and $\theta = \theta^*$ but can assume any value within the friction space depending on the environment, wear particle entrapment and surface roughness. For example, the history of the frictional coefficient change from the static value through the various stages of dynamic friction can be represented as shown in Fig. 12. It starts from the friction surface $\theta = \theta_i$ and moves around in the friction space. A possible path is indicated by the arrows and the heavy line shown in Fig. 12.

An important conclusion of this review of the genesis of friction is that the coefficient of friction is not an inherent material property. It depends on such factors as surface topography, environment and sliding distance in addition to material properties.

Implications of the Genesis of Friction on Delamination Wear

The foregoing discussion of frictional behavior is important to an understanding of delamination wear mechanisms for the following reasons:

(1) Although the overall coefficient of friction can reach a large value, the local surface traction at any one of the contacts cannot have a shear stress exceeding that predicted by the plasticity consideration for a given type of contact. That is, the shear stress at a flat contact cannot exceed 0.4 times the normal stress if all

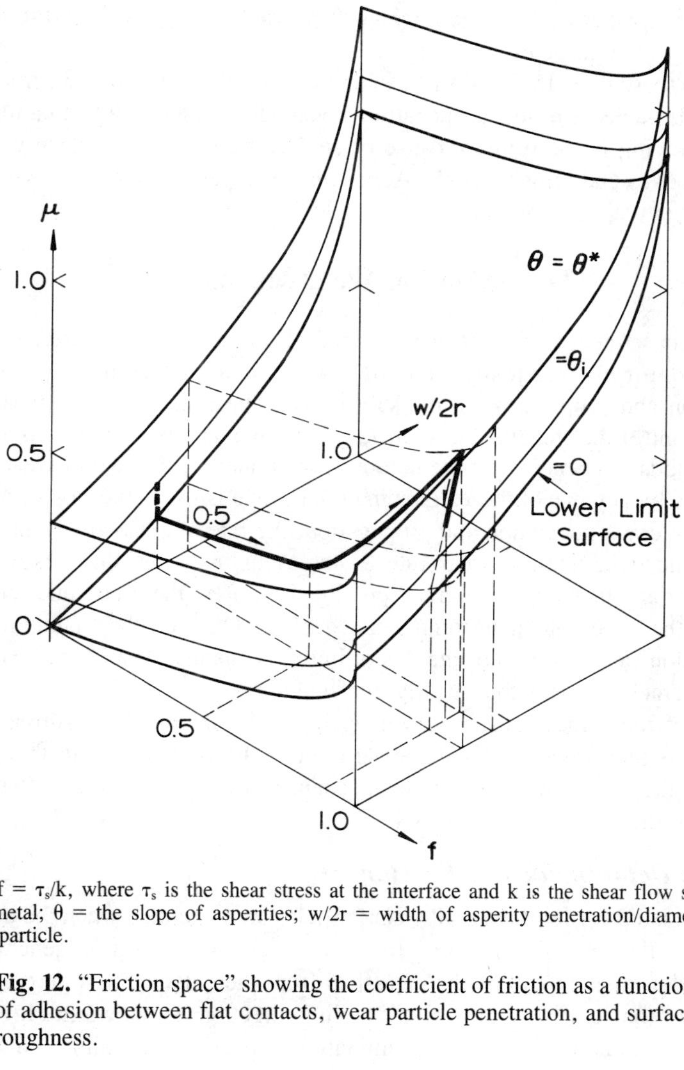

NOTE: $f = \tau_s/k$, where τ_s is the shear stress at the interface and k is the shear flow strength of the soft metal; θ = the slope of asperities; w/2r = width of asperity penetration/diameter of the spherical particle.

Fig. 12. "Friction space" showing the coefficient of friction as a function of adhesion between flat contacts, wear particle penetration, and surface roughness.

the normal load is carried by flat contacts. Likewise, the tangential stress exerted by plowing particles cannot be larger than the normal stress since the maximum value of μ_p cannot be larger than 1, although the overall coefficient of friction may be larger than 1.

(2) The deformation of asperities and the plowing action will often generate wear particles which will accompany delamination wear, and delamination wear often creates surface asperities and wear particles. While the delaminated wear particles will be large sheet-like platelets, the wear particles formed by

flattening of asperities and the furrows of plowed grooves will tend to be long, thin needle-like particles.

(3) All these three forms of surface interaction induce subsurface deformation, and crack nucleation and propagation, eventually causing delamination wear. When plowing is the primary cause of surface traction, the surface will show many grooves and remain rough. Adhesion and asperity interaction will tend to leave a relatively smooth surface.

Delamination Wear Mechanisms

As a consequence of the forces exerted at the surface the material near the surface deforms plastically and ultimately delaminates because of crack nucleation and propagation at the subsurface. These delamination wear mechanisms control the steady-state wear rate of both elasto-plastic and viscoelastic-plastic solids. Depending on the nature of the material, the rate of wear may be controlled by any one of these delamination mechanisms. In the case of materials with rapid crack nucleation and propagation, the rate of plastic deformation will control the delamination wear rate, whereas in materials that easily deform plastically and/or have rapid crack propagation rates, the crack nucleation rate may be the wear rate–determining mechanism. On the other hand, the crack propagation rate can be the rate-determining mechanism in materials with pre-existing cracks or ones that readily nucleate cracks.

Each of these mechanisms was investigated in detail by the author and his co-workers and reported in the special issue of *Wear* in 1977.[6] In this section, some of the recent work on these mechanisms will be reported after briefly reviewing the work done prior to 1977.

Asperity Deformation and Fracture

It was stated earlier that small wear particles are generated when the asperities deform and eventually fracture. In most cases, wear particle generation by asperity deformation can be distinguished from the delamination process by the location of the "crack" relative to the sliding direction, as shown in Fig. 13. The surface cracks associated with delamination appear at the trailing edge of the subsurface crack (relative to the sliding direction), whereas the "cracks" formed by the deformation of asperities appear at the leading edge. Wear particles formed form the asperities are much smaller than the sheet-like wear particles formed by the delamination process.

During steady-state wear, asperities may be generated as a result of the delamination process. In iron, dimples, which often characterize a ductile fracture surface, were observed at the crater surface after delamination.[9] On the other hand, very smooth crater surfaces have also been seen in α-brass.[12] Asperities will

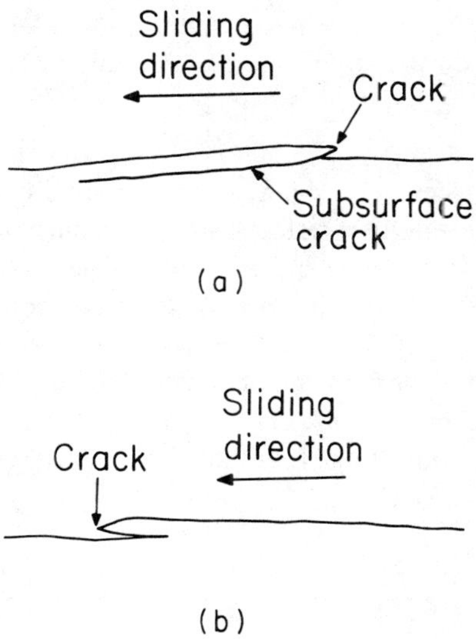

Fig. 13. **(a)** Delamination crack. **(b)** Surface "crack" due to the deformation of asperities.

also be generated as a consequence of delamination of wear sheets because the subsurface cracks are nucleated at different depths and form ridges where they meet. Surface asperities also result because subsurface cracks do not propagate exactly parallel to the surface in most cases.

In some rare situations delamination of wear sheets may occur even before the surface asperities are completely removed. In this case Stages I and II of the frictional behavior are completely absent. In most cases, however, the asperities are deformed and/or removed before the delaminated wear particles are generated.

Crack Nucleation

Subsurface cracks are nucleated as a consequence of plastic deformation of the surface layer under the influence of the surface traction. The mechanism of crack nucleation in materials with inclusions or hard second-phase particles is well known. However, the precise mechanism of crack nucleation in single-phase material without any inclusions is still not clearly known.

In order to nucleate cracks around hard inclusions two criteria must be satisfied. First, the interfacial stress between the matrix and inclusion (or second-phase particle) must exceed the adhesive strength at the interface. Second, the elastic energy released from the inclusion and the matrix surrounding the inclusion must be sufficient to supply the energy required to form new surfaces.

The first criterion is a local stress criterion. Therefore the shape of the inclusion will dictate the magnitude and the location of the maximum normal interfacial stress. Since the maximum stress the material can support is limited by the yield stress σ_y of the matrix material, the local interfacial stress (σ_{rr}) can be expressed as

$$\sigma_{rr} = \sigma_y + \sigma_n \qquad [5]$$

where σ_n is the hydrostatic tensile component of the applied stress. When σ_{rr} exceeds the interfacial stress, decohesion will occur at the inclusion-matrix interface.

Jahanmir and Suh investigated the location of crack nucleation, based on only the stress criterion, for spherical inclusions.[6] The results of the analysis are shown in Fig. 14, which shows the region of possible crack nucleation as a function of the normal and tangential load. The figure shows that as the normal load increases, the depth of nucleation also increases, and that when the normal stress is less than four times the shear yield strength of the matrix, cracks cannot nucleate in the absence of the tangential load. It should be noted that cracks

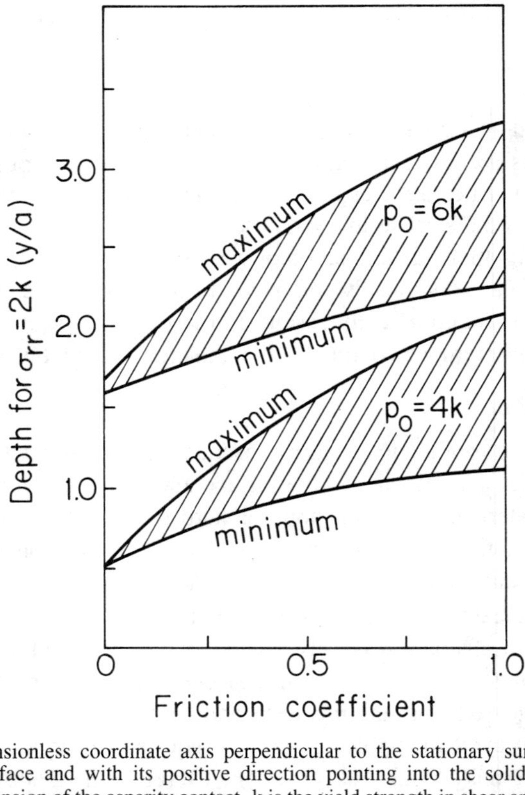

(y/a) is a dimensionless coordinate axis perpendicular to the stationary surface with its origin located at the surface and with its positive direction pointing into the solid. a is a half of the characteristic dimension of the asperity contact. k is the yield strength in shear and p_o is the maximum applied normal stress at the contact which is assumed to have an elliptical stress distribution.

Fig. 14. Depth of void nucleation region as a function of the friction coefficient for two applied local normal stresses.

nucleate only in the subsurface region, and the region of crack nucleation becomes larger as the normal load and the frictional force increase.

The implications of the foregoing results of the stress criterion for crack nucleation are that a very lightly loaded and/or lubricated surface may not wear by the delamination mechanism because of its inability to nucleate cracks and that cracks cannot propagate very near the surface since there are no crack initiation sites. In actual practice even well-lubricated surfaces tend to wear because of the plowing of the surface which eventually leads to delamination wear.

The stress criterion is a necessary but not a sufficient condition. In addition, the energy criterion must also be satisfied to nucleate cracks. The energy criterion may be stated as

$$(E)_b - (E)_a \geq \Delta\gamma A \qquad [6]$$

where $(E)_b$ and $(E)_a$ are the elastic energy stored before and after the void nucleation, A is the surface area of the nucleated void and $\Delta\gamma$ is the surface energy change because of the void nucleation. Since the elastic energy stored depends on the volume of the inclusion, while the surface energy is proportional to its surface area, Eq 6 is inclusion size dependent. There is a critical size below which cracks cannot nucleate even if the stress criterion is satisfied. Until recently there has not been a direct confirmation of this energy criterion.

Su investigated the energy criterion as part of his doctoral thesis at M.I.T.[35,36] The energy criterion was solved analytically using Eshelby's approach.[37] The strain at void nucleation is plotted as a function of the inclusion size in Fig. 15 for the case of epoxy with infinitely hard inclusions; d* is the critical inclusion size at different temperatures of the matrix. This criterion was checked through a tensile test of the filled epoxy which was examined optically after a given uniaxial elongation of the specimen. The separation of the inclusion-matrix interface was directly observed by the change in the reflection pattern of the incident light. The experimental results are plotted in Fig. 16. The figure shows the theoretically predicted elastic strain for void nucleation in glass-bead–filled epoxy at 373°K and 383°K. It also gives the experimentally obtained strains below which no cracks were initiated. The critical inclusion size for crack nucleation in metals is very small, being in the neighborhood of 25 nm. Therefore, most inclusions in metals can become crack nucleation sites.

The crack nucleation mechanism in a single-phase metal without any inclusions has not been definitively established, but various plausible mechanisms have been advanced.[11,19] Although dislocation interactions and the formation of dislocation cells are plausible, the experimental results of Mitchell and Laufer[20] generate questions on the validity of the dislocation cell argument.

Crack Propagation

Once cracks are nucleated, these cracks can propagate under the influence of the applied load. The process of the crack growth from the nucleation stage to a

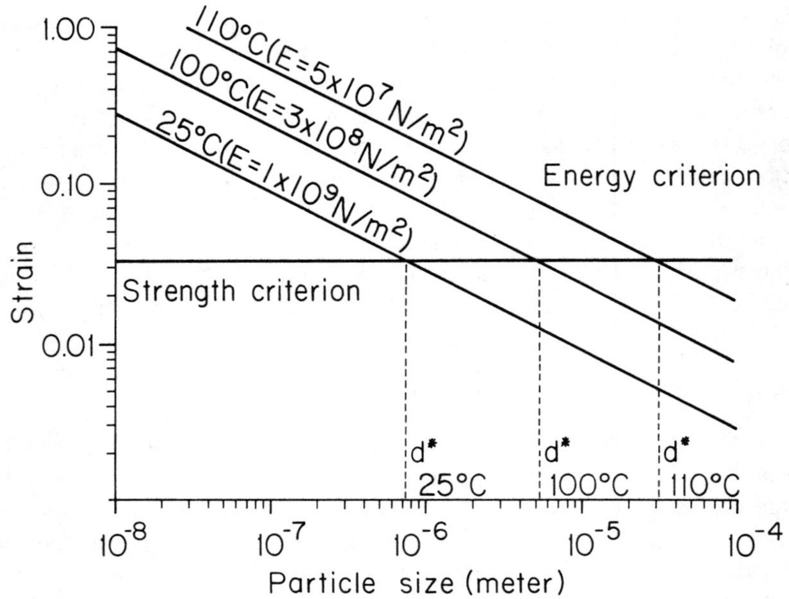

Fig. 15. Energy and strength criteria for void nucleation in epoxy at different temperatures. E is the tensile relaxation modulus at the instant of loading (i.e., tensile elastic modulus).

minimum crack size for steady-state wear has not been analyzed. However, once the crack length reaches a critical crack length, the crack must grow at a constant average rate when the surface is subjected to cyclic loading.

Suh[6] approximately calculated the change in the Mode I and Mode II stress intensity factors as a function of surface traction and contact geometry. Similar analyses have also been done by Rosenfield[38] and Ashelby[39] under different assumptions. The results of these investigations will be reviewed here.

Consider a steady-state crack lying below the surface as shown in Fig. 17. An asperity is moving over the surface and the crack length is longer than the critical size for steady-state crack growth. The wear rate of the material will be dictated by the crack propagation at the ends of the crack, A and B. The crack growth rate under the combined loading of Mode I and Mode II will be assumed to follow Paris's crack growth law for Mode I propagation, although the crack propagation in this case is dictated by Mode II loading.

The phenomena occurring at A and B can be quite different from each other, since A is a region where the state of normal stress is compressive and the plastic zone around A extends to the surface under typical sliding conditions, whereas B is in the tensile region. Fleming and Suh assumed that there is no stress and strain concentration at the forward crack tip A since the crack faces cannot slip

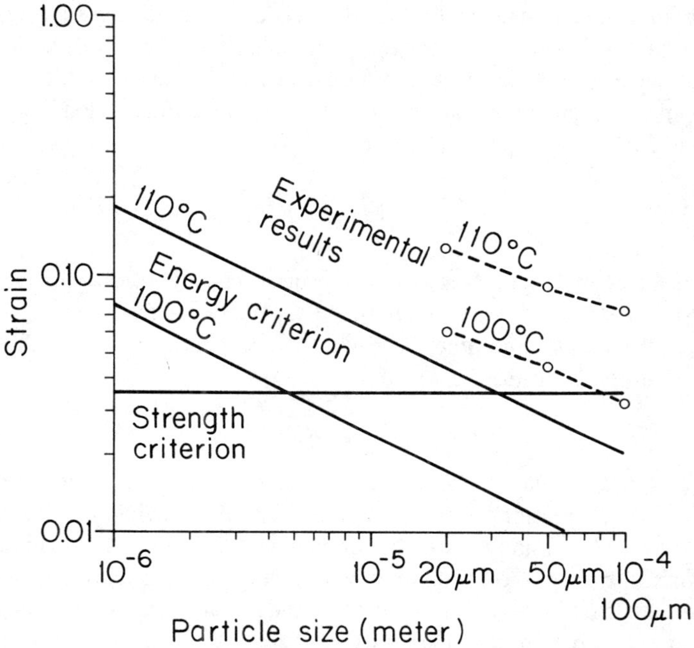

Fig. 16. Results of analytical and experimental investigation of the criteria for void nucleation in glass-bead–filled epoxy.

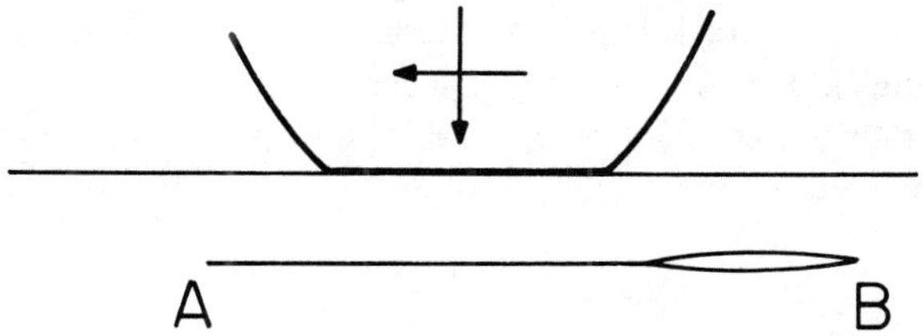

Fig. 17. Asperity moving along the surface over a steady-state subsurface crack.

with respect to each other because of friction in the compression region.[6] On the other hand, Rosenfield considered the stress concentration at A and B by assuming that the crack cannot support shear loading. He concluded that the Mode II stress factor is larger at A than at B and that the friction loading at the surface does not affect the results significantly.[38] Ashelby also found that the maximum stress intensity factor exists at the forward crack tip.[39] The results of these elastic analyses may not be correct for elasto-plastic solids.

Extensive experimental investigation of the wear sheet formation was made by Jahanmir[7] as part of his doctoral thesis work at M.I.T. Micrographs of wear sheets about to delaminate show that the subsurface cracks always reach the surface at the trailing edge of the crack. Although it is not conclusive, it may be an indication that cracks propagate more rapidly at the trailing edge.

If the crack propagation rate is controlled by the Mode II loading, it may be explained as

$$\frac{dC}{dN} = (\Delta K_{II}) = (2K_{IImax}) \qquad [7]$$

where C is the crack length, N is the number of cycles, and $(2K_{IImax})$ is the change in the Mode II stress intensity factor, since K_{II} fluctuates from $-K_{IImax}$ to K_{IImax} under typical loading conditions. The total crack length after the Nth cycle is obtained by integrating Eq 7 as

$$C = f\,(2K_{IImax})\,dN + C_e \qquad [8]$$

where C_e is the critical crack length at which steady-state crack growth is possible. The integration is not possible without knowing the relationship between the function f and the number of cycles N. Since K_{II} is a function of the crack length, the integration would not be normally possible. However, since the wear rate per unit distance slid under the steady-state wear conditions remains nearly constant, Eq 8 may be expressed as

$$C = f_A N + f_B N + C_e \qquad [9]$$

where the functions f_A and f_B are crack propagation rates at A and B, respectively, and C_e is the effective crack length for stable crack growth. Depending on the relative magnitudes of f_A and f_B, one of these terms may be neglected.

If the wear rate is controlled by the crack propagation rate, the wear volume V may be expressed as

$$V = C \cdot d \cdot w \qquad [10]$$

where d is the crack depth and w is the width of the crack. If w is assumed to be about the width of the asperity contact a, then Eq 10 may be expressed as

$$V = \{(f_A + f_B)N + C_e\}\,d \cdot a \qquad [11]$$

The remaining task is to relate f_A and f_B to the changes in stress intensity factors. The exact stress intensity factors can only be determined numerically. This is currently being attempted at M.I.T. This is an important task since the fast-propagating crack determines the wear sheet thickness. It should be noted that cracks do not appear very near the surface, since crack nucleation does not occur very near the surface.

The experimental data on Mode II and combined Modes I and II crack propagation are not currently available for most materials, especially at low values of stress intensity factors. Therefore, it is difficult to estimate the wear volume from first principles.

Wear Sheet Formation

The surface layer with subsurface cracks is under a compressive state of loading because of cyclic loading. The residual stress due to cyclic deformation was analyzed by Jahanmir and Suh.[6] When this compressive stress reaches a critical value for a given crack length (or vice versa), the delaminated sheet will shear, buckle or fracture and separate from the surface.

Extension of the Delamination Theory of Wear to Polymers

Although the delamination theory of wear was originally advanced to explain the wear of metals, it should be applicable to any materials that undergo plastic deformation, nucleate subsurface cracks and allow crack propagation. Crystalline thermoplastics can undergo large plastic deformation and fail in a manner similar to that in metals, whereas glassy thermoplastics and thermosets fail by fracture with limited plastic deformation. Therefore, it is expected that crystalline thermoplastics can wear by mechanisms very similar to those responsible for wear of metals.[40] This was shown to be the case by Clerico,[41] using chopped glass fiber reinforced nylon.

The wear of unfilled thermoplastic was investigated at M.I.T. in order to determine whether the delamination theory can be extended to thermoplastics and minimize the wear of plastic gears and bearings.[42,43]

When viscoelastic-plastic polymers are slid against a slider, the surface layer will undergo shear deformation parallel to the surface along the sliding direction. When the polymer is not filled with reinforcing particles and fibers, subsurface cracks cannot develop readily and, therefore, extensive deformation of the polymer occurs along the sliding direction. As a result, the molecules may be extended and oriented parallel to the plane of the surface, weakening the cohesive force between the molecular "lamellae." The deformation and orientation of molecules will be greater for crystalline polymers. When the attractive force (i.e., adhesion and/or mechanical interlocking force) between the plastic and the slider surface is so strong that the applied stress at the molecular layers is greater than the cohesive force, transfer of films can occur.

The supporting evidence for the foregoing argument is supplied by a large number of experimental results. Thick films of 0.1 to 1.0 μm are seen to transfer to the slider surface in low density polyethylene,[44,45] polypropylene,[44] polyvinylchloride[44,46] and nylon,[47] all of which have low crystallinity or bulky side groups. Very highly oriented thin films of less than 10 to 50 nm transfer to the slider in the case of PTFE,[48,49] high density polyethylene,[50] and polyoxymethylene,[50] all of which are highly linear crystalline polymers. These thin film transfers are seen at low sliding speeds.

Highly glassy polymers (e.g., amorphous polymers below the glass transition temperature) cannot undergo extended plastic deformation. In these polymers cracks are likely to propagate from surface defects (i.e., surface cracks or crazed regions) right behind the slider.[40] These cracks will propagate downward and change their direction in the subsurface region because of the changing direction of the maximum normal stress. The resulting wear particles are expected to be lumpy.

The foregoing description of the wear process suggests that the criteria for wear of plastics can be specified in terms of the interfacial stress between the slider and the surface, τ_s, the shear strength between the molecular "lamellae," K_e, the Hertzian shear stress, τ_h, and the bulk shear strength of plastics, k_b. In crystalline polymers the wear of plastics occurs when the following criteria are satisfied:

$$\tau_s > K_e \quad [12]$$

$$\tau_h < K_b \quad [13]$$

In glassy polymers, the criteria for wear are

$$\tau_s > K_s \quad [14]$$

$$\tau_h > K_b \quad [15]$$

where K_s is the shear strength at the surface. Then film transfer occurs when Eq 12 and 13 are satisfied, whereas thick films or massive lumps transfer when Eq 14 and 15 are satisfied. The assumption is that when plastic deformation does not occur, there cannot be any fracture.

Based on the understanding of how crystalline polymers wear, methods have been devised to improve the wear resistance of these polymers.[43]

Conclusions

(1) The delamination theory of wear can explain the wear of metals and non-metals as long as they can undergo plastic deformation.
(2) The newly developed theory for friction shows that the surface traction that resists sliding motion is caused by a combination of plowing, adhesion and deformation of asperities. Therefore, the magnitude of the surface traction varies from location to location, but the tangential force at each contact point is usually less than that indicated by the overall coefficient of friction.
(3) The crack nucleation criteria, i.e., the stress and the energy criteria, agree with the experimental results.
(4) The delamination theory of wear can be extended to polymeric materials.
(5) Further experimental and theoretical work is required to definitively establish the variation of flow stress at the surface.

Acknowledgment

The work presented in this paper would not have been possible without the generous support of the Office of Naval Research. The author is particularly indebted to Commander Harold P. Martin for his personal support of research on tribology at M.I.T.

References

1. F. P. Bowden and D. Tabor, *The Friction and Lubrication of Solids*, Part I, Oxford University Press (Clarendon Press), 1950, London.
2. N. P. Suh, "New Theories of Wear and Their Implications on Tool Materials," *Wear*, Vol. 62, pp. 1-20, 1980.
3. B. M. Kramer and N. P. Suh, "Tool Wear by Solution: A Quantitative Understanding," *Journal of Engineering for Industry*, Trans. AMSE, Vol. 102, pp. 303-309, 1980.
4. N. P. Suh, "Scientific and Technical Problems in Erosive Wear," Proc. Institute on Scientific Problems Relevant to Coal Utilization, West Virginia University, Department of Energy Symposium Series 46, 1978.
5. H-C. Sin, N. Saka and N. P. Suh, "Abrasive Wear Mechanisms and the Grit Size Effect," *Wear*, Vol. 55, pp. 163-190, 1979.
6. N. P. Suh and co-workers, *The Delamination Theory of Wear*, Elsevier Sequoia, S. A. 1977 (also *Wear*, Vol. 44, pp. 1-162, 1977).
7. S. Jahanmir, "A Fundamental Study of the Delamination Theory of Wear," Ph.D. Thesis, M.I.T., 1977.
8. E. P. Abrahamson II, S. Jahanmir and N. P. Suh, "The Effect of Surface Finish on the Wear of Sliding Surfaces," *CIRP Annals*, Vol. 24, pp. 513-514, 1975.
9. S. Jahanmir, N. P. Suh and E. P. Abrahamson II, "Microscopic Observations of the Wear Sheet Formation by Delamination," *Wear*, Vol. 28, pp. 235-249, 1974.
10. N. Saka and N. P. Suh, "Delamination Wear of Dispersion Hardened Alloys," *Journal of Engineering for Industry*, Vol. 99, pp. 289-294, 1977.
11. N. P. Suh, "The Delamination Theory of Wear," Vol. 25, pp. 111-124, 1973.
12. E. D. Doyle and D. M. Turley, "Delamination Effects in Grinding," *Wear*, Vol. 51, pp. 269-278, 1978.
13. M. V. Swain, "Microscopic Observations of Abrasive Wear of Polycrystalline Alumina," *Wear*, Vol. 35, pp. 185-189, 1975.
14. M. Clerico, "Sliding Wear Mechanisms of Polymers," in *Fundamentals of Tribology*, ed. N. P. Suh and N. Saka, pp. 769-786, M.I.T. Press, 1980.
15. A. L. Zharin and G. P. Shpenkov, "Microscopic Effects of Delamination Wear," *Wear*, Vol. 56, pp. 309-314, 1979.
16. R. B. Waterhouse and D. E. Taylor, "Fretting Debris and the Delamination Theory of Wear," *Wear*, Vol. 29, pp. 337-344, 1974.
17. M. A. Shafia and T. S. Eyre, "The Effect of Surface Topography on the Wear of Steel," *Wear*, Vol. 61, pp. 87-100, 1980.
18. N. Ohmae and T. Tsukizoe, "Analysis of a Wear Process Using the Finite Element Method," *Wear*, Vol. 61, pp. 333-339, 1980.
19. J. P. Hirth and D. A. Rigney, "Crystal Plasticity and the Delamination Theory of Wear," *Wear*, Vol. 39, pp. 133-144, 1976.
20. C. M. Mitchell and E. E. Laufer, "Surface Structure in an Abraded Titanium Alloy," *Wear*, Vol. 61, pp. 111-124, 1980.

21. A. W. Ruff, "Measurement of Plastic Strain in Copper Due to Sliding Wear," *Wear*, Vol. 46, p. 251, 1978.
22. J. A. Kirk and T. D. Swanson, "Subsurface Effects During Sliding Wear," *Wear*, Vol. 35, pp. 63-67, 1975.
23. R. M. Latanision, "Surface Effects in Crystal Plasticity," in *Fundamentals of Tribology*, ed. N. P. Suh and N. Saka, pp. 255-294, M.I.T. Press, 1980.
24. J. A. B. van Dijck, "The Direct Observation in the TEM of the Heavily Deformed Surface Layer of a Copper Pin After Dry Sliding Against a Steel Ring," *Wear*, Vol. 42, pp. 109-117, 1977.
25. J. H. Dautzenberg, "The Role of Dynamic Recrystallization in Dry Sliding Wear," *Wear*, Vol. 60, pp. 401-411, 1980.
26. E. Rabinowicz, "Friction, Wear and Lubrication," Center for Advanced Engineering Study, M.I.T., 1974.
27. E. Rabinowicz, *Friction and Wear of Materials*, Wiley, New York, 1965.
28. A. P. Green, "Friction Between Unlubricated Metals: A Theoretical Model of the Junction Model," Proc. Roy. Soc. (London), Vol. A228, pp. 191-204, 1955.
29. E. Rabinowicz, "The Determination of the Compatibility of Metals Through Static Friction Tests," *ASLE Trans*. Vol. 14k, pp. 198-205, 1971.
30. N. P. Suh, H-C. Sin, M. Tohkai and N. Saka, "Surface Topography and Functional Requirements for Dry Sliding Surfaces," *Annals CIRP*, Vol. 29, No. 1, pp. 413-418, 1980.
31. M. Tohkai, "Microstructural Aspects of Friction," S.M. Thesis, M.I.T., 1979.
32. N. P. Suh and H.-C. Sin, "The Genesis of Friction," *Wear*, submitted for publication, 1980.
33. S. Kato, E. Marui, A. Kobayashi, and T. Matsubayashi, "Characteristics of Surface Topography and Static Friction on Scraped Surface Slideway, Part 1 and 2," *Journal of Engineering for Industry*, Trans. ASME, Vol. 103, pp. 97-108, 1980.
34. J. M. Challen and P. L. B. Oxley, "An Explanation of the Different Regimes of Friction and Wear Using Asperity Deformation Models," *Wear*, Vol. 53, pp. 229-244, 1979.
35. K.-Y. B. Su and N. P. Suh, "Void Nucleation in Particulate Filled Polymeric Materials," to be presented at the Annual Technical Meeting of the Society of Plastics Engineers, Boston, 1981.
36. K.-Y. B. Su, "Void Nucleation in Particulate Filled Polymeric Materials and Its Implication on Friction and Wear Properties," Sc.D. Thesis, M.I.T., 1980.
37. J. D. Eshelby, "The Elastic Interaction of Point-Defects," *Acta Met.*, Vol. 3, p. 487, 1959.
38. A. R. Rosenfield, "A Fracture Mechanics Approach to Wear," *Wear*, Vol. 61, pp. 125-132, 1980.
39. D. H. Hills and D. W. Ashelby, "On the Application of Fracture Mechanics to Wear," *Wear*, Vol. 54, pp. 321-330, 1979.
40. N. P. Suh, "Wear Mechanisms: An Assessment of the State of Knowledge," *Fundamentals of Tribology*, M.I.T. Press, pp. 443-454, 1980.
41. M. Clerico and V. Patierno, "Sliding Wear of Polymeric Composites," *Wear*, Vol. 53, pp. 279-301, 1979.
42. J. R. Youn, "An Investigation of Friction and Wear Behavior of Polymers," S.M. Thesis, M.I.T., 1980.
43. N. P. Suh and J. R. Youn, "Friction and Wear of Surface Treated Polymers," SPE ANTEC, to be presented in 1981.
44. V. K. Jain and S. Bahadur, "Material Transfer in Polymer-Polymer Sliding," *Wear of Materials*, ASME, pp. 487-493, 1977.

45. C. H. West and J. M. Senior, "Frictional Properties of Polyethylene," *Wear*, Vol. 19, pp. 37-52, 1972.
46. D. I. James, "Surface Damage Caused by Polyvinyl Chloride Sliding on Steel," *Wear*, Vol. 2, pp. 183-194, 1958/59.
47. J. H. Warren and N. S. Eiss, Jr., "Depth of Penetration as a Predictor of the Wear of Polymers on Hard and Rough Surfaces," *Wear of Materials*, ASME, pp. 494-500, 1977.
48. K. V. Shooter and D. Tabor, "The Frictional Properties of Plastics," *Proc. Phys. Soc. London*, Vol. B65, pp. 661-671, 1952.
49. W. Bondfield, B. C. Edwards, and A. J. Markham, "Preferred Orientation in PTFE–Carbon Wear Transfer Films," *Wear*, Vol. 7, pp. 383-386, 1976.
50. M. K. Kar and S. Bahadur, "Micromechanism of Wear at Polymer-Metal Sliding Interfaces," *Wear of Materials*, ASME, pp. 482-486, 1977.

Abrasive Wear

MARTIN A. MOORE
Fulmer Research Laboratories Ltd.
Stoke Poges, Slough SL2 4QD
England

Abstract

Abrasive wear is wear by the displacement of materials caused by hard particles or protuberances. Its widespread occurrence accounts for a cost to industrial nations of between 1 and 4% of GNP. The objectives of this paper are to provide a foundation for materials selection and development through review of, illustration of the practical significance of and identification of the gaps in our knowledge of abrasive wear processes.

In practical wear environments it is quite common to encounter more than one type of wear process occurring simultaneously, but also, each wear process may involve a number of wear mechanisms. For abrasive wear, there are two extreme mechanisms of material removal, one in which plastic deformation plays a predominant role and the other in which fracture with limited plastic deformation is dominant. Models for these mechanisms are presented and discussed, and an assessment is made of the contribution of each in practical wear environments, and of the effect of corrosion and other environmental phenomena.

Consideration is given to the influence on wear rates of properties of both the worn material and the abrasive. Variables that are discussed and assessed include abrasive particle size, shape and attack angle, mechanical properties of the abrasive, mechanical properties and response of the worn material and discrete microstructural features of the worn material.

Finally, guidance is given for selection of materials for abrasive wear environments, illustrated by field and laboratory data. Comments are made on the usefulness of laboratory data and the problems associated with simulated testing.

1. Introduction

The Research Group on Wear of Engineering Materials of the Organisation for Economic Cooperation and Development[1] has defined abrasive wear as "wear by displacement of material caused by hard particles or protuberances." Clearly this definition encompasses the intentional removal of material from a surface, as in grinding and polishing of engineering components and cleaning one's teeth, and the unwanted loss of material that occurs when machine components are in relative motion. It includes both the traditional interpretation of abrasion, i.e., wear by hard mineral substance, and wear during metal-to-metal contact where a locally harder surface penetrates and wears a locally softer surface.[2] The latter is often referred to, incorrectly, as adhesive wear.

Other, more specific, terminology has emerged from studies of abrasive wear and is used to describe the severity of wear, the specific combination of material causing wear and material(s) being worn, or the relative properties of the abrasive and worn material. In the United States abrasive wear is often classified as "gouging abrasion," "high-stress abrasion" or "low-stress scratching abrasion," depending on the severity of damage to the worn material. The terms "two-body abrasive wear" and "three-body abrasive wear" are often found in the literature, the first to describe wear caused by abrasive sliding on a material, and the second to describe wear caused by abrasive trapped between two moving surfaces. Richardson[3] differentiated between wear caused by an abrasive of hardness equal to or less than that of the worn surface (a "soft abrasive") and that caused by an abrasive harder than the worn surface (a "hard abrasive").

Abrasive wear causes an enormous annual expenditure by industry and consumers. Most of this is in replacing or repairing equipment that has worn to the extent that it no longer performs a useful function. For many machine components this occurs after only a very small percentage of the total volume has been worn away. For some industries, such as agriculture, as many as 40% of the components replaced on equipment have failed by abrasive wear. Other major sources of expenditure are losses of production consequential upon lower efficiencies and plant shutdown, the need to invest more frequently in capital equipment and increased energy consumption as equipment wears. Estimates[4-6] of the direct cost of abrasive wear to industrial nations vary from 1 to 4% of gross national product, and Rabinowicz[7] has estimated that about 10% of all energy generated by man is dissipated in various friction processes, including abrasion.

A major problem in reducing the widespread occurrence of abrasive wear is that it is almost impossible to exclude from machinery material with abrasive properties, either as a material of construction or as a contaminant. This is hardly surprising since the potentially abrasive materials silica, alumina and iron oxides account for 81% of the earth's crust (Table 1). Even plant matter contains some silica, either as a contaminant or incorporated in the structure. However, the most severe problems are encountered in equipment that is in contact with soils, rocks and minerals as the worked material. Table 2 shows the cost, as a percentage of the total, of abrasive wear to various industries in the Federal Republic of Germany.[6]

The potential savings, the increasingly onerous conditions under which industry requires its machinery to function, and consumers' demands for greater reliability and safety provide considerable incentive for research and development and for education on abrasive wear. Attention to component design may

Table 1. Average Composition of the Earth's Crust

Oxide	Percent	Oxide	Percent
SiO_2	59.07	Na_2O	3.71
Al_2O_3	15.22	K_2O	3.11
Fe_2O_3	3.10	TiO_2	1.03
FeO	3.71	Pb_2O_5	0.30
CaO	5.10	MnO	0.11
MgO	3.45	H_2O	1.30

Table 2. Percentage Costs of Abrasive Wear to Various Industries in the Federal Republic of Germany (Ref 6)

Industry	Percentage cost
Metallurgical extraction and ore mining	40
Coal mining, grading and pulverizing	30
Agriculture, construction, stone gravel and sand quarrying and sundry comminution	20
Production engineering	10

reduce wear by decreasing the frequency of abrasive contact, loads and sliding distances. But much greater improvements in component life have been achieved by the application of knowledge on abrasive wear to the development and selection of materials. Examples are the Cr/Mo cast irons and austenitic Mn steels, originally developed for armor. However, there are no material panaceas to abrasive wear; because of the complexities of the abrasive wear process the optimum material and microstructure may vary considerably for different wear environments. Thus, scientific material development and selection must be founded on a full understanding of how material properties influence abrasive particle contact and material removal. The objectives of this paper are to review, to illustrate the practical significance of, and to highlight the gaps in our current understanding of abrasive wear processes. Emphasis is placed on wear by mineral abrasive particles, but many of the observations made may be applied to wear by hard metallic particles or protuberances.

2. Mechanisms of Material Removal

Most models of abrasive wear are necessarily oversimplifications of a very complex process. In practical wear environments it is quite common to encounter more than one type of wear process occurring simultaneously, but also, each wear process may involve a number of wear mechanisms. For abrasive wear by multi-particle contact there is a wide range of particle loading, particle geometry and attack angle. Each of these has a considerable influence on both the mechanism of material removal and the wear rate. Thus, in order to understand how material properties affect wear it is critical that we understand and recognize the major mechanisms of material removal. There are two extreme mechanisms, one in which plastic deformation plays a predominant role and is the rate-controlling process, and the other in which fracture with limited plastic deformation is dominant and rate controlling. Corrosion or other environmental effects may aid material removal in both cases.

2.1. Plastic Deformation

Two major processes take place when abrasive particles contact the surface of a ductile material:[8]

(1) The formation of grooves which do not involve direct material removal.
(2) The separation of particles in the form of primary wear debris or microchips.

In both cases material is deformed to the sides of the grooves and can become detached to form secondary microchips. Ultimately, material is removed by fracture, but plastic deformation controls the rate at which material is removed.

When a contacting abrasive particle forms a groove by plastic deformation we can represent the volume of material removed, δV, by:

$$\delta V = k_1 \cdot k_2 \cdot A \cdot S \qquad [1]$$

where k_1 is the probability that material is removed, k_2 is the proportion of the groove volume that forms wear debris, as opposed to that which is deformed to the sides of the groove, A is the cross-sectional area of the groove and S is the sliding distance. The cross-sectional area of the groove

$$A = k_3 \cdot p^2 \qquad [2]$$

where the constant k_3 depends on particle shape and p is the depth of indentation of the particle.

If the contacting particle acts like a scratch indenter,[9] the depth of indentation of the particle is given by:

$$p^2 = k_4 \cdot L \cdot H^{-1} \qquad [3]$$

where the constant k_4 depends on particle shape, L is the load on the particle and H is the hardness of the surface. The mean load on the particle is

$$\bar{L} = \sigma \cdot N^{-1} \qquad [4]$$

where σ is the applied load per unit area and the number of particles that contact the surface per unit area is

$$N = k_5 \cdot d^{-2} \qquad [5]$$

where the constant k_5 depends on particle shape and distribution and d is the mean particle diameter.

From Eq 3, 4 and 5,

$$p \propto \sigma^{-1/2} \cdot d \cdot H^{-1/2} \qquad [6]$$

and from Eq 2 and 6,

$$A \propto \sigma \cdot d^2 \cdot H^{-1} \qquad [7]$$

Thus, the sum of δV for N contacts per unit area and unit sliding distance, i.e., the volume wear per unit area per unit sliding distance, is from Eq 1 and 7

$$W = K_1 \cdot K_2 \cdot K_3 \cdot \sigma \cdot H^{-1} \qquad [8]$$

where K_1 is the summed probability of wear debris formation, K_2 is the mean proportion of groove volume removed when debris formation occurs and K_3 is a constant depending on the shape of the particles. In later sections we shall explore the validity of this model and the factors that influence K_1, K_2 and K_3.

Because of strain hardening, plastic deformation is not confined to the grooves. An extensive region below a worn surface is plastically deformed during groove formation, and, because the hydrostatic stress system associated with abrasive particle contact suppresses ductile fracture processes,[10] the strains at worn surfaces can be very high. The depth of the deformed zone and the strain as a function of depth are both proportional to the depth of indentation of the abrasive particles. Thus, data for strain distributions below worn surfaces (of a single material worn under a range of conditions) fall on a common curve when plotted as strain versus the ratio depth, D, to $\sigma^{1/2}d$ (which is related to p from Eq 6) (Fig. 1). Furthermore, the strain at the extreme surface, for one material, is apparently independent of the depth of indentation of the abrasive particles, and may be comparatively insensitive to different methods of achieving surface deformation. Richardson[11] found the hardness of surfaces deformed by abrasive wear (debris formation and rubbing), trepanning (pure rubbing) and shot peening (normal impact) to be remarkably similar (Table 3).

These observations raise a point of considerable practical significance. Since extensive strain hardening occurs during the wear process, wear rates may be affected very little by prior plastic deformation. Avery's results (Table 4)[12] show that pre-hardening austenitic Mn steel by deformation has a negligible effect on its wear rate by wet sand. It has been reported that the highest hardness achieved

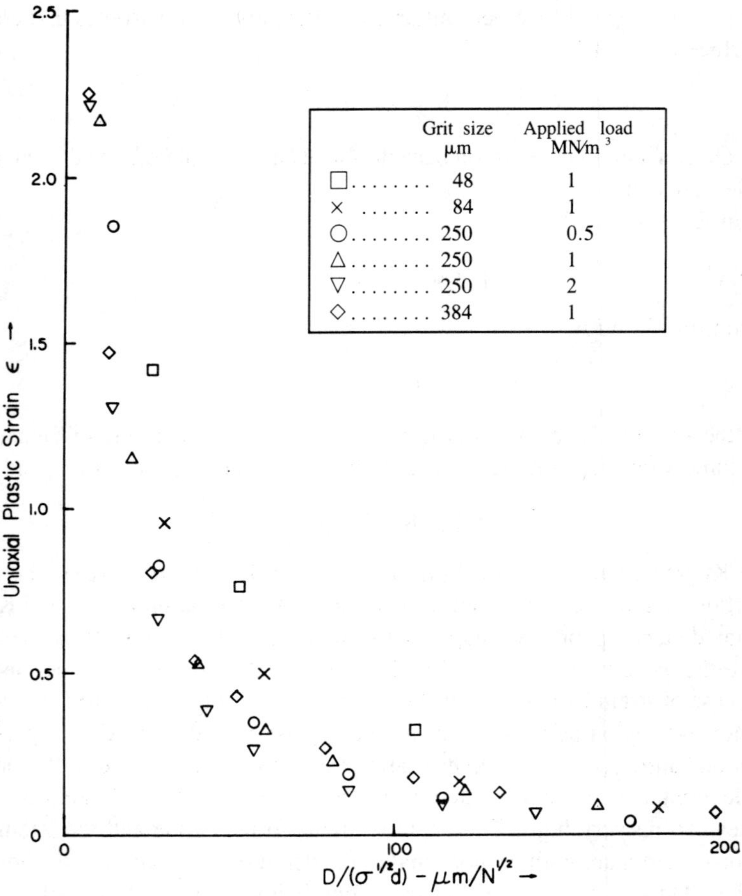

Fig. 1. Strain distribution below worn surfaces of copper-silver solder laminate as a function of $D/(\sigma^{1/2}d)$.

by explosion pre-hardening is only about half that measured at worn surfaces.[11]

In addition, the observations pose several important questions which have yet to be satisfactorily answered. These are examined below:

(1) What are the mechanisms that limit surface strain?

A possible explanation[13] is that plastic deformation is limited by fracture—i.e., material is detached from the surface when some local fracture criterion is satisfied. For example, a microcrack or microstructural feature that is subcritical in the stress system below the surface becomes critical in the stress system at the surface. Such a mechanism would account for the formation of secondary chips. Thus, if fracture did not occur and strain hardening proceeded, the material would continue to deform until it was able to support

Table 3. The Hardness of Strained Surfaces (Ref 11)

Material	Surface hardness (Vickers, kg/mm²)			
	Unstrained	Abraded	Trepanned	Shot peened
0.43%C steel:				
Quenched and tempered	344	726	664	707
Quenched and tempered	590	896	927	852
0.74%C steel:				
Annealed	221	473	684	358
Quenched and tempered	813	967	1250	1143
Austenitic Mn steel	220	885	901	859
Aluminum	30	87	92	
Copper	43.3	136	193	158
Titanium (115)	134	263	313	
Molybdenum	170	530	637	365

Table 4. Effect of Pre-hardening on the Wear of Austenitic Mn Steel (1.17%C, 13.2%Mn, 0.53%Si) (Ref 12)

Pre-treatment	Vickers hardness, kg/mm²	Wet quartz sand abrasion factor(a)
Toughened by water quenching from above 980°C	210	0.79
Explosion hardened	305	0.76
	450	0.73

(a) Wt loss/wt loss 1020 steel control.

elastically the load on contacting particles. Alternatively, as we shall see later, there may be a limit to the energy input at high strains and strain rates, due to thermal and other effects which cause bulk deformation to cease at the extreme surface but continue below the surface.

(2) What are the mechanisms that determine surface and sub-surface hardness?

Clearly, if fracture or other mechanisms limit surface strain then we would expect surface hardness to be determined by this strain and the material stress-strain behavior. Examination of cross sections of worn surfaces [10,14,15] shows that hardness versus depth profiles vary considerably for different materials (Fig. 2). Certain structural features have been identified with perturbations on the hardness/depth profiles, as shown. The unusual combination of high strain,[14] high strain rate[14] and transient temperature rises[16,17] increases the probability of the occurrence of a number of events close to the worn

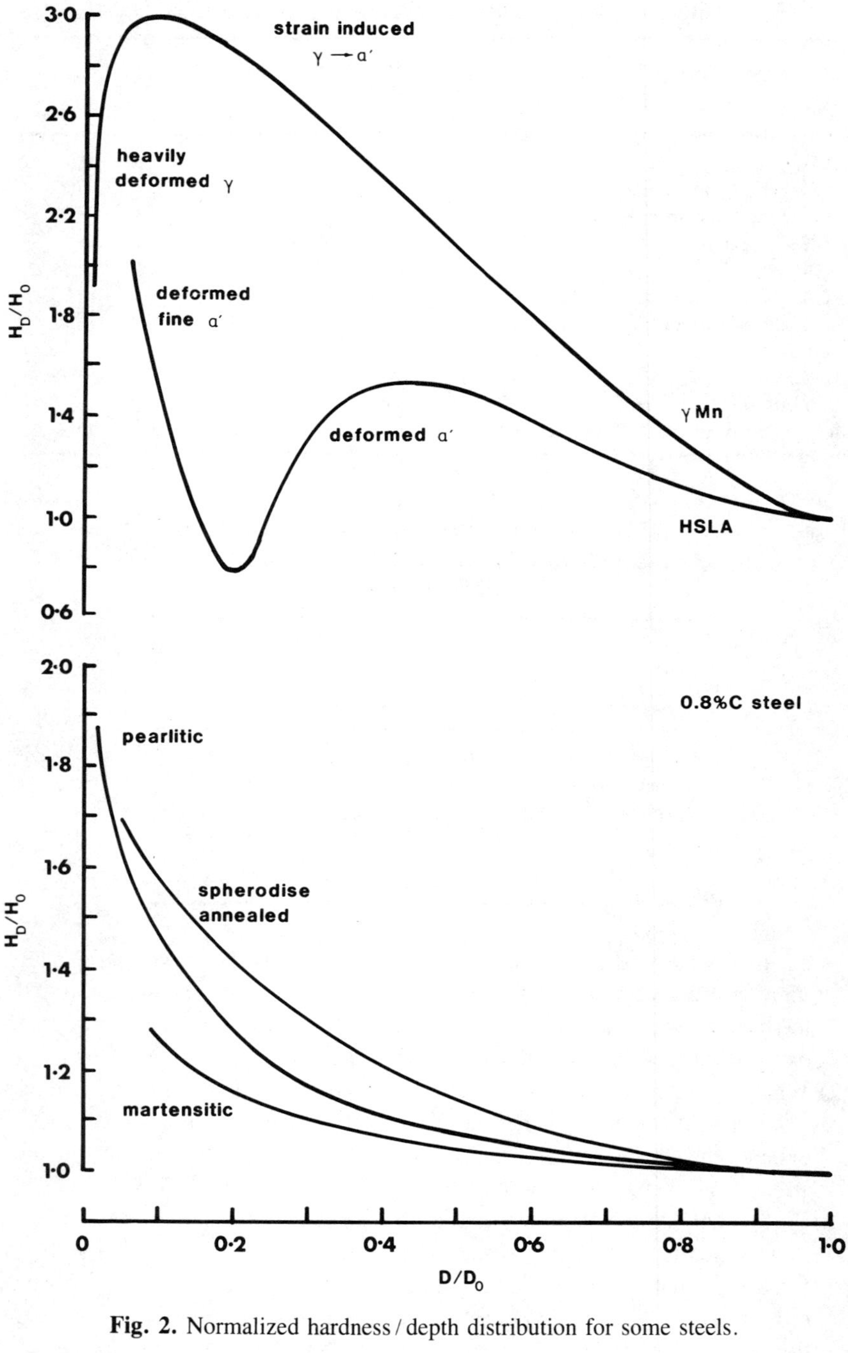

Fig. 2. Normalized hardness / depth distribution for some steels.

surface. These include strain softening due to cyclic deformation,[11] adiabatic shear,[18] dynamic recovery and recrystallization to a very fine grain size,[19,20] strain-induced phase transformation,[15] thermal softening and phase transformation[15,21] and hardening due to solute atom migration and concentration.[22]

Since strain increases rapidly close to worn surfaces we expect hardness to increase but the strain-hardening rate to decrease, from the sub-surface to the surface. However, should any of the phenomena listed above occur also, the strain-hardening rate may increase, become zero or become (dynamically) negative. If the strain-hardening rate is high at the surface, deformation is forced further below the surface, whereas if it is low, deformation is concentrated towards the surface. In the case of locally negative dynamic stress-strain behavior, deformation would be confined to local regions, between which deformation would cease.

(3) Why do different materials strain harden to different degrees during wear?

Examination of the hardness and flow stress of worn surfaces[11,13] has shown that the rate of increase of surface hardness with increase in unstrained hardness varies widely for different groups of materials (Fig. 3). Extremes are shown by the pure metals and by dispersion and precipitation hardened materials. Attempts have been made to explain this systematic behavior[13] by assessing the contribution of dislocation interactions to flow stress at low and high strains. Thus, for the pure metals, with relatively low dislocation densities at low strains and for which the nucleation of cracks is likely to be limited to dislocation interactions, the degree of surface hardening is large. In contrast, the martensitic steels have high dislocation densities and may have pre-existing cracks or crack nucleation at low strains; thus the degree of strengthening by further deformation at worn surfaces is low. For the relatively homogeneous steel structures—martensite, tempered martensite and spheroidal cementite/ferrite—wear resistance, high-strain flow stress and low-strain flow stress are similarly related to structure.[23-26] The implication is that fine carbides in steels make relatively similar contributions to strength, as barriers to dislocation motion, at both low and high strain. Conversely, for alloys strengthened at low strain by coherent precipitates and dispersions the hardness of the worn surface is very similar to that of worn surfaces of the solid solution matrix alone.[13] Thus, the particles in those alloys make little contribution to strength at very high strains.

Because of its predominance in the abrasive wear of ductile materials, plastic deformation of material from grooves at the surface and material below the surface accounts for almost all the external (frictional) work done. From estimates of the strain and flow stress distributions below worn surfaces[10] it is possible to compute the energy expended in plastic deformation (Table 5). Also

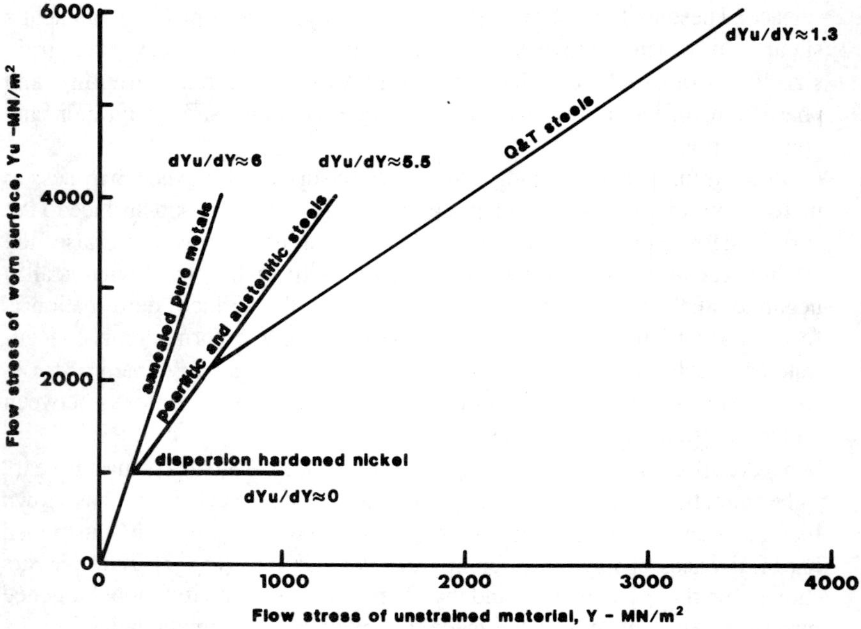

Fig. 3. Flow stress (estimated from hardness) of worn surfaces and unstrained material.

Table 5. Energy Expended in Plastic Deformation and Specific Energy for Removal of Copper During Wear by SiC Bonded Abrasives (Ref 10)

Grit	Wear conditions: Applied load per unit area, MN/m^2	Wear conditions: Estimated abrasive penetration depth, μm	Measured coefficient of friction	Estimated work expended in plastic deformation, J/m · mm^2	Calculated friction work, J/m · mm^2	Specific material removal energy, J/mm^3
320	0.97	2.0	0.60	0.99	0.79	24
180	0.97	3.1	0.64	1.16	0.71	18
60	0.48	6.9	0.82	0.47	0.40	8
60	0.97	8.0	0.82	0.86	0.81	8
60	1.94	10.9	0.84	1.80	1.64	9
40	0.97	12.0	0.99	0.84	0.96	11

given in Table 5 are the coefficients of friction, depth of penetration of the abrasive particles and specific energy for material removal. The data show that the coefficient of friction in abrasive wear, for multiple particle contact, increases as the depth of penetration of the abrasive increases. (Coefficient of friction is not an intrinsic material property, it is determined by the degree and distribution of plastic deformation associated with the sliding surfaces and the mechanical response of the materials.) When plotted against the depth of indentation of the abrasive particles (Fig. 4), the specific energy decreases to a minimum of about 8 J/mm^3 for copper. Similar trends are observed in grinding processes.[27] The practical significance of these observations is that sufficient data would allow the selection of operating parameters for optimum material removal rate and energy utilization in grinding processes.

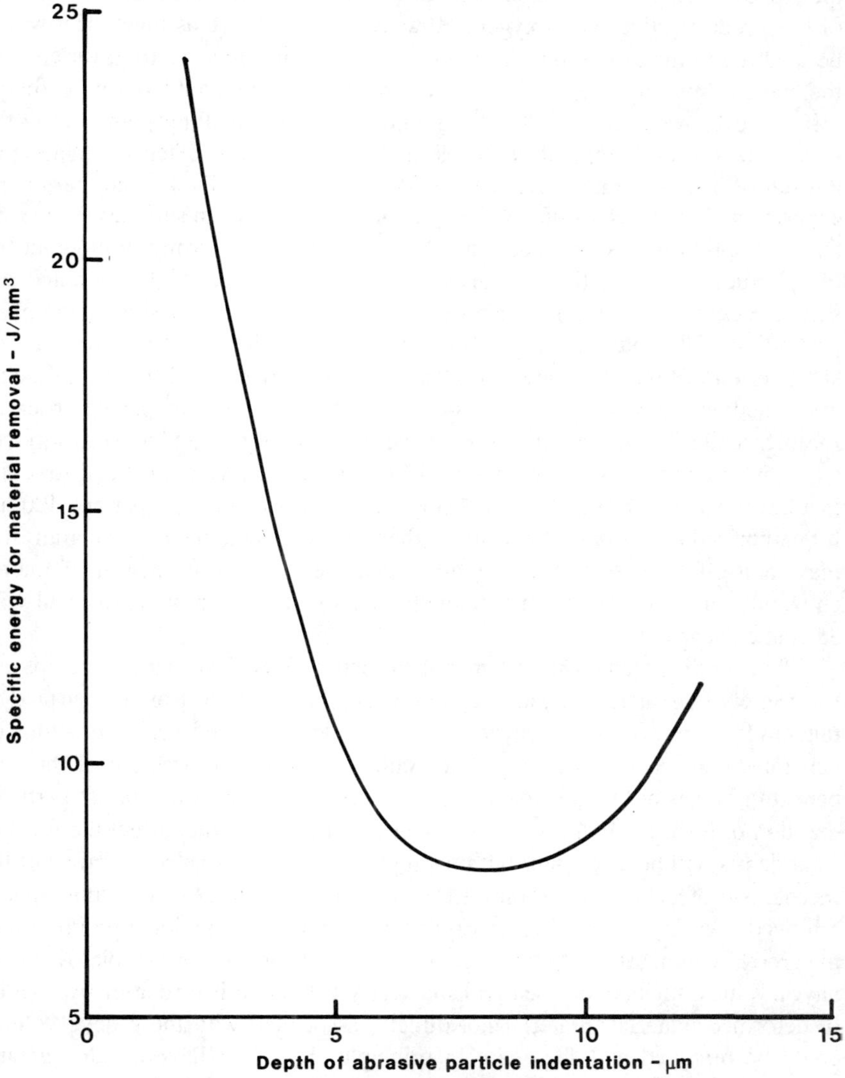

Fig. 4. Specific energy for the removal of copper worn by SiC bonded abrasives, plotted against the mean depth of indentation of the abrasive.

At least 90% of the energy expended in plastic deformation during wear must be dissipated as heat.[28] For abraded copper,[10] the heat generated through the plastic deformation of a wear groove is about 3.5 J/mm^3 of the material displaced from the groove. If this heat remains within the material from the groove, e.g., a microchip, the adiabatic temperature rise would be about 1000°C. Direct and indirect evidence[16,17] supports the occurrence of transient temperature rises of this order on the worn surface and of about 10^{-2} s duration, and we have all witnessed the "sparks" generated during dry grinding, which are wear debris at temperatures of this order reacting with oxygen. However, almost twice as much heat would be required to melt the same volume of metal by adiabatic shearing; for copper, the specific melting energy is about 6 J/mm^3, approximated[27] by the enthalpy difference between the metal in the liquid state at the melting point and in the solid state at room temperature. Thus, as Nabarro[28] states, extensive melting as a result of heat generated by plastic work is extremely unlikely, and there is no experimental evidence to suggest that melting occurs at worn surfaces. However, the assumption has been made that heat is generated uniformly throughout the metal structure. In practice, because a very fine and, often, highly oriented cell structure develops at worn surfaces, [15,20] the energy may be dissipated in only a proportion of the total volume and, because of the high strain rate,[14] over a very short period of time. Adiabatic shear bands would develop,[18,29] having a spacing associated with the size of the cell structure. The interlamellar spacing observed for microchips[19] is of the same order as the fine grain size found at worn surfaces and in wear debris,[19] which is, in turn, of the same order as the fine cell structure that develops at worn surfaces[20,30]—i.e., less than 1 μm for copper and 100 nm for carburized mild steel. Doyle and Aghan[19] have argued that the fine grain size may in itself lead to plastic instability and the lamellar formation of microchips without invoking the high temperatures that are normally associated with adiabatic shear.

Whatever the mechanism of lamellar formation, it is clear that microchips can form solely by plastic flow, although they may be separated from the surface by random fracture events. However, a considerable amount of wear debris may be generated also in the absence of such chip formation[20] by fracture of material beneath grooves or by deformation to the sides of grooves or in front of particles that do not form microchips. In later sections the factors that affect the form of wear debris will be discussed, but it is important to consider also the effect of the mechanism of debris formation on the proportion of the groove volume that is removed, i.e., the constant K_2 in Eq 8. For microchip formation a proportion of the groove volume flows to the sides of the groove and so, in principle, K_2 might have a value lying between nearly 0 and nearly 1. For debris formation by fracture of deformed material, material not directly associated with the groove volume could be removed and so, again in principle, K_2 might have a value greater than 1. Therefore, volume wear is critically dependent on the type of wear debris generated through plastic deformation and fracture processes.

2.2. *Fracture*

During the indentation or scratching of brittle solids by a spherical indenter fracture may occur under apparently elastic contact—Hertzian fracture—with the formation of conical cracks extending into the material. It is unlikely that such fracture would result in material removal since the crack and free surfaces do not encompass a volume of material. However, at small indenter radii of curvature elastic-plastic contact occurs because the radius of curvature has a larger effect on the plastic indentation load than on the Hertzian fracture load. The transition from elastic-plastic indentation, at small radii of curvature, to Hertzian fracture (elastic contact), at large radii, occurs at a critical radius which increases as the hardness of the material decreases and the fracture toughness parameter, K_c, increases.[31]

For sharp indenters, the depth of elastic-plastic indentation increases with increasing load until the indentation reaches a critical size. At this critical size the tensile stresses associated with the indentation are sufficient to cause cracks to propagate from the plastic zone surrounding the indentation.[32,33] For scratching, as opposed to static indentation, the critical indentation size for cracking is significantly lower,[34] presumably because the tensile stresses associated with sliding are higher. The critical size for the onset of cracking increases as the hardness of the material decreases and as its fracture toughness increases. It is also larger for blunt than for sharp indenters. The cracks that develop during scratching are of two main types, median cracks normal to the surface and lateral cracks extending from beneath the indentation towards the surface (Fig. 5). Under such conditions of indentation material removal will occur when the lateral cracks intersect each other or propagate to the surface.[31,32] Even when the indentations are subcritical sub-surface cracking may occur[35] because the stress caused by dislocation pile-up at grain boundaries of brittle solids is not relieved by slip in adjacent grains.

When indentation fracture occurs, the crack lengths scale with the size of the indentation and are larger for material of low fracture toughness. Thus, for abrasive wear in which brittle fracture is the predominant mechanism of material removal we would expect the volume of material removed when a groove is formed to be a function of the fracture toughness and, from Eq 3, the load on the contacting particle and the hardness of the worn surface. Evans and Wilshaw[31] have modelled this process and predict that the upper limit for volume wear per unit sliding area per unit sliding distance is

$$W = N \cdot \bar{L}^{5/4} \cdot K_c^{-3/4} \cdot H^{-3/4} \cdot \qquad [9]$$

On substituting for N from Eq 5 and $\bar{L}$ from Eq 4 this becomes:

$$W = k_5^{-1/4} \cdot \sigma^{5/4} \cdot d^{1/2} \cdot K_c^{-3/4} \cdot H^{-1/2} \cdot \qquad [10]$$

Because, for material removal by indentation fracture, the cracks can extend well

Fig. 5. Median and lateral vent cracks associated with an indentation in glass.

beyond the boundaries of the grooves, the rate of material removal may be greater than for material removal by plastic deformation. Thus, there should be a substantial increase in wear rate as applied load and abrasive particle size are increased such that the size of the indentations increases from subcritical to supercritical. Moore and King[36] studied the wear of a wide range of brittle solids on 60 grit (250 μm) and 180 grit (84 μm) silicon carbide bonded abrasives. By estimating the mean indentation depth of the abrasive particles, from Eq 3, 4 and 5, and the critical indentation depth for fracture,[33] they demonstrated that the indentation depths were mostly subcritical for wear by 180 grit abrasive, but all were supercritical for wear by the 60 grit abrasive, under an applied load per unit area of 1 MN/m^2 in both cases. Wear data from this study and for ductile metallic materials (Fig. 6)[37] clearly show the contribution of indentation fracture to volume wear of the brittle materials on the 60 grit abrasive. There is a rapid increase in the slope of the curve for brittle materials, almost certainly associated with a transition from subcritical to supercritical indentation on the 60 grit abrasive as the general wear rate and depth of indentation increase. In addition, the slope of the linear region of the curve for brittle materials, which is the best fit through data with considerable scatter, is about 20, whereas for the ductile materials it is about 1.5. We can estimate volume wear on the 60 grit abrasive from Eq 10 and on 180 grit abrasive from Eq 8, using values for K_1, K_2, K_3, k_5, H and K_c from the literature[31,36,38-41]—i.e., assuming that the mechanism of removal is by indentation fracture on the 60 grit abrasive and by plastic defor-

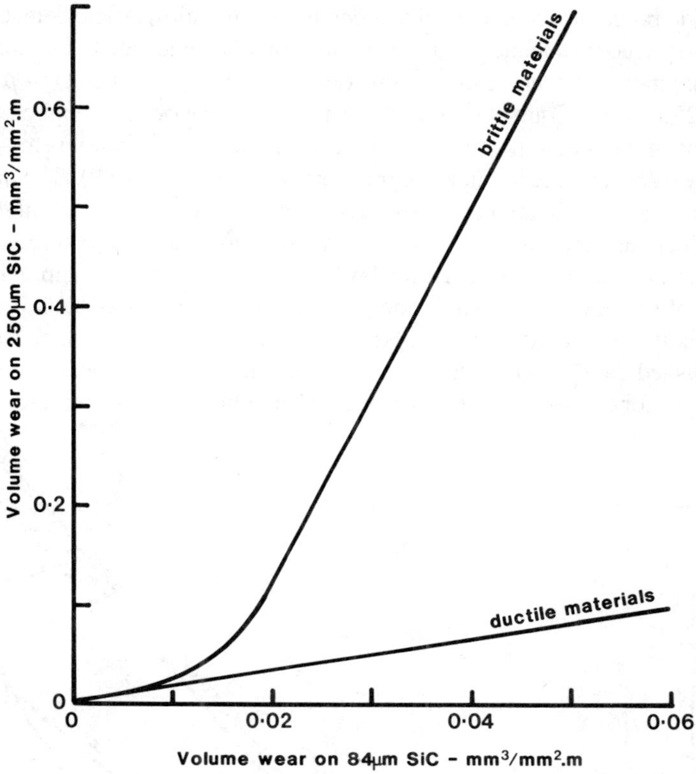

Fig. 6. Wear on 250 μm SiC versus wear on 84 μm SiC (1 MN/m^2 applied load), for ductile and brittle materials.

mation on the 180 grit abrasive. From these estimates, the mean theoretical value of the slope is about 13, with a range from 6.5 to 40.

The practical significance of these phenomena is that brittle materials, such as ceramics, may have very high wear rates in certain wear environments—when the abrasive is angular and "sharp" and when the abrasive size and applied load are high. Also, for rapid stock removal in grinding or for cutting rock and minerals, these should be the operating conditions. Under such conditions the worn surface will contain many small cracks (median cracks from surface indentation), which are generally undesirable for ground engineering components. Thus finishing should be carried out using rounded or "blunt" fine abrasive at low loads. These conditions will promote material removal by plastic deformation mechanisms.

2.3. Material Removal in Practice

The two models of material removal described in the preceding sections are expected to represent the upper and lower limits for volume wear. But because of the wide range of individual contact conditions in multi-particle contact wear, neither may account adequately for wear rates found in practice. To illustrate this, data from laboratory wear tests[3,36,37] on commercial bonded 180 grit (84 μm) and 60 grit (250 μm) SiC and 180 grit (84 μm) flint abrasives at an applied load of 1 MN/m^2 have been plotted as the reciprocal of volume wear—wear resistance—against the hardness of the worn surface (Fig. 7 to 9). The data, for which there is some scatter, are represented by solid lines, but also plotted, as broken lines, are Eq 8 and 10. For Eq 8 no prior assumptions have been made about the values of the constants; the broken lines are the best fit through linear portions of the data, from which the value of $K_1 \cdot K_2 \cdot K_3$ has been estimated subsequently. For Eq 10 it was necessary to assume a value for the constant k_5, and published data[39,40] suggested a value of 0.1. Because the fracture toughness K_c can vary for the same value of hardness H, the broken curve representing Eq

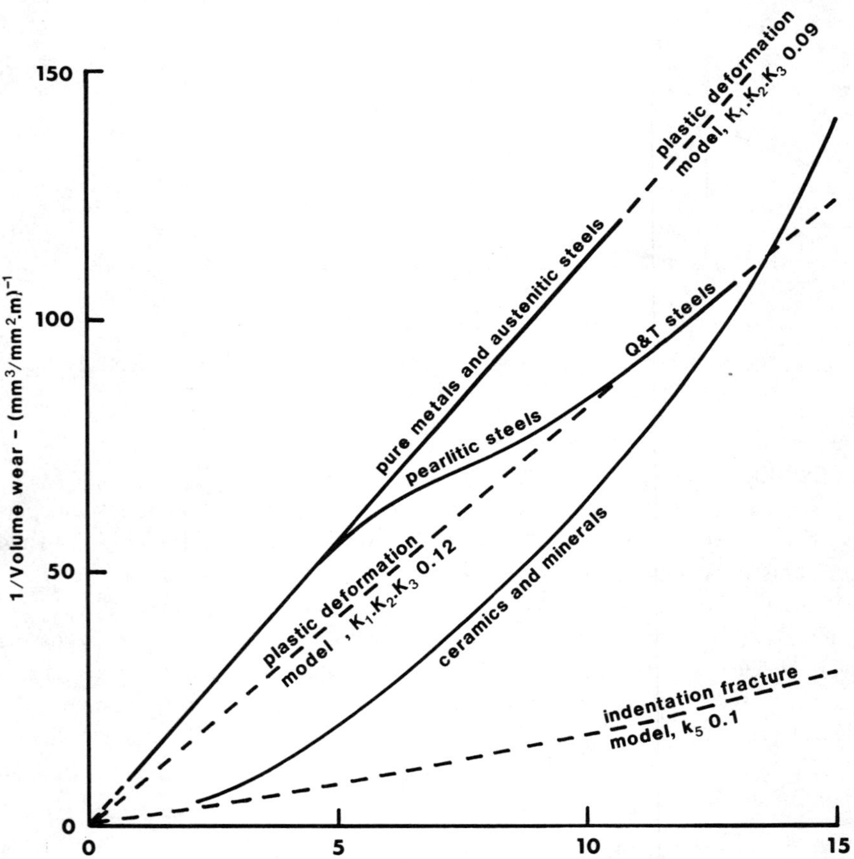

Fig. 7. Data (solid lines) for wear on 180 grit (84 μm) SiC bonded abrasive and theoretical models (broken lines).

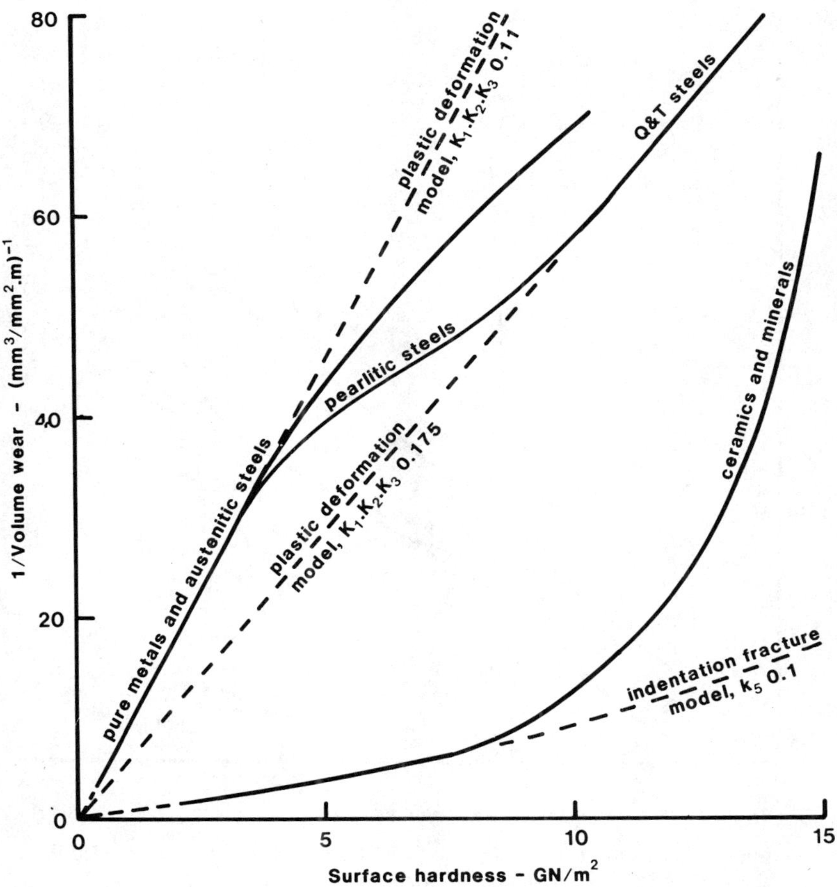

Fig. 8. Data (solid lines) for wear on 60 grit (250 μm) SiC bonded abrasive and theoretical models (broken lines).

10 is the best fit for a range of K_c and H commonly found for brittle materials.[31,38]

For wear on 180 grit SiC (Fig. 7), data for the pure metals and austenitic steels lie on a common line, which from Eq 8 gives a value for $K_1 \cdot K_2 \cdot K_3$ of about 0.09. Published data[38,40,41] suggest that the probability of wear debris formation during particle contact on ductile materials is about 60%, the mean proportion of the groove volume removed when debris formation occurs is about 60%, and abrasive particles approximate to 120° right cones. The value of $K_1 \cdot K_2 \cdot K_3$ calculated from these data is 0.11, which is close to the value obtained from the wear resistance versus hardness plot.

Data for the pearlitic and quenched and tempered steels deviate from the linear behavior shown by the pure metals, but become linear at high values of surface hardness; for this case, Eq 8 gives a value for $K_1 \cdot K_2 \cdot K_3$ of about 0.12.

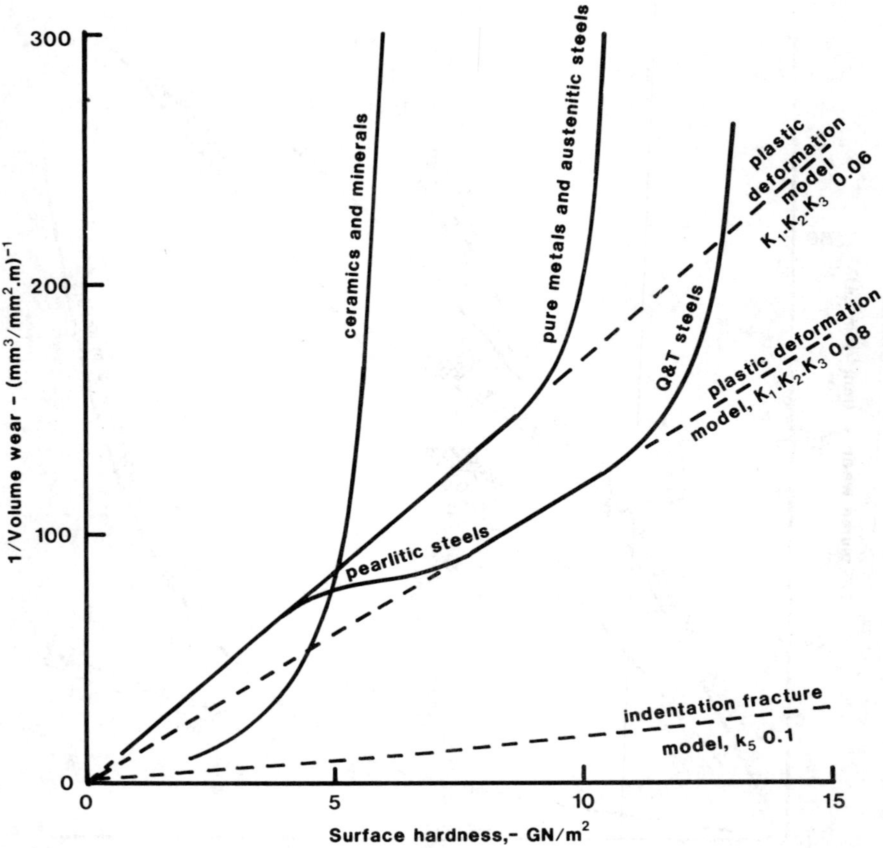

Fig. 9. Data (solid lines) for wear on 180 grit (84 μm) flint bonded abrasive and theoretical models (broken lines).

Assuming that K_3 is the same for both the pure metals and the steels, this increased value of $K_1 \cdot K_2 \cdot K_3$ indicates that either or both the probability of debris formation and the proportion of the groove volume removed depend on material properties other than surface hardness.

Data for ceramics and minerals agree with the indentation fracture model only at very low values of surface hardness; this is expected, as discussed in the previous section. However, the data do not agree with the plastic deformation model either. Experimental evidence[36] suggests that both plastic deformation and fracture mechanisms may operate simultaneously during the wear of brittle materials. Figure 10(a) shows the surface of fused SiO_2, hardness 5.3 GN/m^2, after wear on 180 grit (84 μm) SiC at an applied load of 1 MN/m^2, on which can be identified plastically formed grooves, A, material removal adjacent to the grooves by lateral cracking, B, and median cracks, C. In contrast, Fig. 10(b)

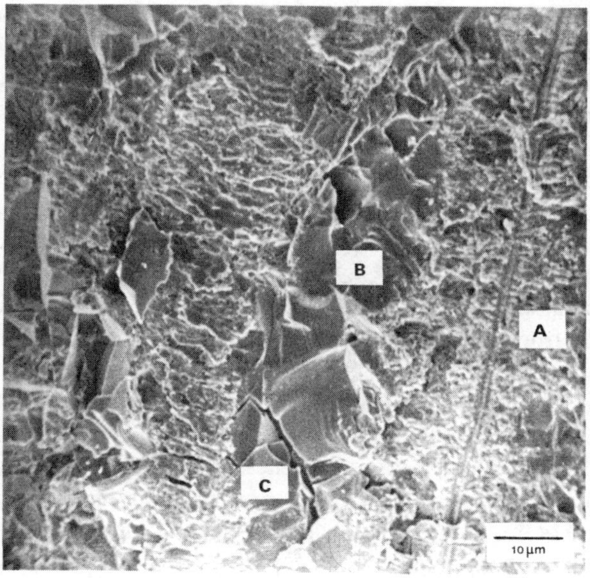

Fig. 10(a). Surface of fused SiO_2 after wear on 180 grit (84 μm) SiC bonded abrasive. See text for identification of A, B and C.

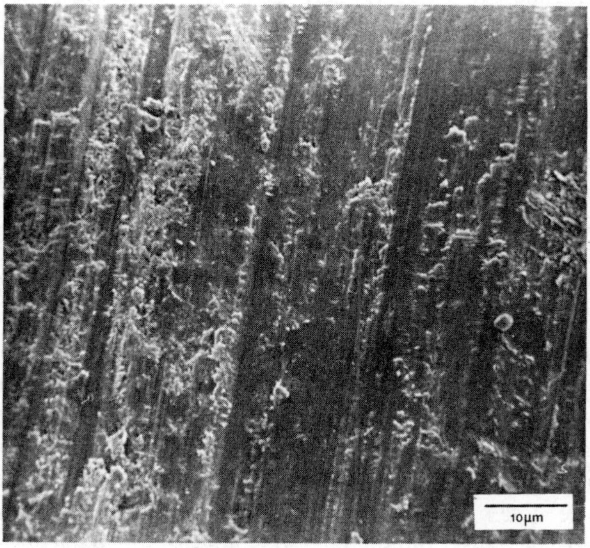

Fig. 10(b). Surface of hot pressed Si_3N_4 after wear on 180 grit (84 μm) SiC bonded abrasive.

shows the surface of hot pressed Si_3N_4, hardness 15.9 GN/m^2, on which the prominent feature is plastically deformed grooves with very little associated fracture. This behavior may be explained in terms of the probability of the indentation depth exceeding the critical identation depth for fracture, being greater for the fused SiO_2 than for the hot pressed Si_3N_4.

For wear on 60 grit SiC (Fig. 8), data for the pure metals and austenitic steels lie on a common curve, which approximates to linear behavior only at low values of surface hardness. For the linear portion of the curve, $K_1 \cdot K_2 \cdot K_3$ approximates to about 0.11—i.e., the 60 grit SiC is a more efficient abrasive than the 180 grit SiC. At high values of surface hardness, data for the pure metals and austenitic steels deviate from linear behavior, indicating that the probability of debris formation and/or the proportion of the groove volume removed increase as hardness increases. Data for the pearlitic and quenched and tempered steels show behavior similar to that on the 180 grit SiC, but the effect on increasing the value of $K_1 \cdot K_2 \cdot K_3$ is greater. Data for ceramics and minerals agree with the indentation fracture model to medium values of surface hardness; again, this is expected because of the increased indentation depth of the abrasive, Eq 6, and, thus, higher probability that the indentation depth will exceed the critical indentation depth. Examination of worn surfaces shows that both plastic deformation and fracture mechanisms account for material removal, but with an increased contribution from fracture.

For wear on 180 grit flint (Fig. 9), the data show significantly different behavior compared with the SiC abrasives. At low values of surface hardness the data for pure metals and steels show similar behavior to that for 180 grit SiC but with much lower values of $K_1 \cdot K_2 \cdot K_3$, indicating that the flint abrasive is less efficient than the SiC abrasive. However, at medium to high values of surface hardness both pure metals and steels show rapid and large increases in wear resistance as hardness increases. Similarly, data for ceramics and minerals deviate rapidly from the indentation fracture model at low values of surface hardness, such that their wear resistance exceeds that of the pure metals. These effects are largely due to modification of the contact conditions and material removal mechanisms by deterioration of the flint abrasive during wear.[3] The effect is largest for brittle materials, because failure and blunting of the abrasive has an effect on both plastic deformation mechanisms of removal, by changing K_3, and on fracture mechanisms, by reducing the indentation depth and increasing the critical indentation depth. For the pure metals and steels, blunting of the abrasive has a similar effect by changing K_3, but, also, it has an effect on K_1 and K_2. This will be discussed in more detail in later sections.

It should be noted that the values of $K_1 \cdot K_2 \cdot K_3$ estimated from the wear resistance versus hardness plots (Fig. 7 to 9) are specific to the wear environment in question, and that these values may vary considerably for different sets of data. In particular, the data are plotted for experiments on commercially bonded abrasives. In many practical wear environments the abrasive is not fixed but is free to move with the worn surface. Loose abrasive particles spend about 90% of the contact time rolling and only about 10% of the time sliding.[42]

2.4. Corrosion and Environmental Effects

In many practical wear situations worn surfaces are subjected to the action of an aggressive liquid or vapor as well as to mechanical damage by abrasives. An example is in pumps handling acidic slurries from coal-fired power station gas scrubbers. Since, during abrasive wear, mechanical damage may remove protective surface films, materials that rely on the development of passive films for protection will tend to have enhanced degradation rates. This is illustrated in Fig. 11 for martensitic high chromium irons tested in a slurry of silica sand for a range of pH values.[43] Similarly, for wood cutting the active liquid environment is acidic, with pH in the range 2.5 to 4, and wear occurs by combined corrosion and mechanical mechanisms.[44] The solution may be not to select a more noble material at the expense of wear resistance, but, where possible, to control the environment by the addition of neutralizing chemicals.

For very lightly loaded abrasive particle contact, it is possible that indentation is limited to surface films, where these exist. Thus, in very fine polishing, for which no surface damage is observed, material may be removed by continuous degradation and formation of surface films.[45] The sensitivity of fine polishing processes to small changes in the chemical activity of the polishing liquid supports this argument.

Even the presence of water vapor may be sufficient to increase wear rates of certain materials.[46] For example, surface deformation of aluminum and fracture of glass are enhanced in the presence of water vapor (Rebinder effect).[47] Similar phenomena may cause the abrasive itself to deteriorate, particularly by fracture,[46,48] which on a minor scale could result in sharpening but on a larger scale could result in blunting of the abrasive. The contribution of environmental effects on both the worn material and the abrasive will depend on the specific wear and chemical environments. This is illustrated by some results on the effects of water and water vapor (Table 6).[46,48-51] In view of their importance in contributing to material removal and their practical significance it is rather surprising that our understanding of these phenomena is so limited.

3. Effects of Properties of the Abrasive

3.1. Abrasive Particle Shape, Orientation and Size

In the preceding sections attention has been drawn to the effect of abrasive particle shape on wear. Shape is an important factor because it influences the transition load from elastic to plastic contact and the critical indentation size for fracture. It also determines groove geometry. If particle load and surface hardness

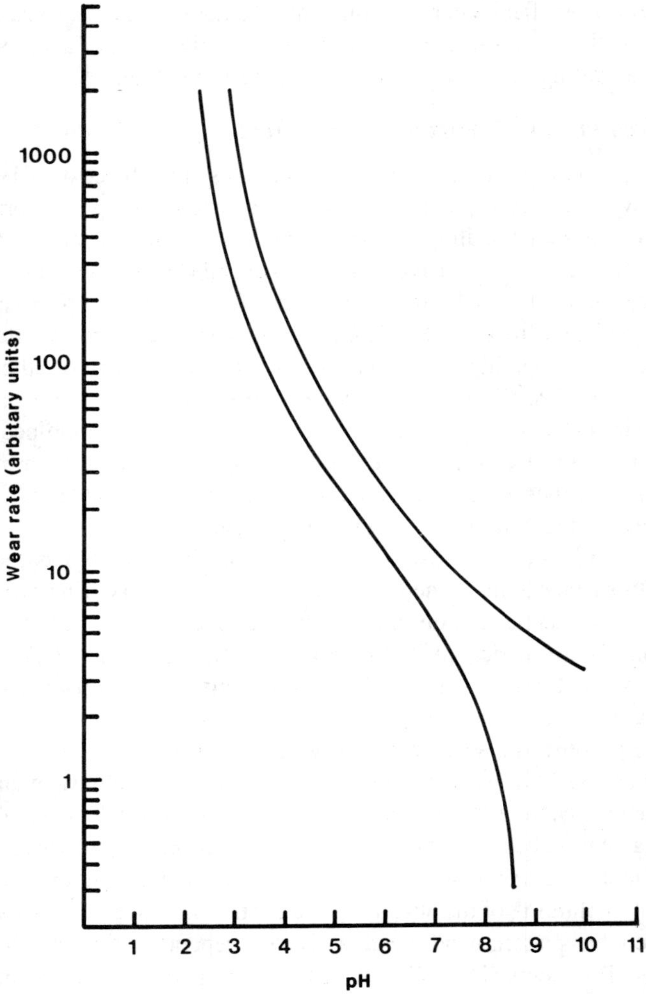

Fig. 11. Effect of pH on the wear of some high chromium cast irons in an abrasive slurry.

are constant, the projected area of contact wil be constant, but the cross-sectional area of a groove resulting from such contact will depend on the particle shape. Simple calculation for pyramidal, conical and spherical contacts shows that the ratio of cross-sectional to projected area of contacts increases as the included angle, cone angle and radius, respectively, decrease. In addition, this ratio is generally less for spherical than for pyramidal or conical contacts. Thus, for both plastic deformation and fracture mechanisms of material removal we would expect the wear rate to be higher for "sharp" pointed abrasives than for "blunt" rounded abrasives. Table 7 illustrates this effect for wear by quartz abrasives.[52]

Table 6. Summary of the Effect on Wear of Water and Water Vapor (Ref 46, 48-51)

Material and wear conditions	Percentage change with respect to "dry" wear 50% RH	100% RH	Wet
Aluminum alloy 6063-T6:			
Fixed SiC abrasive	+20	+10	−10
1040 steel, pearlitic:			
Fixed SiC abrasive	+5	+10	0
Loose SiC abrasive, 3-body	+20	+175	
Pyrex glass:			
Fixed SiC abrasive	+15	+30	0
Loose SiC abrasive, 3-body	−12	+220	
Sintered alumina:			
Fixed SiC abrasive	–	–	−99
Loose Al_2O_3 abrasive, 2-body	–	–	+300
WC/6-10%Co:			
Loose Al_2O_3 abrasive, 2-body	–	–	+54
Cutting sandstone	–	–	−36
Flexible PVC:			
Loose SiO_2 abrasive, 2-body	–	–	+200
Nylon (polyamide) 6.6:			
Loose SiO_2 abrasive, 2-body	–	–	−58

Table 7. Effect on Wear by Quartz of Abrasive Particle Shape (Ref 52)

Abrasive shape characteristic	Wear, arbitrary units: Normalized 1045 steel HV = 130	Ni-hard cast iron HV = 680	High Cr cast iron HV = 700
Sharp crushed			
Wausau quartzite	1.0	0.18	0.04
Round Ottawa sand	0.31	0.03	0.02

In practice it is difficult to characterize and quantify abrasive particle shape, both because of the range found in any one environment and because abrasive particle contact often involves only 10 to 30% of the total surface area. The fine surface structure may be as important as the bulk shape.

Geometry and orientation of abrasive particle contacts have a very significant effect on the formation of wear debris during groove formation by plastic deformation. For spherical particle contact on steels, plastic deformation with no material removal, "plowing," can occur when the ratio of the depth of indentation to width of the groove is between 0.005 and 0.05, but microchip formation occurs when this ratio is greater than 0.05.[53,54] For angular particles microchips are formed even at very small indentation depths.[19,45] A transition from plowing to microchip formation occurs when a critical attack angle—the angle between the leading face of the abrasive particle and ungrooved surface—is exceeded. In simulated abrasive particle tests[39,54-58] the critical attack angle varies considerably for different materials (Table 8). Theoretically[55,56] and in practice[57] the critical attack angle decreases as the coefficient of friction between the leading face of the abrasive particle and worn material increases. The critical attack angle for steels decreases as the hardness increases, and there is a general trend for it to be higher for pure metals than for alloys and steels of similar hardness.

For commercial abrasive particle contacts, which approximate to 120° cones,[41] the median attack angle is about 70°, so the probability that a contact will result in microchip formation, and contribute to K_1 in Eq 8, increases as the critical attack angle decreases. If the critical attack angle is as high as 90°, only 15% of possible particle contacts will form microchips (Fig. 12),[39,41] whereas if the critical attack angle is 30°, about 80% of possible contacts will form microchips. Thus, Murray et al.[58] have suggested that, at the same surface hardness, plowing is predominant in the wear of pure metals and microchip formation is predominant in the wear of hardened steels. This would account for the lower wear resistance of the steels at the same hardness and for the transitional behavior as hardness increases and microchip formation becomes more predominant. (Fig. 7).

In a wide range of abrasive wear environments the size of the abrasive has a significant effect on wear; wear increases as the abrasive particle size increases

Table 8. Summary of Experimental Critical Abrasive Attack Angles for Microchip Formation (Ref 39, 54-58)

Material	HV, kg/mm²	Critical attack angle
Lead	5	50° to 60°
α-brass	180	50° to 60°
Aluminum	35	80° to 90°
Aluminum alloy	145	~ 35°
Copper	120	40° to 50°
Nickel	350	60° to 70°
1040 steel, annealed	180-260	40° to 90°
1082 steel, quenched and tempered	860	~ 30°
Tool steel, quenched and tempered	760	~ 30°

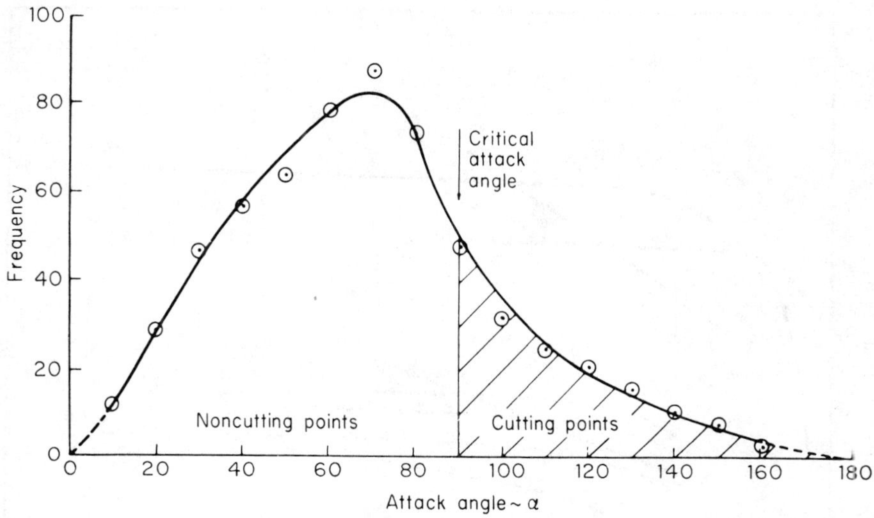

Fig. 12. Frequency distribution of attack angles of contacting abrasive particles for 70 μm SiC bonded abrasive, and proportion of cutting points with a critical angle of 90°.

(Fig. 13). The effect is greatest for non-metals, both ceramics and polymers,[36,59,60] and may be associated primarily with a change in the predominant mechanism of material removal.[36,60] The effect is least for metallic materials. It has been suggested that for fine abrasives a lower fraction of the load is carried by particle contacts that are able to remove material, because (*a*) a larger number of contacts will make elastic contact only,[40,61,62] (*b*) loose wear debris hinders the contact of some particles,[42,62-64] and (*c*) fine abrasives deteriorate much more rapidly than coarse abrasives.[61,62]

3.2. Mechanical Properties

Early investigations of abrasive wear showed that wear decreases as the hardness of the worn material approaches that of the abrasive and drops rapidly when the hardness of the worn material exceeds that of the abrasive. Figure 14 illustrates this effect for metallic materials and ceramics worn on 80 to 400 μm diameter commercial glass, flint, garnet, corundum and SiC abrasives.[3,36,37,65] Richardson[3] found that continuous groove formation ceases for coarse abrasives when the yield stresses of the abrasive and worn material are equal, but groove formation continues for fine abrasives beyond this limit. For the former, material continues to be removed even though groove formation ceases, and for the latter, the grooves deteriorate very rapidly after initial contact. Even when the abrasive is much harder than the worn material the rate of reduction of groove formation with increasing contact distance is very much greater for fine abrasives than for coarse abrasives.[62]

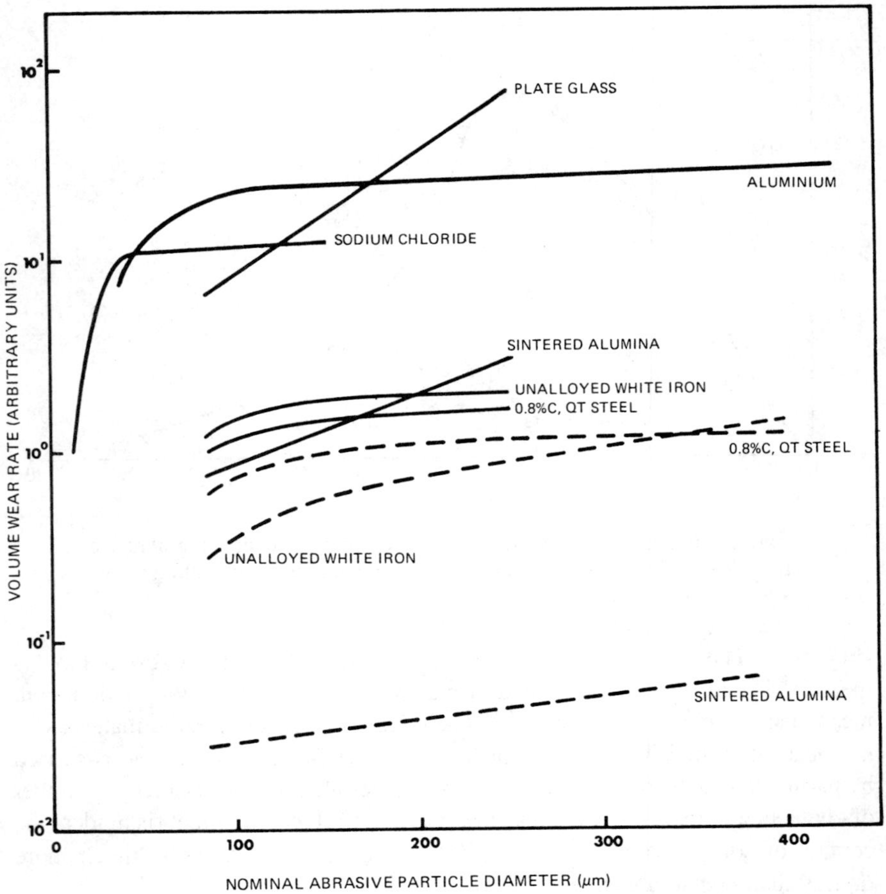

Fig. 13. Effect of abrasive size on wear by bonded abrasives (1 MN/m^2 applied load); solid lines for SiC abrasives, broken lines for flint abrasives.

To account for these phenomena, Richardson[3] proposed that plastic flow of the abrasive occurs when the hardness of the worn surface approaches that of the abrasive, but when the plastic zone in the abrasive particle reaches a critical size[32] fracture occurs. Since coarse abrasives normally indent the worn surface to a greater depth than fine abrasives, the plastic zone is expected to be larger and the probability of fracture higher. In addition, the tensile strength of abrasive particles decreases rapidly as size increases[66-68] because the probability that a particle will have a suitably oriented defect increases.

Since fracture of the abrasive may regenerate sharp cutting facets and produce loose abrasive fragments,[69,70] the effect on wear of fracture is less than the effect of plastic deformation alone. Thus, deterioration of the coarse abrasive by frac-

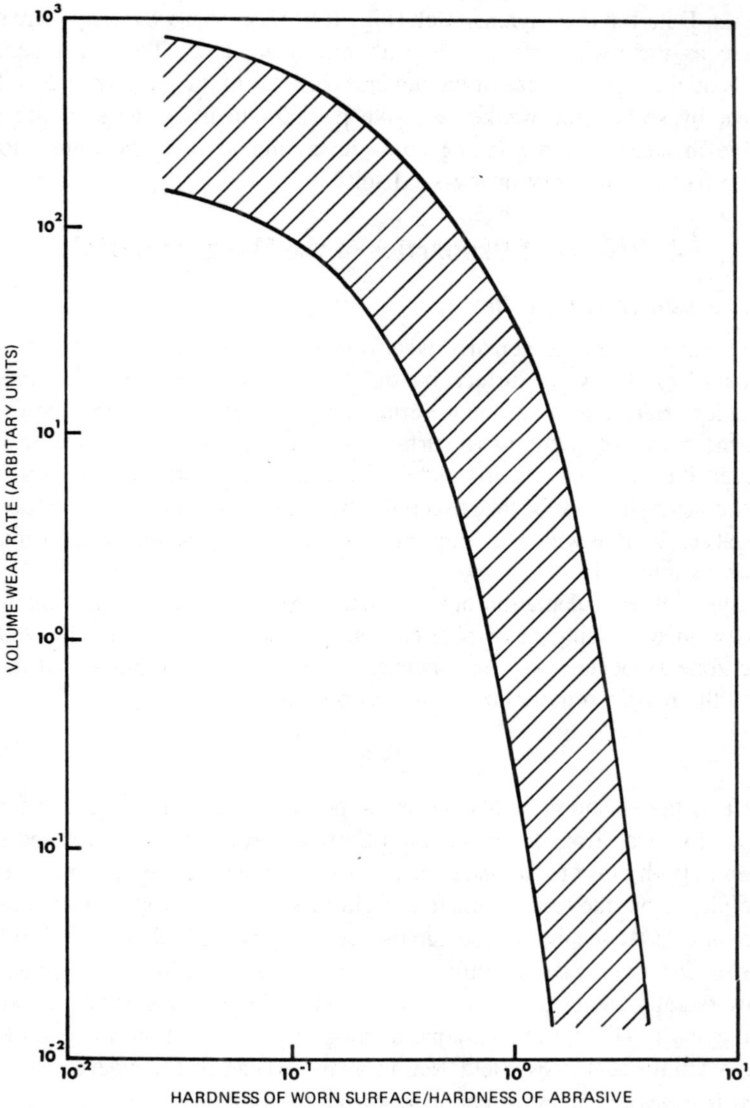

Fig. 14. Effect of abrasive hardness on the wear of metallic materials and ceramics (worn on 80-400 μm bonded abrasives, 1 MN/m^2 applied load).

ture causes discontinuous groove formation but the abrasive is effective up to fracture and, possibly, after fracture. Deterioration of the fine abrasive by plastic deformation inhibits groove formation, presumably because the load can be carried elastically on the blunted contacts. Load is expected to have an effect also, because the depth of indentation of the abrasive particles will increase as load increases. Data for wear against rock (Fig. 15)[71] show that wear may increase or decrease as the rock surface fails with increasing load. The exact behavior depends on the type and size of the mineral particles of the rock mass. Similarly, for wear by soils,[72] the weaker abrasive particles in ironstone soils are more effective in wear than the strong abrasive in flint soils or the lightly loaded abrasive particles in a dry quartz sand soil.

4. Effects of Properties of the Worn Material

4.1. Mechanical Properties

The simple models of abrasive wear discussed in section 2 are based on the indentation of the worn surface by the abrasive particles; thus resistance to indentation (hardness) is a major variable in Eq 8 and 10. In practice (Fig. 7 to 9), the hardness of the worn surface does have a major influence on wear. However, it cannot account completely for wear since it provides no measure of flow characteristics nor of the susceptibility of a material to crack initiation and propagation, both of which are important in determining the mechanism and ease of material removal.

Analyses of the indentation of ideal rigid-plastic material[73-76] liken the indentation to an expanding hemispherical cavity. These analyses predict that the plastic zone associated with the indentation is also hemispherical and that the ratio of the plastic zone radius, c, to indentation radius, a, is

$$c/a \propto (E/\sigma_y)^{1/3}, \qquad [11]$$

where E is the elastic modulus and σ_y is the yield stress. They predict that the material displaced from the indentation flows up around the indenter and forms a raised portion close to the indenter surface. In practice, metallic materials are elastic-plastic but they approximate to rigid-plastic—i.e., elastic strains are negligible for a 120° cone indenter, when the ratio E/σ_y exceeds about 100.[76] At lower values of E/σ_y elastic strains influence the deformation, allowing some material to flow radially rather than up to the surface.[77] In principle these phenomena should apply to sliding indentations, although the deformation will be modified by the asymmetrical stress field, acting mainly ahead of the indenter.[78]

Two features of the deformation associated with indentations are relevant to abrasive wear. First, the depth of the deformed zone at a worn surface should be a function of E/σ_y and the groove size. Second, the proportion of the groove volume that forms ridges along the groove sides should decrease as E/σ_y decreases. From Eq 3 and 11, the depth of the deformed zone is

$$D_{def} \propto (E/\sigma_y)^{1/3}(L/H)^{1/2}. \qquad [12]$$

Estimates of the depth of the deformed zone in copper, iron and steels[14] after wear on 180 and 60 grit SiC are plotted against this function in Fig. 16, where σ_y is

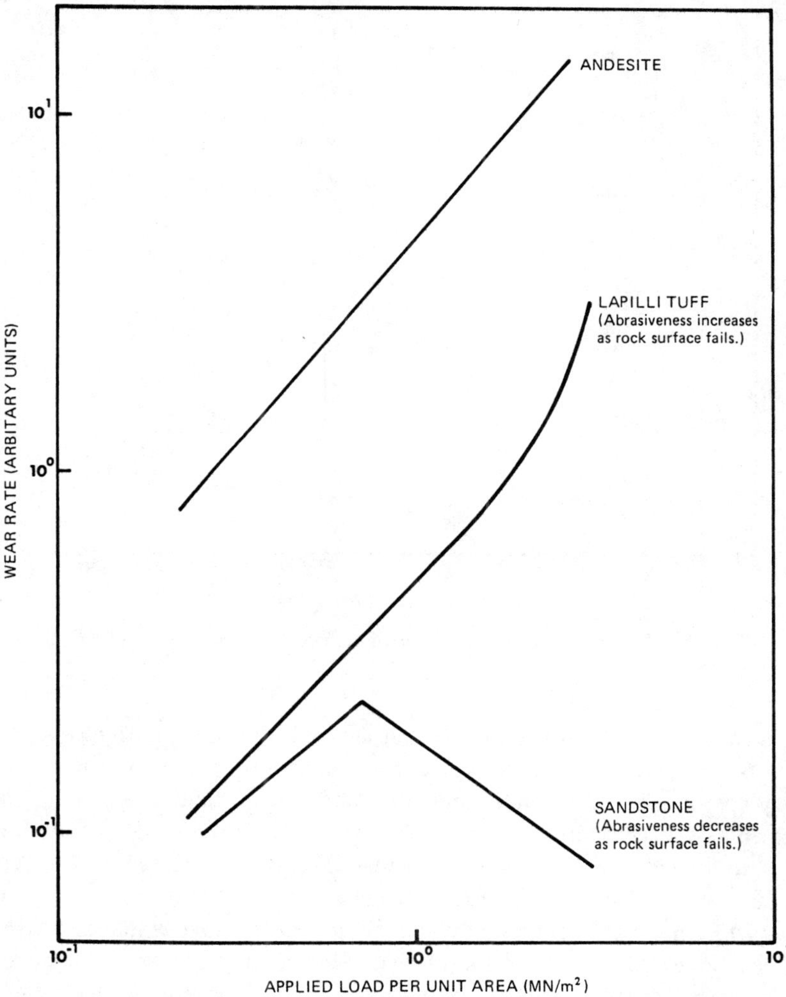

Fig. 15. Effect of applied load on the wear of a 0.3% C low alloy steel against three rock types.

the flow stress at low strain estimated from hardness measurements,[75,76] and H is the surface hardness. While there is considerable scatter of the data, due in part to the inaccuracy of the measurement of depth of the deformed zone and hardness, there is general agreement with Eq 12. Thus, the depth of the deformed zone is dependent on mechanical properties of the material. Since frictional work during the wear of ductile materials is largely associated with plastic deformation, friction might be reduced if the depth of the deformed zone decreases—i.e., as E and L decrease and σ_y and H increase. In addition, when cracks initiate and

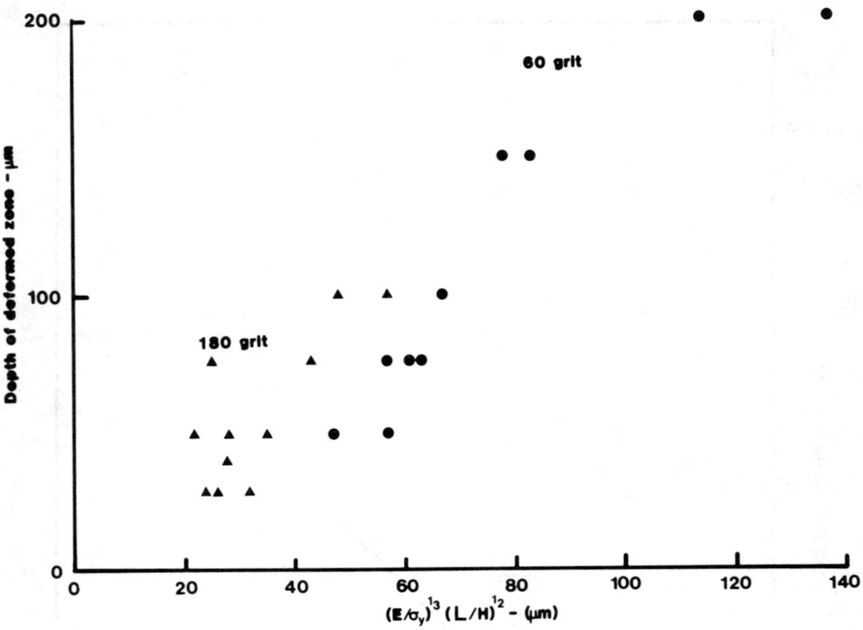

Fig. 16. Dependence of the depth of deformed zone on the function $(E/\sigma_y)^{1/3} \cdot (L/H)^{1/2}$.

propagate beneath the worn surface and contribute to material removal, wear might be reduced as plastic deformation is forced closer to the surface.

In experiments with single simulated abrasive particle contacts[79,80] measurements of the proportion of the groove volume removed and proportion displaced to the sides of the groove show that the former increases, and the latter decreases, as hardness of the surface increases and as E/σ_y decreases (Fig. 17). Two factors probably contribute to this behavior. The first is that at low E/σ_y material flows less readily to the groove sides because of the contribution of elastic strains, and thus material displaced from the groove more readily forms wear debris. The rapid fall in proportion of the groove volume removed over the low E/σ_y range and the general trend for materials of the same hardness to have higher wear resistance at higher E/σ_y values[11,37] support this. The second is that as hardness increases, resistance to crack propagation decreases, so fracture might occur during flow of material around particle contacts and contribute to the proportion of groove volume removed. The dependence of flow geometry on mechanical properties of the surface material offers an explanation for the variation of critical attack angle with material and contact friction. At high values of the contact friction rigid-plastic behavior, i.e., flow to the surface around the indenter, is less likely to occur,[75] so radial deformation and microchip formation are more probable.

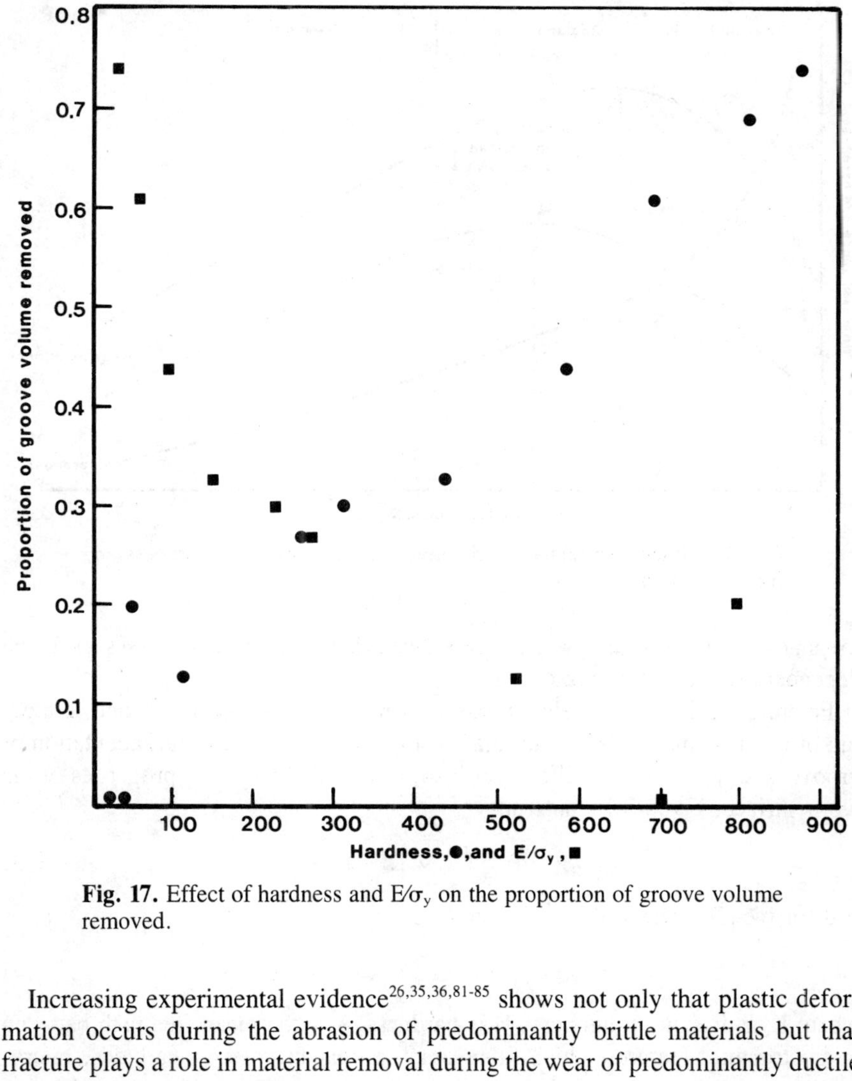

Fig. 17. Effect of hardness and E/σ_y on the proportion of groove volume removed.

Increasing experimental evidence[26,35,36,81-85] shows not only that plastic deformation occurs during the abrasion of predominantly brittle materials but that fracture plays a role in material removal during the wear of predominantly ductile materials. The contribution of fracture, which as discussed in section 2 can significantly increase wear, is dependent on the size of the wear grooves and on mechanical properties of the material. Zum Gahr[86] has rationalized this behavior for wear by a hard abrasive by plotting the upper limit to wear resistance and hardness against fracture toughness, shown schematically in Fig. 18. As fracture toughness increases, wear resistance increases to a maximum, which tends to a higher fracture toughness value for increasing groove size, i.e., increasing applied load and/or abrasive particle size. Over this range, region I, wear is very dependent on fracture toughness, but decreases as hardness decreases. As fracture

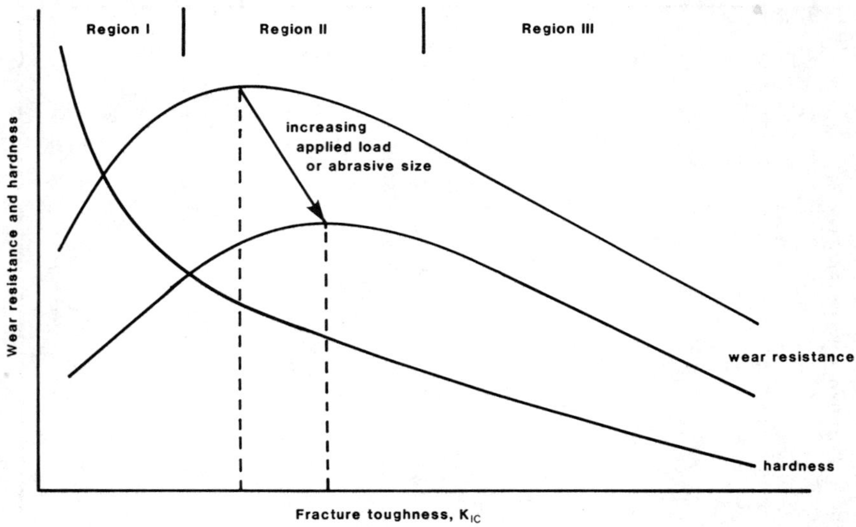

Fig. 18. Schematic relationship between wear resistance, hardness and fracture toughness.

toughness increases further, wear resistance decreases and wear is seen to be dependent on hardness, region III.

Fracture mechanics models for indentation of brittle materials[33,87] and scratching of metallic materials[85] yield relationships between the critical indentation or groove size, p_{crit}, the critical contact load, L_{crit}, and mechanical properties of the worn surface. For brittle materials[87]

$$p_{crit} \propto (K_c/H)^2;\ L_{crit} \propto K_c \cdot (K_c/H)^3, \qquad [13]$$

and for metallic materials[85]

$$p_{crit} \propto \lambda^{1/2} \cdot K_c/s \cdot \mu \cdot H;\ L_{crit} \propto \lambda \cdot K_c^2/s^2 \cdot \mu^2 \cdot H, \qquad [14]$$

where K_c is fracture toughness, H is hardness, λ is the mean free path between microstructural features which favor crack propagation (e.g., carbides, microcracks, prior austenite grain boundaries, etc.), s is the distance between abrasive particle contacts, and μ is the coefficient of friction between abrasive particles and the material. Other parameters, including shape of the abrasive particle contact, also enter into these relationships, but are assumed constant in the following. When p_{crit} is exceeded fracture associated with indentations or grooves is favored.

From Eq 13 and 14 it is clear that the ratio K_c/H is critical in determining the groove size and contact load at which fracture occurs. By reference to Fig. 18 it can be seen that the generalized value of K_c/H and, thus, p_{crit} decreases as hardness increases. The actual depth of indentation of an abrasive particle, p, also decreases as hardness increases (see Eq 6), but at a much slower rate than does p_{crit}. Thus, if, at high values of fracture toughness and low values of hardness, p_{crit} is greater than p, they will converge as fracture toughness decreases and hardness increases. When p is just smaller than p_{crit}, fracture does not occur, and material removal is entirely controlled by plastic deformation, but when p is just larger than p_{crit} fracture occurs and we expect wear to increase. This region of close proximity of p and p_{crit} corresponds to the maximum wear resistance, region II in Fig. 18. Here hardness is at a maximum for plastic deformation-controlled material removal.

It should be noted that Fig. 18 is schematic only and that materials having the same fracture toughness may have quite different values of hardness. Furthermore, the values of these mechanical properties appropriate to this analysis are those of the worn surface. While the hardness of worn surfaces may be estimated from direct measurements there is no method at present of estimating the fracture toughness of worn surfaces. In making use of the analysis and its predictions to optimize wear it is possible only to use the low strain values of hardness and fracture toughness as guides. From their analysis, Lawn and Marshall[87] predict that wear resistance is maximized:

(1) In region I, when $p_{crit} \ll p$ and damage is essentially fracture controlled, by maximizing K_c;
(2) In region II, when $p_{crit} \approx p$, by maximizing K_c/H and, to a lesser extent, K_c;
(3) In region III, when $p_{crit} > p$ and damage is essentially deformation controlled, by maximizing H.

In Zum Gahr's analysis,[85] the effects of microstructural features and particle contact spacing are included. Thus in addition to point 1 above, wear resistance should increase as λ increases, and should be greater for fine than for coarse abrasives. Figures 7 and 8 show that the wear behavior of materials is generally in agreement with these predictions. The higher wear resistance of the pure metals compared with steels of the same hardness is probably associated with lower λ values for the latter. This effect and other microstructural effects are discussed in the following section.

4.2. Microstructural Features

The role of microstructure in abrasive wear may be divided, simply, into two major categories. The first is when microstructure influences bulk properties, including the initiation of cracks. This is generally the case when the size of microstructural features is smaller than the depth of indentation of abrasive particles. The second is when microstructural features become effective as discrete components and their individual properties assume increasing importance. This is generally the case when the size of microstructural features is about equal to or larger than the depth of indentation of abrasive particles. Wear data amassed over the last thirty years have provided a good understanding of the qualitative influence of microstructure on abrasive wear, but only recently have concerted efforts been made to provide sufficient quantitative data to enable specification of the optimum microstructure and metallurgical production techniques. Currently, this objective has been met only for a limited number of practical materials; there remain large gaps in our knowledge. The following is intended to provide a summary of the qualitative effects of microstructure on wear and to draw attention to quantitative data, where they exist. For further information, the reader is referred to reviews by Hornbogen[88] and Zum Gahr.[89]

Because bulk mechanical properties are determined by microstructural parameters—such as volume fraction and distribution of dispersed phases and inclusions, their coherency and hardness—these parameters also have an effect on wear. The effect of microstructural parameters on strain hardening at worn surfaces has been discussed in section 2.1, and, depending on the role that microstructure plays at high strains, wear may correlate with microstructural features. For example, the wear of steels containing finely dispersed carbides and of aluminum alloys containing hard incoherent particles is linearly related to the square root of the mean free path between particles,[23-26,90] whereas the wear of alloys containing soft coherent dispersions is little affected by the presence of or distribution of the particles.[83,91]

In considering the effect of microstructure it is important to appreciate that certain structures and features may be beneficial in some wear environments but detrimental in others. This is clearly demonstrated by the effect of microstructure on the wear of ferrous materials. Data for steels worn by "hard" abrasives[25,37,92] show that wear resistance increases with increasing hardness, but that austenitic and bainitic structures are superior to ferritic, pearlitic and martensitic structures of the same hardness (Fig. 19). This is generally attributed to the higher strain-hardening capacity and ductility of austenite (see Fig. 3) and lower bainite, which contains about 20% retained austenite. For martensitic structures wear resistance may increase or decrease as the retained austenite content increases (Fig. 20).[26,93] The beneficial effect of retained austenite on wear by relatively hard abrasives or at high abrasive particle contact loads (Eq 4) is associated with suppression of the initiation and propagation of microcracks close to the surface,[26] i.e., an increase in the values of p_{crit} and L_{crit} in Eq 14. This may arise because volumetric changes associated with transformation of austenite to martensite close to the surface introduce compressive stresses in the martensite, and because cracks propagating through martensite beneath the surface are blunted and stopped in the ductile austenite. These effects are greatest at retained austenite contents of between about 20 and 50%,[26,93,94] above which the bulk softening effect of the austenite causes a decrease in wear resistance. The detrimental effect of retained austenite on wear by relatively soft abrasives or at low abrasive particle loads is associated with the reduction in bulk hardness.

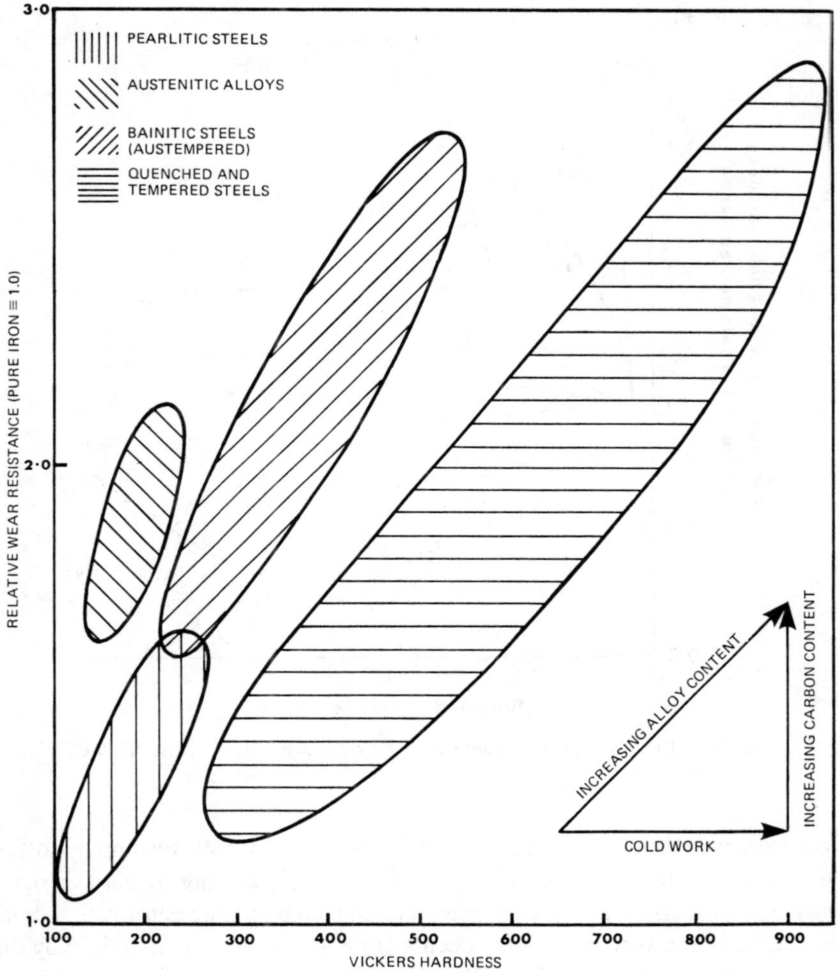

Fig. 19. Effect of structure, heat treatment and alloy content on the wear of steels (worn on 90 μm Al_2O_3 abrasive, 1 MN / m^2 applied load).

Austenite close to the worn surface will transform to martensite or strain harden to a hardness about equal to the martensite, but below the surface will remain soft. Thus, because for soft blunted abrasives and low particle loads $p_{crit} > p$, wear is predominantly dependent on hardness and increases (region III in Fig. 18). Again, this effect is greatest at retained austenite contents up to about 50%, above which the increase in strain-hardening capacity and rate with increasing austenite content has a favorable effect.

Taken to the extreme of fully austenitic structures, e.g., austenitic Mn steels, the data in Fig. 20 illustrate a different trend: fully austenitic structures have

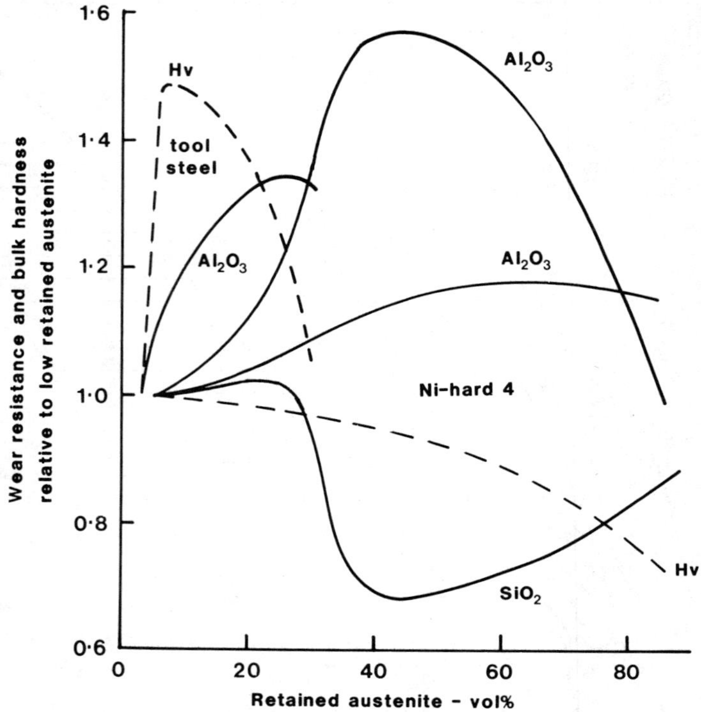

Fig. 20. Effect of retained austenite on the wear of a tool steel and Ni-hard 4 cast iron.

higher relative wear resistance under conditions of low particle loading by soft or blunt abrasives than under conditions of high particle loading by hard or sharp abrasives[12] (see also Fig. 8 and 9). Material removal by plastic deformation, with fracture playing a very minor role, occurs in the former and is minimized by the high strain-hardening capacity and ductility of the austenite. In contrast, with the more efficient abrasive, the increased probability of fracture at the larger indentation depths and the large depth to which deformation is forced (Eq 12) account for the relatively higher wear.

Many practical engineering materials, selected for abrasive wear environments, have structures of hard carbides in either hardened or soft matrices. In such materials, the size, spacing, volume fraction and mechanical properties of the carbides and the mechanical properties of the matrix, relative to both the carbides and abrasive, have significant effects on wear. Figure 21 shows the dependence of wear resistance on carbide content for steels, cast irons and cobalt alloys.[26,94-96] When the carbides are small relative to the abrasive particle indentation depth, and when the matrix is of about the same hardness — e.g., M_3C type carbides in martensitic matrix tool steels — wear by hard abrasives increases as

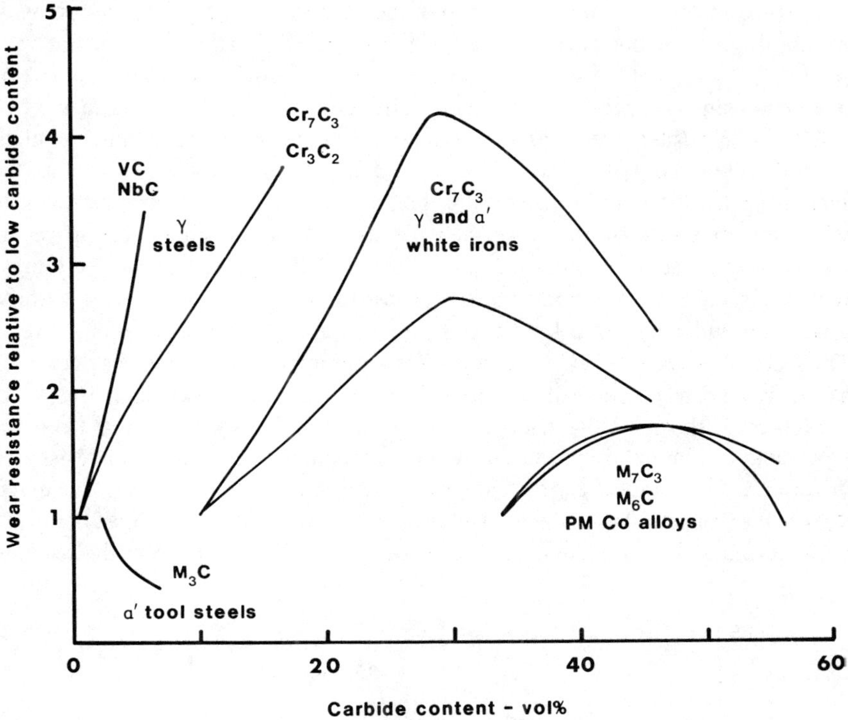

Fig. 21. Effect of carbide content on the wear of steels, cast irons and cobalt alloys.

the carbide content increases. The carbides act as stress raisers at which cracks initiate and propagate through the low toughness matrix.[26] However, when the carbides are large relative to the abrasive particle indentation depth, they represent a powerful obstruction to indenting abrasive particles, particularly for relatively soft abrasives. Thus, wear resistance increases with increasing carbide content up to a maximum and then decreases. The transition is associated with the relative contributions of fracture of the hard carbides and plastic deformation of the matrix to the total wear. As carbide content increases, material removal by fracture of the carbide is more than compensated by reduced material removed from the matrix and of whole carbides by plastic deformation.[94,95] The largest increases in wear resistance with increase in carbide content occur for tough carbides, e.g., VC and NbC compared with Cr_7C_3, and low abrasive particle loading, i.e., minimal fracture. As the carbide content increases still further the contribution of fracture of the carbides to total wear increases while that due to plastic deformation mechanisms of the matrix decreases. Above the carbide content for which fracture becomes predominant, in terms of volume removal, wear increases.

In some materials, a necessary step in the fracture of carbides is the removal, by plastic deformation mechanisms, of the supporting matrix.[97,98] Thus at very high carbide contents, e.g., WC/Co composites, a further transition may occur, associated, in part, with the increasing difficulty of removal of the matrix. In WC/Co composites wear resistance increases rapidly as the carbide content exceeds about 80 vol% (Fig. 22).[97,99,100] Part of this increase is due to the increasing hardness and to increasing degradation of the abrasive, particularly when the hardness of the material exceeds that of the abrasive. However, during wear by soft abrasives the binder is extruded between the carbide grains, which subsequently fracture or are loosened and plucked out.[98] Thus carbide grain size and mean free binder path are major factors in determining wear. The wear resistance of WC/Co composites on relatively soft abrasives decreases as the mean free binder path increases[98] and as the carbide size increases.[99]

Figure 23 illustrates the effect of carbide size on the wear of a WC/Co composite and a cobalt alloy produced by powder metallurgy. While wear resistance of the WC/Co composite decreases with increasing carbide size, as discussed above, the opposite behavior is displayed by the Co alloy, which has about 40 vol% carbides. For the latter, the ability of the carbides to obstruct the passage

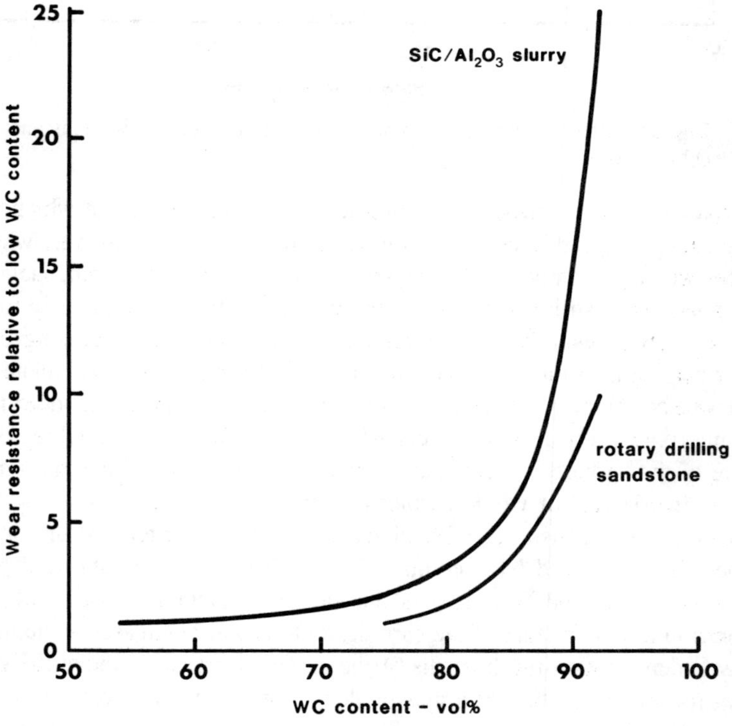

Fig. 22. Effect of carbide content on the wear of WC/Co composites.

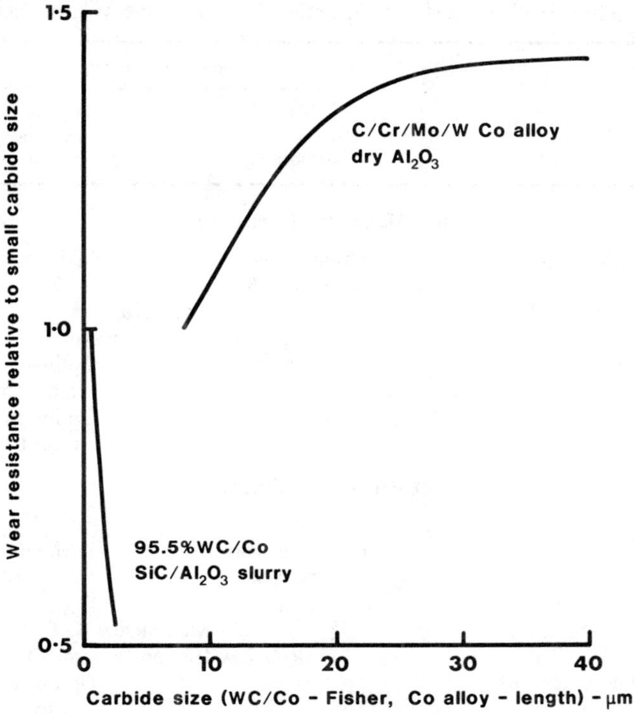

Fig. 23. Effect of carbide size on the wear of a WC/Co composite and a cobalt alloy containing M_7C_3 and M_6C type carbides.

of indenting abrasive particles and cause abrasive degradation may increase as the carbide size increases, particularly over a range about equal to the range of depth of indentation of the abrasive particles.

5. Selection of Materials for Abrasive Wear Resistance

Because neither wear resistance nor friction coefficient are intrinsic material properties, the first step in selecting a material for abrasive wear resistance is to determine the environment and define the properties required of the material. The required material properties will include those required for other, possibly primary, functions of the equipment or structure in question. For example, in mining and quarrying, equipment may be required to withstand high impact loads. Table 9 shows the required material properties for two simple wear environments, and illustrates how these may vary. This highlights the dangers in selecting materials on the basis of performance data from tests under conditions that dc not closely simulate the service conditions. If performance data do not exist, and it is necessary to devise a test procedure, a good guide to the proximity of the test

Table 9. Required Material Properties for Abrasive Wear Resistance

Required material properties	Abrasive environment: High load, large angular, hard abrasive	Abrasive environment: Low load, small rounded, soft abrasive
Bulk Mechanical Properties		
H	Optimize with respect to K_c	Maximize, >0.6 $H_{abrasive}$
E/σ_y	Maximize (for low friction, minimize)	
K_c	Maximize	Optimize with respect to H
K_c/H	Maximize	Optimize with respect to H
Microstructural Properties		
Matrix:		
H_{matrix}	Optimize with respect to toughness	Maximize
Strain-hardening capacity	Maximize (for low friction, minimize)	
Ductility and toughness	Maximize	Optimize with respect to H_{matrix}
Dispersed hard phase:		
H	Maximize, but with respect to $H_{abrasive}$ and H_{matrix}	
Size	Maximize, >p	
Content	Optimize in range 30-50 vol%	

environment to the service environment is the similarity of surface damage resulting from the two environments—i.e., size, shape and length of scratches, degree and depth of sub-surface deformation, degree of fracture damage, etc. The performance data[3,36,37,72,101-104] in Table 10 show that the relative wear resistance and ranking order can change from one environment to another. Actual wear rates vary even more widely; for example, wear rates of about 0.001 μm/hr for flotation impellers in abrasive mineral slurries and 25 μm/hr for hammers in impact pulverizers are typical. Thus, factors such as tolerable wear and ease of replacement enter into the selection of material. Inevitably, the final choice is a compromise, particularly when both wear resistance and toughness are required.[12,94,105,106]

The preceding sections have shown that abrasive wear involves unique high strain, high strain rate deformation and fracture of microvolumes of material. It

Table 10. Examples of Abrasive Wear Performance Data (From Ref 3, 36, 37, 72 and 101-104)

Material	Wear resistance relative to 0.4% C low alloy steel, quenched and tempered to about 500 HV, in various wear environments: Sliding of chutes				Laboratory wear tests		
	By coke	By blast furnace sinter	Wear of ball mill media grinding quartz ores	Wear of deep cultivators, flintstone sand loam soil	Jaw crusher, siliceous ore	Pin-on-disc, commercial bonded abrasives: 384 μm flint	Pin-on-disc, commercial bonded abrasives: 84 μm glass
Cast Irons							
Low alloy 2.5-2.8%C, 800 HV	0.85	1.25	1.0	3.5	–	1.65	~∞
HT nodular graphite, 700 HV	1.0	6.5	–	0.65	–	–	–
15/3 Cr/Mo martensitic	–	–	1.25	–	3.5-25	1.25-1.4	–
High Cr, 25-30%, martensitic	1.25-1.8	6.5	1.25	3.5	1.65-13	2.0	–
Ni-hard (3C4Ni2Cr)	1.65	14	1.0	2.5	13	1.65	–
Cast and Wrought Steels							
0.4%C low alloy, 500 HV	1.0	1.0	1.0	1.0	1.0	1.0	1.0
0.8%C, 800 HV	–	–	1.0	2.0	–	1.8	~∞
2%C, 12%Cr, 700 HV	–	–	1.1	2.0	–	1.8	–
1%C, 12-14% Mn, austenitic	0.9-1.4	–	0.85	1.1	0.7-2.8	1.25	6.0
Hard Facings							
3.5%C, 20-30%Cr, Fe base	1.25-2.2	11	–	2.5-4	1.4	1.4-4	130
WC/ferrous base	–	5	–	3.5	–	3	–
3.5%C, 33%Cr, 13%W, Co base	–	–	–	2-5	–	1.25-2.2	–
Non-Metallics							
Sintered 95-99% Al_2O_3	–	5	–	–	–	3.5-25	–
WC/6%Co	14	–	–	–	–	140	–
Aluminous cement concrete	0.3	0.07	–	–	–	–	–
Polyurethane	0.05	0.4	–	–	–	–	–

is very difficult to quantify the response of materials in such environments and the relative importance of mechanical and structural properties. We have tended in the past to rely on bulk properties as a guide for materials selection. Certainly, bulk hardness and fracture toughness, etc., have an influence on wear, but they provide only very coarse guides to performance. Similarly, wear performance of ferrous materials in many environments correlates reasonably well with carbon content, but such correlations should be used with caution, since they do not take account fully of the different mechanisms of material removal that may operate during the wear process. Because of the complexity of the abrasive wear process, materials selection is itself a complex and time-consuming task, but proper attention to detail is essential and the potential savings are usually sufficient to justify the investment of resources.

References

1. Research Group on Wear of Engineering Materials, Glossary of Terms and Definitions in the Field of Friction, Wear and Lubrication—Tribology, OECD Publications, Paris, 1969.
2. T. R. Bates, K. C. Ludema and W. A. Brainard, A. Rheological Mechanism of Penetrative Wear, Wear, 1974, vol. 30, pp. 365-375.
3. R. C. D. Richardson, The Wear of Metals by Relatively Soft Abrasives, Wear, 1968, vol. 11, pp. 245-275.
4. Department of Education and Science, Lubrication (Tribology), Education and Research—A Report on the Present Position and Industry's Needs, Her Majesty's Stationery Office, London, 1966.
5. H. P. Jost, in Principles of Tribology, ed. J. Halling, Macmillan & Co., London, 1975, p. xii.
6. Ministry for Research and Technology, Tribologie, Research Report T76-38, Zentralstelle für Luft-und Raum-fahrtdokumentation und-information, Munich, 1976.
7. E. Rabinowicz, in "Materials Technology—1976" (APS New York Meeting), ed. A. G. Chynoweth and W. M. Walsh, American Institute of Physics Conference Proceedings No. 32, New York, 1976, pp. 165-174.
8. M. M. Khrushchov and M. A. Babichev, Research on the Wear of Metals, Moscow, 1960, chap. 18.
9. C. A. Brookes, P. Green, P. H. Harrison and B. Moxley, Some Observations on Scratch and Indentation Hardness Measurements, J. Phys. D, 1972, vol. 5, pp. 1284-1293.
10. M. A. Moore and R. M. Douthwaite, Plastic Deformation below Worn Surfaces, Met. Trans., 1976, vol. 7A, pp. 1833-1839.
11. R. C. D. Richardson, The Maximum Hardness of Strained Surfaces and the Abrasive Wear of Metals and Alloys, Wear, 1967, vol. 10, pp. 353-382.
12. H. S. Avery, Work Hardening in Relation to Abrasion Resistance, Materials for the Mining Industry Symposium, 1974, Climax Molybdenum Co., pp. 43-77.
13. M. A. Moore, R. C. D. Richardson and D. G. Attwood, The Limiting Strength of Worn Metal Surfaces, Met. Trans., 1972, vol. 3, pp. 2485-2491.
14. M. A. Moore, The Strain-Hardening and Fracture of Ferritic Materials under Abrasive Wear, Ph.D. thesis, University of Newcastle-upon-Tyne, 1974.

15. U. Bryggman, S. Hogmark and O. Vingsbo, Studies of Abrasive Wear in a Modified Impact Testing Machine, Institute of Technology, Uppsala University, Report UPTEC 7879R, 1978.
16. M. A. Moore, A Preliminary Investigation of Frictional Heating During Abrasive Wear, Wear, 1971, vol. 17, pp. 51-58.
17. R. McPherson, Transient Temperature Produced by Abrasion of Aragonite, Wear, 1973, vol. 23, pp. 83-86.
18. R. E. Winter and I. M. Hutchings, The Role of Adiabatic Shear in Solid Particle Erosion, Wear, 1975, vol. 34, pp. 141-148.
19. E. D. Doyle and R. L. Aghan, Mechanism of Metal Removal in Polishing and Fine Grinding of Hard Metals, Met. Trans., 1975, vol. 6B, pp. 143-147.
20. L. K. Ives, Microstructural Changes in Copper Due to Abrasive, Dry and Lubricated Wear, Wear of Materials 1979, ed. K. C. Ludema, W. A. Glaeser and S. K. Rhee, ASME, New York, 1979, pp. 246-256.
21. R. R. Agarwala and H. Wilman, Deformation of Iron Crystals by Unidirectional Abrasion, JISI, 1955, vol. 179, pp. 124-131.
22. K. Nakajima, N. Yamamoto and A. Isogari, Interaction of Solute Atoms with Dislocations in the Surface Layers of Abraded Cu-Al Alloys, Trans. JIM, 1969, vol. 10, pp. 223-226.
23. J. Larsen-Badse, The Abrasion Resistance of Some Hardened and Tempered Carbon Steels, Trans. AIME, 1966, vol. 236, pp. 1461-1466.
24. J. Larsen-Badse and K. G. Mathew, Influence of Structure on Abrasion Resistance of a 1040 Steel, Wear, 1969, vol. 14, pp. 199-206.
25. M. A. Moore, The Relationship Between the Abrasive Wear Resistance, Hardness and Microstructure of Ferritic Materials, Wear, 1974, vol. 28, pp. 59-68.
26. K-H. Zum Gahr, The Influence of Thermal Treatments on Abrasive Wear Resistance of Tool Steels, Z. Metallkde., 1977, vol. 68 (12), pp. 783-792.
27. S. Makin and N. Joseph, Minimum Energy in Abrasive Processes, Wear, 1975, vol. 32, pp. 15-23.
28. F. R. N. Nabarro, Theory of Crystal Dislocations, Oxford University Press (Clarendon), London, 1967, pp. 696-698.
29. S. Ramalingam and J. T. Black, An Electron Microscope Study of Chip Formation, Met. Trans., 1973, vol. 4, pp. 1103-1112.
30. D. M. Turley, Dislocation Substructures and Strain Distributions Beneath Machined Surfaces of 70/30 Brass, J. Inst. Metals, 1971, vol. 99, pp. 271-276.
31. A. G. Evans and T. R. Wilshaw, Quasi-Static Particle Damage in Brittle Solids—I. Observations, Analysis and Implications, Acta Met., 1976, vol. 24, pp. 939-956.
32. B. Lawn and R. Wilshaw, Indentation Fracture: Principles and Applications, J. Mat. Sci., 1975, vol. 10, pp. 1049-1081.
33. B. Lawn, T. Jensen and A. Arora, Brittleness as an Indentation Size Effect, J. Mat. Sci., 1976, vol. 11, pp. 573-575.
34. B. Bethune, The Surface Cracking of Glassy Polymers under a Sliding Spherical Indenter, J. Mat. Sci., 1976, vol. 11, pp. 199-205.
35. M. V. Swain, Microscopic Observation of Abrasive Wear of Polycrystalline Alumina, Wear, 1975, vol. 35, pp. 185-189.
36. M. A. Moore and F. S. King, Abrasive Wear of Brittle Solids, Wear of Materials 1979, ed. K. C. Ludema, W. A. Glaeser and S. K. Rhee, ASME, New York, 1979, pp. 275-85, and Wear, 1980, vol. 60, pp. 123-140.
37. R. C. D. Richardson, The Wear of Metals by Hard Abrasives, Wear, 1967, vol. 10, pp. 291-309.
38. M. F. Stroud and H. Wilman, The Proportion of Groove Volume Removed as Wear in Abrasion of Metals, Brit. J. Appl. Phys., 1962, vol. 13, pp. 173-178.

39. T. O. Mulhearn and L. E. Samuels, The Abrasion of Metals: A Model of the Process, Wear, 1962, vol. 5, pp. 478-498.
40. J. Larsen-Badse, Influence of Grit Size on Groove Formation During Sliding Abrasion, Wear, 1968, vol. 11, pp. 213-222.
41. M. Y. Friedman, S. M. Wu and P. T. Suratkar, Determination of Geometric Properties of Coated Abrasive Cutting Edges, J. Eng. Ind., 1974, vol. 96, pp. 1239-1244.
42. E. Rabinowicz, L. A. Dunn and P. G. Russell, A Study of Abrasive Wear under Three-Body Conditions, Wear, 1961, vol. 4, pp. 345-355.
43. J. Dodd, Further Progress in the Development and Utilisation of Abrasion Resistant Alloys in the U.S., Fourth International Abrasion Colloquium, Grenoble, 1979.
44. W. M. McKenzie and W. C. Hillis, Evidence of Chemical Acceleration of Wear in Cutting Plant Material, Wear, 1965, vol. 8, pp. 238-243.
45. R. L. Aghan and L. E. Samuels, Mechanisms of Abrasive Polishing, Wear, 1970, vol. 16, pp. 293-301.
46. J. Larsen-Basse, Influence of Atmospheric Humidity on Abrasive Wear—I. 3-Body Abrasion, Wear, 1974, vol. 31, pp. 373-379.
47. M. Ciftan and E. Saibel, Rebinder Effect and Wear, Wear of Materials 1979, ed. K. C. Ludema, W. A. Glaeser and S. K. Rhee, ASME, New York, 1979, pp. 659-664.
48. J. Larsen-Basse, Influence of Atmospheric Humidity on Abrasive Wear—II. 2-Body Abrasion, Wear, 1975, vol. 32, pp. 9-14.
49. J. E. Hines, Abrasive Wear of High Density Alumina, M. S. thesis, Pennsylvania State University, 1974.
50. A. Latin, The Properties of Tungsten Carbide-Cobalt Alloys Used for Mineral Cutting Tools, Metallurgia, 1961, vol. 64, pp. 211-216, 267-273.
51. W. Dawihl and B. Frisch, Wear Properties of Tungsten Carbide and Aluminium Oxide Sintered Materials, Wear, 1968, vol. 12, pp. 17-25.
52. H. S. Avery, The Nature of Abrasive Wear, 1975 SAE Off-Highway Vehicles Meeting, Milwaukee, 1975, SAE Paper No. 750822.
53. I. V. Kragelskii, Friction and Wear, Butterworth, London, 1965, pp. 20-26, 80-110.
54. D. Graham and R. M. Baul, An Investigation into the Mode of Metal Removal in the Grinding Process, Wear, 1972, vol. 19, pp. 301-304.
55. A. J. Sedriks and T. O. Mulhearn, Mechanics of Cutting and Rubbing in Simulated Abrasive Processes, Wear, 1963, vol. 6, pp. 457-466.
56. A. J. Sedriks and T. O. Mulhearn, The Effect of Work Hardening on the Mechanics of Cutting in Simulated Abrasive Processes, Wear, 1964, vol. 7, pp. 451-459.
57. S. K. Dean and E. D. Doyle, Significance of Grit Morphology in Fine Abrasion, Wear, 1975, vol. 35, pp. 123-129.
58. M. J. Murray, P. J. Mutton and J. D. Watson, Abrasive Wear Mechanisms in Steels, Wear of Materials 1979, ed. K. C. Ludema, W. A. Glaeser and S. K. Rhee, ASME, New York, 1979.
59. G. R. Davies, An Abrasion Test for Use in Agricultural Engineering, Proc. Inst. Mech. Engrs., 1965/66, vol. 180(2A) pp. 237-247.
60. J. K. Lancaster, Abrasive Wear of Polymers, Wear, 1969, vol. 14, pp. 223-239.
61. J. Larsen-Badse, Influence of Grit Diameter and Specimen Size on Wear During Sliding Abrasion, Wear, 1968, vol. 12, pp. 35-53.
62. S. W. Date and S. Malkin, Effects of Grit Size on Abrasion with Coated Abrasives, Wear, 1976, vol. 40, pp. 223-235.
63. B. W. E. Avient, J. Goddard and H. Wilman, An Experimental Study of Friction and Wear During Abrasion, Proc. Roy. Soc., 1960, vol. A258, pp. 159-180.

64. E. Rabinowicz and A. Mutis, Effect of Abrasive Particle Size on Wear, Wear, 1965, vol. 8, pp. 381-390.
65. G. K. Nathan and W. J. D. Jones, Influence of the Hardness of Abrasives on the Abrasive Wear of Metals, Proc. Inst. Mech. Engrs., 1966/67, vol. 181 (30), pp. 215-221.
66. S. Hata and T. Muro, Mechanism of Friction and Wear of Steel Plate Against Solidified Sandy Soil, Mem. Fac. Eng. Kyoto Univ., 1969, vol. 31(4), pp. 456-489.
67. R. Komanduri and M. C. Shaw, J. Eng. Mater. Technol., 1974, vol. 96, pp. 145-156.
68. R. Komanduri and M. C. Shaw, Attritious Wear of Silicon Carbide, J. Eng. Ind., 1976, vol. 98, pp. 1125-1134.
69. R. W. Johnson, The Use of the Scanning Electron Microscope to Study the Deterioration of Abrasive Papers, Wear, 1968, vol. 12, pp. 213-216.
70. G. W. Patterson and T. O. Mulhearn, The Fracture of Idealised Abrasive Particles, Wear, 1969, vol. 13, pp. 175-182.
71. S. Hata and T. Muro, private communication.
72. R. C. D. Richardson, The Wear of Metallic Materials by Soil—Practical Phenomena, J. Agric. Eng. Res., 1967, vol. 12(1), pp. 22-39.
73. R. F. Bishop, R. Hill and N. F. Mott, The Theory of Indentation and Hardness Tests, Proc. Phys. Soc., 1945, vol. 57(3), pp. 147-159.
74. R. Hill, The Mathematical Theory of Plasticity, Clarendon Press, Oxford, 1950.
75. K. L. Johnson, The Correlation of Indentation Experiments, J. Mech. Phys. Solids, 1970, vol. 18, pp. 115-126.
76. C. J. Studman, M. A. Moore and S. E. Jones, On the Correlation of Indentation Experiments, J. Phys. D: Appl. Phys., 1977, vol. 10, pp. 949-956.
77. D. S. Dugdale, Cone Indentation Experiments, J. Mech. Phys. Solids, 1954, vol. 2, pp. 265-277.
78. S. Ramalingam and L. L. Lehn, A Photoelastic Study of Stress Distribution During Orthogonal Cutting, Part I, J. Eng. Ind., 1971, vol. 93, pp. 527-537.
79. T. C. Buttery and J. F. Archard, Grinding and Abrasive Wear, Proc. Inst. Mech. Engrs., 1970/71, vol. 185, pp. 537-551.
80. J. C. Buttery and M. S. Hamed, Some Factors Affecting the Efficiency of Individual Grits in Simulated Grinding Experiments, Wear, 1977, vol. 44, pp. 231-245.
81. I. A. Cutler and R. McPherson, Plastic Deformation of Al_2O_3 During Abrasion, J. Amer. Ceram. Soc., 1973, vol. 56, pp. 266-269.
82. T. F. Page, G. R. Sawyer, O. O. Adewoye and J. J. Wert, Hardness and Wear Behaviour of SiC and Si_3N_4 Ceramics, Proc. Brit. Ceram. Soc., 1978, vol. 26, pp. 193-208.
83. K-H. Zum Gahr, Friction and Wear of a Precipitation Hardenable Austenitic Stainless Steel under Abrasive Conditions, Z. Metallkde., 1977, vol. 68, pp. 381-389.
84. N. Nocker and K-H. Zum Gahr, Abrasive Wear of Cast Iron with Lamellar Graphite and with Spheroidal Graphite, Arch. Eisenhüttenwes., 1978, vol. 49, pp. 155-160.
85. K-H. Zum Gahr, Influence of Internal Notches on Abrasive Wear, Z. Metallkde., 1978, vol. 69, pp. 313-319.
86. K-H. Zum Gahr, Relation Between Abrasive Wear Rate and the Fracture Toughness of Metallic Materials, Z. Metallkde., 1978, vol. 69, pp. 643-650.
87. B. R. Lawn and D. B. Marshall, Hardness, Toughness and Brittleness: An Indentation Analysis, J. Amer. Ceram. Soc., 1979, vol. 62, pp. 347-350.
88. E. Hornbogen, Metallurgical Aspects of Wear, Fortschr.—Bed. VDI-Z, Reihe 5,

No. 24, 1976.
89. K-H. Zum Gahr, Relationship Between Abrasive Wear Rate and the Microstructure of Metals, Wear of Materials 1979, ed. K. C. Ludema, W. A. Glaeser and S. K. Rhee, ASME, New York, 1970, pp. 266-274.
90. J. Larsen-Badse, Abrasion Resistance of Some SAP-type Alloys at Room Temperature, Wear, 1968, vol. 12, pp. 357-368.
91. D. S. Lin and H. Wilman, The Relation of the Friction and Wear in Abrasion of Al-4 wt% Cu Alloy to the Estimated Precipitate Particle Size and Separation During Age Hardening, Wear, 1969, vol. 14, pp. 337-346.
92. M. M. Khrushchov and M. A. Babichev, The Effect of Heat Treatment and Work Hardening on the Resistance to Abrasive Wear of Some Alloy Steels, Friction and Wear in Machinery, 1965, vol. 19, pp. 1-15.
93. N. F. Fiore, Microstructural Effects in Abrasive Wear, Final Report, Phase I, University of Notre Dame, Dept. of Metallurgical Engrg. and Materials Science, Report COO-4246-9, 1979.
94. K-H. Zum Gahr and D. V. Doane, Optimizing Fracture Toughness and Abrasion Resistance in White Cast Irons, Met. Trans., 1980, Vol. 11A, pp. 613-620.
95. V. S. Popov and P. L. Nagorny, Influence of Carbides on the Abrasive Wear Resistance of Alloys, Russ. Cast. Prod., 1969, vol. 8, p. 377.
96. N. F. Fiore and T. H. Kosel, Microstructural Effects in Abrasive Wear, University of Notre Dame, Dept. of Metallurgical Engrg. and Materials Science, Reports COO-4246-10 and -11, 1979.
97. J. Larsen-Basse, Wear of Hard-Metals in Rock Drilling: A Survey of the Literature, Powder Met., 1973, vol. 16(31), pp. 1-32.
98. J. Larsen-Basse, Some Mechanisms of Abrasive Wear of Cemented Carbide Composites, Fourth International Abrasion Colloquium, Grenoble, 1979.
99. E. Carboni, A Contribution to Research on the Method of Evaluating Abrasion Resistance of Sintered Carbides, Metal. Ital., 1969, vol. 61, pp. 593-601.
100. H. Feld and P. Walker, Investigation on Abrasive Wear of Cemented Carbides, Z. Werkstoffkde, 1976, vol. 7, pp. 300-303.
101. T. E. Norman and E. R. Hall, Abrasive Wear of Ferrous Materials in Climax Operations, in Evaluation of Wear Testing, ASTM STP 446, 1969, pp. 91-114.
102. F. Borik and W. G. Scholz, Gouging Abrasion Test for Materials Used in Ore and Rock Crushing: Part II, J. Materials, 1971, vol. 6, pp. 590-605.
103. W. Fairhurst and K. Rohrig, Abrasion Resistance High Chromium White Irons, Foundry Trade Journal, 1974.
104. H. Hocke, Wear Resistant Materials for Coke and Sinter Handling Plant, Iron and Steel Int., 1977, pp. 361-371.
105. H. S. Avery and H. J. Chapin, Selection of Grinding Mill Liner Alloys for Optimum Wear Resistance, AIME Annual Meeting, Paper No. 69-B-6, 1969.
106. D. E. Diesburg and F. Borik, Optimizing Abrasion Resistance and Toughness in Steels and Irons for the Mining Industry, Materials for the Mining Industry Symposium, 1974, Climax Molybdenum Co., pp. 15-41.

Dislocation Concepts in Friction and Wear

D. KUHLMANN-WILSDORF
Department of Materials Science
University of Virginia
Charlottesville, Virginia 22901

Abstract

Using dislocation concepts, primarily as relevant to geometrical considerations, and other basic facts concerning plastic properties of crystalline materials, a number of qualitative and quantitative relationships in the area of friction and wear are developed:

(1) The accepted model for friction and wear is expanded to include the micro-roughness which must necessarily arise at contact spots on account of inhomogeneities in elastic and plastic deformation.

(2) A simple formula is proposed to describe the conditions under which adhesion is important.

(3) The concept of the "basic" case is introduced, which is sliding wear between average-rough and average-clean surfaces such that adhesion as well as surface films can be neglected.

(4) A closer examination of the deformation processes at indentations and underneath wear tracks shows that the reactive force balancing the normal force between the two sliding surfaces originates at considerably larger distances from the interface than the reaction to the tangential traction.

(5) Based on this insight, an expression for the coefficient of friction in the "basic" case is derived. This is found to predict decreasing values of the coefficient of friction with decreasing crystal symmetry and, hence, decreasing numbers of "easy" crystallographic glide systems.

(6) This expression for the coefficient of friction predicts a number of effects which appear to be consistent with observations, including dependence on surface textures, surface hardness, and surface temperature.

(7) The conditions under which surface films may be neglected are more closely examined, together with some effects of surface platings.

(8) Delamination wear is found to follow necessarily whenever the flow-stress of the material near the surface is a function of the cumulative strain.

(9) Expressions for the wear coefficient and for the incubation period preceding "steady state" delamination wear are derived and compared with available data.

(10) The subsurface shear strain distribution underneath surfaces exposed to "steady state" sliding wear is derived for the case that the applied shear stress, due to the tangential tractions, is balanced by the local shear strength as raised by work hardening. The result depends on the prevailing work hardening law. It is shown that one well-investigated case is explained in this manner, using a work hardening law that appears to be common in wire drawing.

(11) Dislocation cells underneath worn surfaces are shown to be very similar to those found in drawn wire, and evidence is presented indicating that these cells are in agreement with theoretical predictions based on dislocation theory.

(12) While either a hard or soft very thin surface layer may be present, it is unlikely to be due to any esoteric dislocation reactions. However, it is important to realize that dislocation cell walls are mobile by nature. Therefore dislocation cells do not deform homologously with the material. Therefore, also, the dislocation cells may generate the appearance of recrystallization when in fact none has taken place.

Introduction

The area of friction and wear is so vast that any single treatment or model cannot explain all, or even the majority of, measurements and observations. Further, for most purposes it is highly unlikely that detailed dislocation mechanisms can throw much light on friction and wear, for the reason that the sizes or volumes involved, e.g., the size of the wear debris, the depth of the deformed surface layer, and the diameter of the typical contact spot, are very large compared to the individual dislocation. The value of dislocation theory in the present context will therefore mostly be in the light it sheds on materials properties and their effects on friction and wear. The following discussion, correspondingly, does not attempt to analyze specific dislocation motions or reactions but uses knowledge of materials properties, as known in connection with dislocation behavior, to extend understanding of friction and wear in what is considered to be the "basic" case. This is the case of dry sliding after repeated passes between two parallel surfaces, neither extremely rough nor smooth, whose nominal area of contact is much larger than the load-bearing area of contact, and in which the two contacting materials are homogeneous, have no unduly thick or strong surface layers, and undergo no crystallographic or chemical transformations.

What Makes a Contact Spot Strong and How Are Asperities Maintained?

Very commonly, friction and wear are considered in terms of two apparently contradictory models, indicated in Fig. 1. Figure 1(A) exemplifies the model of the adhesion theory. According to Bowden and Tabor[1] this theory is ultimately founded on the work of T. Desaguliers in 1784. Its modern development was pioneered mainly by R. Holm[2] in connection with electrical contacts and by

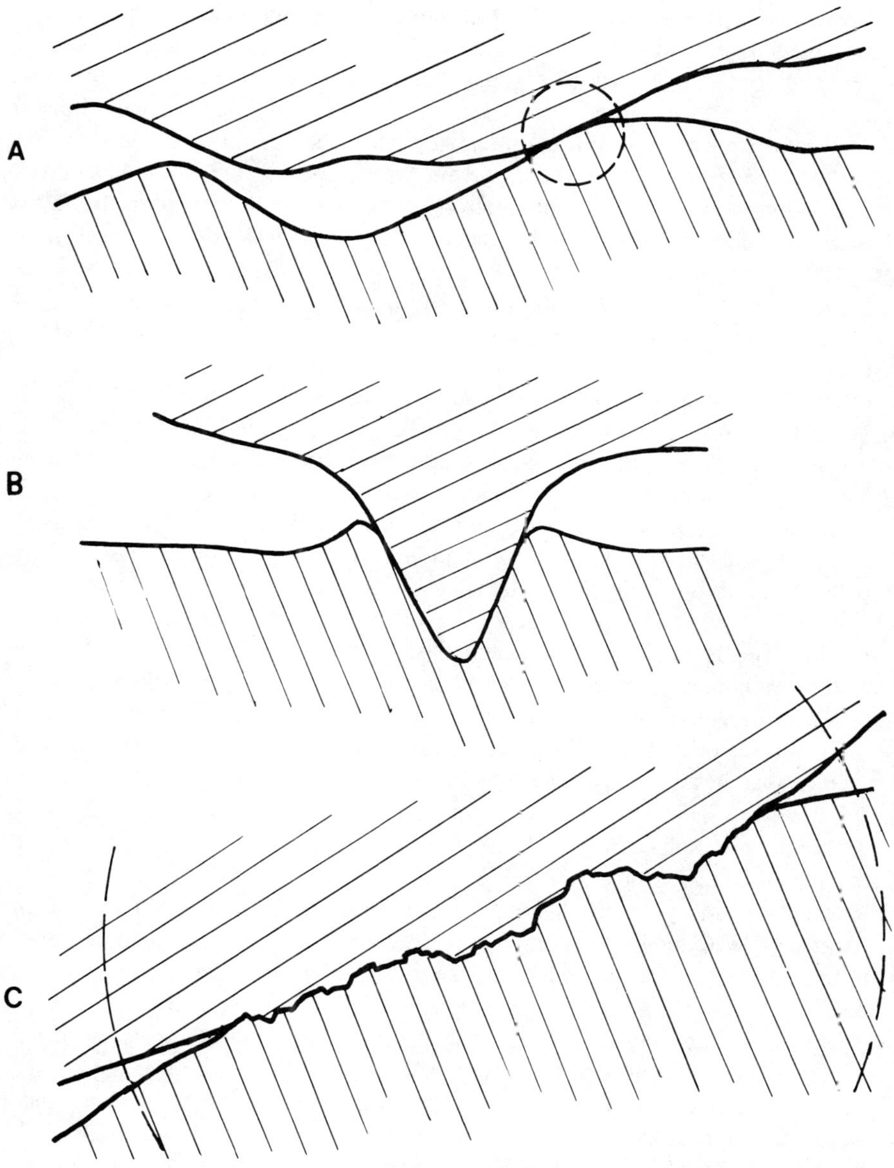

Models are based on adhesion (A), on "plowing" of sharp asperities through the softer material (B), and on roughening of contact spots by micro-roughness generated through inhomogeneities of elastic and plastic deformation (C). A complete contact spot may consist of two or more interacting "hills." If so, (C) depicts one such hill rather than a complete contact spot. Also, the micro-roughness is not believed to provide complete atomistic contact throughout, but more likely only over ⅓ to ½ of the area. Gap widths are almost certainly no more than a few hundred angstroms, and typically much less.

Fig. 1. Commonly accepted (**A** and **B**) and improved (**C**) models for friction and wear.

Bowden and Tabor as summarized in their books.[3,4] In the adhesion theory of friction it is recognized that two sliding surfaces touch only at a number of separate contact spots which together amount to a small fraction of the geometrical contact area, as indicated in Fig. 1(A). These contact spots are considered to be very strong, causing friction damage rather than permitting simple sliding of the two surfaces past each other, on the hypothesis that atomistic contact gives rise to adhesion, or local welding, stronger than the shear strength of the softer material underneath. This view has recently been vigorously criticized, especially by Bikerman[5] and, in regard to the wear mechanism, by N. P. Suh and coworkers.[6] Rabinowicz,[7] although generally in favor of the adhesion theory, had already pointed out its shortcomings. Most obviously, simply pressing or rubbing metals together does not give rise to sticking, as it would if the adhesion at contact spots was as general and strong as assumed in the adhesion theory of friction. If in fact the postulated strong adhesion of metals at their contact spots were to take place, metals could not be handled without some protective surface treatment, lest they stick together on mere contact, much like cellotape.

On the plus side, it is true that cold welding (or frictional welding) is often observed. Indeed, it was its demonstration by Desaguliers for the case of lead balls which originally launched the adhesion theory as outlined by Bowden and Tabor.[1] Further, the adhesion theory is effective in explaining the basic laws of friction. This is due to two fundamental assumptions of the theory: first, that the total area of actual atomistic contact is proportional to the normal force between the two surfaces because, at the contact spots, plastic deformation of the softer of the two materials typically takes place to the point of saturation hardening; and second, that the adhesion or local welding between the two surfaces at the contact spots is always so strong that surface sliding is inhibited, and, instead, shearing or fracture takes place in the subsurface layers, at least of the softer material. In this manner, the coefficient of friction is found independent of load as well as of apparent (overall geometric) area of contact and, within fairly wide limits, of roughness. With the further fact that the flow stress of crystalline materials tends to be little dependent on strain rate, the effective velocity independence of the coefficient of friction, again within fairly wide limits, is explained.

Figure 1(B), by contrast, is a model for the "plowing" of asperities of the harder of the two materials through the softer. Indeed, it seems clear that some "plowing" or, perhaps a better term, "grooving," takes place in friction, since wear tracks almost uniformly show fine grooves. Examples are shown in Fig. 2, due to L. K. Ives.[8] From these micrographs it may be seen that the asperities or particles which are generating the grooves are evidently rather long-lived, since the line pattern does not change significantly over the length of tracks shown. Specifically, the right part of Fig. 2 is included because it shows more clearly than the left part that the asperities or particles causing the grooves are on the micron size scale, at least in this example, and that the grooving is for the most part quite shallow.

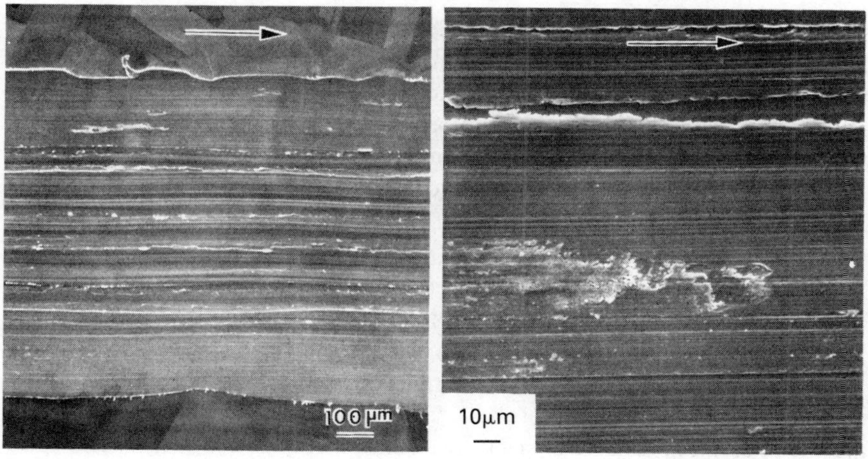

The grooving (which can be seen to consist of rather long scratches on the left part of the figure, and to be very shallow as seen on the right) is believed to be caused by micro-roughness of the kind indicated in Fig. 1(C). Although at actual contact spots, as in Fig. 1(C), there is interlocking and relative rest between the two sides, grooving must occur at the sides of the contact spots and interacting "hills," and where surfaces are sufficiently separated to prevent locking but still within range of the highest asperities. The latter condition will routinely occur when contact spots of the kind in Fig. 1(C) are in the process of being released. "Grooving" is considered to contribute on the order of between 10% and 20% to the coefficient of friction, and only very little to wear.

Fig. 2. Wear track on copper ring, tested under lubricated sliding condition, seen with two different magnifications. *(Courtesy of L. K. Ives, Ref 8)*

The question which arises in this connection is how the asperities or particles responsible for this grooving are formed. Are they the same asperities which make contact spots of the type illustrated in Fig. 1(A)? To answer that question, one must try to evaluate the size of the average contact spot. This approach is not entirely straightforward because of uncertainty as to the number of contact spots which remains in spite of some superb research on the geometry of contacting surfaces (e.g., Ref 9-11). What, then, is the typical size of a contact spot? According to adhesion theory, and in accordance with much experimental evidence, the average load-bearing area, A_b per contact spot, with N contact spots between two surfaces pressed together with the force P, is

$$A_b \simeq P/HN \simeq \pi d^2/4 \tag{1}$$

if H designates the hardness of the softer of the two members. Much more will have to be said below about the "hardness" or "flow pressure," H, introduced here. At this point it is sufficient to assume that H is a fairly well defined materials constant and that Eq 1 is obeyed independent of the magnitudes of N and P.

For the majority of cases of interest, in practice as well as in the laboratory, P is on the order of 3 kg weight within, say, two orders of magnitude. Next, N is almost never a large number. For electrically conducting contact spots, Holm[12] found values of N up to 18 in a particular run of experiments, while recently N has been given by McNab[13] as between 10 and 50; and direct observations by means of infra-red microscopy by H. Kongsjorden, J. Kulsetås and J. Sletbak[14] revealed several separate temperature peaks over an apparent contact area of a few square millimeters. In fact, quite commonly N is still smaller than the quoted values—namely, three, or two, or even one. The reason for this is thermal instability, a topic that is the subject of increasingly intense investigation, largely pioneered by Burton and co-workers.[15] Namely, frictional heat raises the temperature of contact spots and their surroundings, causing them to bulge out, thereby increasing the frictional force, with a resultant further rise in temperature. This process acts more powerfully on larger than on smaller contact spots, eliminating the latter in favor of the former. Subsequent deformation and wear either cause a slow migration of these contact spots or reduce the local stress at the one or several operative contact spots below a critical level when other spots take over. The result, by theory as well as by observation,[16,17] is that only one or a very few contact spots exist per sliding interface, the so-called hot spots, at least beyond some critical loads and speeds. Figure 3 gives an example due to J. P. Netzel.[17]

To obtain some very rough impression of the size of a typical contact spot let, then, N be ten, and assume $H = 50$ kg*/mm^2 at a load of three kilogram weight, i.e., $P = 3$ kg*. With these fairly typical numbers, Eq 1 yields $A_b = 6 \times 10^{-3}$ mm^2 or $d \simeq 90$ μm. Similarly, in the experiments by Holm referred to above, d ranged between 70 and 126 microns. Such numbers are evidently much larger than the width of the average plowing asperity inferred from Fig. 2. In fact, considering Eq 1 it will be seen that, with $N \leq 50$ and with H typically ≤ 100 kg*/mm^2, contact spots of the size of the grooving in Fig. 2 could only arise with loads of several grams or less, whereas for the case of Fig. 2 the computed contact spot diameter is about 100 microns.

Another problem, besides size considerations, with the model of a "plowing" asperity is that in plowing, such as in a farm field, the plowed material is cut and heaped up on either side of the furrow, not transported towards the end of the furrow. This kind of deformation is indeed often observed alongside grooves and furrows at the appropriate small scale. However, severe dragging in the direction of relative motion, to a depth much larger than the grooving, is very commonly observed in the subsurface regions, at least in the softer member. Figures 4 and 5 give examples, due to Dautzenberg[18] and Hirst and Lancaster,[19] respectively. Thus we see that experimental evidence regarding subsurface strain distribution as given in Fig. 4 and 5 supports the adhesion model of friction according to Fig. 1(A) and not the plowing model of Fig. 1(B), whereas the topography of the wear track as shown in Fig. 2 supports the plowing model, albeit with asperities on a much smaller scale than the size of the contact spots.

At this point a more detailed consideration of the deformation processes, in terms of dislocations, becomes useful. As seen, the strain distribution under wear

Fig. 3. "Hot spot," representing one persistent contact spot which is possibly composed of several interacting "hills," in the mating ring of a face seal after 45 minutes of operation. *(Courtesy of J. P. Netzel, Ref 17)*

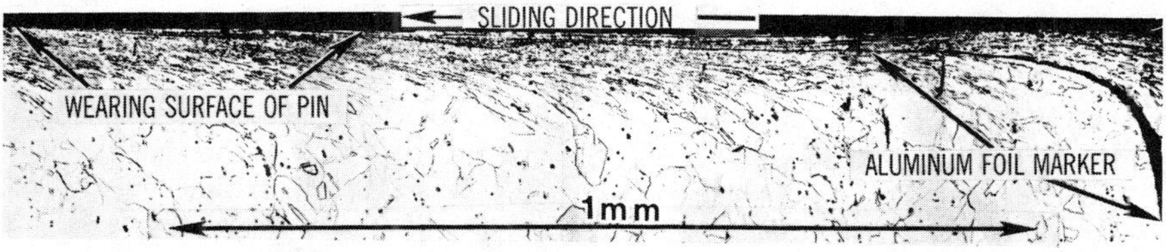

Fig. 4. The displacement field of a worn copper pin, made visible by placing an aluminum foil between the two halves of the pin which was cut perpendicular to the sliding direction and pressed together before the wear test against SAE 1045 steel. 4 kg* load. (*Courtesy of J. H. Dautzenberg, Ref 18)*

tracks, e.g., as illustrated in Fig. 4 and 5, extends to a distance comparable to the estimated diameter of the average contact spot. This is in agreement not only with stress analysis such as performed by Davies[20] and Jahanmir and Suh,[21] but also with observations by L. D. Dyer[22] of strains and dislocation distributions about indentations (modelling the asperities at contact spots) in highly perfect single crystals. Figure 6 due to Dyer shows the shape of an indentation made by a sapphire ball on the cube face of a copper single crystal, and Fig. 7 and 8 by Dyer show the resulting dislocation distribution. Admittedly, in this particularly soft material with its very low yield stress, the indentation size as well as the zone through which dislocations spread were larger for the same load than in an ordinary metal. Yet, the geometry is believed to be well represented, because the

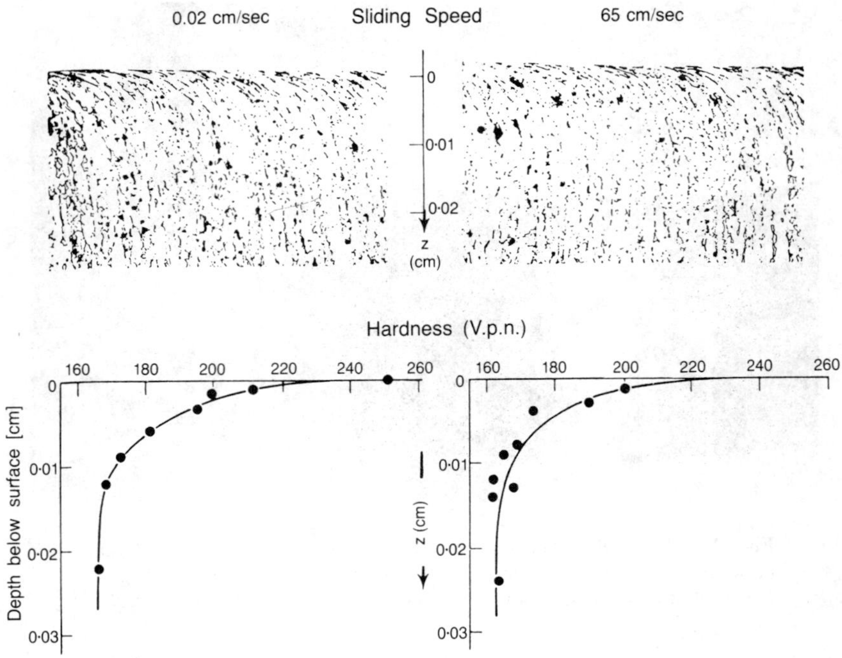

Fig. 5. Microstructure under wear track of 60/40 brass pins **(top)** and correlated hardness measurements **(bottom)** for two different sliding speeds under otherwise the same conditions. Note that the wear layer is roughly 120 μm thick, which is considered to be comparable to the diameter of the average contact spot in this case, and at the same time to be a fairly representative value for contact spot diameters. Note that the thickness of the wear fragments was only ~3 μm. *(After W. Hirst and J. K. Lancaster, Ref 19)*

ratio of these two sizes is presumably largely independent of yield stress and, hence, choice of material, provided that the number of available slip systems is adequately large.

It is gratifying that in Fig. 8 the most intense dislocation concentration (indicative of the highest operative stress and hardening) is found well below the surface, in line with continuum theory.[20,21] Indeed, there is acceptable agreement between the observed dislocation distribution in Fig. 8 and stress maps obtained from theoretical stress analysis, as will be seen by comparison of Fig. 8 with Fig. 9. Clearly, stress analysis and observations on dislocations both yield the result that the highest shear stresses and shear strains under an indenter or asperity occur somewhat below the surface and extend for a distance comparable with the size of the indentation. In turn, the range of the strong surface shear strain (Fig. 4 and 5) is of the same magnitude as the size of the contact spots, namely,

$$d \simeq \sqrt{A_b} \simeq \sqrt{P/HN} \simeq \sqrt{(P/H)}/3, \qquad [2]$$

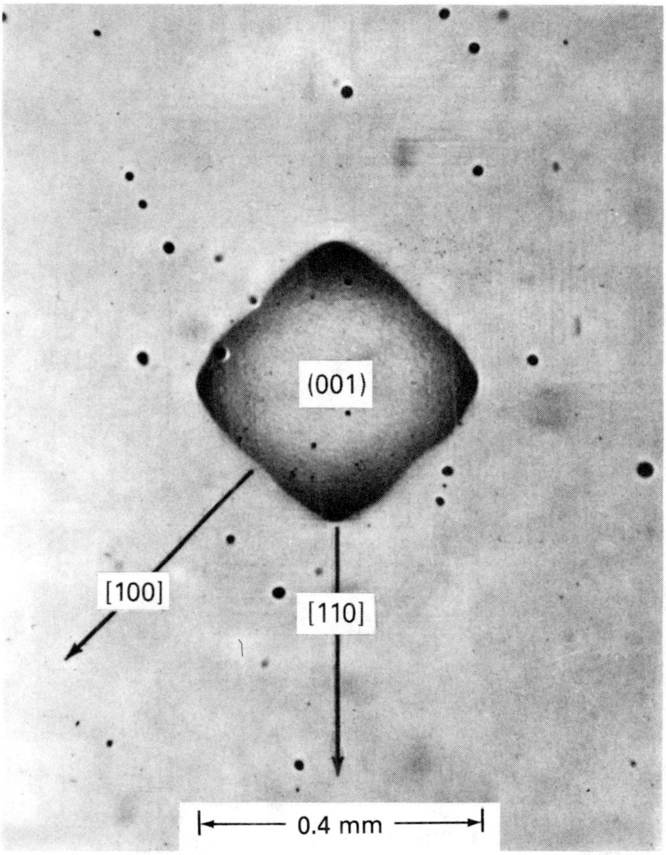

Fig. 6. Indentation made by a ¼ in. (6.35 mm) polished sapphire ball on the highly polished cube face of an especially perfect copper crystal, under 1 kg* load. *(Courtesy of L. D. Dyer, Ref 22)*

found to be in the neighborhood of 0.1 mm in typical cases as explained. Therefore, there can be little doubt left that in friction and wear the surface layers are dragged along by strongly adhering contact spots, and that the asperities responsible for the grooving are of a quite different kind and of a much smaller size.

In order to obtain a possible answer as to the nature of the smaller asperities, consider yet another aspect of the deformation processes at contact spots of metals. To begin with, assume that, although mildly curved so as to produce localized plastically deformed contact spots, the two surfaces are ideally smooth. An examination of the two surfaces directly after pressing them together without sliding might show that the harder of the two materials has retained a smooth surface. The same could not possibly be true for the softer one, however, because

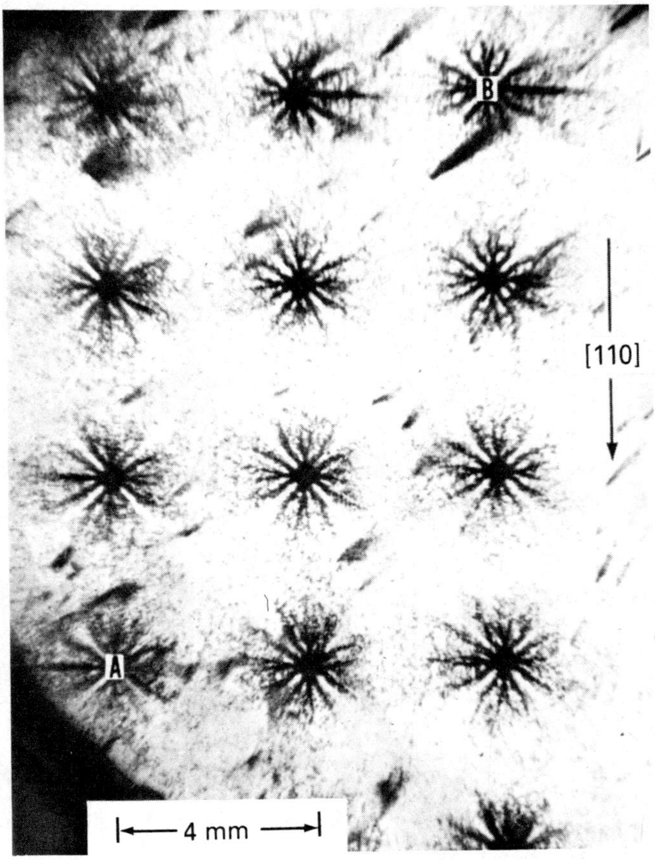

Fig. 7. Regularly arranged indentations of the kind shown in Fig. 6, made on the same crystal and etched to reveal dislocation intersections with the surface. *(Courtesy of L. D. Dyer, Ref 22)*

there is absolutely no deformation process in crystalline materials which does not cause inhomogeneous distortions. If both objects had been single crystals, crystallographic slip would have caused slip lines and the corresponding surface roughening. Had there been twinning, this would have caused even greater roughening. Had there been grain boundaries in the softer material, this would further have added roughening because of differences in the hardness of the different grains depending on orientation. Figure 10, also due to Dyer,[23] shows this anisotropy by inference, plus an effect perhaps yet more unexpected—namely, that even for the same surface, (rolling) friction and hardness depend on the direction of relative motion. Correspondingly, different directions of relative motion give rise to different slip line patterns and different surface roughness, and incidentally also different coefficients of friction.

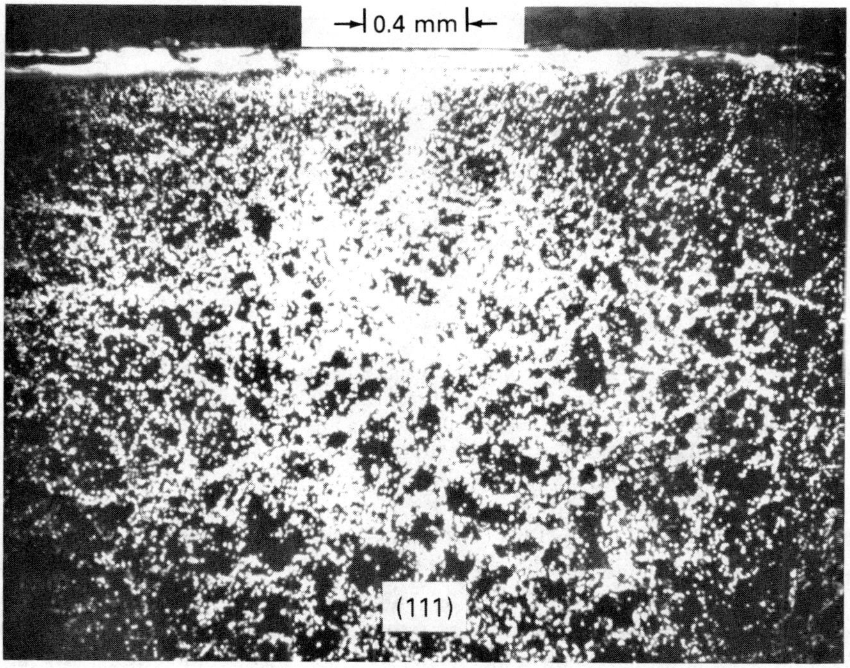

Fig. 8. Dislocation distribution beneath an indentation such as in Fig. 6 and 7, revealed by etching a polished section parallel to a {111} plane. It passes not through the center of the indentation but through the center of the dislocation concentration underneath, thereby somewhat underrepresenting the dislocation density directly at the surface at the indentation. Note also that, because of the obliqueness of the cut, vertical distances are expanded by about 20%. *(Courtesy of L. D. Dyer, Ref 22)*

To all of these effects one must add elastic inhomogeneity: specifically, elastic anistropy in the softer material means that, depending on orientation, the elastic part of the deformation differs from crystallite to crystallite, along with peak hardness. As a result, the crystallites of the harder material are subject to different elastic strains, and hence reactive pressures, from one to the next crystallite of the softer partner, besides responding differently even to the same pressure on account of their own elastic anisotropy. Thus the elastic compression in both materials differs from grain to grain on their own side, as well as from grain to grain on the opposite side, and these differences translate into the corresponding variations of the plastic permanent strain in the softer material. In addition, compatibility stresses at the grain boundaries must necessarily cause further rather abrupt changes in the plastic deformation of the softer material from one side to the other of any grain boundary, in both of the opposing surfaces.

Consider further that real materials are not ideally homogeneous, and that they often contain second and third phases of greatly different hardness and elasticity.

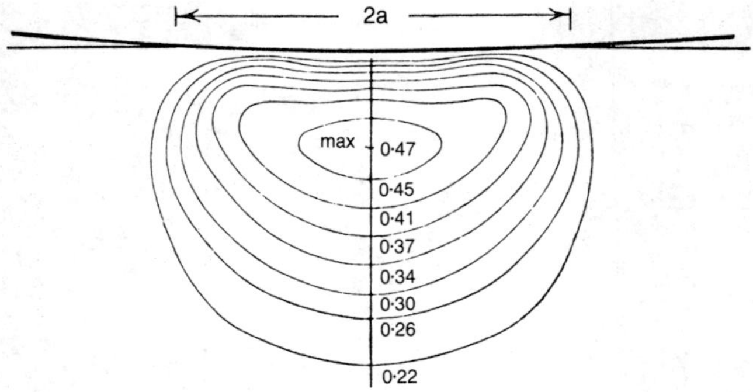

Fig. 9. Elastic distortion of a flat surface by an indenting sphere, and contour lines showing the values of the maximum shear stresses beneath the (elastic) indentation in units of the mean pressure exerted by the ball over the contacting area, according to D. Tabor (Fig. 24 of Ref 31), based on R. M. Davies, Ref 20.

Also to be considered is the effect of entrapped extraneous particles such as dust. Last, but not least, if either or both of the materials are covered by a brittle surface film, this may crack in a type of crazing pattern, with the correspondingly varied abrupt changes of the normal pressure transmitted across the interface and the plastic deformation in response thereto. Altogether, then, compressing two surfaces together must automatically and necessarily lead to micro-roughness. At the least, the depth of this micro-roughness must be on the scale of slip lines, and of the differences in plastic deformation normal to the interface that are complementary to the differences in the elastic deformation of grains. Note that such micro-roughness can never be rubbed off or otherwise vanish during wear, since it reestablishes itself in ever different configurations as wear occurs, and as through sliding the local conditions change continuously.

The depth of slip lines is very variable. It depends on strain, surface orientation, material, surface films, and loading conditions, including strain rate and temperature. Typically, the scale of roughness from this cause will be in the ten to a few hundred nm range. To estimate the roughness due to elastic anisotropy, let Young's modulus, E, vary by the amount of ΔE as a function of orientation. The elastic compression of a layer of thickness d under pressure H is $h = Hd/E$. Therefore, Δh, the height difference resulting from the elastic deformation of a layer thickness d under compressive stress H at elastic anisotropy $\Delta E/E$, is given by

$$\Delta h \simeq d(H/E)\,(\Delta E/E), \qquad [3]$$

provided the grains are of a size larger than, or comparable to, d. Taking $H/E \simeq 5 \times 10^{-3}$, $\Delta E/E \simeq 0.2$, and $d = 0.1$ mm as fairly typical numbers, one thus obtains $\Delta h = 0.1$ μm from this source. Note that this is just about the scale of the roughness which appears to have caused the grooving in Fig. 2.

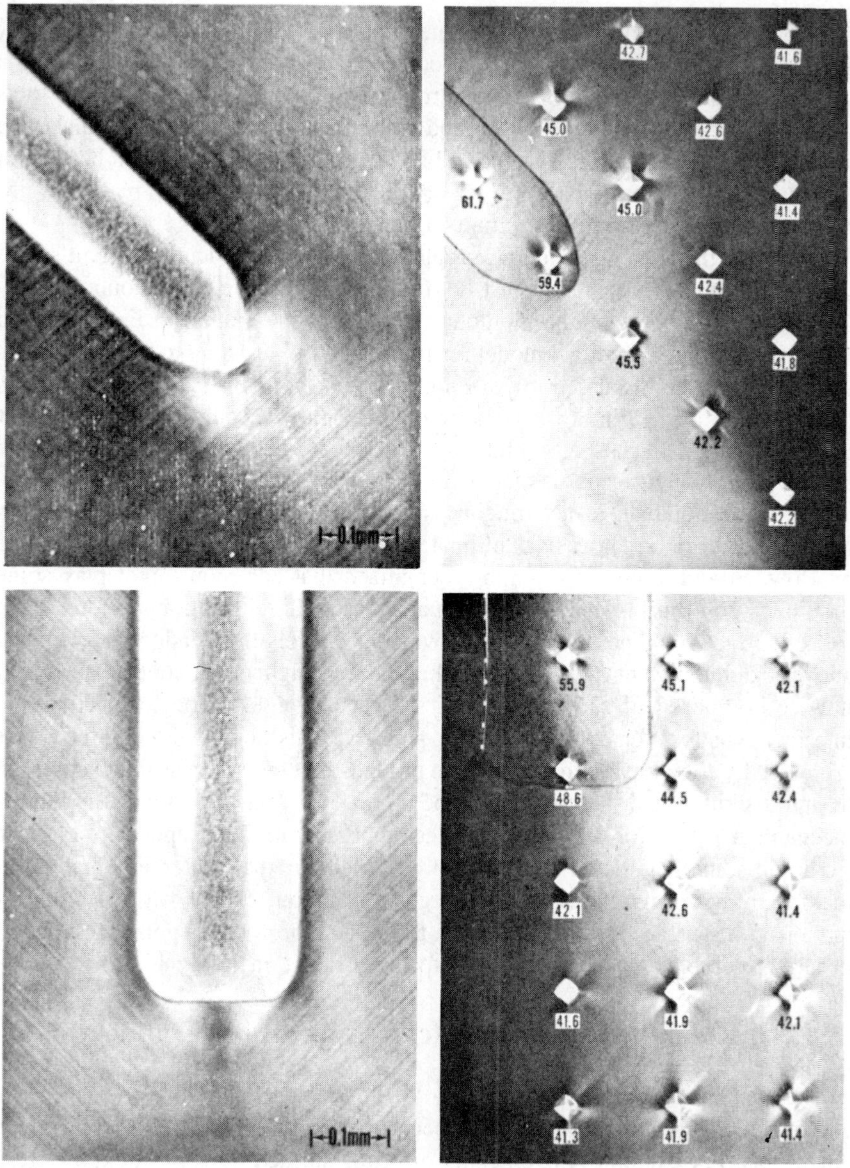

Fig. 10. Micrographs showing surface features **(left)** and associated micro-hardness values **(right)** about the ends of rolling tracks made by a ¼ in. (6.35 mm) polished sapphire ball under a load of 200 g* on the highly polished cube surface of an exceptionally perfect copper single crystal. At the top the rolling direction is [110] and the measured coefficient of rolling friction in this direction was 0.0059. For the [100] rolling direction at the bottom the coefficient of rolling friction was 0.0045. *(Courtesy of L. D. Dyer, Ref 23)*

That micro-roughness is indeed generated at plastically deformed contacts, even if between two initially quite smooth surfaces free of grain boundaries, is directly documented by Fig. 11. This micrograph compares to Fig. 6 but for a much smaller load (namely, 20 g*) and using interference contrast to make the slightest surface roughness visible. In Fig. 11, traces of slip lines can be seen in the indentation, plus small hillocks which might be due to materials inhomogeneities, or perhaps to entrapped dust.

In light of the above considerations and experimental evidence, then, it seems certain that micro-roughness must be a ubiquitous feature at all contact spots. This conclusion permits a harmonious synthesis of the models in Fig. 1(A) and 1(B): namely, the theoretical model for friction and wear ought to be modified by including in it the discussed micro-roughness, as shown in Fig. 1(C). With this modification, the origin of the strength of the contact spots, and the reason why the subsurface material below the contact spots shears, in preference to simple sliding along the surfaces, is clarified.

The suggestion that surface roughness may lead to interlocking at contact spots is not new but, rather, goes back ultimately to Coulomb (1736-1806). As outlined by Bowden and Tabor,[1] Coulomb "recognized that adhesion might play some part in friction [but] rejected it as the main cause. He felt friction was due to the role of surface roughness . . ."—all of which is entirely the standpoint developed above. Coulomb went on to consider the surface roughness at contact spots as if it were composed of ramps inclined, say, at an angle φ to the macroscopic interface whence, for uphill sliding $\mu = \tan \varphi$. This notion of Coulomb's was criticized by John Leslie[1] in 1804 on the grounds that there was as much downhill as uphill sliding, and that there could be no net effect on the basis of Coulomb's mechanism on μ for that reason. The approach in this paper differs from Coulomb's in that the present model is based on the contention that interlocking by micro-roughness prevents relative motion, except via plastic deformation in the surface layer, at least over the area of the contact spots. This model, in turn, resembles the hypothesis put forth by Leslie in 1804,[1] it seems.

Micro-Roughness and Persistent Asperities of Contact Spots

The application of the model of Fig. 1(C) does not mean that local welding and/or noticeable adhesion does not occur at all, or is restricted to very clean conditions. For example, the cracking of brittle surface films must expose fresh surfaces at the contact spot interface. If that happens among materials of similar hardness, true adhesion can occur where film cracks on the two sides intersect, provided that they are wide enough for the underlying material to touch, perhaps after squeezing through the cracks by means of plastic flow. According to J. B. P. Williamson,[24] that latter effect can be microscopically verified for two pieces of aluminum after pressing them together; much the same effect was earlier suggested by Whitehead.[25] Also, clean surfaces are continuously exposed in the

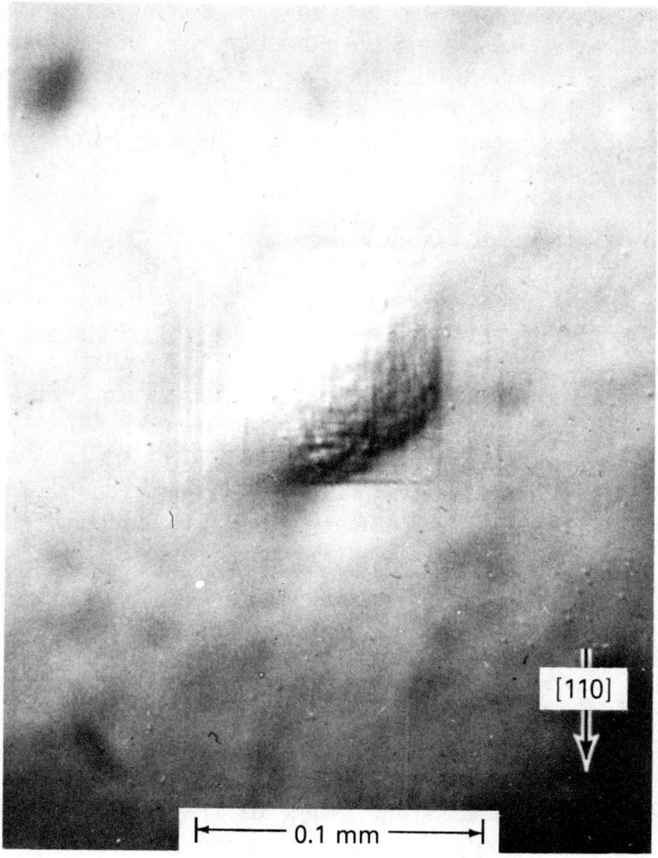

Fig. 11. As Fig. 6 but for a 20 g* load and using interference contrast to reveal finest details in surface topography. Note the micro-roughness in the indentation, caused partly by slip lines, and partly due to other causes, including perhaps minute surface contamination or irregularities in the unseen oxide layer. *(Courtesy of L. D. Dyer, Ref 22)*

course of wear, and these are capable of adhesion, especially at high pressures which have the effect of relatively increasing contact area to geometrical area and thus retarding re-contamination. At any rate, whenever and to whichever degree adhesion and/or welding arise, they supplement the mechanical locking due to surface roughness and, if strong, they can be dominant and cause corresponding effects.

Regardless of the relative strength of adhesion (provided it is not dominant, a point to which we shall return later), the achievements of the adhesion theory of wear remain intact by the use of the model of Fig. 1(C), since the essential assumptions remain unchanged: namely, even more than in the unmodified adhesion theory, the typical load-bearing spot is not subject to sliding between the surfaces but deforms and/or shifts through subsurface plastic deformation. Secondly, as before, the total load-bearing area between the two surfaces remains (nearly) proportional to the normal force between them, subject to the limitations that the load be large enough to cause adequate local plastic deformation and small enough that only a minor fraction of the geometrical surface is in load-bearing contact.

The above statement that with and without micro-roughness the contact area is proportional to the load, and the more detailed properties of contact spots, deserve some elaboration. To begin with, it was shown by Archard[26] and Greenwood and Williamson[9] that, quite commonly, for large N a proportionality between NA_b and P comparable to Eq 1 may be expected even if the contact spots are not plastically deformed, whereas for a single asperity in elastic loading A_b is proportional to $P^{2/3}$ according to the Hertzian solution.[27] This was found to be true especially when smaller protuberances were located on individual asperities. It would thus be expected to be true also for the model of Fig. 1(C). Further, N was found to be proportional to P and, for Gaussian distributions, A_b to be independent of load.

Furthermore, micro-roughness on asperities has received repeated attention and Fig. 1(C) has to be interpreted in light of preceding work. Specifically, Fig. 1(C) is not meant to imply that microscopic contact is complete over all of the micro-roughness features indicated in Fig. 1(C). From the viewpoint of materials science this is clear, assuming a very thin or no surface film, because of the inability of glide deformation along slip lines to produce smoothness, or true conformity to an opposing surface, on a scale less than the distance between, and depth of glide on, active slip lines. While load is applied, elastic deformation will cause better conformity than plastic deformation alone can achieve. However, as will be discussed with a somewhat different viewpoint in connection with Eq 4 to 9, a local pressure H, as exists at a contact spot, will not even out surface undulations beyond an angle of $\sim 2H/\pi E$. Assuming $H/E \simeq 5 \times 10^{-3}$ as before, this means that only undulations equal to, or less than, about 0.2° will be obliterated, which will be a small minority.

If, then, micro-roughness asperities due to inhomogeneous deformation are superimposed on the contact asperities, the considerations of Pullen and Williamson[11] apply. Their theory showed that peaks of interacting asperities on an indented region are flattened to the point that actual contact is close to 50%, provided that the surface and subsurface have the same hardness. That theory explains the intuitively most surprising persistence of roughness in macroscopic ball indentations that was first convincingly documented by A. J. W. Moore's[28] classical paper, as seen in Fig. 12.

Moore[28] had suggested that the hills in macroscopic indentations were persistent on account of preferential work hardening. Quite in line with our own

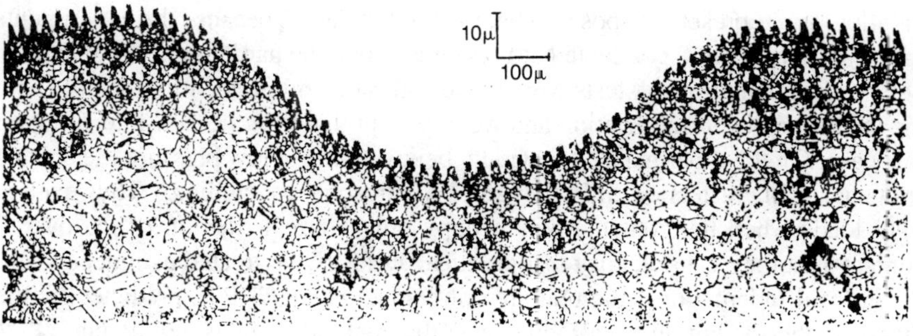

Fig. 12. Taper section of a deep indentation in annealed copper that had had a grooved surface made via turning on a lathe. Note the persistence of the "hills" in the indentation. However, it was noted that these hills were soon obliterated through sliding wear at the same load. *(After A. J. W. Moore, Ref 28)*

preceding argument Moore realized that the indentation hardness controlling the compressing of the hills is that at a distance of their diameter below the surface, while the macro-indentation conforms to the hardness at a distance of the macro-indentation's diameter below the surface, more or less. Yet, although such hardness differences will be real enough, they are not the major reason for the persistence of hills in indentations, as was shown by Williamson and Hunt.[10] Rather, according to the theory and experiments by Pullen and Williamson,[11] flattening of hilltops is compensated by simultaneous rising of the valleys in the surface profile. This accounts for the volume displaced, and causes equilibrium to be established when the local pressure at the flattened hilltops is twice the average contact pressure over the indentation, while real contact occurs over only one-half the area of the indentation, provided, that is, the hardness is the same everywhere. When the hardness at the surface is H_s as compared to the bulk hardness H_b, Pullen and Williamson[11] find that the fraction of surface making real contact is $H_b/(H_b + H_s)$, which reduces to ½ for $H_b = H_s$.

Pullen and Williamson[11] also consider the possible existence of a micro-roughness on the flattened hilltops as follows (Ref 11, p. 170): "If a large number of discrete micro-asperities exist on each asperity then in principle these also should act cooperatively and it can readily be shown that the degree of contact within the 'real' contact area (now the nominal contact area of the micro-asperity population) would be two thirds"—assuming constant hardness throughout, that is.

This latter speculation was quite likely prompted by a very fine roughness on the flattened hilltops, on the order of, say, 0.3 μm, visible on the surface profile tracings published by Pullen and Williamson[11] and Williamson and Hunt.[10] It is this which is here identified as the micro-roughness considered in this paper and indicated in Fig. 1(C). The hills in the indentations may or may not have a true parallel for our present purposes. That uncertainty arises because the experiments in Ref 10 and 11 were conducted at pressures above the hardness of the specimen before the indentations or tests were made. Thus one might identify the hills with the contact spots during friction and wear, except that they are non-interacting in that case. On the other hand, it is widely believed that at low pressure the contact spots occur in clusters, with Pullen and Williamson[11] concluding: "Thus even at light loads when the contact spots occupy only a small fraction of the overall contact area they cannot, for most practical surfaces, be treated as independent. As in local indentations, the real area of contact is not W/P as predicted by the classical theory but just half this; and the real pressure is not P but 2P."*

In the further discussion in the present paper, it will be argued that the surface shearing deformation takes place below the level of the valleys in the micro-roughness or the valleys between the clusters of contact spots, as the case may be. The reason for this is that no systematic shear deformation is possible above that level. Therefore, the results of Pullen and Williamson do not affect Eq 1, provided A_b is taken to be the (average) area over which correlated hills and micro-asperities make contact with the opposing member, while N is the (average) number of spots at which systematic shear takes place at any one moment. A "contact spot" will therefore mean that local region over which interacting contact spots are distributed and participate in what is in effect a single larger indentation. A_b will be the average area of such regions and N their average number.

The above details may be less important for wear than might initially be thought because the persistence of hills and valleys observed by Moore,[28] which triggered the subsequent work by Williamson and co-workers,[10,11] is typical for stationary contacts only. In relative sliding the hills are rapidly obliterated, but it appears that the micro-roughness is preserved (see Ref 28, p. 237 and Fig. 11 and 12). Even so, it is possible, if not probable, that on average there is more than one hill per contact spot also during sliding.

Having thus clarified the meaning of A_b and N as used throughout this paper, a similar clarification is needed in regard to the hardness, H. Indeed, this is a thornier problem, in spite of the fact that it is usually agreed among researchers that hardness is a materials property which, in wear, may be assumed to be independent of the load, the number of contact spots, and their shape. One particular difficulty encountered for our present purposes is that the hardness (defined as the "Meyer hardness," e.g., Ref 29—namely, $H = P/NA_b$ with A_b the projected area of the indentation as in Eq 1) depends on the relative depth of indentation. Measured values of H, obtained from single ball indentations, rise rapidly from zero to a kind of saturation value while the ratio of depth of indentation to ball radius rises from zero to about 3%.[30-33] This initial rise of

*The symbols W and P in the above quotation correspond to P and H, respectively, in this paper.

hardness with increasing depth of indentation coincides with the stress range before all requisite slip systems are activated. Beyond this range, the hardness is equal to $2.8\ \sigma_y$[30,31] or $H \simeq 3\ \sigma_y$, if σ_y is the yield stress in tension (say, the stress for 0.2% plastic strain), more or less. However, the plastic deformation made by the ball indenter (and equivalently by the harder side meeting an asperity on the softer) work hardens the material. As a consequence the measured value of H continues to rise.

Tabor[30,31] has shown that the measured hardness approximates the hardness which the material would have if it were subjected to a tensile strain of $\epsilon = 0.1\ d/r$ beyond the existing state, wherein d is the diameter of the indentation and r is the radius of the indenting ball. It is therefore possible to infer the tensile stress–tensile strain curve, i.e., $\sigma(\epsilon)$, of an annealed material from hardness measurements made with different ball sizes and loads, substituting 2.8 H for $\sigma(\epsilon)$ and 0.1 d/r for ϵ.

Translating this result into the conditions in sliding friction, one might thus say that H equals $2.8\ \sigma_y^*$, where σ_y^* is the yield stress of the material (at a distance roughly $\sqrt{A_b}$ below the surface) after it has been work hardened beyond its original state as if by a tensile strain of $0.1\sqrt{A_b}/r$.

As a further complication, H is somewhat rate dependent, increasing with decreasing time of indentation. This effect is strong for soft materials, but for rather hard materials the dynamic hardness is practically constant[20,30] and is equal to (or very close to) $2.8\ \sigma_y^*$ as defined above.

Adhesion and Surface Roughness

About the rim of any contact spot an annular region exists in which the opposing surfaces are separated by only a very small gap.[34] In Fig. 1(C), for example, the region at issue is that where the two surfaces include a small angle, just inside the circle. Electric tunneling in that annular zone, believed to span a gap on the order of 0.5 nm, has recently been investigated.[35,36] Similarly, one must expect adhesive forces of interaction to act across that gap. The extent of that annular zone, relative to the load-bearing area A_b, increases with the number of contact spots and decreases with increasing load. Above all, it is a strong function of surface smoothness, increasing with decreasing surface roughness. As the load per contact spot decreases, so that the individual contact spots shrink in size and the surface distortion due to local plastic deformation decreases, so, evidently, the annular gap up to some fixed distance of separation increases in relative area. Similarly, decreasing the pre-existent surface curvature causes a size increase in the discussed annular zone.

One may make a very rough estimate of the strength of the adhesive forces across the annular gap as follows: If the total interfacial energy of the two sides is reduced by $\Delta\gamma$ when they are in atomistic contact, as compared to their being separated by a distance s equal to the range of the adhesive forces, then the average attractive stress between the two sides at distance s is

$$p_\gamma \simeq \Delta\gamma/s = \Delta\gamma/nb, \qquad [4]$$

expressing s as a multiple of b, the atomic diameter. Presumably, the range of adhesive forces is a few or several angstroms, so that n = 2 appears to be a reasonable estimate. Now, in a material of elastic modulus E, a sine wave of displacement

$$u = A \sin\frac{2\pi}{\lambda}x \qquad [5]$$

is elastically flattened by a pressure[16]

$$p_A = \pi EA/\lambda. \qquad [6]$$

From Eq 4 and 5 one may find the maximum surface waviness such that it would be flattened by adhesive forces and, hence, give rise to complete contact between the two surfaces. In this, it is helpful to employ a rough empirical relationship between the surface tension γ, the atomic diameter b, and Young's modulus E, namely

$$\gamma \simeq Eb/20. \qquad [7]$$

The relative change of total surface energy may therefore be written

$$\Delta\gamma/2\gamma \simeq 10\Delta\gamma/Eb, \qquad [8]$$

assuming that the surface energy and elastic constants are similar on the two sides.

From equating the two pressures, p_A and p_γ, and using Eq 8, one finds the maximum value of the average angle of surface inclination due to roughness such that adhesive forces will establish complete atomistic contact over the surface,

$$(2A/\lambda)_{crit} \simeq \theta_{crit} \simeq (\Delta\gamma/\gamma)/10\pi n. \qquad [9]$$

With n = 2, as introduced above, and $\Delta\gamma/\gamma \leq 1$, it correspondingly follows that very strong adhesive effects must arise for surface waviness with average surface inclinations between the two surfaces smaller than, or about equal to,

$$\theta_{crit} \simeq 0.015\ (\Delta\gamma/\gamma) \lesssim 0.9°. \qquad [10]$$

For surface roughnesses much larger than the limiting value given by Eq 10, which are indeed very common because of preceding surface preparation as well as sliding and micro-roughness, the annular zones of adhesion must necessarily be rather small compared to the load-bearing interface, A_b. In fact, the typical surface waviness is $\theta \simeq 5°$. Therefore adhesive forces can be neglected in very many cases, especially since $(\Delta\gamma/\gamma)$ is generally quite small, as shown by the fact that it is not possible to mend fractured objects by simply pressing the two sides together, however perfect may be the geometrical match between the two surfaces.

The reason for the low values of $\Delta\gamma/\gamma$ commonly experienced is, of course, that the adhesive forces are screened by surface films. Such films form almost instantaneously in atmospheric air. Obvious conditions under which strong adhesion arises are, therefore, twofold. The first is when Eq 9 and/or 10 are fulfilled—i.e., in the case of very planar, highly polished, clean surfaces making contact under almost any pressure. The second is when surface films are eliminated faster than they can re-form. This may readily occur under moderate to high pressures, especially in protective atmospheres or in vacuum when, during wear, new surfaces are made and quickly covered by relatively large contact spots in which air access is inhibited.

For strong adhesion to occur, surface waviness in accordance with Eq 9 and 10 is most important when the surfaces are relatively at rest. However, it is intuitively clear that adhesion as manifested by seizing can spread via local plastic deformation when surface films are destroyed through relative motion, especially when motion is slow and/or occurs under high pressures, even if the initial surface should be rough and Eq 10 not be obeyed. These expectations appear to be in general agreement with observation.

For the sake of completeness, a few more details regarding adhesion may be in order: the concept of compatibility, linking the expected degree of friction and wear among pairs of materials to their alloying behavior, championed by Rabinowicz,[37,38] may be theoretically linked to epitaxial behavior. Namely, the value of $\Delta\gamma/\gamma$ in Eq 9 and 10 would seem to be the same as that which controls epitaxy and plays a considerable role there.[39-41] One may thus expect that materials which are known to form good epitaxial systems, namely, those with large $\Delta\gamma/\gamma$ in the clean state, are those which are "compatible" in the sense of Rabinowicz, and are those in which adhesive effects are strong. To the extent that sliding wear under high pressure lays bare new uncontaminated surfaces which are at least partly and temporarily shielded from contamination, "compatibility" is thus very relevant if "seizing" is experienced.

Turning to the other extreme, namely, contacts under light loads, one may estimate the contribution made by adhesive forces on the basis of the Hertzian contact as discussed by Holm:[2] the average diameter of N plastically deformed load-bearing contact spots, modelled as formed by lightly indenting, locally spherical asperities of radius r_c contacting a flat surface under load P, is, according to Eq 1,

$$d = \sqrt{4P/\pi HN}. \qquad [11]$$

However, as explained, the average contact spot is surrounded by an annular zone whose outer diameter may be given by the symbol d_t, within which adhesive forces act. If, again, the gap width within which this effect is significant is taken to be $s = nb$, then by simple geometry[34]

$$d_t = \sqrt{d^2 + 8r_c nb} = \sqrt{(4P/\pi HN) + (8r_c nb)}. \qquad [12]$$

As seen before, $d \simeq 0.1$ mm within a factor of ten or so, i.e., $10^{-6}\text{m}^2 \lesssim d^2 \lesssim 10^{-10}\text{m}^2$ in the majority of cases. By comparison, with $10^{-3}\text{m} \lesssim r_c \lesssim 10^{-5}\text{m}$ as a reasonable range, with $n \simeq 2$ and with $b \simeq 3 \times 10^{-10}\text{m}$, one obtains $5 \times 10^{-14}\ \text{m}^2 \lesssim 8r_c nb \lesssim 5 \times 10^{-12}\ \text{m}^2$, i.e., orders of magnitude smaller than d^2. Therefore, spreading of contact spots is not going to occur to any noticeable extent except under the conditions of high pressure, and/or cleanliness and surface smoothness as outlined above.

The Number of Slip Systems and the Coefficient of Friction

According to the model of Fig. 1(C), with or without the action of adhesive forces, sliding requires shear deformation of the weaker of the two materials, which is sheared off or carried along by the stronger at least for short distances. However, as clarified by Fig. 6 to 8 and 10, it is not possible to shear or drag a surface element without propagating the shear deformation into the surrounding material. In other words, the cross-sectional area over which shearing must occur is larger than the actual load-bearing area of the contact spot and indeed larger than A_b. Certainly it is impossible to drag along a surface element without deforming also surrounding material. This is dramatically obvious in cases of rather strong roughness in which material may be pulled over parts of the original surface, as in Fig. 13 and 14, in extreme cases giving rise to the burying of extraneous material under material dragged along by contact spots, as shown in Fig. 14. Thus, the case of no relative motion and great simplicity (Fig. 6 to 8), as well as the case of surface turbulence of rather extreme magnitude (Fig. 13 and 14), suggest that the area sheared in sliding is larger than the area bearing the normal load by some constant c. This constant c, which is estimated to lie between 1.5 and 5, is likely to be subject to variations depending on geometry and load.

Using the above result we may write for μ, the coefficient of friction,

$$\mu = \frac{F_t}{P} = \frac{NcA_b\tau_s}{NA_bH} = \frac{c\tau_s}{H} \qquad [13]$$

where F_t is the tangential force required for sliding, and τ_s is the flow stress of the softer material with regard to shear parallel to the surface.

From dislocation theory it is known that plastic deformation of crystalline materials deforming in glide occurs if, and as soon as, the momentary critical

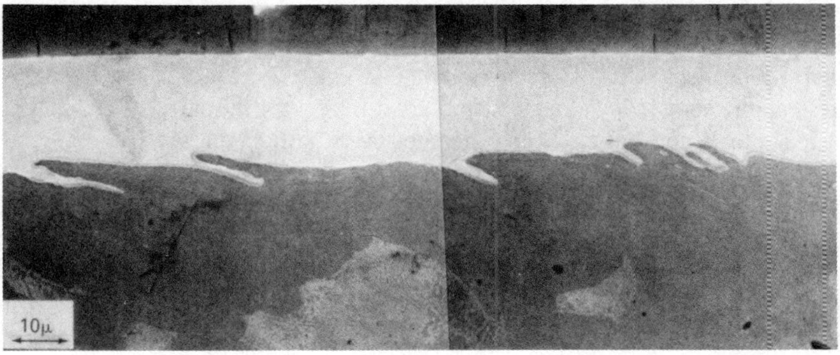

Fig. 13. Highly irregular deformation at the interface between a thick gold plating and its AISI 1020 steel substrate underneath the wear track of an AISI 52100 steel slider. *(Courtesy of N. P. Suh, Ref 6, p. 4)*

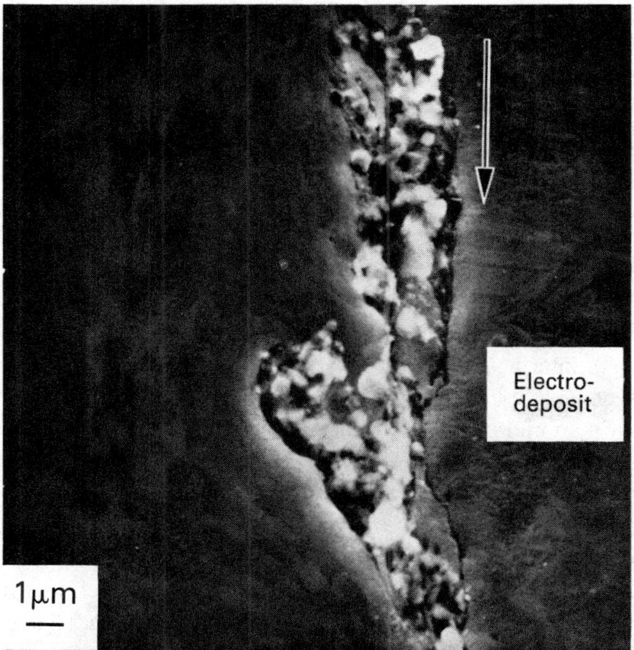

Fig. 14. Abraded copper surface showing highly irregular surface deformation to the point that some of the abrading sand particles have become embedded. *(Courtesy of L. K. Ives, Ref 6)*

resolved shear stress is exceeded on any crystallographic slip system. It appears to have been overlooked in the past that both the magnitude of the critical resolved shear stress and the factor relating the applied stress to the resolved shear stress on the controlling slip system(s) are different for the surface, at which the shear strain is concentrated, and the subsurface region, in which the hardness is controlled.

Fortunately, for purposes of theory, the plastic strains concerned (in the subsurface region establishing load-bearing indentations and near the surface during sliding) are frequently so large, and/or work hardening is so small, that one may assume that saturation flow stress is attained everywhere. If so, τ_s and H are related in a simple manner—namely, in agreement with the discussion in the preceding section,

$$H = 2.8\ \sigma_y, \qquad [14]$$

where σ_y is the saturation tensile stress (for the present assumed to be constant). Further, in single glide the saturation resolved shear stress, τ_y, is correlated with σ_y via the so-called Schmid factor, m_b, as

$$\tau_y = \sigma_y\ m_b = \sigma_y\ (\cos \lambda \cos \chi), \qquad [15a]$$

where λ and χ are the angles made by the compression axis with the active slip directions and slip plane normals, respectively. In Eq 15, meant to be used in connection with Eq 14, the subscript b has been used as a reminder that the relevant deformation takes place somewhere in the interior and thus may be taken to reflect the orientation in the bulk. Specifically, in regard to indentations (i.e., Eq 14 and 15 together), m_b of course depends on crystal orientation or texture. Its value is not easily found in view of the complexity of the deformation, as exemplified in Fig. 7 and 8.

The conditions at the surface are different. In that case the applied shear stress, parallel to the interface and in the sliding direction, must be resolved into the active slip planes and slip directions carrying the surface shear. However, if single rather than multiple glide occurs, i.e., if slip continues on only one set of parallel glide planes and in only one glide direction, the active slip plane and slip direction reorient to be parallel to the plane and direction of shear, respectively. This has been well known ever since Schmid and Boas' fundamental work.[42] Directly at the surface, where the shear deformation is very large as seen in Fig. 4 and 5, a crystal texture should therefore develop such that the "easiest" crystallographic glide plane is parallel to the surface, and the easiest glide direction therein is parallel to the direction of sliding. The factor relating applied stress to resolved shear stress in the top layer should therefore be nearly unity when, or if, shear in the top layer is confined to only one single slip system. In that case $\tau_s = \tau_y$, whence with Eq 13 to 15

$$\mu = c m_b / 2.8. \qquad [16]$$

One other important consideration intervenes here: namely, that it is as yet uncertain under which circumstances and for what materials single glide is realized in the surface layers during sliding wear. The most simple geometrical concepts outlined above would lead one to expect single glide at the surface but in its totality the deformation in wear is very complicated, as discussed above, and multiple slip would seem to be rather more likely than single glide. In that case, on account of the very large shear strains achieved at the surface, one expects the development of some symmetrical orientation in which two or more glide systems are equally highly stressed.

The simplest example in this connection is that of layer-type crystals. These have only one crystallographic slip plane, namely, the basal plane, within which, however, there commonly are three non-parallel glide directions according to three directions of atomic close packing. In single glide, the surface texture would be that described already, namely, with the basal plane parallel to the surface and a close-packed direction parallel to the direction of wear. However, even in the absence of any alternative mode of plastic deformation it is equally well possible that the strain at the surface settles down to double glide on two equivalent glide directions, thus generating a surface texture in which the direction of wear bisects the angle between the two.

Similarly, in crystal systems with more than one "easy" glide plane (meaning all but layer-type and hcp crystals with a c/a ratio larger than ideal) multiple glide on more than one glide plane could well be expected. The resulting texture, if consisting of but one component, should then be one in which the surface is symmetrical in regard to two, or perhaps three or even four, different slip systems.

The discussed principles may be illustrated by experimental evidence for the case of wire drawing. In wire drawing the wire axis of bcc metals is parallel to <110>, i.e., not parallel to the <111> slip direction but symmetrical to two pairs of two <111> directions each. Fcc metals have a double fiber texture in which different grains have either <100> or <111> parallel to the wire axis. Since <110> is the slip direction in fcc metals, as already mentioned, again the texture axes are symmetrical to more than one of the <110> slip directions—namely, symmetrical to four <110> directions in the case of <100>, and three for <111>.

Low stacking fault energy causes a distinct dissociation of $\frac{1}{2}$<110> dislocations into two $\frac{1}{6}$<112> partials, to the effect that cross slip as well as dislocation intersections are inhibited to a degree depending on the width of dissociation. This, in turn, has the effect of impeding the simultaneous action of different slip systems. In this light, it is understandable why low stacking fault energies shift the fcc wire drawing texture so as to favor the <111> component, as was first pointed out by N. Brown.[43] Similarly, therefore, one may suspect that the wear texture, and hence wear behavior, of fcc and hcp metals depend on stacking fault energy, a point to which we shall return later.

In rolling textures of metals the same principles may be perceived, albeit with some complications. The rolling texture of bcc metals is <110>, {100}, naming first the rolling direction and then the plane of the sheet. That of hcp metals can be <11$\bar{2}$0>, {0001}, as would be expected on the most simple model for single glide, but generally the hcp rolling texture is variable, depending on the number of slip systems and twinning modes participating.[44] In fcc metals, <112> is parallel to the rolling direction (i.e., midway between two <110> slip directions) but the texture with regard to the plane parallel to the rolling plane is not well defined. For pure fcc metals the average orientation of many grains is best represented as {123} parallel to the sheet, it seems,[45] while for fcc alloys (i.e., low stacking fault energies) it is more nearly {110}. However, especially for pure fcc metals, the texture has at least two components, much like the fcc drawing texture.

One may ask why nature should prefer multiple glide, entailing short mean free dislocation paths and profuse dislocation intersections, before single glide with potentially much longer free paths and thus fewer dislocations generated per unit of glide and many fewer dislocation intersections. It is here suggested that the reason may be found in the specifics of sub-boundary formation as follows. With gratifying regularity, under a host of different deformation conditions, the dislocation density, ρ, is found to depend on $(\tau - \tau_o)$, i.e., the resolved shear stress in excess of the intrinsic frictional stress, τ_o, as

$$\rho = [(\tau - \tau_o)/\alpha G b]^2, \qquad [17]$$

with $\alpha \simeq 0.4$ (within a factor of two, say), b the Burgers vector, and G the shear modulus. The great increase of dislocation density during plastic deformation, which in accordance with Eq 17 occurs while τ rises in the course of work hardening, arises via mutual dislocation trapping into low-energy configurations. In unidirectional deformation, whenever several independent crystallographic slip systems exist and dislocations are not significantly dissociated into partials, these low-energy configurations typically are, or approach, low-angle boundaries arranged into characteristic networks, commonly referred to as "cell walls" forming a "cell structure."

It can be shown theoretically that the energy of cell walls decreases with increasing number of participating non-parallel Burgers vectors, meaning dislocations belonging to different slip systems,[46] especially in strongly elongated cells in which tilt walls predominate. The effect of twist components in the cell wall structure modifies this result to the effect that there is not much additional gain in specific dislocation energy stored in the cell walls when the number of non-parallel Burgers vectors rises above three.[47] It follows, therefore, that an accumulation of dislocations with just one type of Burgers vector strongly attracts dislocations of other types. Among these, the dislocations belonging to highly stressed glide systems are evidently favored by the externally imposed forces. The discussed tendency for the simultaneous operation of two or more glide systems counteracts the impediment against intersecting glide that otherwise exists for all types of dislocations, and it overcomes it unless the dislocations are substantially extended and intersect only with great difficulty, thereby inhibiting dislocation cell formation.

Theory further shows that the number of possible low-angle boundaries, and hence of cell walls, is truly stupendous, and that cell walls in any arbitrary orientation can be made of any three or more independent Burgers vector directions, provided only that the dislocation densities are free to adjust.[48] Therefore it is difficult to predict the cell structure that will form under different circumstances, but the proclivity for multiple glide, whenever several independent "easy" crystallographic slip systems exist and dislocation intersections are not impeded through unduly large dislocation dissociation, is readily understood from the above principles.

Later on, more will be said about dislocation cells in relation to friction and wear. The topic has been introduced here mainly in preparation for the later discussion. The macroscopic effect of the discussed dislocation and cell wall properties is simply that multiple glide takes place readily, and that the resolved shear stresses on different, simultaneously acting glide systems are quite similar. On this basis—i.e., that, locally, all simultaneously acting glide systems continue to have the same critical resolved shear stress throughout work hardening, plus the fact first pointed out by Taylor[49] that five independent glide systems are necessary and sufficient for all grains in an aggregate to deform homologously with the external strain—most textures can be generated via computer modelling.[44] However, the fact that deforming single crystals also break up into crystallites, separated by cell walls with gradually increasing angles of misorientation, cannot be so simply explained but requires understanding of the dislocation mechanisms introduced above. In this light, the effect of dislocation dissociation into partials is to selectively increase the critical resolved shear stress of otherwise equivalent crystallographic slip systems.

Returning to the problem at hand, namely, to find the coefficient of friction in accordance with Eq 13 to 15 for the general case, we need to relate τ_s, the shear flow stress at the surface, to ${}_s\tau_y$, the critical resolved shear stress of the slip systems operating at the surface, via the appropriate Schmid factor. To do this, we make use of the general relationship between the applied stress, σ_{ij}, and the resolved shear stress on the various (operative) slip systems, of which Eq 15a is but a specific case. Namely,[44]

$$\tau_y = \tfrac{1}{2}\Sigma\sigma_{ij}(\cos\alpha_i \cos\beta_j + \cos\alpha_j \cos\beta_i) = \Sigma\sigma_{ij}m_{ij}, \qquad [18]$$

with α and β referring to the respective slip directions and slip plane normals. For the case of the surface shear stress generated by sliding, say, along the xy plane (normal z) in the x direction, Eq 18 simplifies to one single term. In that case

$\sigma_{zx} = \sigma_{xz} = \tau_s$ is the only stress component, and

$$_s\tau_y = m_s\tau_s = \tau_s \cos \alpha_x \cos \beta_z, \qquad [19]$$

where α_x is the angle which the slip direction includes with the sliding direction, and β_z is the angle which the slip plane makes with the plane of sliding. Equation 19 is a simplification in that the resolved shear stress is in truth a function of z, a point to be discussed below. For the present purpose $_s\tau_y$ stands for the resolved shear stress at the level of maximum strain rate.

The Schmid factor in Eq 19 has been given by the symbol m_s in order to remind the reader via the subscript s that it is governed by the conditions in the topmost surface layers, in which the critical resolved shear stress is $_s\tau_y$. By contrast, in the subsurface layers the critical resolved shear stress is $_b\tau_y$, i.e.,

$$_b\tau_y = {_b\sigma_y}\, m_b = H_b\, m_b/2.8. \qquad [15b]$$

By combining Eq 13, 14, 15b and 19, one thus obtains for the coefficient of friction

$$\mu = \frac{c\,\tau_s}{H} \simeq \frac{c\,{_s\tau_y} \cos\lambda \cos\chi}{2.8\,{_b\tau_y} \cos\alpha_x \cos\beta_z} \simeq \frac{c\,{_s\tau_y}\, m_b}{2.8\,{_b\tau_y}\, m_s} \qquad [20a]$$

for all cases in which slip is the sole mode of plastic deformation.

In order to obtain a quantitative value for μ, one must consider the likely values of the relevant Schmid factors m_b and m_s. As was shown by Dyer,[22,23] most crystallographically possible "easy" glide systems do in fact operate below and about an indentation, which is not surprising considering the complexity of the imposed strain and G. I. Taylor's rule that five independent slip systems are required for the general deformation.[49] Further, as was discussed, at any one spot all simultaneously acting equivalent slip systems have about the same Schmid factor and critical resolved shear stress value, while at the same spot no equivalent slip system operates whose Schmid factor is much smaller. It follows that m_b rises with M, the number of crystallographically possible "easy" slip systems: if M is larger than five, the value of m_b decreases when M drops—i.e., as the choice of "easy" slip systems decreases—because increasingly those in less favorable orientations are forced to act. If M is less than five, "easy" slip systems must be supplemented by slip systems with intrinsically larger values of $_b\tau_y$, or else twinning or cracking will result. Thus m_b in Eq 20 depends not only on orientation, as was demonstrated by Fig. 6, for example, but even more on crystal structure.

Reliable determinations of m_b for asperity contacts, at different degrees of flattening and for different crystal types and orientations, would require arduous calculations and/or experiments of the kind conducted by Dyer,[22,23] a task obviously entirely out of the range of the present paper. However, some very simple arguments permit clarification of the basic concepts involved as follows.

For the case of polycrystals, m_b is the larger the higher the number of independent glide systems is. To the extent that an indentation parallels a compression test, $(\lambda + \chi) \geq 90°$ in Eq 20a, and the Schmid factor m_b cannot be larger than 0.5. This value of $m_b = 0.5$ would be realized if every crystallographic plane could act as slip plane, and every direction as slip direction. In essence, amorphous materials and glassy metals have that property so that, for them, $m_b = 0.5$. Among the metals, it seems that aluminum at elevated temperatures comes closest to this ideal, when it can glide on all {111} and {100} planes in the twelve<110> directions. Therefore aluminum, when sliding so as to elevate surface temperature to, say, 300°C and above, should have an m_b value near ½, and thus the highest coefficient of friction among the crystalline metals, given the same value of the other parameters in Eq 20a.

The next largest m_b values are expected for bcc metals, since in them ⟨111⟩ serves as slip direction, and, in pencil glide, most if not all planes in the ⟨111⟩ zone may act as slip planes, especially in non-refractory bcc metals. From the research on textures referred to before, $m_b = 0.44$ appears to be the most likely estimate for the case of pencil glide.* A somewhat lower value of m_b is predicted for fcc metals with their ½⟨110⟩, {111} slip systems—namely, $m_b \simeq 0.33$ for polycrystals under the constraints of "polyslip."[50]

In typical hcp metals the basal plane is the only strongly preferred glide plane, with the three closest-packed directions in it serving as slip directions, i.e., M = 3 in this case. Depending on c/a ratio, twinning or glide on pyramidal planes will supplement these glide systems or cracking will occur. In any event, lack of at least five "easy" slip systems will cause the hardness to be relatively higher than indicated by Eq 15b, to the same effect as if m_b were lower than geometrically determined. If, then, ${}_b\tau_y$ is taken to be the critical resolved shear stress of the basal plane, $m_b \simeq 0.25$ may be a reasonable estimate for hcp metals.

Conditions are still more restrictive for layer-type materials. In these, cracking on account of insufficiency of glide systems is presumably widespread since for them glide on the basal plane is the only mode of plastic deformation. The most extreme case is presented by very hard materials such as carbides and nitrides, which lack any effective glide systems or other modes of plastic deformation and thus are brittle. For these, therefore, the hardness to shear strength ratio is abnormally high, which is formally expressible as abnormally low m_b values.

Turning next to the m_s values in Eq 20 it is evident that these are apt to be larger and to be less dependent on crystal type than m_b, because the applied surface tractions and resulting strains are close to simple shear, and therefore the number of participating slip systems is smaller, i.e., one, two or three. If single glide obtains, the final texture would cause $m_s = 1$, with $\alpha_x = \beta_z = 0$ in Eq 19.

---

*The inverse of the Schmid factor, m_b, is the Taylor factor for polyslip. That Taylor factor has been tabulated for various cubic slip modes by G. Y. Chin in "Work Hardening in Tension and Fatigue," ed. A. W. Thompson (AIME, Warrendale, PA, 1977), p. 45.

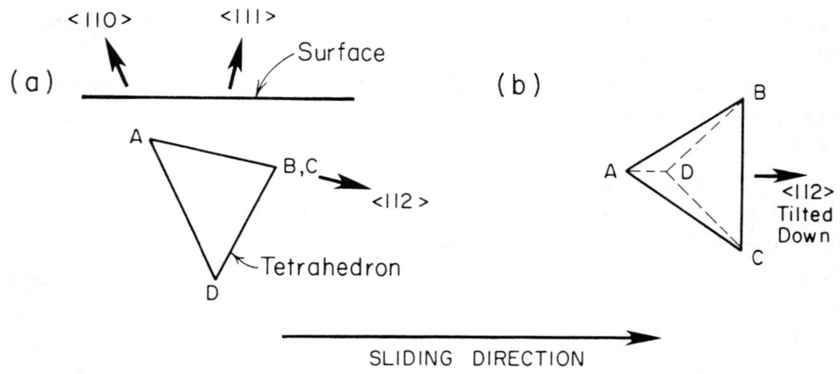

Fig. 15. Surface texture geometry reported for sliding wear in fcc metals, clarified by means of Thompson's tetrahedron, composed of the four {111} slip planes as bounded by the six <110> slip directions. Also for hcp metals and for layer-type materials the slip plane, namely, the basal plane in that case, is nearly parallel to the surface in the wear texture. **(a)** Longitudinal section normal to surface; **(b)** top view. (*Courtesy of J.P. Hirth and D.A. Rigney, Ref 51*)

Hirth and Rigney[51] have recently made a literature survey regarding surface textures due to sliding wear or abrading. As might be expected, these resemble, or may be identical with, the corresponding rolling textures. Figure 15 due to Hirth and Rigney[51] indicates the surface wearing texture for fcc metals. The textures for hcp and bcc metals, and for layer-type crystals, can be derived from Fig. 15 by substituting, for {111}, the pertinent crystallographic slip plane (namely, the basal plane for hcp and layer-type crystals, and {110} for bcc metals). In this light, considering the texture of Fig. 15, it seems clear that once the surface texture is established the value of m_s cannot be much lower than $m_s = 0.8$ in accordance with Eq 19, with $m_s = 1$ as the upper limit.

Altogether, therefore, with m_s not very variable in the fully "run-in" specimen, the coefficient of friction is found roughly equal to m_b, *if* ${}_s\tau_y = {}_b\tau_y$ as would be true for very hard or not work hardenable materials. With ($c/2.8\ m_s$) not far from unity in accordance with the above discussion, Eq 20a then reduces to

$$\mu \simeq m_b \qquad [20b]$$

for the "standard" case after the run-in period. Correspondingly, the m_b values outlined above are indicative of the value of the coefficient of friction in steady state, namely, $m_b \simeq 0.5$ for aluminum at mildly elevated temperature, $m_b \simeq 0.44$ for bcc metals deforming in pencil glide, $m_b \simeq 0.33$ for fcc metals, $m_b \simeq 0.25$ for hcp metals, and perhaps $m_b \simeq 0.15$ for layer-type materials. Finally, for ultra-hard brittle materials, the coefficient of friction on this basis should be quite low, e.g., 0.15 or even less, because formally m_b would be very low (indeed perhaps zero). From a different viewpoint one would rephrase this latter result, saying that brittle materials have a low coefficient of friction because they lack any efficient mechanism to convert mechanical work into heat.

The above values presumably play the role of lower limits because, generally, ${}_s\tau_y > {}_b\tau_y$ on account of work hardening in the course of surface shear, as seen for example in Fig. 5. Secondly, m_s rises during the run-in period before the surface texture is established and may start out at much lower values than 0.8. Third, the work expended in grooving must be added, which is generally presumed to account for 10% or 20% of the coefficient of friction. Finally, on new surfaces, "prow formation" as discussed recently by Antler[52] will add to the sliding resistance and thus to μ. Conversely, c may have been overestimated. This question is difficult to resolve at this point: It is considered that the effect of the microroughness to modify local pressures and actual surface contact by a factor of perhaps 2 or even 3 in accord with the investigations by Williamson and coworkers[10,11] does not influence our considerations because, unnoticed, microroughness was there all along and was implicitly incorporated into the factor $2.8 = H/\sigma_y$ without this having been realized. The same cannot be said of the possible clustering of interacting hills into any one contact spot. The effect of such clustering would be to put a lower limit on c, namely, $c = 2$ in soft materials and up to 3 in harder ones.[10,11] These are lower limits because, to the thus enlarged average contact spot area, the previously considered plastically deformed zone about the spot must be added.

Actual values of the coefficient of friction appear to follow the order of the dependence of m_b values on number of slip systems[51,53] outlined above. Specifically, it has often been noted that hcp metals have a lower coefficient of friction than fcc metals (e.g., Ref 51, 53, 54), and bcc metals have a higher friction coefficient than these, while layer-type crystals, foremost among them graphite and molybdenite, have such low coefficients of friction (on the order of 0.15) that they are widely used as lubricants.

A correlation between the friction coefficient and the number of slip systems and/or stacking fault energy was previously suggested by Heilmann and Rigney,[53] Rigney and Glaeser[55] and Hirth and Rigney.[56] In the context of the present theory the connection of coefficient of friction with stacking fault energy has already been explained. It comes about because low stacking fault energies, in fcc as well as hcp metals, act to reduce the effective number of slip systems, and thus m_b and μ.

One is hard put to give actual experimental values for μ so as to test Eq 20a and thereby to place some limits on the constant c. The following appear to be somewhat representative for the "standard" case to which Eq 20a applies: for layer-type crystals a lower limit value of the friction coefficient on the order of 0.1; about 0.4 for fcc metals; about 0.5 for bcc metals; and, finally, closer to 0.25 for hcp metals.

One must hedge in quoting such values because we restricted our considerations to the average-clean, average-rough, well-run-in sample, and disregarded lubrication, strong or thick surface layers, adhesion (which is expected to be strong in vacuum), exceptionally planar and smooth surfaces, or high pressures. Regrettably, measurements of the friction coefficient are frequently given without clarifying the conditions under which they were obtained. Furthermore, in line with the prevailing focus on the adhesion theory, measurements under very clean conditions were usually believed to be more meaningful, and therefore these are mostly reported. In fact, values of the friction coefficient below unity are often believed to be artifacts due to an assumed lubricating effect of surface impurities or surface films. The above discussion will have shown why a re-examination of friction data for dry sliding between average-clean, average-rough surfaces would be highly desirable, indeed essential, if Eq 20a is to be tested, with the possible outcome that these conditions will be established as the standard case, from which all others may be derived by considering the various modifying factors.

Effects of Sliding Distance, Sliding Direction, and Bulk Texture on the Coefficient of Friction

The preceding interpretation of friction coefficients in terms of slip geometry accounts for or predicts some special effects. First, as long as surface texture is not fully established, m_s, the Schmid factor for the shear deformation at the surface, changes. Generally m_s tends to rise, perhaps up to unity as was explained. This should cause the friction coefficient to drop during run-in by up to almost a factor of two even in cubic crystals, and more in materials of lower symmetry. Conversely, work hardening through surface deformation increases ${}_s\tau_y$ while ${}_b\tau_y$ remains rather stable, causing the coefficient of friction to rise proportionately. Therefore one must expect substantial changes of μ during run-in which could go either way, and which could (indeed often should) show an intermediate extremum.

Secondly, for the fully run-in condition there should exist an orientation dependence on the direction of wear. That is, if a texture has been established in which the active slip direction is parallel to the direction of sliding, the friction coefficient is expected to have a minimum value (say, μ_o) when testing in the initial direction of sliding, while in other directions μ should be inversely proportional to the corresponding (lower) direction cosine in accordance with Eq 20a. In case of the perfect texture, i.e., $\alpha_x = \beta_z = 0$, this means that $\mu = \mu_o/\cos\varphi$ where φ signifies the angle between the direction of testing and the original sliding direction, or the nearest equivalent slip direction in the slip plane parallel to the surface. Of such equivalent slip directions there are two (besides the original one) on the hexagonally close-packed slip planes of fcc, hcp and layer-type substances, but only one on the {011} slip plane of bcc metals slipping parallel to <111>.

For any texture in which the sliding direction is midway between two slip directions, e.g., <112> in fcc metals with {111} nearly parallel to the surface as in Fig. 15, the coefficient of friction is predicted to have a maximum in the original sliding direction. In agreement with that prediction Moore[28] found the coefficient of copper that had been grooved by turning on a lathe (which presumably generated the discussed <112>{111} texture) to be 0.6 in dry sliding parallel to the grooves, but 0.57 at 45° to the grooves. For an accurate <112>{111} texture the expected ratio of μ in the two directions is cos 30°: cos 15° = 0.60 : 0.54, and for {111} somewhat tilted against the surface, as in the actual texture, the predicted coefficient at 45° should correspondingly be moderately higher than 0.54 but below 0.60, in good agreement with Moore's measurement.

Equation 20a further indicates a dependence of the friction coefficient on any pre-existent texture in the bulk of the material. This offers the hypothetical opportunity of reducing friction coefficients via suitable choices of textures. Some indirect support for this contention is provided by Fig. 10, in conjunction with Dyer's[23] observation that the coefficients for rolling friction were 0.0045 for rolling in the [100] direction, and 0.0059 for rolling in the [110] direction.

The Role of Surface Films

Much evidence exists in support of the contention that the coefficient of friction and/or adhesion are greatly influenced by surface films. Thus substantial effects can result even from very minor contaminations of either or both of the two contacting surfaces, as elaborated by Bowden and Tabor[3,4] and Rabinowicz[7] as well as numerous other workers, e.g., J. R. Whitehead,[57] K. L. Johnson and D. V. Keller,[58] and D. H. Buckley.[59] Figure 16 from the book by Bowden and Tabor[4] gives an example, including in the legend the interpretation advanced by these authors.

In view of the wide acceptance of the adhesion theory of friction, it is not surprising that authors commonly ascribe decreases in the coefficient of friction following the introduction of some contaminants to the lubricating effect of surface layers, including those which are only several atomic layers thick, much as Bowden and Tabor did in their interpretation of Fig. 16. Not at all infrequently, however, major changes in the nature of the surface films do not cause any significant, reproducible changes in the coefficient of friction. Examples may be found in Ref 60 and 61. Figure 17, presenting data extracted from Ref 60, illustrates the latter phenomenon.

The model of Fig. 1(C) and the preceding considerations are clearly incompatible with the assumption that very thin surface films act as lubricants. In fact, the theory presented in this paper does not leave room for any major effects of surface films, whether pre-existent or generated through chemical reactions or transfer of

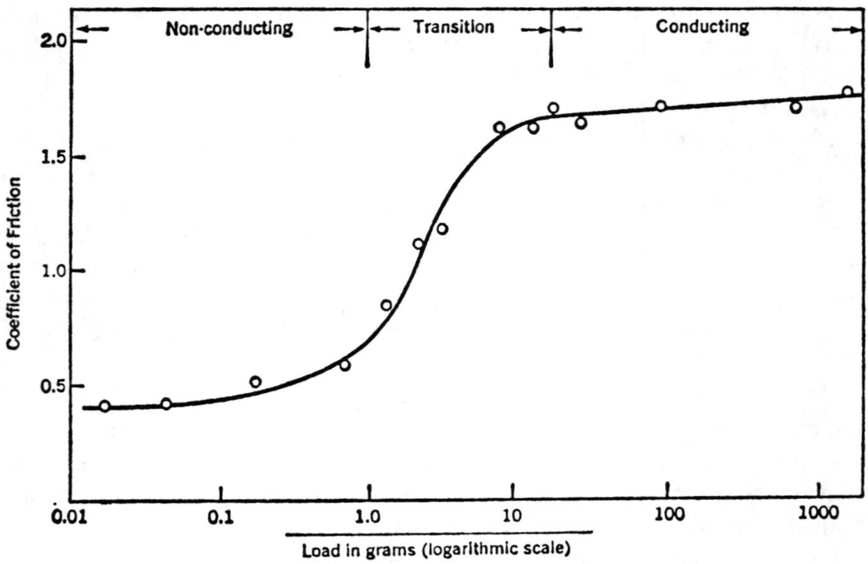

This is Fig. 38 in Ref 4, to which the authors, F. P. Bowden and D. Tabor, write: "Below 1 gm. the friction is low ($\mu \simeq 0.4$) and the surfaces are electrically insulating; sliding occurs within the oxide film on the copper surface. Above 100 gm. the friction is high ($\mu \simeq 2$) and the surfaces become electrically conducting. The oxide is penetrated giving metallic contact and increased interfacial adhesion." According to the theory in this paper the above explanation is modified to exclude sliding within the oxide layer, and seeing the "basic" case realized at low loads.

Fig. 16. Coefficient of friction of copper versus a hard steel slider as a function of load, using slow speed and well-controlled conditions.

material in the course of wear, except under a number of specific circumstances, including at least one of the following:

1 Films of a thickness comparable to, or larger than, the micro-roughness
2 Films so strong that they compete with the strength of a layer thickness comparable to $\sqrt{A_b}$ of the softer material
3 Films of a kind to inhibit the formation of micro-roughness
4 Films which form hard wear debris particles.

Note that films falling under point 1 may significantly reduce the coefficient of friction, and thus be truly lubricating, provided their shear strength is low. Soft platings such as discussed by Peterson and Ramalingam in this book provide examples. Films of group 3 can also reduce the coefficient of friction, especially if they are very hard, and those in group 4 can reduce μ by providing particles which separate the two surfaces while either rolling between them like balls in a ball race, or plowing through either or both sides. The wear rate in the last case could be affected either way, but more likely it is increased. Hard platings such as discussed by Peterson and Ramalingam in this book tend to belong to both 2 and 3.

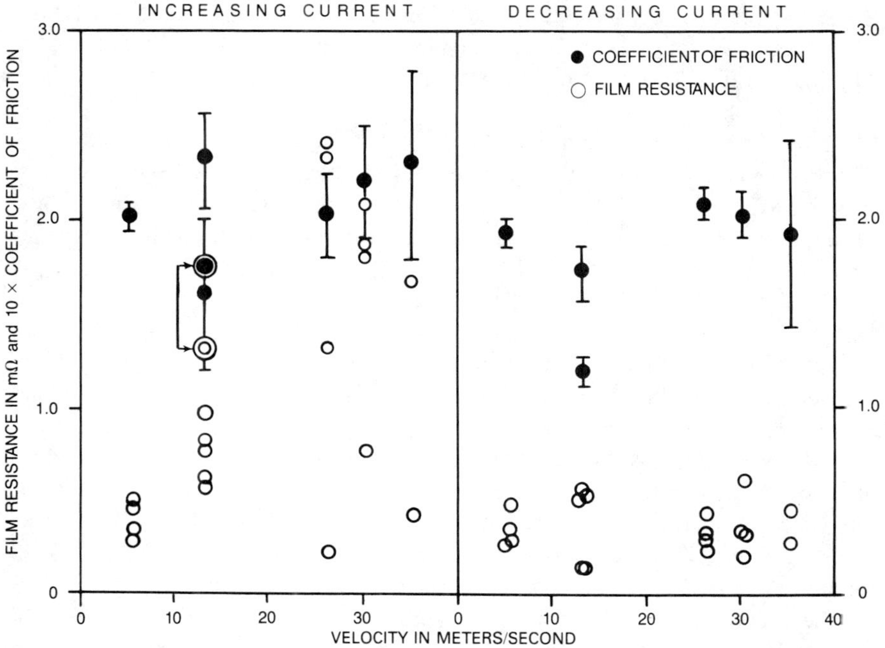

Note that the film resistance varies irregularly by as much as a factor of ten on increasing current, indicating correspondingly large changes in the film thickness and/or structure, while the coefficient of friction changes only within about ±25% and without any correlation with the film resistance. On decreasing the current, the film resistance changes much less but the coefficient of friction more than on increasing current. Clearly, therefore, there is no direct correlation between coefficient of friction and film character, at least not in this case.

Fig. 17. Values of the coefficient of friction and of the electrical resistance of the surface film for an electrical silver graphite brush (75w/o Ag) sliding on a polished copper ring, measured during the same experiment (see Ref 60).

Any of the above possibilities, plus perhaps a number of others not considered, singly and in combination, will surely be realized under a bewildering variety of different circumstances and will thus give rise to the correspondingly varied effects. The range of effects that one may expect in connection with films of types 1 to 4 is further increased on account of the surface temperature changes which, at constant pressure and velocity, accompany changes in the coefficient of friction, and which at the same time can influence surface films in various ways.

Ordinary thin surface films on nominally clean surfaces, meaning films not visible to the naked eye (i.e., of thickness between, say, twenty and two hundred angstroms) are unlikely to fall into any of the above categories, except perhaps 4. Therefore, materials with surface films not thick or strong enough to be classed under any of the four points enumerated above will be included in the "basic" case. Conversely, it requires the local absence of any film whatever, including monatomic adsorption layers, to permit strong adhesion,[58,59] and furthermore, all "ordinary" films down to monatomic layers on the two surfaces have a significant electrical resistance.

In light of the above considerations, and in agreement with Bowden and Tabor's own interpretation, the transition shown in Fig. 16 is seen as one between the "basic" case including an invisible surface film, and surface film removal through wear which exposes new surface under conditions in which re-contamination is inhibited. Then true adhesion takes place, and the measured coefficient of friction correspondingly rises steeply.

It requires very well controlled conditions of loading in order to obtain a virtually constant coefficient of friction when there is a significant adhesive force between the two sliding objects, e.g., as was seemingly the case in Fig. 16 and similar measurements by Whitehead[57] that were performed in the same laboratory. However, at least in the latter case,[57] the data points indeed were averages of a large number of individual measurements which showed wide scatter, including stick-slip. More importantly yet, the sliding took place only on fresh tracks and at a very slow speed (namely, 0.008 cm/sec), and the load was kept constant in any one test. Similar conditions were observed by Buckley.[59]

These conditions are important because the meaning of the coefficient of friction is ambiguous when true adhesion is present . Namely, the coefficient of friction is given by $\mu = F_t/P$ according to Eq 13. If adhesion is dominant, F_t is presumably controlled by the highest value of P attained while the surfaces are relatively at rest, since any increase of P increases the true area of contact irreversibly. Albeit, it does not require that F_t depend reversibly on P in order to collect "coefficient of friction" data. Therefore, using samples which cling together not only when $P = 0$ but even if under slight tension, meaning negative values of P, while the adhesive junctions provide a finite value for F_t, permits obtaining values for μ between minus and plus infinity, including zero. Bikerman[5] has previously made this same point.

Cases like that of Fig. 16, in which μ is found constant within a wide range of P, even though adhesion is significant, may arise when the force of adhesion is proportional to P. While this situation is sometimes observed (e.g., Fig. 45 of Ref 62) it is by no means general.[58]

The converse phenomenon of the above, i.e., independence of measured coefficients of friction (in the $0.2 \pm .05$ range as in Fig. 17, and in the 0.4 ± 0.1 range in Ref 61), in the presence of drastic changes in the nature of the surface film, is also easily understood. As long as the films do not belong to categories 1 to 4 above, they are unlikely to have any significant influence on the coefficient of friction. At the same time, the fact that in both Ref 60 and 61 (and in Fig. 17) the coefficients of friction, although not subject to systematic changes, yet show substantial irregular variations is a further argument in favor of the earlier conclusion that the number of contact spots can be quite low. In both Ref 60 and 61 conditions were such as to cause considerable local heating, i.e., favoring a small number of contact spots. Further, in both cases local materials differences at the contact spots would have been erratic and not insignificant because composite materials were used. It is thus believed that the erratic behavior of the friction coefficient in Fig. 17 reflects local materials differences as the few acting contact spots begin, grow, and decay.

Some Thoughts on Geometrical Effects, Surface Stress/Strain Distribution, Wear, and Delamination

Repeated reference has been made to the elevated temperature that may exist at the contact spots, and to the resultant temperature gradients towards the interior. The analytical problems involved in determining the local temperatures on the two sides of the sliding contact are formidable and no detailed treatments are so far available. Similarly, only scattered measurements of contact spot temperatures are available. Much excellent work has indeed been done in this most vital area (e.g., Ref 2, 15, 33). Even so, much remains to be done.

Some aspects of temperature effects become especially evident when the coefficient of friction is considered in terms of the conversion of mechanical energy into heat rather than in terms of forces. That viewpoint has a distinguished history which was recently reviewed by Rigney and Hirth,[63] and a new treatment has recently been proposed by Hirth and Rigney.[51,63] These authors, as well as Dautzenberg,[18] emphasize that plastic deformation is the principal mechanism by which the observed conversion of mechanical energy into heat takes place during friction. They also emphasize that this conversion does not take place at the interface but mainly in the subsurface regions of the softer of the two materials. One may add that the less thermally conductive tends to be hotter than the more thermally conductive and tends to develop the hot spots, given similar hardnesses and an extended geometrical interface.[15,16] Besides, for geometrical reasons, in a pin and ring combination only the pin can develop hot spots, which is no doubt one important reason that the same materials combination can show different coefficients of friction and wear when pin and ring materials are interchanged.

These considerations are highly significant because of the general rule that hardness decreases with temperature. As a consequence the softer material (as judged from tests at ambient temperature) must be expected to wear rather faster when used as the pin than as the ring. Further, material transport, while primarily going from the softer to the harder as judged at ambient temperature, should have a superimposed tendency to go from the pin to the ring. Thus, while Fig. 1(C) leads to the conclusion that the softer material always transfers faster to the harder material than vice versa, it is also true that, on account of geometry and coefficients of thermal conductivity, and the differences in local temperature at the contact spots resulting therefrom, the roles may be inverted as judged from relative hardness at ambient temperature. According to Holm (Ref 12, footnote to p. 238), Bowden and Hughes were the first to emphasize that fact.

Hirth and Rigney have made a qualitative analysis of the coefficient of friction based on steady-state stress and strain distribution in the subsurface regions, primarily in the softer material.[51,63] The standpoint in the present paper differs from their treatment in focusing on cases in which only negligible recovery takes place and surface hardening continues to fracture. Consider, then, the resolved shear stress causing the surface strain, such as depicted in Fig. 4 and 5. It is due to the tangential tractions applied at the contact spots. Let us assume that it is adequately represented by

$$\tau(z) = \{z_o/(z + z_0)\}{}_s\tau_y, \tag{21}$$

[although the subsequent argument does not depend on the specific form of $\tau(z)$], where z_o is comparable to d, i.e., $\sqrt{A_b}$ of Eq 1. Here, as before, ${}_s\tau_y$ is the resolved shear stress at the surface. For the case of homogeneous materials, we shall specify its position, identified as $z = 0$, more precisely as the level of maximum strain rate. Thus $z = 0$ will be located at the level of the valleys between interacting hills making up the contact spot, and for the case of only one hill per contact spot $z = 0$ will be at the level of the valleys of the micro-roughness.

The resolved shear stress $\tau(z)$ must be in equilibrium with the local flow stress of the material which, to a first-order approximation at small strains, depends on shear strain γ and shear strain rate $\dot{\gamma}$ as, say,

$$\tau(\gamma, \dot{\gamma}) \simeq \tau_o + \theta_{II}\gamma + a\dot{\gamma} \tag{22}$$

where θ_{II} is the linear work hardening coefficient and a is a constant which typically decreases with temperature.

On the assumption that one may treat the different layers of z = constant as gliding and work hardening independently (vindicated by the results to be discussed later), one may therefore write

$$\tau(z) = {}_s\tau_y - \theta_{II}\{\gamma_s - \gamma(z)\} - a\{\dot{\gamma}_s - \dot{\gamma}(z)\} \tag{23}$$

where the subscript s refers to values at $z = 0$. Combining Eq 21 and 23 yields

$$\dot{\gamma}(z) = \dot{\gamma}_s - {}_s\tau_y\, z/\{a(z + z_o)\} - \theta_{II}\{\gamma(z) - \gamma_s\}/a. \tag{24}$$

Now, if steady state were reached, meaning that the strain rates at the various values of z were independent of time, the respective strains would be $\gamma = \dot{\gamma}t$ with t the elapsed time of testing, presumed to be large compared to the run-in period.

Correspondingly, in steady state Eq 24 would render the clearly time dependent expression

$$\dot{\gamma}(z) = \dot{\gamma}_s - {}_s\tau_y\, z/\{a(z + z_o)\} - \theta_{II}\{\dot{\gamma}(z) - \dot{\gamma}_s\}\, t/a, \quad [25]$$

an evident contradiction.

This paradoxical result points up the impossibility of reaching steady state in sliding friction as long as the local hardness depends on the cumulative local strain. This conclusion is actually quite independent of the specific forms of $\tau(z)$ and $\tau(\gamma,\dot{\gamma})$. In this light, it is evident that any steady-state equilibrium in sliding wear can only be dynamic in either of two ways, or a combination thereof. Namely, hardening can cease without fracture, or fracture can occur. The former can be realized only via continuous recovery or annealing to exactly balance the hardening resulting from the shear deformation. However, the rate of recovery and the hardness of the recovered state are functions of temperature. Practically no recovery takes place below some threshold temperature which decreases with hardness, and while the rate of recovery strongly rises with temperature, the resulting hardness in the annealed state drops only slowly with temperature and is largely independent of the initial hardness. It would seem that these relationships cannot yield steady state under a shear strain rate which monotonically increases towards the surface and is accompanied by a monotonically increasing temperature. Rather, if dynamic recovery were strong, it should give rise to cyclic hardening and annealing in the top layer, below which there would be a layer hardened to saturation under the prevailing conditions and no longer participating in the deformation. On average, therefore, maximum hardness would exist somewhere below the surface, and the top surface layer itself would be relatively soft. However, monotonic hardness gradients such as in Fig. 5 seem to be the rule rather than the exception. Besides, there is no strong effect of ambient temperature on the coefficient of friction (e.g., Fig. 18) as should be the case if dynamic recovery played a major role. Altogether, therefore, it is unlikely that steady state is due to dynamic recovery.

This, then, means that fracture must occur. Such fracture must take the form of delamination for the reason that, statistically, cracking will always occur where stress/strain conditions of fracture are first reached, which will always be at some more or less well defined value of z.

The Effect of Changes in Surface Hardness on the Coefficient of Friction

The temperatures at the interface of contact spots can be quite high; indeed, they can easily reach the melting point. Figure 19 gives a recent example due to Saka, Eleiche, and Suh.[64] In the preceding discussion, repeated reference was

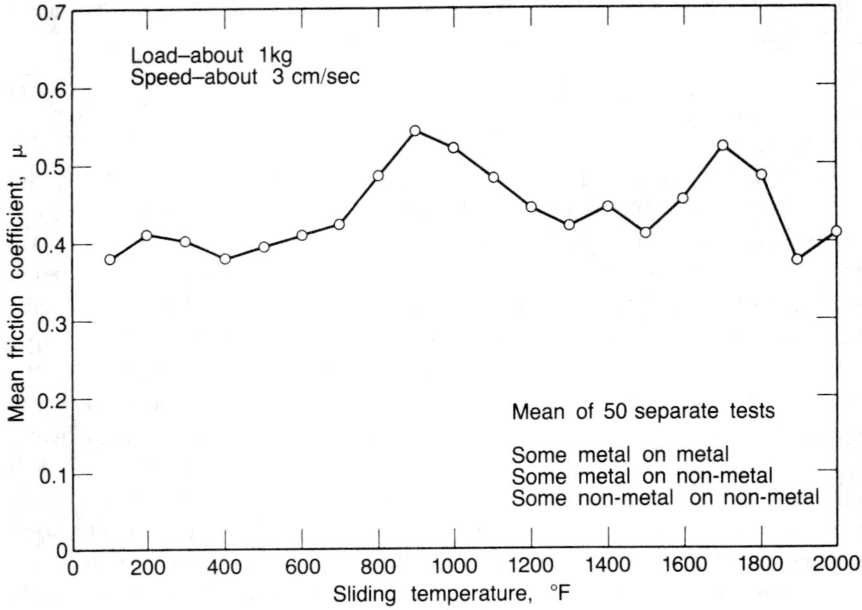

Fig. 18. The friction coefficient is insensitive to externally applied temperature changes, as was demonstrated by E. Rabinowicz with this figure representing measurements of over fifty widely varying materials combinations. *(Fig. 4.33, Ref 7)*

made to the mechanical effects of such temperature increases at the contact spots. Also, they can cause transformations, especially important in the case of steels (compare the paper on "Wear of Steels" by Vingsbo and Hogmark in this book). Correspondingly, as a function of depth below the surface, zones of different phases can be formed with their own characteristic microstructures and hardnesses, leading on occasion to very pronounced effects on wear rates.[65] However, in the present paper no further attention will be paid to this complication, since it is viewed as a deviation from the "basic" case as defined above.

A very illuminating study of the effect of temperature on the area of contact spots and the materials hardness derivable therefrom has recently been published by Tamai and Tsuchija.[66] Figures 20 and 21 present their results. The temperature variation was generated by passing current between the contacting materials while these were at rest and the contact spot was under direct observation, thus permitting measurements that would otherwise not have been possible.

The data in Fig. 20 and 21 are not surprising. The purpose of presenting them is to document that the softening expected to accompany temperature rises does indeed occur, and that it causes the corresponding increase in the load-bearing area A_b. It will hardly be doubted that in this connection it is immaterial that the temperature increase was due to Joule heat rather than frictional heat. However,

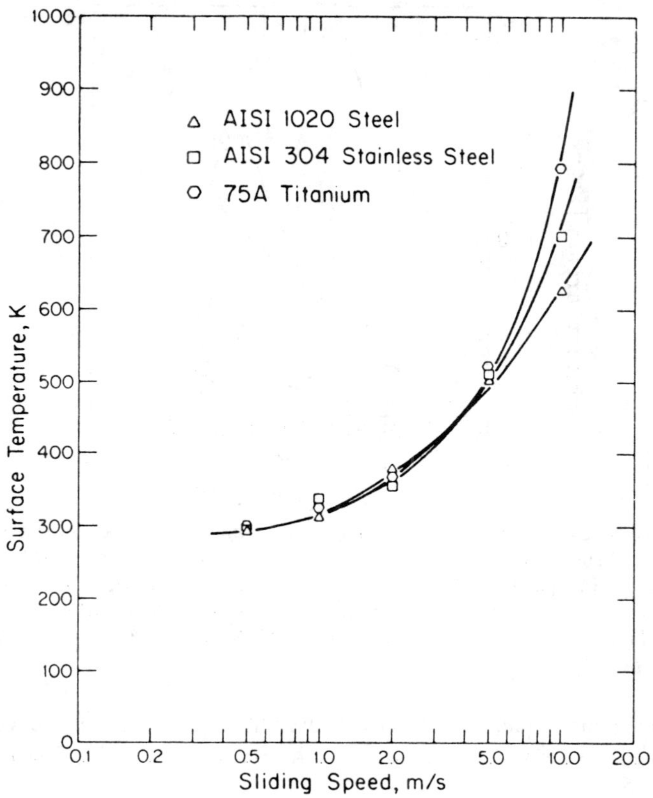

Fig. 19. Average temperatures at the interface between different materials as indicated and a 52100 AISI steel ring on which they slide under a 49 N load, as dependent on sliding speed. *(Courtesy of N. Saka, A. M. Eleiche and N. P. Suh, Ref 64)*

whether heating is due to electrical currents or friction, or both, there will be a strong temperature gradient away from the surface.

At significant values of peak temperature the gradient will not be as strong in the case of friction heat compared with Joule heat,[63] but it can certainly be very strong in both cases, e.g., presumably on the order of 5,000°C/mm in the case of Fig. 19. Correspondingly, there can be strong gradients in (momentary) flow stress. This will cause an additional concentration of the shear strain parallel to the sliding direction in the topmost surface layers beyond what would be expected from considerations such as those given in Eq 21 to 25, and yet further accentuated because greater strain rates in turn lead to greater heating. Therefore, it must be understood that ${}_s\tau_y$ and ${}_b\tau_y$ in Eq 20a refer to their values during sliding, which are not necessarily identical with those determined afterwards, such as in Fig. 5, especially not when the temperature differences between surface and

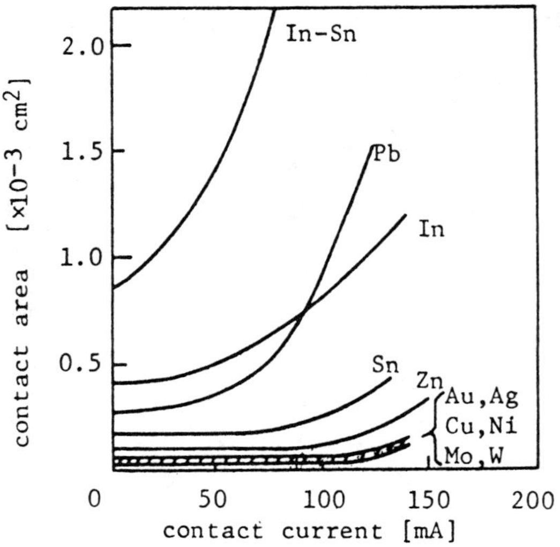

Fig. 20. Size of current-conducting contact spot (one "a-spot") formed at the polished end of a wire, pressed against a glass surface provided with a thin, transparent conducting layer, under a 100 g* load, as a function of current. With increasing current, the interfacial temperature is raised and the contact spot size increases correspondingly. *(After T. Tamai and K. Tsuchija, Ref 66)*

slightly lower subsurface regions are so large that significant differences of flow stress values are caused thereby. To emphasize this point, Eq 20a may be rewritten as

$$\mu = \frac{c\, m_b\, {}_s\tau_y(\gamma_s, \dot{\gamma}_s, T_s)}{2.8 m_s\, {}_b\tau_y(\gamma_b, \dot{\gamma}_b, T_b)} \simeq m_b \frac{{}_s\tau_y(\gamma_s, \dot{\gamma}_s, T_s)}{{}_b\tau_y(\gamma_b, (\gamma_b, \dot{\gamma}_b, T_b)}, \qquad [20c]$$

where, again, the subscript s refers to the layer at $z = 0$ as defined and b refers to the subsurface region at $z \simeq \sqrt{A_b}$. Equation 20c qualitatively includes the dependence of the coefficient of friction on sliding speed and also some of the changes which occur during run-in.

On account of Eq 20c, for "steady state" conditions, one expects a gradual decrease of friction coefficient with sliding speed, by a factor of two or so between very slow relative motion and a sliding speed just below that causing superficial melting, commensurate with the decrease of critical resolved shear stress between ambient temperature and several degrees below melting. The estimated factor of "two or so" is meant to be a rough guide, with actual figures hopefully determinable with the aid of Eq 20c. At still higher speeds, when surface melting occurs, the coefficient of friction will drop to quite low levels since then ${}_s\tau_y(\gamma_s, \dot{\gamma}_s, T_s) = 0$ and dynamic effects of the liquid metal take over.

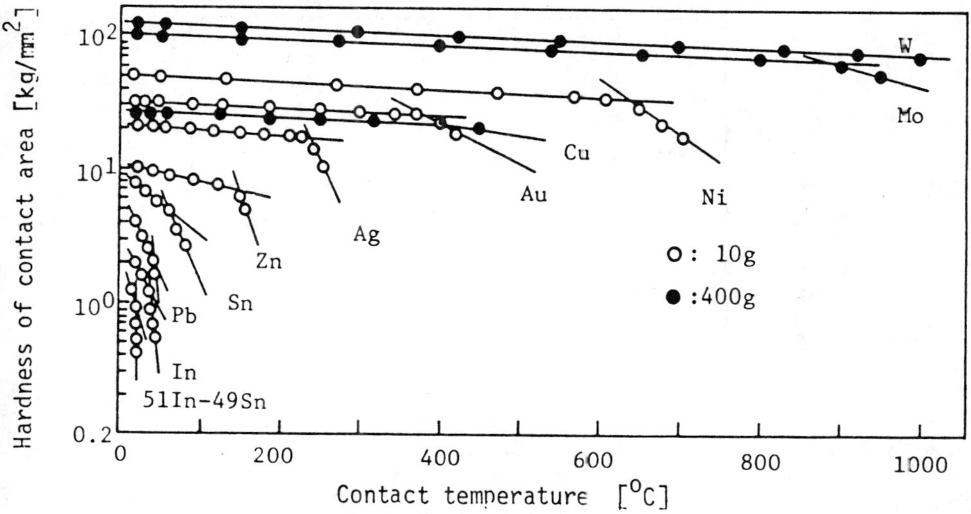

Fig. 21. Dependence of hardness of the wire materials in Fig. 20 derived from the data in that figure. *(After T. Tamai and K. Tsuchija, Ref 66)*

Available data are in qualitative accord with the above prediction. An example is given in Fig. 22 for the range in which the surface remains solid and in Fig. 23 for extremely high speeds (achieved by Beams' method, Ref 67) up to surface melting. Already Bowden and Tabor[1,3] and Rabinowicz[68] clearly realized that different effects of temperature on F_t and H (Eq 13) are responsible for the typical decline of the friction coefficient with sliding speed, and Rabinowicz pointed out explicitly that temperature changes per se were not the cause, giving Fig. 18 as proof.

One aspect of Eq 20a and 20c is the prediction of a rapid increase in μ due to work hardening in the surface, raising ${}_s\tau_y$ as compared to ${}_b\tau_y$ during run-in provided that (*a*) the softer of the two materials is initially in a condition to permit strong work hardening, and (*b*) the surface temperature does not quickly rise into the softening range. Now, the initial rise of ${}_s\tau_y$ is rather more rapid than the increase of m_s as a result of surface texture formation. Therefore, during run-in a rapid rise of μ could be followed by a mild decrease before equilibrium is established. It seems possible that at least some of the $\mu(t)$ curves during run-in reported by Feinle and Feller[69] record this phenomenon.

Wear Rate and "Wearing-In"

There is no universal relation between wear rates and coefficient of friction. Intuitively, one would expect that high friction coefficients are correlated with

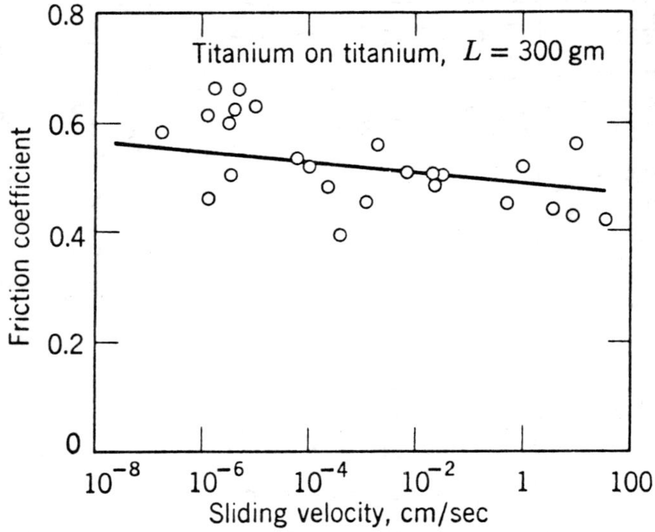

Fig. 22. For hard materials, the coefficient of friction declines slowly with speed, as in this example. *(After E. Rabinowicz, Fig. 4.12, Ref 7)*

high wear rates on the grounds that much energy is expended when friction is high. However, the derivation of the coefficient of friction leading to Eq 20a and 20c does not give any grounds for a correlation between wear rates and friction.

The question of wear has so far been touched upon only once, in connection with Eq 25, which showed up the fact that dynamic steady state not accompanied by fracture cannot be reached in sliding friction unless strain hardening ceases as a result of continuous annealing, which as indicated already is not expected to happen. Typically, therefore, hardening will proceed until fracture takes place, presumably within that zone parallel to the surface where fracture conditions are reached first. This, then, brings up the question of where fracture will take place.

To begin with, one should distinguish between the interface at which actual contact takes place between the two materials and the zone (mainly in the softer of the two) within which strong shear in the direction of sliding takes place. In recognition of that distinction, in the preceding considerations the origin of the z-axis (pointing downwards into the softer material as in Fig. 5) was placed at the average level of the valleys in the micro-roughness for an otherwise continuous contact spot, or at the level of the valleys between the interacting hills if the contact spot is composed of such. This was done because above those levels, on account of the constantly changing geometry, the slip planes are broken up into a multitude of disjointed and mutually misaligned regions, permitting no systematic glide. For the same reason, this zone cannot be the source of the typical flake-like wear debris that is so very generally observed. On the other hand, small

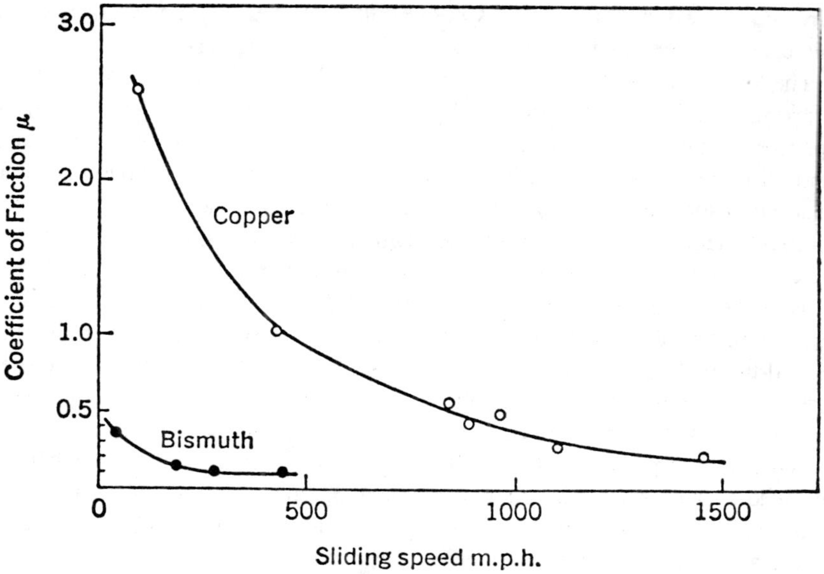

Fig. 23. Friction coefficient of a steel ball rubbing against copper and bismuth at extremely high speeds. Modest vacuum. With copper the friction falls to a relatively low value at sliding speeds of 1,500 m.p.h. because of the formation of a thin molten surface film of copper. With bismuth, where the melting point is much lower, the drop-off in friction occurs at a correspondingly lower speed. *(Figure and legend after F. P. Bowden and D. Tabor, Ref 4, Fig. 42)*

pieces of material projecting from either of the two materials into valleys (between hills and/or the micro-roughness) of the other, are doubtlessly sheared off in large numbers even after very slight relative movements, including those occurring coincidentally with normal loading and unloading without sliding.

Evidently, more particles of the softer should be sheared off and cling to the harder than vice versa, but both transfers should take place profusely in terms of numbers of particles, albeit involving only very small volumes. Also, once the two surfaces are macroscopically separated, and the configurations of the micro-roughnesses have changed in response to the load reduction, many of the sheared-off particles will simply fall off, or be attracted back to their own original side by adhesive forces. Further, pre-existing roughnesses will have much the same effect until obliterated.

It appears that all of these conclusions are in adequate agreement with observations. Specifically it is a well-known rule that preferentially the softer material moves onto the harder, while traces of the harder material can usually be found on the softer. For example, Moore[28] has noted that iron can be found on a copper surface over which hard steel has been sliding. However, the wear debris from this cause will be small, on the scale of the micro-roughness.

The larger wear particles which cause the bulk of wear are "delaminated" via cracking parallel to the surface.[6] Systematic crack formation in the zone of positive z-values (using our previous coordinate system) will typically require some time. Shear strain will have to accumulate until the deformability of the material in shear has been exhausted. Whether there is an additional requirement for delamination because shear stresses by themselves may not be sufficient cause for crack nucleation and propagation is not considered in the present paper. That question is even now the object of doubt and disagreement. Fleming and Suh[70] have emphasized the role of subsurface tensile stresses normal to the surface behind the moving contact spot, while Rosenfield[71] believes compressive/shear stresses to be much more important, and Jahanmir and Suh[21] have concentrated attention on inclusions as likely sites for crack nucleation.

In the present paper the stand will be taken that cracking will occur whenever and wherever a critical shear strain is reached. The value of that critical shear strain will depend on numerous parameters including perhaps the local normal stress, and it will require considerable further research before it can be theoretically predicted. The salient point is that it is not believed that crack nucleation sites are ever scarce, with and without inclusions. Not only can cracks start at grain boundaries,[72] but as was shown by Gardner, Pollock and Wilsdorf[73-75] for the case of tensile stress, and postulated by Rigney and Glaeser[51,55] for the case of wear, cell walls, or joints between them, provide crack nucleation sites.

With this, a few simple conclusions regarding the dependence of wear on hardness and surface coatings emerge. First, the shear strain is largely concentrated in the softer material so that delamination-type wear (which by present consensus tends to account for the bulk of wear, as indicated already) will be concentrated mainly in the softer partner, whence the rule that wear decreases sharply with increasing hardness. The same protective effect of hardness can be achieved by thin hard platings, especially if they inhibit the formation of micro-roughness so that in favorable cases true sliding may be expected. Soft platings, by contrast, are beneficial when they concentrate the shear in themselves and are of a kind either not to work harden at all, like indium, or to continuously anneal under the prevailing temperature at the surface.

For homogeneous samples, some simple quantitative relationships regarding wear rates and induction times derive directly from the model as follows: Fracture is expected to occur by cracking parallel to the surface, within the shear-stressed layer, of thickness $\sqrt{A_b} \simeq z_o$ according to Eq 21, once a critical local shear strain, γ_{crit}, is reached, thereby generating wear particles of thickness

$$t = \sqrt{A_b}/n = \sqrt{P/NH}/n. \qquad [26]$$

Here n is a number which depends on the position of the critically deformed layer below the interface, but is insensitive to changes in P. The critical shear strain will be some multiple of $\bar{\gamma}$, the average shear strain in the deformed zone, say, $\gamma_{crit} = \bar{\gamma}/f$ with f smaller than unity. In turn, the average shear strain is

$$\bar{\gamma} = (NA_b/A_t)S/\sqrt{A_b} = N\sqrt{A_b}S/A_t = (S/A_t)\sqrt{PN/H} \quad [27]$$

where A_t is the total, macroscopic area of contact and S is the distance of sliding. Equation 27 follows because the displacement S of the contact spots of total area NA_b is averaged over the whole area of macroscopic contact, whereas the thickness of the deformed zone is taken to be $\sqrt{A_b}$.* Once steady state is reached, the volume of wear particles produced when the deformed layer undergoes a shear deformation $\bar{\gamma}$ may thus be written

$$V_w = A_t t\bar{\gamma}/f\gamma_{crit} = \sqrt{PN/H}\, tS/f\gamma_{crit} \quad [28]$$

since the volume of wear particles will total $A_t t$ when $\bar{\gamma} = f\gamma_{crit}$. Combining Eq 26 and 28 we thus obtain

$$V_w = (PS/H)/(nf\gamma_{crit}) = ZPS/H. \quad [29]$$

The last formulation gives the empirical formula for wear rate in the form preferred by Holm (Ref 12, pp. 233 ff.).

Equation 29 refers to steady state, whereas there must be an incubation period during which V_w/S, on account of the formation of delaminated wear particles, rises from zero to its steady-state value. The sliding distance to reach steady state is that for which $\bar{\gamma} = \gamma_{crit}/f$, i.e., from Eq 27 and Eq 1

$$S_{inc} = A_t\sqrt{H/NP}\ \gamma_{crit}/f. \quad [30]$$

During the incubation period, smoothing of surface roughnesses with the associated shearing-off of particles goes on, leading to the overall transfer of softer material to the harder, which then comes to a dynamic equilibrium. Depending on the material loss due to this process the wear rate in the incubation period can be larger or smaller than in steady state.

It is evident that Eq 29 represents the well-known law of wear which says that the wear volume is proportional to the load as well as the sliding distance and is inversely proportional to the hardness. The wear coefficient deduced from Eq 26 to 29 is related to the formulations of Holm (Ref 32, p. 233, using Z), and of Rabinowicz (Ref 7, p. 138, using k) as

$$(nf\gamma_{crit})^{-1} = Z = k/3. \quad [31]$$

Experimental data yield a very wide range of values for Z. Holm (Ref 12, pp. 237-239) lists values for the softer contact (metal) member between

*As will be further shown below, the deformed zone is in fact significantly thinner than $\sqrt{A_b}$. In the present context the effect of this is to underestimate the parameter n. However, the lower limit of n in the present consideration was obtained from data in which the deformed layer thickness was insufficiently well specified. Therefore it seemed safer to proceed as above.

$Z = 2 \times 10^{-6}$ and 2×10^{-3} in air, and up to 2.7×10^{-2} for two like metals. Correspondingly, one must compare reasonable values of $(nf\gamma_{crit})$ with that range, i.e., one must check whether

$$30 \lesssim (nf\gamma_{crit}) \lesssim 5 \times 10^5. \qquad [32a]$$

For the particular cases of nickel, copper, and gold each sliding on itself, Soda, Kimura and Tanaka[76] found that the thickness of the wear fragments was proportional to the thickness of the plastically deformed surface layer in the range from 12 μm $\lesssim t \lesssim$ 43 μm with $n = 4$. This is most welcome support for Eq 26 and at the same time provides what may be presumed to be the lower limit of n. Namely, as has been known for a long time, and is also apparent from the quoted Z values, the heaviest wear occurs when two like materials slide on each other. Holm explains this phenomenon (Ref 12, p. 240) on the grounds that "... the contact area [is] prescribed by the strength of the softer member. But the harder member prescribes the structure of the surface." Therefore, when both members are the same, both deform and "the contact surface is in a labile state and will attain a wavy structure" (Holm, p. 240). Consequently, when like material slides on like, the gradient of shear deformation from the surface inward is lower, and the region in which systematic glide and fracture can occur begins at a relatively greater depth below the interface. For all other cases n is bound to be larger. Thus, for example, in the investigation by Hirst and Lancaster,[19] to which Fig. 5 pertains, the thickness of the wear particles was only about three micrometers, i.e., $n \simeq 40$. It is, of course, extremely unlikely that this should coincidentally also be the maximum value for n, and $n = 100$ would appear to be a very conservative value for the maximum.

The lower limit of f is about 0.2, and the value of γ_{crit} is typically comparable to γ_s, as will be shown presently. Therefore Eq 32a might be read to imply

$$40 \lesssim \gamma_{crit} \lesssim 5000. \qquad [32b]$$

These values seem to be entirely reasonable, seeing that Dautzenberg[18] and Dautzenberg and Zaat[77] determined shear strain values of $\gamma_s \simeq 120$ and $\gamma_s \simeq 700$, respectively, for copper sliding on steel. (Note that their "effective strain" $\bar{\delta}$ is $\gamma/\sqrt{3}$.) However, the limits of Eq 32b are too conservative. Some additional insights may be gained by the use of Fig. 24 by Dautzenberg and Zaat,[77] into which the interpolation curve

$$\gamma = \gamma_s \exp(-mz/z^*) \qquad [33]$$

has been entered with $\gamma_s = 70 \times \sqrt{3}$ and $m = 5$. Equation 33 is the type of interpolation suggested by Heilmann and Rigney.[53] In effect Eq 33 defines z^*, the thickness of the strongly sheared layer, as that distance at which $\gamma(z^*) = \gamma_s \exp(-m) \simeq \gamma_s/150$, in this case yielding $z^* = 100$ μm as indicated in

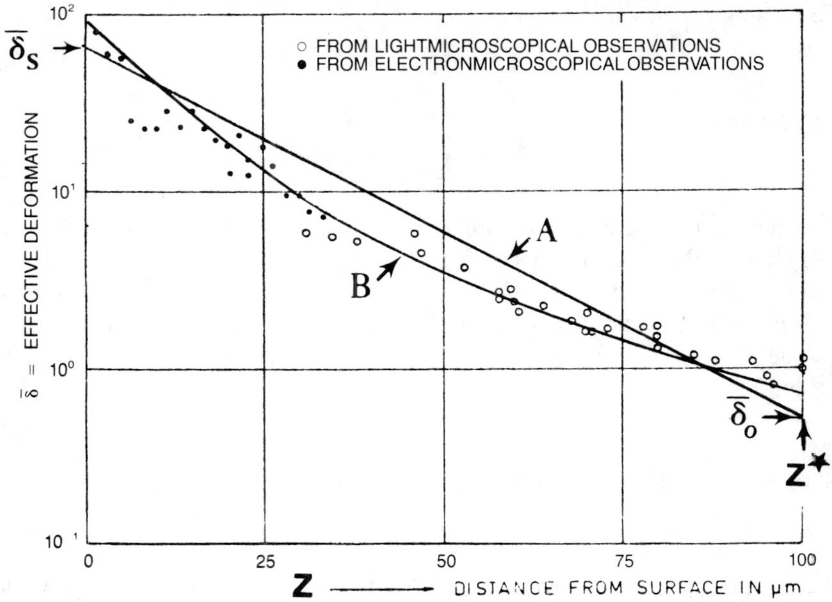

The interpolation curve A represents Eq 33 for the case of m = 5 with z* = 0.1 mm and $\bar{\delta}_s$ = 70, whence $\bar{\delta}_o = \bar{\delta}_s/150$. The interpolation curve marked B represents Eq 48a, again with m = 5 and with $\bar{\delta}_s$ = 100. Thus, curve B is the predicted strain dependence when the applied tangential stress drops off as $(1 + z/z_o)^{-1}$ and a logarithmic work hardening law prevails according to Eq 45b.

Fig. 24. Subsurface shear strain (in terms of "effective" strain $\bar{\delta} = \gamma/\sqrt{3}$) in an 8 mm diam polished OFHC copper pin slid on a normalized polished SAE 1045 steel ring at 2 m/sec under 4 kg* load. *(Diagram, excepting interpolation curves, due to J. H. Dautzenberg and J. H. Zaat, Ref 77)*

Fig. 24. Evidently the fit is acceptable, and from here on we shall continue to employ the same definition of z*, namely, via $\gamma_s/150$, whether or not the dependence of γ on z is exponential. This has the disadvantage that the shear strain $\gamma_o = \gamma(z^*)$ may not be insignificant, namely, $\gamma_o = 30$, for example, if m = 5 and $\gamma_s = 4500$, but it simplifies the subsequent considerations. Making a different choice for the thickness of the deformed layer is equivalent to changing the value of m, perhaps by as much as a factor of two in Fig. 24. However, it seems more logical to define z* in terms of $\gamma(z^*)/\gamma_s$ than in terms of some arbitrary value of strain and thereby making it also dependent on the value of γ_s.

From Eq 33 follows

$$\bar{\gamma} = \gamma_s/m = 0.2\gamma_s \qquad [34]$$

and

$$\gamma_{crit} = \gamma(z^*/n) = \gamma_s \exp(-m/n) = \bar{\gamma}/f. \qquad [35]$$

Combining Eq 34 and 35 renders

$$f = \exp(m/n)/m, \tag{36a}$$

i.e., for $m = 5$ and $n \geq 4$

$$0.2 \lesssim f \lesssim 0.7. \tag{36b}$$

While γ_{crit} is always smaller than γ_s, it is evidently not very much smaller, and will typically be comparable with it, e.g., for the data by Hirst and Lancaster,[19] $\gamma_{crit} = 0.88\gamma_s$ when $n = 40$. When n is lowest, i.e., if $n = 4$, $\gamma_{crit} = 0.29\gamma_s$. Therefore, and because γ_s is more accessible to measurements than γ_{crit}, we rewrite the wear coefficient Z in terms of the surface shear strain, γ_s, using Eq 31 and 35 as

$$Z = [f\,\gamma_s\, n \exp(-m/n)]^{-1}. \tag{37a}$$

With $m = 5$ and $n \geq 4$ one thus obtains theoretically

$$0.015 \lesssim \gamma_s Z \lesssim 4.4. \tag{37b}$$

Comparing this with the experimental limits for metals, extracted from Tables 41.07 and 41.09 in Ref 12 as quoted above, namely,

$$2 \times 10^{-6} \lesssim Z \lesssim 2.7 \times 10^{-2} \tag{37c}$$

therefore requires

$$0.89 \lesssim \gamma_s \lesssim 160 \tag{37d}$$

for $Z = 2.7 \times 10^{-2}$ and

$$7500 \lesssim \gamma_s \lesssim 2.2 \times 10^{6} \tag{37e}$$

for $Z = 2 \times 10^{-6}$ in steady state if the theory is to be acceptable. Similarly, from Eq 30, employing the same limiting values, one obtains

$$1 \lesssim S_{inc}/(A_t\sqrt{H/NP}) \lesssim 10^{6} \tag{38}$$

for delamination wear in metals.

The above relationships will permit testing the theory presented here. At this point little more can be done than to check general trends and order of magnitude values. In that spirit it may be pointed out that the theory does not provide for any connection between the coefficient of friction and delamination wear, for it is only delamination wear which is considered in Eq 26 to 38. Intuitively, one does expect a correlation between Z and the friction coefficient such that if one rises then so should the other, but experimentally there is no real evidence that such a correlation exists. In fact, recently Rabinowicz and Chan[78] found the reverse relationship between μ and Z, albeit with very much scatter, as shown in Fig. 25. Note that in this case the Z values are partly lying well below the lower limit indicated in Eq 37c. The reason is that Fig. 25 primarily concerns the wear of composite silver-graphite brushes, i.e., it involves graphite, a non-metal. It remains to be seen whether the wear also in that case can be treated by the above theory. In any event, one expects higher upper limits for γ_{crit} and n values in layer-type materials than in metals, and thus, optimally, lower Z values. This, then, is tentatively offered as an hypothesis to account for the excellent wear

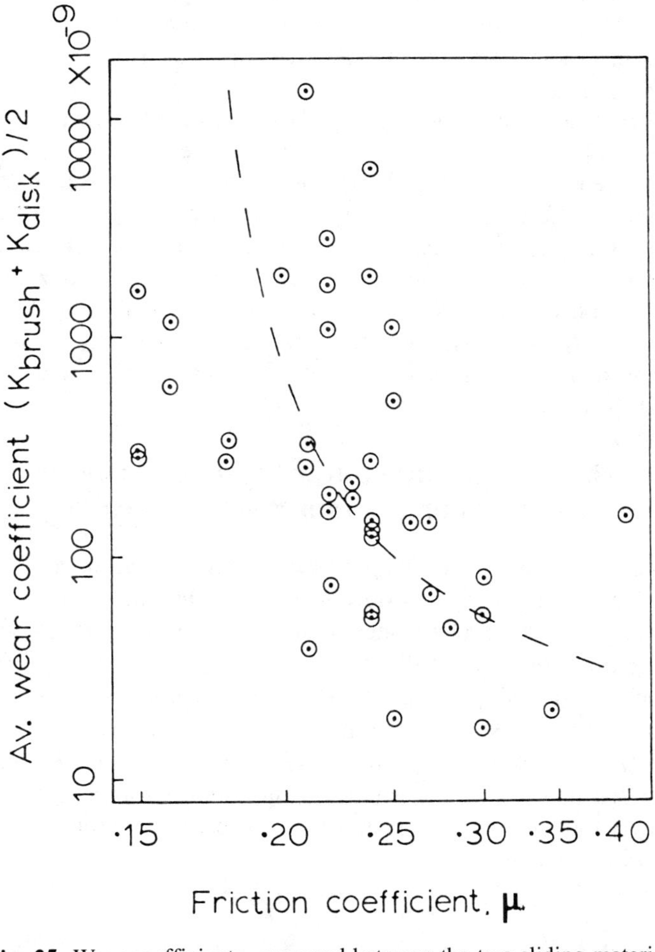

Fig. 25. Wear coefficients, averaged between the two sliding materials, namely a silver graphite (50 w/o Ag) brush and a disk made variously of Ag, Au, Cu, Pd, Pt, Ni, Rh, Re, Ru and Ir, plotted against the respective coefficients of friction as determined in different runs. *(After E. Rabinowicz and P. Chan, Ref 78)*

characteristics of layer-type materials, foremost among them graphite and molybdenite, to the point that they are often used as lubricants, as is well known.

One other relevant set of data concerns the variation of wear rates with sliding velocity. Hirst and Lancaster[19] have shown that, at least in the case they investigated, the sliding speed does not act directly on wear rate but via the concomitant rise in surface temperature. Figure 26, due to Hirst and Lancaster, shows the typical resultant dependence of wear rate of homogeneous metals (not subject to phase transformations) on sliding speed, including an initial significant decrease in wear rate followed by a rise. In terms of the present theory, the decrease is entirely to be expected because the strain to failure rises with increasing temperature, so that one should expect also γ_{crit} to increase with increasing temperature.

More complicated behavior, presumably responsible for the later increase in wear rates, will be expected when the surface temperature rises strongly. Even well below the point of surface melting one must expect chemical attack, principally oxidation for tests in the atmosphere, to raise the wear rate. Also, where freshly exposed surfaces touch, local welding will increasingly occur and increase wear. A further deviation from the simple theory outlined here occurs when phase transformations take place, a subject treated by Vingsbo and Hogmark in this book. At melting, finally, the wear rates could become high or low, depending on the opportunities for healing of wear damage by resolidifying material.

The Near-Surface Shear Strain Distribution As Determined From Work Hardening

Hirst and Lancaster[19] found that sliding velocity has no direct influence on strain distribution and hardening (see Fig. 5), and that the velocity dependence of wear rate is due to surface temperature changes. These results suggest that within wide limits and presumably for a wide range of metals strain rate effects per se can be neglected. If so, plastic shear deformation will continue until the local flow stress as raised by work hardening equals the applied stress, i.e., $\tau = {}_s\tau_y\, z_o/(z + z_o)$ according to Eq 21. That this is so is strongly indicated by Fig. 5, in that much the same hardness distribution is caused by the different velocities. It follows that for the most simple case, i.e., linear hardening as in Eq 22 with a negligible $a\dot{\gamma}$ term,

$$\tau_o + \theta_{II}\gamma(z) = {}_s\tau_y\, z_o/(z + z_o), \qquad [39]$$

i.e., that

$$\gamma(z) = \{z_o/(z + z_o)\}\, {}_s\tau_y/\theta_{II} - \tau_o/\theta_{II} \qquad [40]$$

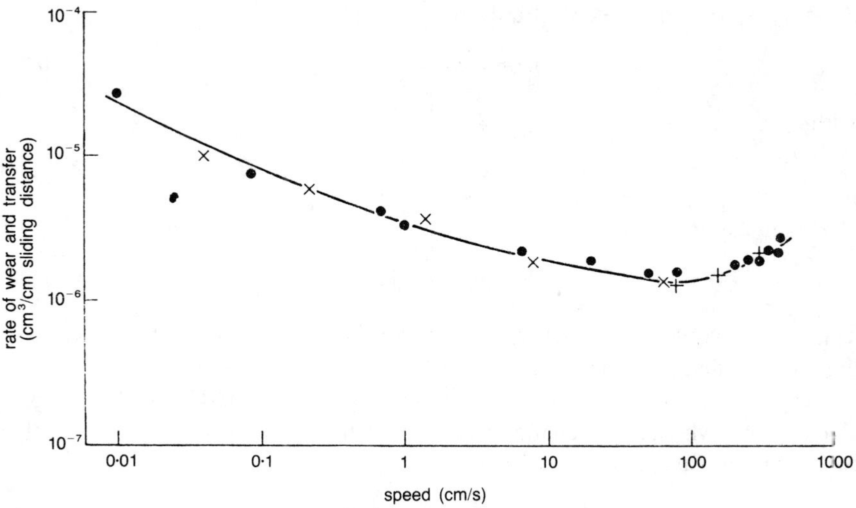

Fig. 26. Dependence of the wear rate on sliding speed for a 60/40 free cutting brass at 22.5 kg* load. A distinction is made here between "rate of wear" (•) and "transfer" (× and +) which may be neglected for the present purposes. This dependence of wear rate on speed, namely an initial drop followed by a rise, is common for non-transforming metals. *(Figure due to W. Hirst and J. K. Lancaster, Ref 19)*

and

$$\gamma_s = ({}_s\tau_y - \tau_o)/\theta_{II}, \qquad [41]$$

while $\gamma = 0$ for

$$z(\gamma = 0) \geqq \frac{{}_s\tau_y - \tau_o}{\tau_o} z_o \simeq \frac{{}_s\tau_y - \tau_o}{\tau_o} K\sqrt{A_b} \simeq z^* \qquad [42]$$

with, say, $0.1 \leq K \leq 10$.

In this case of linear hardening with finite τ_o, the assumption that z^*, the width of the deformed layer as introduced above in connection with Eq 33, is a measure of the size of the contact spot, i.e., $\sqrt{A_b}$, is evidently not true, except by occasional coincidence. Thus the distinction between z_o, a measure of the width of the layer subject to significant shear stress (roughly of size $\sqrt{A_b}$) defined by Eq 21, and z^*, the width of the strongly shear strained layer defined by Eq 33, is very real for linear hardening, in which case z^*/z_o strongly depends on the prevailing conditions.

Depending on the choice of θ_{II}, ${}_s\tau_y$ and τ_o, the shear strain distribution represented by Eq 40 can vary widely. Thus, for very low work hardening, i.e.,

${}_s\tau_y = \tau_o + \delta\tau$ with $\delta\tau \ll \tau_o$, Eq 40 reduces to

$$\gamma \simeq [1 - z/z^*]\delta\tau/\theta_{II} = \gamma_s[1 - z/z^*], \quad [42a]$$

with

$$z^* = (\delta\tau/\tau_o)z_o, \quad [42b]$$

i.e., with Eq 26,

$$n \simeq z_o/z^* = \tau_o/\delta\tau. \quad [42c]$$

This, then, would be the strain distribution for the ideally plastic material so widely considered in continuum theory.

For ${}_s\tau_y = 2\tau_o$, $z_o = z^*$, i.e., the deformed zone would have a width comparable to $\sqrt{A_b}$, which has been the assumption in the earlier qualitative parts of this paper. In that case, Eq 40 becomes

$$\gamma = (\tau_o/\theta_{II})/(1 + z/z^*). \quad [43]$$

Finally, for ${}_s\tau_y \gg \tau_o$ one obtains

$$\gamma = \frac{{}_s\tau_y/\theta_{II}}{1 + z/z_o} = \frac{\gamma_s}{1 + z/z_o} \quad [44a]$$

and, from $\gamma(z^*) = \gamma_s/150$,

$$z^* = 149\, z_o. \quad [44b]$$

It is perfectly well possible (although as we shall see presently it is not common) that linear work hardening occurs during wear. At any rate, linear work hardening does not explain the data in Fig. 24. Namely, when the function $\gamma(z)$, Eq 40, is plotted logarithmically as in Fig. 24, it is characterized by a very steep decline of shear strain near $z*$, the edge of the deformed zone. Also, the shear strains in the deformed zone are typically much larger than expected from ordinary linear work hardening.

To put the values found in [37d] and [37e] into perspective, it may be pointed out that in wire drawing, for slip planes oriented at 45° to the wire axis, the shear strain, γ, is related to the tensile strain, $\Delta\ell/\ell$, as $\gamma = \sqrt{2}\,\Delta\ell/\ell$, and to the "true" strain ϵ as

$$\epsilon = \ell n(\gamma/\sqrt{2}) \quad [45a]$$

Hence the limit $\gamma = 7500$, as in [37e], corresponds to $\epsilon = 8.5$ or less in wire drawing (less because in wire drawing textures the slip planes and directions do not have the ideal 45° orientations).

Such strains, while entirely unheard of in tensile tests, are commonplace in wire drawing, and are similarly expected to be realized in sliding wear. Namely, in wear even more than in wire drawing, compressive stresses are superimposed on the shear stresses and retard cracking. This would not be so if n, defined by Eq 26, were near to, or less than, say, 2, because the delamination-type cracking would then take place at low compressive stress or even under dilatation. However, as we saw, $n \geq 4$, *and* the strain is strongly concentrated near the surface.

Taking, then, the standpoint that worn surface layers closely resemble highly drawn material, we turn to the work by Langford and Cohen,[79] who investigated the flow stress and microstructure of iron wires as a function of true strains up to $\epsilon = 7.4$ without intermediate annealing, which for the <110> texture in their case corresponds to $\gamma \simeq 4000$. Note that even this large strain can be easily exceeded in iron, while pure copper and platinum can be drawn practically indefinitely without annealing.

Langford and Cohen[79] observed a work hardening law of the type

$$\tau = \tau_o + \theta'_{II}\,\ell n\,\gamma. \qquad [45b]$$

This law is approximated by Eq 22 for small changes in strain, provided strain rate effects can be neglected. It should indeed be quite prevalent, being expected whenever the work hardened state at any strain is essentially the same as the original state, i.e., whenever it is the potential beginning state for further work hardening, as is the case in so-called stage II behavior. Namely, stage II behavior goes hand in hand with a microstructure which shrinks in scale but remains substantially unchanged geometrically,[79-81] as is demonstrated in Fig. 27.

Combining Eq 45b with Eq 21 renders

$$\gamma = \exp(-\tau_o/\theta'_{II})\exp[({}_s\tau_y/\theta'_{II})/(1 + z/z_o)] \qquad [46a]$$

and

$$\gamma_s = \exp[({}_s\tau_y - \tau_o)/\theta'_{II}]. \qquad [46b]$$

By the use of the previous convention that

$$\gamma(z^*)/\gamma_s = e^{-m} \qquad [47]$$

where, specifically, $m = 5$ was chosen above, one may rewrite Eq 46a into

$$\gamma = \gamma_s \exp[-2m/(1 + z^*/z)] \qquad [48a]$$

with

$$z_o = z^* \qquad [48b]$$

provided that

$$m = {}_s\tau_y/2\theta'_{II}. \qquad [48c]$$

Equation 48 has been entered into Fig. 24 as curve B, choosing $m = 5$ as

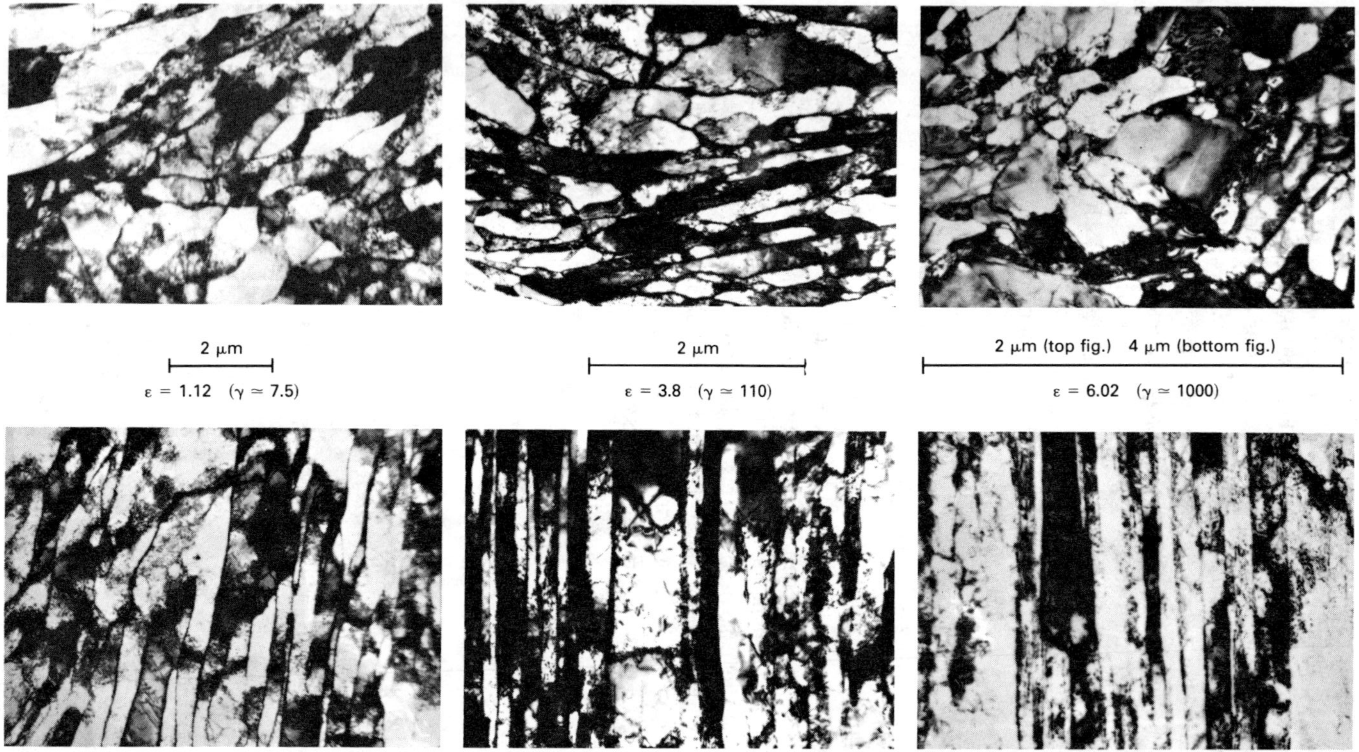

Fig. 27. Cross-sectional **(top row)** and longitudinal **(bottom row)** micrographs of the dislocation cell structure in drawn iron wire. The true strains (ϵ) and the shear strains (γ) to which these would correspond in the surface deformation under tracks are as indicated, using the specific wear texture in the case concerned. Note that, because of "similitude," the cells shrink in size without changing shape, yet no recrystallization took place. *(Courtesy of G. Langford and M. Cohen, Ref 79)*

before and $\bar{\delta}_s = 100$ (i.e., $\gamma_s = \sqrt{3}\ 100 = 173$). The agreement with the data is excellent, and it could still be bettered by a minor adjustment in the value of m. Therefore the only remaining question is whether ${}_s\tau_y/\theta'_{II} \simeq 2m = 10$ is a reasonable value and is compatible with the requirements of Eq 46b with Eq 48c, namely,

$$\ell n\,\gamma_s = ({}_s\tau_y - \tau_o)/\theta'_{II} = 2m(1 - \tau_o/{}_s\tau_y). \quad [49a]$$

With $\gamma_s = 173$, i.e., $\ell n\,\gamma_s = 5.15$, and with $m = 5$, Eq 49a yields

$$\tau_o/{}_s\tau_y = 1 - (\ell n\,\gamma_s)/2m = 0.485, \quad [49b]$$

i.e., ${}_s\tau_y/\tau_o = 2.1$. This result compares with the measurements by Langford and Cohen[79] in which at the true strain of $\epsilon \simeq 4.3$, to which γ_s would correspond in the <110> wire drawing texture, ${}_s\tau_y/\tau_o \simeq 3$ was found. It thus seems that the strain distribution of Fig. 24, and by inference of a wide range of pure metals under sliding wear, is in all respects consistent with a logarithmic work hardening law of the type of Eq 45b without noticeable strain rate dependence.

For the sake of completeness it may be added that Eq 33 used for the interpolation curve A in Fig. 24 in conjunction with Eq 21 corresponds to the work hardening law

$$\tau = \tau_o/[1 - (\text{const})\,\ell n(\gamma/\gamma_o)] \quad [50]$$

or more specifically for our case,

$$\tau = {}_s\tau_y/[1 - (z^*/mz_o)\ell n(\gamma/\gamma_s)]. \quad [51]$$

Such a work hardening law has never been proposed but, given skillful choice of the disposable parameters, will also permit reasonable approximations to actual work hardening curves—as may indeed be judged from curve A in Fig. 24 compared with curve B.

In summary of this section: For metals not subject to phase transformations, under a wide range of sliding conditions, flow stress gradients due to temperature and strain rate gradients can be neglected. Therefore the surface layer responds to the imposed shear stress (assumed to decrease as $z_o/(z + z_o)$ in accordance with Eq 21) with plastic shear until the local flow stress has reached the local value of the applied stress. Stress and strain are related via work hardening laws. The work hardening law which from the theoretical viewpoint is most likely to be widely applicable has been observed in wire drawing of iron[79] as well as aluminum alloys.[82] According to this law, the flow stress rises linearly with the logarithm of the strain. Available data on the shear strain distribution near a worn copper surface (Fig. 24) are completely explained with it.

It may be added that the logarithmic work hardening law is not the only one found in wire drawing, and in fact work softening is occasionally observed.[82] It would be most interesting to test the wear rates of materials that work soften in wire drawing. For these, one would expect an extreme concentration of shear strain at the surface (i.e., large n values in Eq 37a) in conjunction with very large surface shear strains, γ_s, and thus low wear coefficients, Z.

For the ideal plastic material, without effective work hardening, one expects γ to decrease linearly with z and to be extremely concentrated at the surface, meaning a very large value of n and correspondingly a very low wear rate according to Eq 37a. Also, all other work hardening laws will concentrate the shear strain near the surface, but not as extremely, with the correspondingly expected moderate to large values of n.

The often debated, suspected soft surface layer would appear to be impossible in the framework of the preceding analysis unless dynamic effects were very large, for which there is no evidence. However, all previous considerations concerned positive z-values, and the origin of our coordinate system, i.e., $z = 0$, was placed at the level of the valleys between cooperating hills, or of the micro-roughness, respectively. For negative z-values conditions are far more complex. When a contact spot is released, while the two sides part, the roughened zone at negative z-values, which has not undergone systematic shear deformation and thus will be softer, will be subject to grooving which causes strong local deformation and hardening. Yet, even after grooving, the top layer, of one micron or less, could still be softer than the layers below, and in that case would have the correspondingly lower dislocation density. However, if so, this softer layer is softer on account of a quite complex deformation history and not as the result of any particular dislocation mechanism.

The question of the suspected soft surface layer has been reviewed by Hirth and Rigney[51,56] from the standpoint of dislocation theory, and even more extensively by Nabarro.[83] Rather earlier, and somewhat less searchingly, the same question has been addressed by the writer,[84] and it was further illuminated by some remarkable experiments.[85,86] On balance, it seems clear that, on account of dislocation mechanisms, the top surface layer could be harder or softer, depending on detailed conditions, but that esoteric dislocation effects do not seem to play any appreciable role in modifying the hardness of surface layers in friction and wear.

Dislocation Behavior and Microstructures

The picture of regularity and simplicity of mechanical behavior during sliding wear that has been developed above for what we have termed the "basic" case leaves little room for surprises with regard to microstructures. For homogeneous metals, at least, one will expect the same kind of dislocation cell formation as in wire drawing, but with a texture much as in rolling.

The microstructure of drawn wire has previously been examined for the case of iron[79] (Fig. 27 gives examples) and for the case of aluminum,[82] and it has been interpreted theoretically with considerable success,[80] albeit still leaving some unanswered problems.[82,87]

The basic reasons for the formation of sub-boundaries have been given earlier; they involve energy reduction of the trapped dislocations which in the presence of more than one Burgers vector direction leads to the formation of dislocation cells. The cell interiors are almost free of dislocations and are rigidly rotated relative to the interiors of the neighboring cells.

"Similitude," as previously defined and illustrated in Fig. 27, causes stage II behavior. Included in stage II behavior are linear work hardening at small strains, and work hardening linear in log strain (i.e., linear from moment to moment with respect to the existing state) for large strains, as in Eq 45b. While similitude obtains, the dislocation cell size and mesh length in the cell walls are both nearly inversely proportional to the acting stress. This has indeed been found to be true in iron, in the work by Langford and Cohen,[79,80] but a more complicated, though generally similar, behavior was reported for aluminum alloys.[82,87]

The similarity between surface shear in wear, and the deformation in wire drawing, is sufficiently close that below wear tracks one expects to see dislocation cells resembling those in wire drawing, which follows stage II behavior, except perhaps in the layer at $z < 0$ in our coordinate system. Correspondingly, the dislocation cell diameter, D, should rise roughly linearly with z, i.e., distance from the surface, since the shear stress near the surface is $\tau \simeq {}_s\tau_y z_0/(z + z_0)$ while $D \propto 1/\tau$ in similitude, so that $D \propto (z + z_0)/z_0$ or $D = D_s + (\text{const})(z)$. Figure 28 due to L. K. Ives[8] shows an increase of D of just about this type but, regrettably, micrographs are as yet too few to establish this dependence definitely. Should similitude break down in the topmost layer, this could well give rise to a zone of constant minimum cell size there, commensurate with expected behavior in stage III.[81] Further, since Ψ, the angle of relative misorientation across cell walls, is approximately inversely proportional to the meshlength of the dislocations in the cell walls, one expects $\Psi \propto \tau$, i.e., $\Psi \propto z_0/(z + z_0)$, more or less. This, too, should be tested. The qualifier "more or less" comes about because of a minor deviation from simple proportionalities by the intrusion of a logarithmic term in the dislocation line energy.[80,81]

Dislocation cell walls are principally composed of mobile dislocations and are themselves mobile, always adjusting towards minimum energy configurations. Therefore it would be a mistake to expect dislocation cells to deform homologously with the shear of the surface material, and for the case of wire drawing Fig. 27 shows clearly that they do not; namely, the wire lengths in Fig. 27 are in the ratios of 1 : 14.6 : 134—i.e., a much wider spread than that of the cell lengths or cross sections. However, because of the awesome variety of possible configurations and our present inability to compute the energy of even some of the most simple ones to an accuracy sufficient for relative comparisons, no quantitative predictions can be made as to the size of the cells in the deformed surface zone.

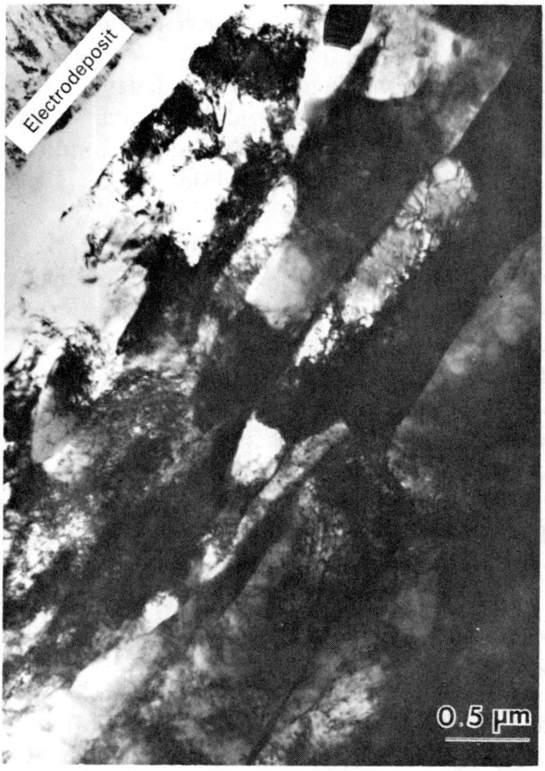

Fig. 28. Cell structure beneath wear track in copper tested under lubricated sliding conditions. Note the qualitative similarity with Fig. 27, taking into consideration the relatively higher magnification in this figure. According to the theory in the present paper, the cell diameter should rise linearly with distance from the surface down to the thickness of the sheared layer, of roughly 0.1 mm in extent. *(Courtesy of L.K. Ives, Ref 8)*

In drawn iron wire,[79] the cell diameters shrink in line with the rising stress (Eq 46) but, as evident from Fig. 27, less fast than the wire diameter, so that cell walls vanish continuously. It seems highly improbable that the eliminated cell walls simply dissolve, and equally improbable that, by and large, they leave through the free surfaces. Rather, they presumably annihilate pairwise, i.e., by the elimination of whole cells.[81] That effect is unlikely to be found in sliding wear, because the surface layer, though shearing, does not change its macroscopic area or thickness. Thus no cell walls need to be eliminated during cell refinement but cell walls must be formed instead. Again, that process is believed to consist of the pairwise formation of opposite cell walls.[81] Furthermore, during sliding wear the average cell diameter as a function of z will not change beyond run-in. Thereafter, the average cell length and shape is presumably maintained by the statistical pairwise formation of new cell walls, in whichever new areas become exposed as wear via delamination removes surface layers.

Going in the direction of smaller z-values, i.e., towards the surface, the cell walls will look more and more like grain boundaries, because as Ψ and the dislocation density in the walls increase, so the wall thickness, being roughly proportional to the dislocation spacing, shrinks. For this reason the cells may often look like, and be mistaken for, recrystallized grains.

Measurements of shear strain below the surface by analyzing crystallite shapes[18,77] are bedeviled by the discussed mobility of cell walls, the resulting impossibility of deducing the macroscopic shape change from them, and the difficulty of distinguishing cell walls from grain boundaries. The belief that "grain boundary" rearrangements are necessarily caused by recrystallization or dynamic recrystallization[18] is only too understandable and it is occasionally correct, but, in connection with cell walls, it is generally erroneous as indicated. Figure 27 demonstrates this clearly. From the dislocation cells one might gain the impression that the material is annealed, but the experimentally observed dependence of cell diameter on stress and strain,[79] as well as the theoretical interpretation,[80] show without doubt that there was no recrystallization, dynamic or otherwise. This does not mean that recrystallization cannot take place at all during sliding wear: Fig. 29 by Ives[8] proves that it can. In this case it is unmistakable on account of the presence of annealing twins.

Rigney, Hirth and Glaeser[51,55,56] have recently suggested that the dislocation cells below the wear track are roughly rectangular, with the largest face parallel to the surface, and the cells elongated in the direction of sliding. In addition to the arguments by Rigney et al., this is also a most reasonable possibility when viewed from the basis of a computational theory of dislocation cells[47,88] which came to the following conclusions: The minimum energy configuration attainable with three equivalent, coplanar Burgers vectors forming an equilateral triangle, such as on close-packed planes, is a set of rectangular cells whose relative misorientations are about a common axis and alternate left, right, left, right as if in a three-dimensional checkerboard pattern.

In sliding wear, the sliding direction, as well as the normals to the sliding surface and to the sliding direction, are preferred directions. For reasons of symmetry, either of the three could serve as the mentioned common axis of relative misorientation among the cells that should exist according to theory.[38] If it is the surface normal which so serves, there would be twist boundaries parallel to the surface, while the two sets of boundaries at right angles to the surface would be tilt walls. If, on the other hand, the axis of misorientation is the sliding direction, all cell walls containing the sliding direction (i.e., two-thirds of the walls, and the larger ones at that) would have tilt character, while only the walls normal to the sliding direction would be twist boundaries. Finally, if the common axis of misorientation is parallel to the surface and normal to the sliding direction,

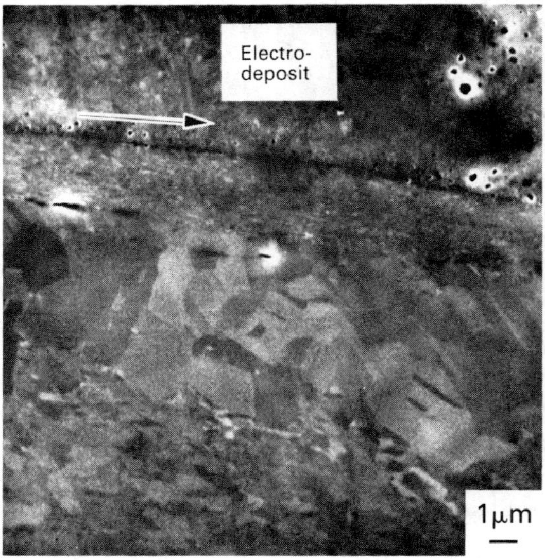

Recrystallized grains can be clearly recognized by the presence of annealing twins in them. Note also fissures, partly in the recrystallized grains and partly in the fine-grained material — which incidentally could also be recrystallized, a question which cannot be definitely decided without further information.

Fig. 29. Cross section through a wear track of copper tested under dry sliding condition, 1 kg* load, 38 mm/sec sliding speed, after 200 m of sliding. *(Courtesy of L. K. Ives, Ref 8)*

the cell walls parallel to the surface and those normal to the sliding direction would have tilt character, whereas the cell walls parallel to the sliding direction would be twist boundaries.

Hirth and Rigney[51,56] expect tilt walls at right angles to the sliding direction, composed of primary dislocations, which would imply a rotational axis normal to the sliding direction, parallel to the surface. However, tentatively, Ives[89] has identified the majority of cell walls as of tilt type for the case of copper. If so, the sliding direction is the common axis. At any rate, now that the different simple possibilities have been stated, it becomes readily possible to search for the answer in a systematic manner.

We may conclude this discussion with two tantalizing micrographs due to Lawless[90] (Fig. 30). These give an example of that ideal cell geometry which was predicted by Rigney, Hirth and Glaeser[51,55,56] and which is theoretically expected for the case of highly symmetrical conditions, including at least three coplanar Burgers vectors in a symmetrical orientation to surface and sliding direction.[88] These micrographs of the cell structure underneath a wear track were obtained from a 1000 Å thick (100)-oriented copper film that had been epitaxially grown on cleaved rock salt, and over which a very lightly loaded hemispherical sapphire was slid in the [011] direction.[90]

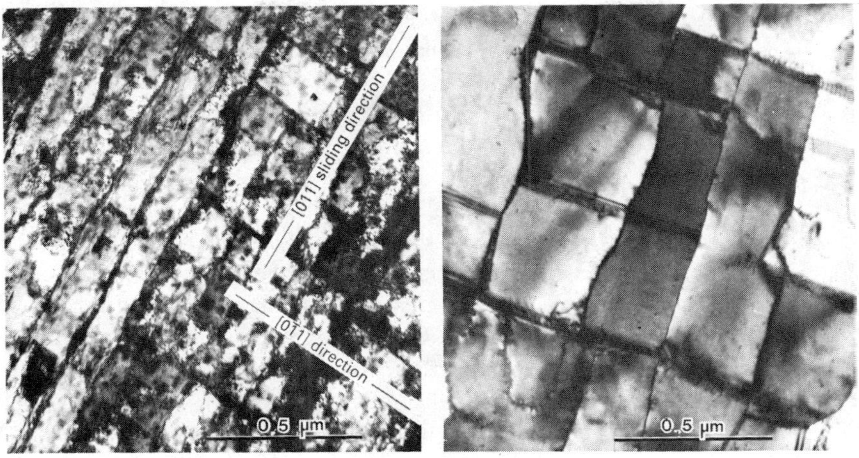

As deformed **After heating**

By geometry and orientation, the cells are exactly as had been predicted previously (Ref 51, 55, 56 and 88). However, it cannot be determined whether the cells have a common axis of misorientation, as predicted by theory (Ref 88) and, if so, which one. Part of the difficulty here arises because it is not entirely clear whether some of the features parallel to $[0\bar{1}1]$ in the micrograph at right are sets of dipolar edge dislocation pairs, or twin boundaries, or stacking faults. More experiments to clarify this point are needed.

Fig. 30. Dislocation cell structure under the wear track of a very lightly loaded sapphire ball slid over a 1000 Å thick, (100)-oriented copper film on cleaved rock salt. *(Courtesy of K. R. Lawless, Ref 90)*

Evidently, in this geometry, with ½<110> serving as Burgers vectors, the stipulated conditions for high symmetry are met, and evidently, too, the cell shapes are precisely those that were predicted. Yet it is not possible to decide from Fig. 30 which direction served as the common axis of misorientation for the cells that should be present according to theory, for the reason that not all of the short boundaries normal to the sliding direction in these micrographs are necessarily cell walls. Specifically, the short boundaries, which show no diffraction contrast change from one side to the other and are composed of just one set of lines parallel to the surface, are probably stacking faults. If the dark lines in these features were dislocations, they could only be dipolar walls composed of equal numbers of positive and negative primary (i.e., [011]) edge dislocations, and would thus not be associated with any relative rotation. Alternatively, these features could be twin boundaries.

The fact that another set of dislocations, steeply inclined to the surface, is appearing to move into those structures or to move towards extending them is an argument in favor of interpreting them as dipolar walls which are in the process of converting into tilt walls with the tilt axis parallel to [100]. Also, the cell walls parallel to [011] appear to be threaded with dislocations which are steeply inclined to the surface. It is thus possible that all of the walls either are, or are in the process of developing into, tilt walls with the rotation axis normal to the surface. This interpretation would be consistent with the lack of diffraction contrast in the left micrograph of Fig. 30 but inconsistent with the rather pronounced diffraction contrast across the middle cell in the right one. Perhaps, at the price of a further complication, that contrast could then be interpreted as due to an initial rotation about an axis parallel to the [011] sliding direction.

Alternatively, the common axis of misorientation could be parallel to the sliding direction. The lack of diffraction contrast in the left micrograph in that case would be explained as due to a too great deviation from the Bragg condition. Altogether this hypothesis of [011] as the rotation axis is in better agreement with the micrographs. The features across which there is no evidence of misorientation, if they are not stacking faults or twin boundaries, could in that case be interpreted as close pairs of walls which are in the process of moving apart, thereby generating new dislocation cells as explained before in connection with cell size refinement in wear layers.

Altogether, therefore, Fig. 30 provides tantalizing support for the theoretical considerations presented in this paper but falls short of providing proof. Thus more experimental work also in this area is highly desirable.

Summary and Outlook

The various ideas, predictions, and quantitative relationships in this paper provide much opportunity for experimental testing, which no doubt will lead to considerable refinements if not major modifications. As has been shown, all of the above are in adequate or better agreement with existing data, so that the theory is certainly promising. If it is then assumed that it is basically correct, one may be impressed with the underlying simplicity of the mechanisms involved in friction and wear. A small number of relatively restricted contact spots, usually representing only a very small fraction of the geometrical area, interlock from one side of the sliding couple to the other, to enforce momentary relative rest. This enforces sliding, which is concentrated in the top layer of the softer of the two materials concerned, or is shared equally among two like materials. At the edges of these spots, as well as while spots part, the interlocking micro-roughnesses, generated by inhomogeneities of local elastic and plastic deformation, plow a shallow grooving pattern into the wear track. The shear stresses, resulting from the tangential tractions applied at the contact spots, extend into the materials to a depth comparable to the contact spot diameters, which typically are in the 0.1 mm range, whereas the micro-roughness has only about 1% of that depth, or less.

The two materials respond to the applied tractions with shear deformations. In the majority of technologically interesting dry sliding situations, those shear deformations cease when the local flow stress, as raised by work hardening, equals the applied resolved shear stress on the active glide systems, wherein each layer parallel to the surface may be assumed to act independently of its adjoining layers, and in which, typically, strain rate effects are negligible.

During the initial period of "running-in," statistically the contact spot migrations generate a rather uniformly sheared and work hardened surface layer. Local shearing-off of asperities, which are caught in the "valleys" of the opposing surface contour, continues throughout sliding. It acts to preferentially transfer particles of the softer material to the harder, but not exclusively so. Transfer is initially stronger in the case of initially rough surfaces and is initially slower for the case of initially smooth surfaces. It should come to an equilibrium when the micro-roughness appropriate to the prevailing sliding conditions is established throughout. Shearing-off of asperities should then continue at some constant rate which, however, is slower than "delamination wear," which sets in gradually and reaches a constant rate after the running-in period.

"Delamination wear," consisting of the flaking-off of a surface layer of more or less constant thickness, in flakes whose area is of little particular significance for the process, must occur whenever the flow stress of the softer (or both) material(s) is a monotonically rising function of cumulative strain. If so, a limit must be reached beyond which the material cannot be deformed, and cracking will result. This cracking takes the form of "delamination" because of the essential uniformity of stress/strain history and mechanical properties of layers parallel to the surface which, however, act rather independently of the neighboring layers. The simple formulae which describe this process give a wear rate parameter and length of induction path for wearing-in which are in general harmony with observations.

Complications of the above basic processes arise because the flow stress of the surface material subject to shearing, and the flow stress of the bulk material which determines the total extent of the contact spots, are not the same and are subject to different effects with regard to crystallographic texture and temperature gradients such as arise in the course of sliding. However, the semi-quantitative predictions of coefficient of friction and its dependence on crystal structure, sliding direction, development during running-in, and sliding velocity are all in good accord with experimental observations.

Dislocation concepts are important in the understanding of the above derivations and conclusions, especially with regard to geometrical considerations. That the dislocations behave according to the local conditions of stress and strain, without the obvious intervention of any peculiar effects such as are otherwise known to occur at surfaces, is established to a high degree of confidence through comparing dislocation structures underneath wear tracks with those in drawn wires. The working conditions in the two cases are comparable, similarly high shear strains are attained, and the dislocation structures formed are also similar.

Since those in wire drawing have previously been successfully interpreted in terms of basic dislocation properties, it is considered probable that the structures under wear tracks are also in accord with theoretical expectations, at least in cases of symmetry sufficiently simple that predictions can be made. No such cases are at this point available with enough detail provided to permit a proof of this contention. However, in one specific case of sliding on an oriented copper film intriguing partial confirmation was obtained.

It is much to be hoped that the present theory will stimulate experimentation to test and expand it. If successful, the theory may become very useful in the prediction of coefficients of friction and wear rates. One particular application of the theory may be found in the opportunity it offers to improve wear behavior through purposefully selected crystallographic textures.

Acknowledgments

The financial support of this research by the Office of Naval Research (Materials Sciences Division, Arlington, VA) is gratefully acknowledged. Sincere thanks are due to colleagues who provided micrographs and/or stimulation and advice, foremost among them L. D. Dyer, J. P. Hirth, L. K. Ives, K. R. Lawless, E. Rabinowicz, D. A. Rigney, N. P. Suh and J. B. P. Williamson.

References

1. F. P. Bowden and D. Tabor, "Friction — An Introduction to Tribology" (Anchor Press/Doubleday, New York, 1973), pp. 16-24.
2. R. Holm, "Electric Contacts" (Springer, New York, first published 1958).
3. F. P. Bowden and D. Tabor, "Friction and Lubrication" (Methuen, London, 1967, first published 1956).
4. F. P. Bowden and D. Tabor, "Friction — An Introduction to Tribology" (Anchor Press/Doubleday, New York, 1973).
5. J. J. Bikerman, *Wear*, Vol. *39* (1976) 1.
6. N. P. Suh and co-workers, "The Delamination Theory of Wear" (Elsevier Sequoia, Lausanne, 1977).
7. E. Rabinowicz, "Friction and Wear of Materials" (John Wiley and Sons, New York, 1965).
8. L. K. Ives, "Microstructural Changes in Copper Due to Abrasive, Dry and Lubricated Wear," in "Wear of Materials — 1979," ed. K. C. Ludema, W. A. Glaeser and S. K. Rhee (ASME, New York, 1979), p. 246.
9. J. A. Greenwood and J. B. P. Williamson, Proc. Roy. Soc. (London) Vol. *295A* (1966) 300.
10. J. B. P. Williamson and R. T. Hunt, Proc. Roy. Soc. (London) Vol. *A327* (1972) 147.
11. J. Pullen and J. B. P. Williamson, Proc. Roy. Soc. (London) Vol. *A327* (1972) 159.
12. R. Holm "Electric Contacts" (Springer, New York, 1967), p. 34.
13. I. McNab, "Recent Advances in Electric Current Collection," *Wear,* Vol. *59* (1980) 259.
14. H. Kongsjorden, J. Kulsetås and J. Sletbak, "Electrical Contacts — 1978" (Proc. Twenty-fourth Holm Conference on Electrical Contacts, Chicago, IL), p. 477.

15. "Thermal Deformation in Frictionally Heated Systems," ed. R. A. Burton (Elsevier Sequoia, Lausanne/New York, 1980).
16. R. A. Burton, *Wear*, Vol. *59* (1980) 1.
17. J. P. Netzel, *Wear*, Vol. *59* (1980) 135.
18. J. H. Dautzenberg, *Wear*, Vol. *60* (1980) 401.
19. W. Hirst and J. K. Lancaster, Proc. Roy. Soc., Vol. *A259* (1961) 228.
20. R. M. Davies, Proc. Roy. Soc., *A197* (1949) 416.
21. S. Jahanmir and N. P. Suh, *Wear*, Vol. *44* (1977) 17.
22. L. D. Dyer, Trans. ASM Quarterly, Vol. *58* (1965) 620.
23. L. D. Dyer, Acta Met., Vol. *9* (1961) 928.
24. J. B. P. Williamson, private communication.
25. J. R. Whitehead, Proc. Roy. Soc., Vol. *A201* (1950) 109.
26. J. F. Archard, Proc. Roy. Soc., Vol. *A243* (1957) 190.
27. H. Hertz, cf. Ref 2, 4th ed., 1967, pp. 367-368.
28. A. J. W. Moore, Proc. Roy. Soc. (London) Vol. *A195* (1948) 231.
29. E. Holm, R. Holm and E. I. Shobert II, J. Appl. Phys., Vol. *20* (1949) 319.
30. D. Tabor, Proc. Roy. Soc. (London), Vol. *A192* (1948) 247.
31. D. Tabor, "The Hardness of Metals" (Clarendon Press, Oxford, 1951).
32. R. Holm, "Electric Contacts" (Springer, New York, 1967), pp. 373-375.
33. E. I. Shobert, "Carbon Brushes—The Physics and Chemistry of Sliding Contacts" (Chemical Publ. Co., New York, 1965), p. 24.
34. R. Holm, loc. cit., p. 132.
35. C. M. Adkins and D. Kuhlmann-Wilsdorf, "Electrical Contacts—1979" (Proc. Twenty-fifth Holm Conference on Electrical Contacts, Chicago, IL), p. 171.
36. C. M. Adkins and D. Kuhlmann-Wilsdorf, "Electrical Contacts—1980" (Proc. Twenty-sixth Holm Conference on Electrical Contacts, Chicago, IL), p. 67.
37. E. Rabinowicz, ASLE Trans., Vol. *14* (1971) 198, 206.
38. S. A. Barber and E. Rabinowicz, "Electrical Contacts—1980" (Proc. Twenty-sixth Holm Conference on Electrical Contacts, Chicago, IL), p. 33.
39. J. H. van der Merwe, Proc. Phys. Soc., Vol. *A62* (1949) 315.
40. J. H. van der Merwe, in "Single Crystal Films," ed. M. H. Francombe and H. Sato (Macmillan Co., N.Y., 1964), p. 139.
41. W. A. Jesser and D. Kuhlmann-Wilsdorf, Phys. Stat. Sol., Vol. *19* (1967) 95.
42. E. Schmid and W. Boas, "Plasticity of Crystals" (F. A. Hughes and Co., London, 1950), p. 58; first published in German, 1934.
43. N. Brown, Trans. AIME, Vol. *221* (1961) 236.
44. G. Y. Chin, in "Constitutive Equations in Plasticity" ed. A. S. Argon (M.I.T. Press, Cambridge, MA, 1975), p. 431.
45. R. E. Smallman, J. Inst. Met., Vol. *84* (1955) 10.
46. J. T. Moore and D. Kuhlmann-Wilsdorf, Surface Science, Vol. *31* (1972) 456.
47. M. N. Bassim and D. Kuhlmann-Wilsdorf, Crystal Lattice Defects, Vol. *4* (1973) 9.
48. D. Kuhlmann-Wilsdorf, J. Appl. Physics, Vol. *33* (1962) 648.
49. G. I. Taylor, J. Inst. Met., Vol. *62* (1938) 307.
50. U. F. Kocks, Acta Met., Vol. *6* (1958) 85.
51. J. P. Hirth and D. Rigney in "Dislocations in Solids," ed. F. R. N. Nabarro (North-Holland Publ. Co., Amsterdam), in press.
52. M. Antler, "Electrical Contacts—1980" (Proc. Twenty-sixth Holm Conference on Electrical Contacts, Chicago, IL), p. 1.
53. P. Heilmann and D. A. Rigney, "Sliding Friction of Metals" (Leeds-Lyon Symposium on Tribology, Leeds, England, Sept. 9-12, 1980).
54. E. Rabinowicz, "Friction and Wear of Materials" (John Wiley and Sons, N.Y., 1965), p. 67.

55. D. A. Rigney and W. A. Glaeser, *Wear*, Vol. *46* (1978) 241.
56. J. P. Hirth and D. A. Rigney, *Wear*, Vol. *39* (1976) 133.
57. J. R. Whitehead, Proc. Roy. Soc., Vol. *A201* (1950) 109.
58. K. L. Johnson and D. V. Keller, J. Appl. Phys., Vol. *38* (1967) 1896.
59. D. H. Buckley, J. Appl. Phys., Vol. *39* (1968) 4224.
60. S. Dillich and D. Kuhlmann-Wilsdorf, "Electrical Contacts — 1979" (Proc. Twenty-fifth Holm Conference on Electrical Contacts, Chicago, IL), p. 185.
61. P. K. Lee, "Electrical Contacts — 1979" (Proc. Twenty-fifth Holm Conference on Electrical Contacts, Chicago, IL), p. 133.
62. D. Tabor, "The Hardness of Metals" (Clarendon Press, Oxford, 1951).
63. D. A. Rigney and J. P. Hirth, *Wear*, Vol. *53* (1979) 345.
64. N. Saka, A. M. Eleiche and N. P. Suh, *Wear*, Vol. *44* (1977) 109.
65. K. L. Hsu, T. M. Ahn and D. A. Rigney, *Wear,* Vol. *60* (1980) 13.
66. T. Tamai and K. Tsuchija, "Electrical Contacts — 1978" (Proc. Twenty-fourth Holm Conference on Electrical Contacts, Chicago, IL), p. 469.
67. J. W. Beams, Physics Today, Vol. *12* (1959) 20.
68. Ref 7, p. 91.
69. P. Feinle and H. G. Feller, Z. Metallkunde, Vol. *68* (1977) 9.
70. J. R. Fleming and N. P. Suh, *Wear*, Vol. *44* (1977) 39.
71. A. R. Rosenfield, *Wear*, Vol. *61* (1980) 125.
72. P. Schwellinger, Z. Metallkunde, Vol. *71* (1980) 520.
73. R. N. Gardner, T. C. Pollock and H. G. F. Wilsdorf, Mat. Sci. Eng., Vol. *29* (1977) 169.
74. H. G. F. Wilsdorf, in "Strength of Metals and Alloys" (Proc. ICSMA5, Aachen, Germany, Aug. 1979, ed. P. Haasen, V. Gerold and G. Kostorz, Pergamon, N.Y., 1979), p. 169.
75. H. G. F. Wilsdorf, Kristall und Technik, Vol. *14* (1979) 1265.
76. N. Soda, Y. Kimura and A. Tanaka, *Wear,* Vol. *40* (1976) 23.
77. J. H. Dautzenberg and J. H. Zaat, *Wear*, Vol. *23* (1973) 9.
78. E. Rabinowicz and P. Chan, "Electrical Contacts — 1979" (Proc. Twenty-fifth Holm Conference on Electrical Contacts, Chicago, IL), p. 123.
79. G. Langford and M. Cohen, Trans. ASM, Vol. *62* (1969) 623.
80. D. Kuhlmann-Wilsdorf, Met. Trans., Vol. *1* (1970) 3173.
81. D. Kuhlmann-Wilsdorf, in "Work Hardening in Tension and Fatigue," ed. A. W. Thompson (AIME, Warrendale, PA, 1977), p. 1.
82. D. J. Lloyd and D. Kenny, Acta Met., Vol. *28* (1980) 639.
83. F. R. N. Nabarro, in "Surface Effects in Crystal Plasticity," ed. R. M. Latanision and J. T. Fourie (Noordhoff, Leyden, 1977), p. 49.
84. D. Kuhlmann-Wilsdorf, in "Environment-Sensitive Mechanical Behavior," ed. A. R. C. Westwood and N. S. Stoloff (Gordon and Breach, N.Y., 1966), p. 681.
85. G. E. Ruddle and H. G. F. Wilsdorf, Appl. Phys. Lett., Vol. *12* (1968) 271.
86. H. G. F. Wilsdorf and G. E. Ruddle, in "Surface Effects in Crystal Plasticity," ed. R. N. Latanision and J. T. Fourie (Noordhoff, Leyden, 1977), p. 565.
87. D. J. Lloyd, Metal Science, May 1980, p. 193.
88. M. N. Bassim and D. Kuhlmann-Wilsdorf, Phys. Stat. Sol. (a) Vol. *17* (1973) 281; Vol. *17* (1973) 379; Vol. *19* (1973) 335.
89. L. K. Ives, 1980, private communication.
90. K. R. Lawless, in "Interdisciplinary Approach to Friction and Wear," ed. P. M. Ku (NASA SP-181, 1968), p. 433.

The Role of Fatigue in Sliding Wear

YOSHITSUGU KIMURA
Institute of Interdisciplinary Research
Faculty of Engineering
The University of Tokyo
Tokyo 153, Japan

Abstract

Adhesive wear has been widely accepted as the type of wear most frequently encountered in practical sliding systems. However, with developments in theoretical research and analysis of wear particles, the concept of adhesive wear has been re-examined, and the possible role of fatigue has acquired increasing importance in recent work on wear processes.

Theories of wear are reviewed in which fatigue of surface asperities is mainly considered. The classical linear wear law is interpreted in terms of mechanics and materials science. Further, a substrate fatigue theory is introduced which provides a generalized treatment of wear and accounts for its complex behavior.

Experimental findings are then reviewed. Wear is caused by forces acting on the surfaces, and accumulation of damage is confined within a thin layer adjacent to the surfaces. Deterioration of the layer can appear in several ways: plastic deformation, increase of hardness, formation of cell structure, etc. Some mechanical interpretations are also given.

During the final stage of deterioration, cracks initiate, propagate, and eventually separate particles from the surface. Observation of surface cracks and wear particles and quantitative dependency of particles on sliding and lubricating conditions are discussed. In particular, results of close examination of worn surfaces are presented which show features similar to those of fatigue fracture.

Finally, quantitative correlation between wear and fatigue is discussed in lubricated and unlubricated cases, though published evidence is rather sparse. The poor correlation usually observed shows the necessity of analyzing the details of wear processes.

Introduction

More than thirty years ago, Ragnar Holm wrote in his textbook on electric contacts as follows: "Of course an extensive literature about wear exists, but it lacks fundamental principles, and the theory of wear has to be built up from the very beginning and without too great expectation."[1] Though thirty years is not a long period when we recall the history of tribology, which is said to compare almost with recorded history,[2] we must realize that much the same state of things exists in this field even at the present time. Accordingly, it should be considered reasonable that there is confusion in recognizing the possible roles of fatigue in wear.

For example, a lubrication handbook edited by the Japan Society of Lubrication Engineers lists four basic types of wear—adhesive wear, abrasive wear, corrosive wear and fatigue wear.[3] In this classification, fatigue wear implies the so-called rolling contact fatigue, namely, pitting, spalling or flaking. There is a problem, however, in terming these phenomena fatigue "wear." A glossary of terms in tribology published by OECD defined wear as "the progressive loss of substance from the operating surface of a body occurring as a result of relative motion at the surface."[4] Rolling contact fatigue does not satisfy the necessary condition of progressiveness and, strictly speaking, it should be excluded from wear.

The role of fatigue to be considered here is that in sliding wear. Of course this does not mean to exclude wear which accompanies the microslip at rolling contacts; this is known to show essentially similar characteristics to those observed in pure sliding systems if properly interpreted.[5]

Though it is often said that the number of classification systems of wear is not less than that of the researchers, it can be reasonably assumed that the basic types of sliding wear are abrasive, adhesive and corrosive wear. Abrasive and corrosive wear are rather clearly defined: the former is caused by cutting action with hard particles or hard protuberances, while chemical or electrochemical reaction with the environment predominates in the latter. However, the majority of wear problems we encounter in practical sliding systems are neither abrasive nor corrosive, and the vague term "adhesive wear" has been adopted to describe them, as illustrated in Fig. 1(a). Considerable effort has been devoted to evolving adhesion theories.

Adhesive wear can be defined as "wear by transference of material from one surface to another during relative motion, due to a process of solid-phase welding."[6] This concept, however, has been criticized in two aspects.

First, it has been argued that the adhesion theory fails to explain the presence of the loose wear particles as a logical consequence.[7] If the welding occurring at a junction is not strong enough to prevent separation along the original interface, neither metal transfer nor a loose particle results. On the other hand, if it is strong enough, there would be transfer but no loose wear particles.

The second criticism has come from practice, or observation of particles, to be exact. Analysis of wear particles is the most rapidly developing field in recent wear research. In particular, ferrography has succeeded in resolving several important types of wear in actual lubricated sliding systems, but, curiously enough, adhesive wear particles have seldom been reported. Instead, it has turned out that the most frequently observed particles appear as thin flakes of metal,[8] as

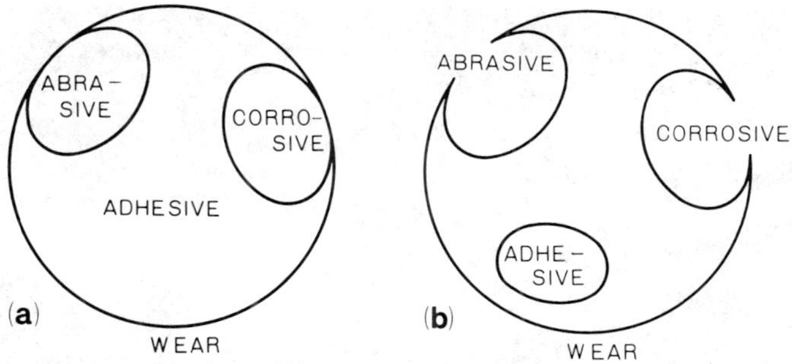

Fig. 1. Classification of wear by **(a)** vague and **(b)** strictly defined adhesive wear concepts.

exemplified by those shown in Fig. 2. These have been hypothesized to be caused by fatigue within a Beilby layer, and have been termed "rubbing wear" particles.[8]

Let us make the problem clear. It might be premature to deny the existence of adhesive wear, but as illustrated in Fig. 1(b), adhesive wear may be less common than is generally assumed. It is the remaining majority of wear cases, neither abrasive, nor corrosive, nor adhesive in the strict definition, for which the author thinks fatigue mechanisms are responsible.

Fatigue Theories of Wear

To relate fatigue to wear is not a new concept, but it had limited popularity among tribologists until recently. One of the reasons for this is in the shade of meaning of the term *fatigue*. This term used to be understood to mean a type of fracture caused by many loading cycles of stresses under the elastic limit of a given material. On the other hand, it was commonly recognized that interaction of mating surfaces takes place only at the real contacts, which form as the result of plastic deformation of asperities. Hence, not until low-cycle fatigue was considered has fatigue been regarded as a potential mechanism of the so-called adhesive wear of metals.

The adhesion theory of wear developed by Archard[9] assumes that the area of real contact is determined by a plasticity relation, that it consists of a number of circular contact points, and that hemispherical wear particles are formed at the contact points with a probability k. The amount of wear W under the load L during sliding distance S is then given by a simple linear law as

$$W = \frac{kL}{3p_m} S \,, \qquad [1]$$

where p_m is the flow pressure of the material. Because the microscopic quantities used for the deduction unfortunately do not appear in the resultant equation,

Fig. 2. Ferrogram showing "rubbing wear" particles. *(Courtesy of Research Laboratory, Mitsubishi Oil Company)*

the assumptions escaped re-examination, and the wear law was applied conveniently to cases in which linearity holds at least approximately.

The introduction of the fatigue concept started with interpreting the probability factor in terms of fatigue fracture. This can be illustrated after Rozeanu[10] as in Fig. 3, in which the states of an asperity are shown schematically. Figure 3(a) is a non-fatigue model. After an interaction between asperities, the probability k results in wear or transfer, and $(1-k)$ leaves no effect on the asperity. On the contrary, Fig. 3(b) shows a fatigue model by Rozeanu, where repeated interaction between asperities leads to their successive deterioration and to the final fracture. It should be noticed here that, after sufficient damage is accumulated, a weak interaction may be enough to break the weakened asperity resulting in detachment of a particle; thus one of the essential shortcomings of the adhesion theory was eliminated.

Admitting that wear is a result of fracture at microscopic asperities, some argue about the "necessity" of fatigue failure. They claim that wear can occur even in a single-pass experiment. This apparently simple comment implies some interesting problems. Many engineering wear problems arise from systems in which asperity interaction occurs repeatedly. The predominating mechanism operating in a practical wear process may or may not be common to that found in the case of a single-pass experiment. Therefore, if we focus our attention on problems of practical importance, such an exceptional case could be tentatively ignored. The problem is then reduced to a probabilistic consideration. If fatigue is simply considered as a process in which fracture occurs after repeated cycles of stresses or strains, it includes the fracture after two, three, . . . , to infinite cycles. On the other hand, non-fatigue fracture implies a limiting case in which fracture occurs

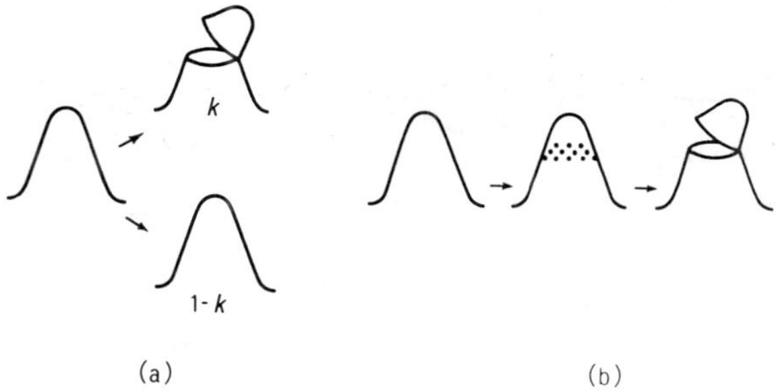

Fig. 3. Schematic illustration of **(a)** non-fatigue model and **(b)** fatigue model of wear. *(After Rozeanu, Ref 10)*

with a single application of forces. Probability theory tells us the former is far more probable, and what needs proof is the latter. Further, there is another problem about the nature of surfaces in a single-pass experiment which will be discussed in a later section.

A mathematical expression of wear was proposed by Kraghelsky for the case where elastic or plastic contact prevails.[11] He supposed "that wear results from stresses and deformation, and their values depend on the geometrical factors and adhesion forces." In the case of plastic contact, for example, penetration of conical asperities is assumed, and plastic deformation of each asperity is calculated from the coefficient of friction, the penetration depth and some other parameters. The deformation is then related to the number of cycles to failure, the reciprocal of which is assumed to be the probability factor, through the well-known Manson-Coffin low-cycle fatigue relation. The penetration was calculated on the basis of a complex contact model peculiar to U.S.S.R. works, but the concept itself is rather simple.

The asperity fatigue theory was further refined by Halling[12] and then by Finkin.[13] Their ideas will be briefly reviewed in order.

An important contribution made by Halling is the introduction of Greenwood and Williamson's probabilistic treatment of asperity contact.[14] As shown in Fig. 4, it is assumed that both surfaces are covered with asperities with spherical summits of radius β. One surface is rigid and its asperity peaks are in a plane; the other is elasto-plastic and the height of its asperities, z, follows a distribution function, typically Gaussian. Deformation and fracture are considered for the latter surface only, and are approximated by those encountered in normal approach of the surfaces. This model is not necessarily realistic, but it should be realized that it was based on a sound foundation of mechanics, and thus Halling showed a novel interpretation of the linear wear law.

Finkin appreciated the importance of this approach, and tried to extend and generalize it by eliminating some of the simplifying assumptions.[13] Halling had

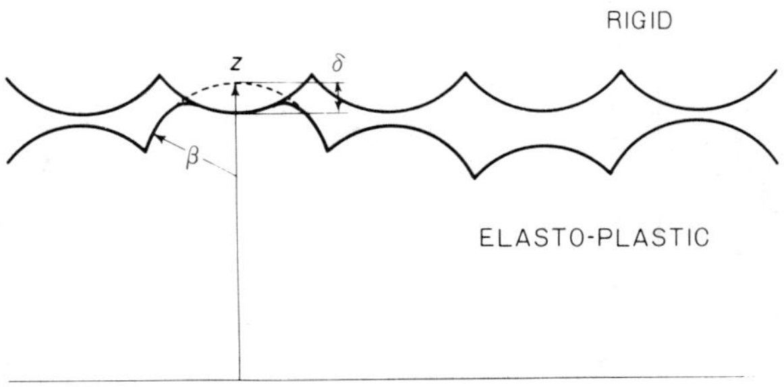

Fig. 4. Greenwood and Williamson's model of contact between nominally flat surfaces. *(Ref 14)*

assumed that all asperity interactions, elastic or plastic, contribute to wear through a strain ratio fatigue failure law. It would be more reasonable, however, to assume that only those interactions the strain of which exceeds a critical minimum value contribute to wear particle formation, whereas all asperity contacts support load, elastically or plastically. With some other minor modification Finkin derived an expression as follows:

$$W = \frac{LS}{K}\frac{\rho v}{\bar{\epsilon}_1^m}\left(\frac{K}{\pi Bc}\right)^{m/n}\frac{\lambda^{m/2-1-n/2}}{\beta^{m/2+1-n/2}}\frac{\Gamma(1+m/2,\ \epsilon_p^2/\varphi^2)}{\Gamma(2+n/2)}, \qquad [2]$$

where W is the amount of wear in volume under the load L during the sliding distance S as in Eq 1, K is the product πBcD^n in which B is the stress/strain ratio, c is the hardness/yield stress ratio, D is a strain/geometry constant and n is the index describing work hardening. Further, ρ is the asperity density, v is the mean volume of a single wear particle, $\bar{\epsilon}_1$ is the strain to failure in one loading cycle, m is the low-cycle fatigue constant, being 2 typically, λ is a constant for normal approach, being 1 for elastic and 2 for plastic cases, β is the radius of the spherical summit of each asperity, ϵ_p is the strain for beginning plastic deformation, and φ is the proportional constant relating the maximum strain in an asperity with the square root of normal deformation, δ in Fig. 4.[14]

Equation 2 is obtained for the case of interaction between asperities, or contacts like those of Archard. Finkin also generalized his theory to include the case of the failure of a hill moving against a plane and the failure occurring within the plane under the moving hill. He showed that a wear equation like Eq 2 could be derived for both cases.

Of course, the author does not intend these equations to be understood at a glance, but to show that they inevitably become complex even though a number of simplifying assumptions are made. Seemingly, these would provide a limiting case one can reasonably achieve by trying to give a fundamental significance to a single wear equation like the linear law.

An entirely different approach to wear was undertaken by IBM investigators. They proposed an "engineering model for wear" based on the experimental results obtained in a Bowden-Leben type reciprocating ball/flat machine.[15] As the experiment was conducted generally in the elastic contact stress range, their theory has nothing to do with the microscopic contact mechanics between asperities, and aims at relating wear to the Hertzian stresses.

They assumed the maximum shear stress criterion for fracture, and considered the maximum value of the shear stress, τ_{max}, which the substrate material experiences as the quantity representing the severity of the contact. It was then proposed that for a given number of passes no appreciable wear, or significant change in surface topography, occurs if

$$\tau_{max} < \gamma\tau_y, \quad [3]$$

where τ_y is the critical shear stress for yielding and γ is a constant, which depends upon the materials, the lubricating conditions and the number of passes.

An interesting feature of the theory in connection with fatigue is the last dependency. Here, Lundberg and Palmgren's ninth power law[16] is hypothesized, which was originally obtained for the fatigue life of SKF ball bearings. When the values of γ are written as γ_1 and γ_2 for the numbers of passes N_1 and N_2, the relation becomes

$$\gamma_1/\gamma_2 = (N_1/N_2)^{-1/9}. \quad [4]$$

Substrate Fatigue Theory

The fatigue theories of wear described above have opened up the possibility of discussing so-called adhesive wear in sensible terms; hopefully the terms can be common to those in materials science. It should be realized, however, that the available theories are still too simple to fully explain the complicated behavior of wear.

One important but poorly understood phenomenon is the nonlinear dependency of wear amount on sliding distance. The well-known break-in phenomenon and the presence of initial and steady states of wear provide familiar examples. As is usual with tribological phenomena there are a number of factors which have potential influence on this behavior, but the fatigue process itself can cause the nonlinearity in some cases.

In connection with this problem there are at least two important points which have been ignored in the theories introduced in the preceding section. One is the initial state of the sliding surfaces. Usually these surfaces are prepared by machining, grinding or other machine work. Such preparation procedures are essentially sliding processes, and damage is expected to accumulate in the subsurface layer during them. That is, the process of wear seldom starts with the material at the surface having the structure and properties of the bulk. Rather, it is likely to be deformed differently and may be in a somewhat fatigued state.

A similar situation prevails in the steady state of wear. As will be discussed in the following section in some detail, a deteriorated layer of a certain thickness is usually observed near the wear surfaces. This means that the damage done during sliding can extend into the substrate far deeper than the thickness of a single wear particle. Consequently, when a layer of particles is removed, some damage has already accumulated in the surface which is newly exposed.

Thus it is insufficient to describe the career of a single particle; we must also consider the history of the surface. A mathematical model based on this idea[17] will be introduced below.

A cross section of a rubbing surface is idealized in Fig. 5. Material is regarded as an aggregation of rectangular blocks with area δA and height δh, and wear is assumed to proceed when these blocks detach one by one. The size of these blocks should be determined by mechanical conditions and material properties, but tentatively it is assumed to be constant. The coordinate Oz is defined as shown with the origin O at the initial surface. Progressive wear is expressed by the change of the position of the surface, h, with sliding distance, which is obtained from calculation with the "sample column."

It is assumed that the amount of damage, E_s, caused on the surface of the sample column during sliding over a unit distance is a random variable having a distribution function $W(E_s)$, and that when the surface suffers damage of E_s the material at z also receives damage which amounts to

$$E(z) = E_s \epsilon (z - h), \qquad [5]$$

where h is the position of the surface at the moment.

As shown in Fig. 6(a), the total amount of damage the material at z can withstand, $E_f(z)$, may be lowered near the initial surface by its preparation; the bulk value, $E_{f\infty}$, is approached as z tends to infinity.

Now the probability p_1 of the detachment of a block during sliding over the first unit distance is assumed to be given as the probability that E_s exceeds $E_f(0)$,

$$p_1 = \int_{E_f(0)}^{\infty} dW(E_s). \qquad [6]$$

The resulting wear, that is, the expected position of the surface, h_1, after that sliding becomes

$$h_1 = p_1 \delta h. \qquad [7]$$

When it is further assumed that the damage is accumulated only when the detachment does not take place, the residual damage in the same sample column at this moment, $E_1(z)$, is given by

$$E_1(z) = \int_{E_f(0)}^{\infty} 0 \times dW(E_s) + \int_{0}^{E_f(0)} E_s \epsilon(z) dW(E_s). \qquad [8]$$

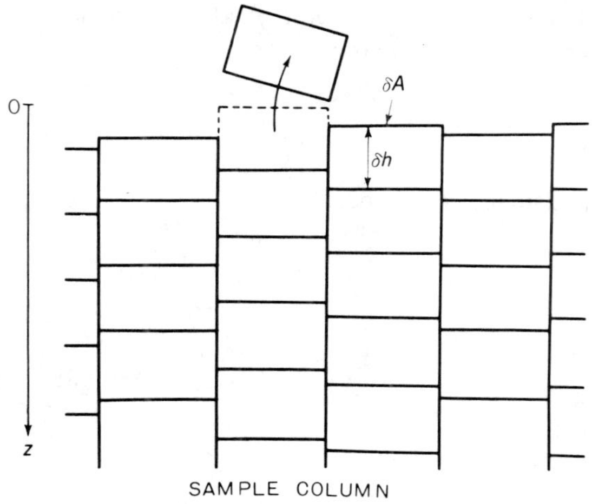

Fig. 5. Cross section of an idealized rubbing surface.

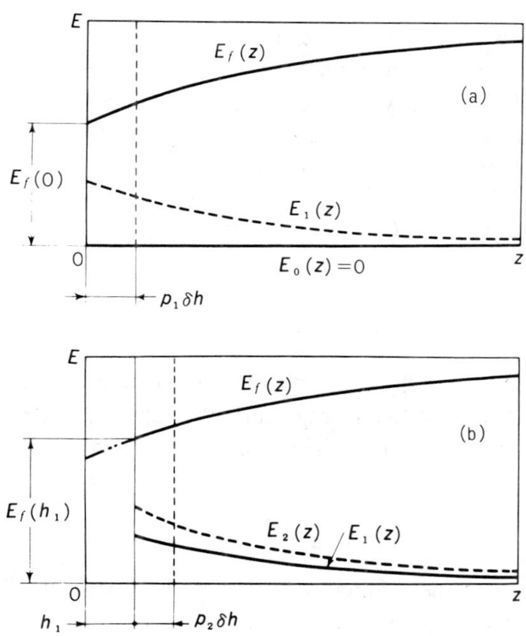

Fig. 6. Accumulation of damage in the substrate and the change of expected position of the surface: **(a)** before sliding and **(b)** after sliding over the first unit distance.

In reference to Fig. 6(b), the probability p_1 of the detachment in the second unit distance is given by the same reasoning as

$$p_2 = \int_{E_f(h_1) - E_1(h_1)}^{\infty} dW(E_s), \qquad [9]$$

and the residual damage, $E_2(z)$, becomes

$$E_2(z) = \int_{E_f(h_1) - E_1(h_1)}^{\infty} \{0 + E_1(z)\} dW(E_s) + \int_0^{E_f(h_1) - E_1(h_1)} \{E_s \epsilon(z - h_1) + E_1(z)\} dW(E_s). \qquad [10]$$

The position of the surface is given by

$$h_2 = \sum_{i=1}^{2} (p_i \delta h) = h_1 + p_2 \delta h. \qquad [11]$$

By repeating this procedure, the quantities after sliding over the nth unit distance can be determined as recurrence formulae, by which progressive wear behavior is described.

In order to show the quantitative examples, some numerical calculation will be given by assuming appropriate functions for $E_f(z)$, $W(E_s)$ and $\epsilon(z - h)$. All these functions are defined only for non-negative values of the parameters and approach zero as the parameters tend to infinity. As a first approximation, the following exponential forms are assumed:

$$E_f(z) = E_{f\infty}(1 - \alpha e^{-z/h_f}), \qquad [12]$$

$$W(E_s) = 1 - e^{-E_s/\beta E_{f\infty}}, \qquad [13]$$

$$\epsilon(z - h) = e^{-(z - h)/h_s}, \qquad [14]$$

where α, β, h_f and h_s are the constants. The initial damage is governed by α and h_f, β denotes the ratio of the average surface damage of the unit sliding to $E_{f\infty}$, and h_s indicates the depth which the given surface damage influences. Then p_n is written as

$$p_n = \exp[-\{E_f(h_{n-1}) - E_{n-1}(h_{n-1})\}/\beta E_{f\infty}]. \qquad [15]$$

Typical results of the parametric study are given in Fig. 7 to 10. Figures 7 to 9 show changes of p_n with n when α, h_f or β is varied, which correspond to the changes in tangent of the curves in wear versus sliding distance plot, or the changes of wear rate. Various wear-distance relations are obtained from the combinations of these parameters, though the assumed values are rather arbitrary.

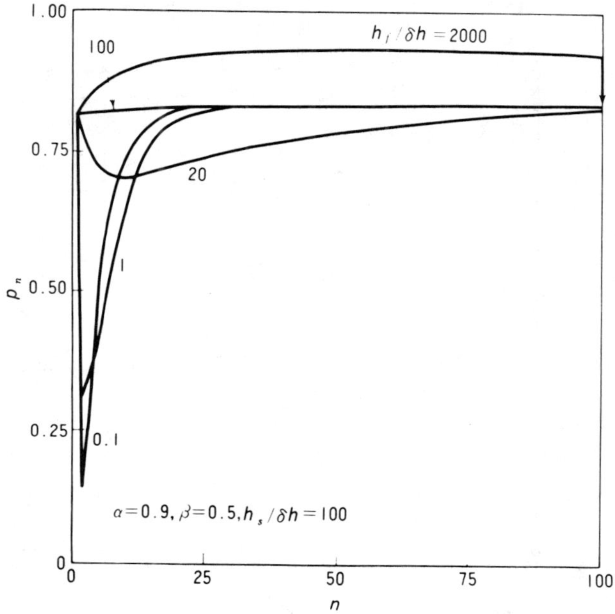

Fig. 7. Changes of wear rate with sliding distance: effect of h_f on p_n.

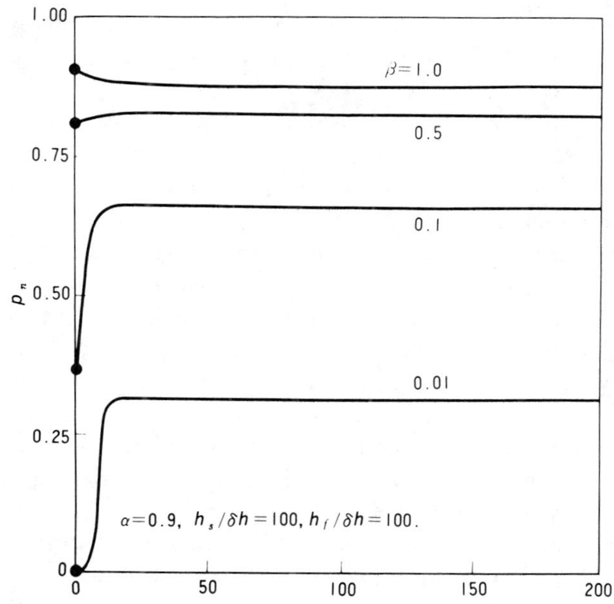

Fig. 8. Changes of wear rate with sliding distance: effect of β on p_n.

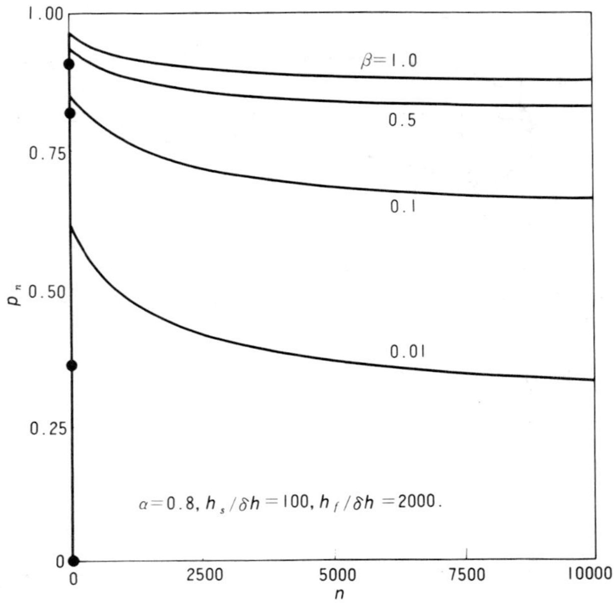

Fig. 9. Changes of wear rate with sliding distance: effect of β on p_n (further examples).

This means that the interaction of the damage caused by sliding with the initial damage can result in nonlinear relations even though there are no changes in the sliding conditions.

Figure 10 shows the probability p_n when n tends to infinity, which is related to the steady wear rate, as a function of β. Previous fatigue wear theories predicted that a parameter corresponding to β would be in proportion to the wear rate, but the present result shows that it is not the case when β is varied over a wide range. There are two main reasons behind this apparent disagreement. First, E_s is given as a random variable in the present analysis, which means that what is effective in wear is an extraordinarily large value of E_s rather than its mean. Roughly speaking, the presence of large values of E_s raises the curves in the low β region, while fractional low E_s values lower them in the high β region. Secondly, the accumulation of damage within the substrate generally raises the curves. As mentioned earlier, a block detaches from the material, leaving some residual damage behind it, which lessens the effort required to make successive blocks come off.

It should be noted here that this theory effectively includes the linear wear law if the parameters are suitably arranged, and that it makes no further assumptions than the previous ones, but rather relaxes the restrictions for the cases to which the theory should apply.

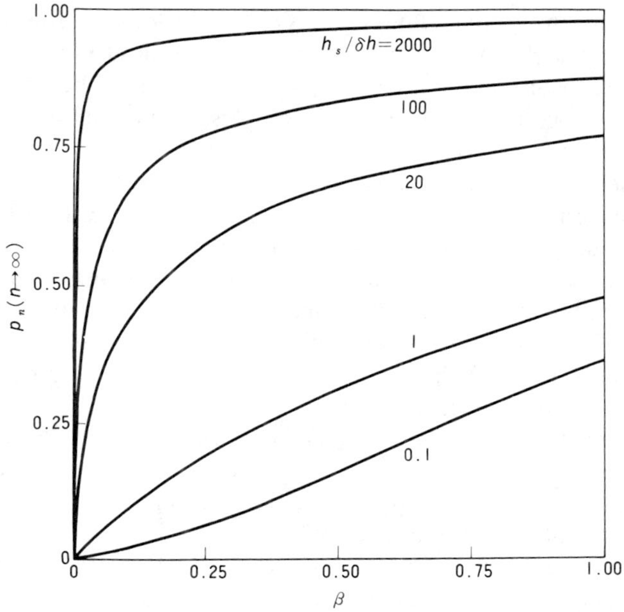

Fig. 10. Wear rate in the steady state: effect of β on $p_{n(n \to \infty)}$.

Changes in Subsurface

Fatigue failure can be characterized as occurring as a result of accumulation of damage in materials caused by repetitive stresses and strains. Though a number of studies have reported various microscopic changes in materials being fatigued, seemingly no simple means are available by which the changes could be correlated with quantitative scales of damage. There is no reason to expect that the matter has been carried further in the process of wear. Therefore, it should be pointed out first that little quantitative correlation can be established at this time between fatigue theories of wear as outlined above and experimental findings.

In the case of wear, forces that provide the ultimate cause of the whole process act on the surface or, generally speaking, at the points of real contact between surfaces. The diameter of a real contact point is typically several tens of micrometers in sliding systems of machinery. Hence stresses are induced, and strains are introduced, in very thin layers just below the surface. Consequently the thickness of the damaged layer is limited when compared with other dimensions of sliding systems—for example, the diameter of a shaft, the width of a bearing, or even the thickness of bearing metals.

Various features of the damaged layer have been reported: for example, plastic deformation, hardening (or, rarely, softening), bending of crystal planes, increase in dislocation density, formation of cell structure, selective diffusion of certain constituents, phase transformation or formation of unusual intermetallic compounds, and eventual initiation and propagation of cracks. Of course, it may not be the case that all these features are always present in the subsurface layer, nor are all responsible for fracture in any circumstance.

The most commonly reported feature is plastic deformation, as this takes place in ductile materials except at the lightest loads, and it is readily observable by optical microscopy of sections. Figure 11(a) is a longitudinal section of a nickel pin slid against a disk of similar material in vacuum, and Fig. 11(b) is that of a similar pin slid in dry air under the same conditions in other respects.[18] On the other hand, Fig. 11(c) shows a longitudinal section of a cylindrical specimen of a plain carbon steel, which was slid against a disk of similar material in the presence of a mineral oil lubricant.[19]

The sliding direction is left to right in all cases, and plastic flow in that direction is evident. Furthermore, a similar plastic flow pattern can be observed under the surfaces after lubricated or unlubricated sliding, though the material and sliding conditions differ considerably.

As stated above, such deformation is caused by the forces acting at the discrete points of real contact. They induce a three-dimensional stress field around each of them, and their full elasto-plastic analysis does not seem straightforward. Therefore, the total thickness of the deformed layer is considered first.

The base of this layer can be taken as the elastic/plastic boundary as suggested by Samuels,[20] and the critical condition for the deformation can be predicted by the theory of elasticity irrespective of the phenomena occurring in the deforming layer. If the forces acting at a single contact point are estimated, the maximum shear stress τ_m which a point at any depth experiences in the direction of sliding on a plane parallel to the surface can be calculated as the curve in Fig. 12.[21] The point of intersection of the curve with the level of critical shear stress for yielding of the material gives a prediction of the deformed layer thickness, which shows reasonable agreement with the observed thickness.

If the thickness is determined in such a way, it would not grow with sliding distance. Figure 13 shows that this is indeed the case; the thickness of the deformed layer in a steel specimen formed during lubricated sliding is plotted against the distance slid,[19] which remains substantially constant.

The hardness in the subsurface of similar specimens is shown in Fig. 14,[19] which was determined on a section. As is often seen, the hardness has its maximum at the surface and decreases monotonically to the bulk value. The thickness of the layer in which appreciable hardening occurs generally coincides with that of the deformed layer.

Recently, strains around the wear track were determined by Ruff.[22] He conducted a wear experiment in which a bearing ball 2.4 mm in diameter recip-

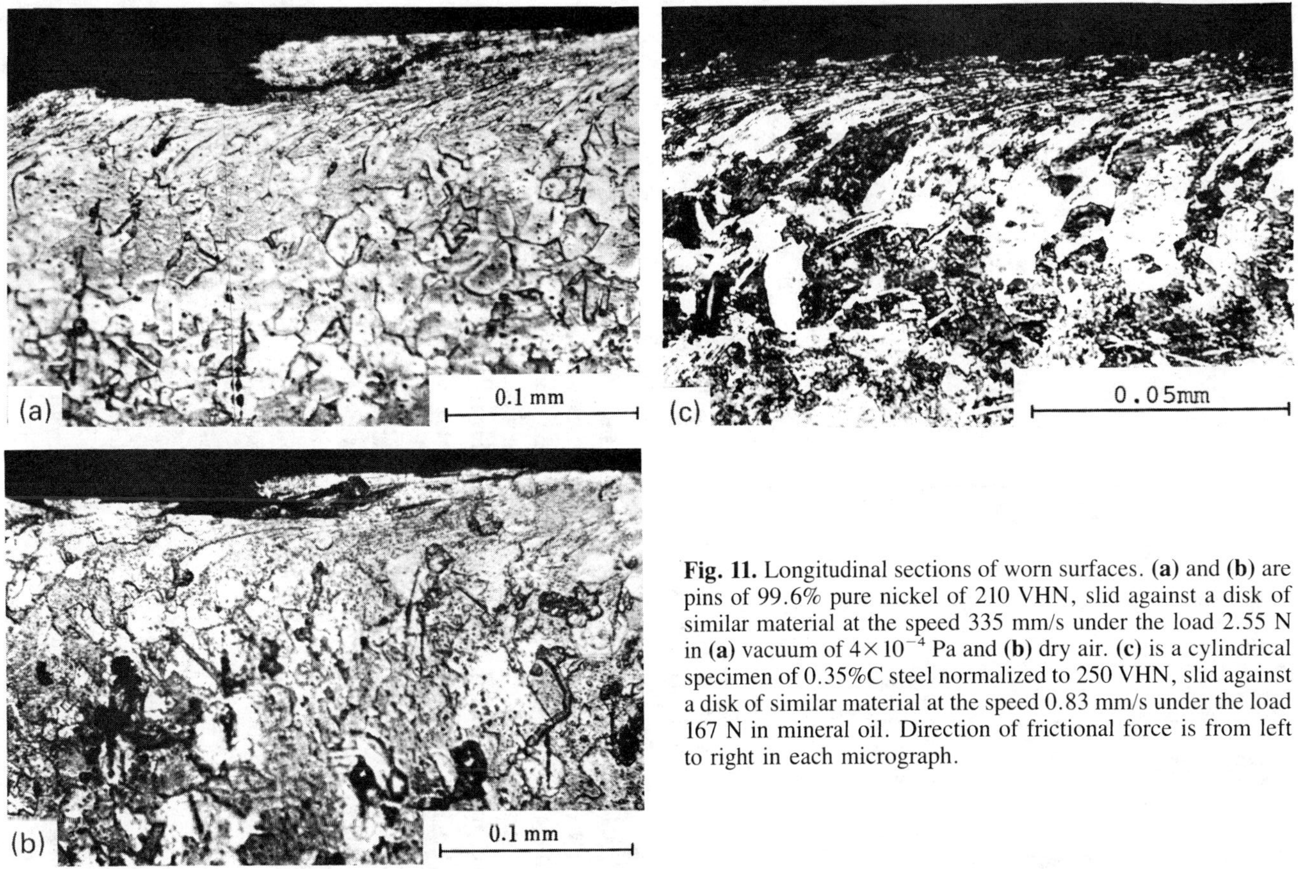

Fig. 11. Longitudinal sections of worn surfaces. **(a)** and **(b)** are pins of 99.6% pure nickel of 210 VHN, slid against a disk of similar material at the speed 335 mm/s under the load 2.55 N in **(a)** vacuum of 4×10^{-4} Pa and **(b)** dry air. **(c)** is a cylindrical specimen of 0.35%C steel normalized to 250 VHN, slid against a disk of similar material at the speed 0.83 mm/s under the load 167 N in mineral oil. Direction of frictional force is from left to right in each micrograph.

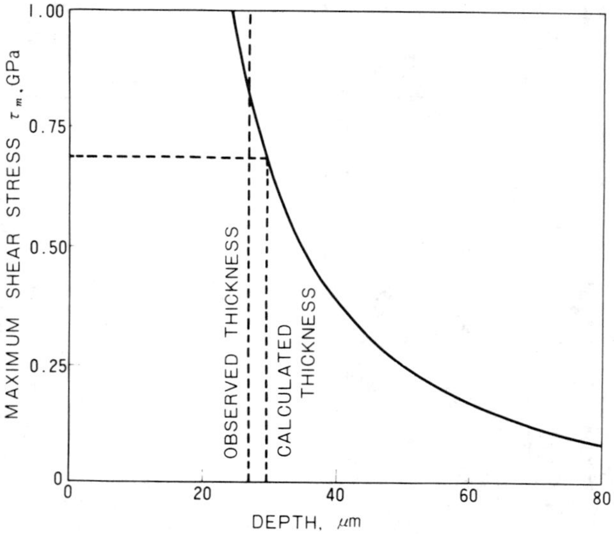

Fig. 12. Maximum shear stress, τ_m, in the substrate as a function of depth. Concentrated normal load is 2.55 N, and the coefficient of friction is 1.3. Critical shear stress of the material is 0.69 GPa(70 kgf/mm^2).

rocated on a copper plate prepared by annealing and electropolishing to make it free from residual strain. After 500 cycles of reciprocating sliding under an unlubricated or lubricated condition, the copper plate was placed in a scanning electron microscope and the strain value was determined by selected area electron channeling pattern technique. Measurements were also made of the subsurface by successively removing material with an electropolishing procedure.

The results are shown in Fig. 15. The abscissa is the distance on the surface across the track. It is observed that the surface strains extend considerable distances beyond the edges of the wear track, and that the strains have their maxima at the surface and decrease smoothly with depth. Such general distribution patterns are common to both lubricated and unlubricated conditions, and moreover, similarity could be found to the distribution of hardness of the subsurface as shown in Fig. 14.

The deformed layer discussed above can be divided further. For example, by optical microscopy of taper sections of mechanically polished brass surfaces Samuels found three layers exhibiting different degrees of deformation.[20] The third layer, that exhibiting the least severe deformation, was characterized by slip lines after etching, and the second by well-developed slip bands. On the other hand, the first layer was very thin and its structure could not be resolved under the metallurgical microscope.

Similar distinctions can be seen in other papers. Tsuya made a microscopic observation of longitudinal sections of worn copper surfaces.[23] The wear experi-

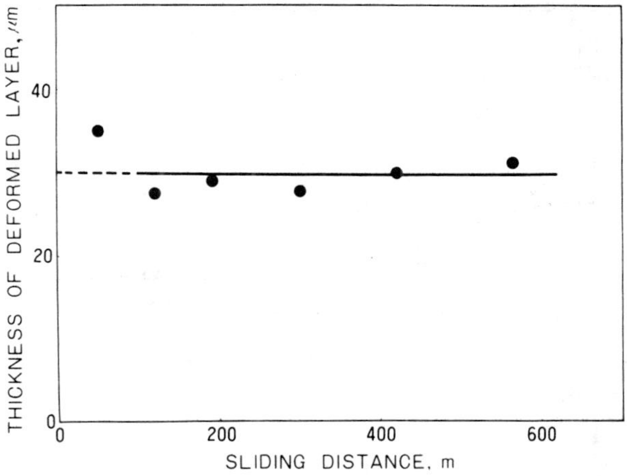

Fig. 13. Thickness of the deformed layer in the end surface of a cylindrical specimen of 0.53%C steel normalized to 250 VHN and slid against a disk of similar material at a speed of 8.3 mm/s under a load of 245 N (lubricated with mineral oil).

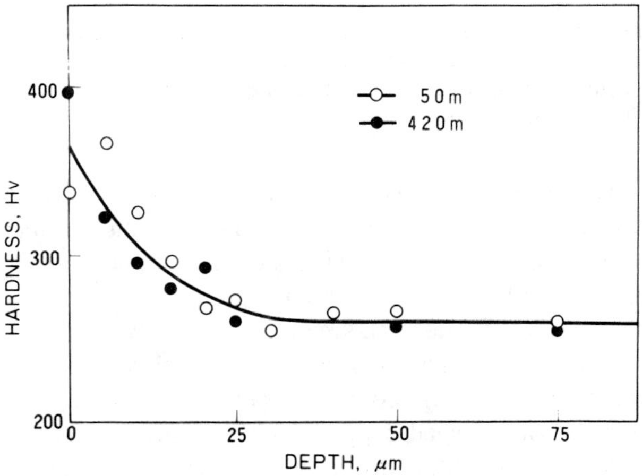

Fig. 14. Distribution of Vickers micro-hardness in the subsurface region of the specimen described in Fig. 13. Distances slid are indicated in the figure.

ment was conducted under an unlubricated condition with a ring/flat machine, and with specimens annealed before testing. In her terminology, there were three layers—namely, a layer having highly micronized crystals, a platelike crystalline region and a region with flexed, distorted crystals, in order of increasing depth. Nakajima and Mizutani investigated by electron microscopy the surface

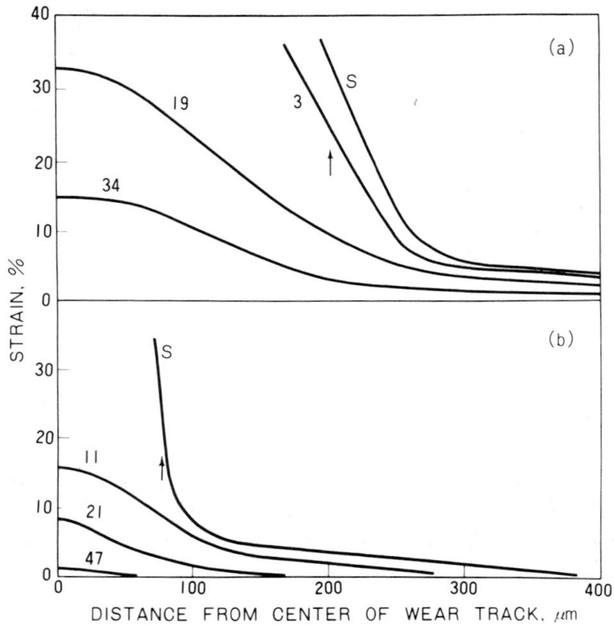

Fig. 15. Strains at the surface (S) and below the surface (depths indicated in μm) of a 99.999% pure copper plate slid against a bearing ball 2.4 mm in diameter at the speed 20 mm/s under the load 0.49 N **(a)** without and **(b)** with SAE 30 automotive oil.

layer of low carbon steel specimens after wear experiments with another ring/flat machine.[24] They distinguished between a region of very high dislocation density and a less severely deformed layer, and found that the thickness of these layers changed differently with a change in sliding speed.

Three-dimensional dislocation cell structure around a wear track was observed by Ohmae and Tsukizoe with transmission electron microscopy.[25] A single-pass experiment was conducted with a conical copper pin slid in the [111] direction on an (001) surface of an annealed copper single crystal. Sections were prepared normal and parallel to the sliding direction and photomicrographs were taken as shown in Fig. 16. Heavily tangled dislocations are readily seen. Moreover, a very fine cell structure, elongated along the sliding direction, is observed adjacent to the wear track. The cell size increases with increasing depth.

Rigney and Hirth reviewed experimental results on microstructural changes during wear and during fatigue,[26,27] and they discussed the case of steels.[28] They suggest that a process can take place in which the carbon atoms in the matrix migrate to the cell walls. As a result, martensite and carbides may disappear from this region, leaving cells of α-ferrite, while cell walls with a higher carbon concentration are formed.

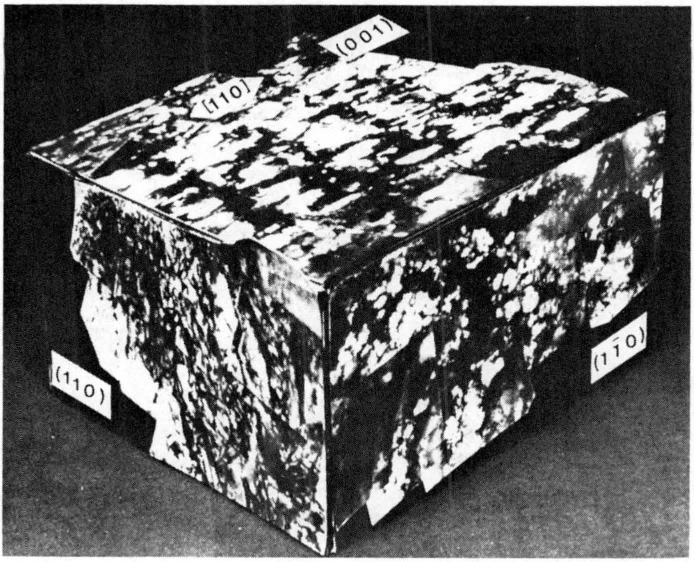

Fig. 16. Three-dimensional dislocation cell structure observed near an (001) surface of a 99.99% pure copper single crystal after a single pass of a conical copper pin at the speed 10 μm/s under a normal load of 0.15 N in dry air. Thickness of the rectangular block represents about 13 μm in the actual solid. *(Courtesy of N. Ohmae and T. Tsukizoe)*

The author is not sure whether or not the descriptions of the layers given by various authors refer to identical structures below the surface. It seems probable, however, that a highly deformed layer of very fine structure with constant average thickness develops adjacent to the sliding surface. Rigney and Hirth noticed this layer and tried to correlate its thickness with friction,[28] referring to a previous approach made by Tsuya.[29] Using a simplified model they proposed a relation,

$$F = wt\tau\epsilon, \quad [15]$$

for the case where plastic work is confined within the near-surface region of one of the materials. Frictional force is F, w and t are the width and the thickness of the highly deformed region, τ and ϵ are the shear stress in the sliding direction and the strain per cycle in that region. When it is assumed that τ ranges from 70 to 1000 MPa, which should be approximately half the ultimate tensile strength, and the values of 5 to 20 for ϵ and 0.1 to 30 for the coefficient of friction are used for assumed unlubricated sliding, the thickness t becomes 5 to 50 μm, which seems to be in a reasonable range. More recently, Heilmann and Rigney have extended this plastic-deformation model of friction by including continuous distributions for both stress and strain and appropriate boundary conditions.[30,31] The thickness of a highly deformed layer remains important in that work.

Formation of Wear Particles

Although the presence of cracks in the sliding surfaces is not a newly discovered feature, not a long time has passed since it captured the particular attention of tribologists.

In an earlier paper describing a sliding experiment between copper surfaces,[32] Tsuya showed some transmission electron photomicrographs of etched cross sections using a two-stage replica technique, in which cracks were clearly visible. She found that the cracks extended along etched bands which were almost parallel to the surface, and suggested that the cracks initiated at inclusion particles and propagated at the base of the highly micronized crystal layer mentioned in the previous section.

Cracks on the sliding surface were reported by Quinn when low-alloy medium-carbon steel (AISI 4340) pins were slid against disks of similar material under unlubricated conditions.[33] From observation of the disk surfaces with transmission electron microscopy of replicas, he reported that the wear tracks were covered with roughened regions and smooth flat regions, and that generally cracks were also found whenever a very smooth surface was found. The cracks tended to lie perpendicular to the direction of sliding, and Quinn pointed out that these cracks were very similar to those obtained in fatigue failure.

Evidence has accumulated that cracks are present in sliding surfaces, and that the cracks participate in the process of wear particle formation. A typical crack is evident in Fig. 11(a),[18] which shows a potential wear particle ready to separate from the surface. Further, some clear pictures of subsurface cracks taken with a scanning electron microscope[34] formed a basis of the delamination theory, which is discussed in other parts of this book.

Before discussing the process of wear particle formation further, some basic quantitative characteristics will be described briefly.

Even though fatigue mechanisms are responsible for wear processes, techniques similar to those used in studying fatigue cannot always be employed in studying wear. The major reasons are that fracture occurs on a microscopic scale in wear, and that a tremendous number of particles are formed during sliding. These make observation of an individual fracture process extremely difficult.

For example, Fig. 17 shows a size distribution of wear particles obtained in a lubricated wear experiment of a 0.35%C plain carbon steel in a three-pin and disk machine; the load was 147 N and sliding speed was 1.57 mm/s. Under these conditions, sliding over 100 m produced some 10^8 wear particles of 2.2 μm in mean diameter. The distribution shows a generally monotonic decrease of relative frequency with increasing diameter. In unlubricated sliding, the particles become comparatively coarse; typically, the mean diameters are of the order of several tens of micrometers as is illustrated in Fig. 11(a) or (b).

Effects on the particle size of various environments were examined by Finkin.[35] He compared data from R. G. Foster's experiments on sliding copper on

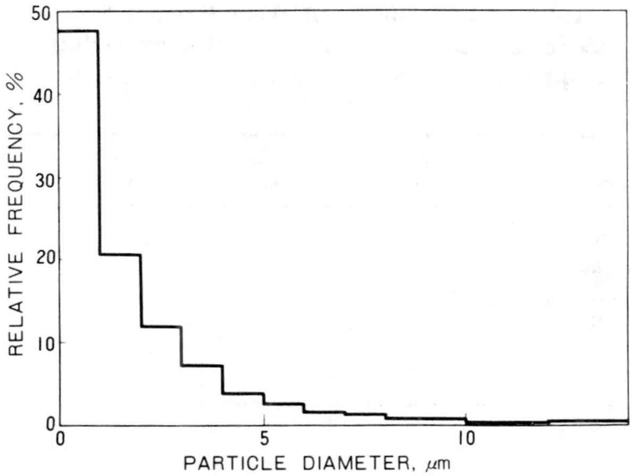

Fig. 17. Example of size distribution of wear particles obtained from a wear experiment with 0.35%C steel in a three-pin/disk machine, run at a speed of 1.57 mm/s under a normal load of 147 N in mineral oil.

copper with various lubricants, and pointed out a good correlation between wear particle size and surface roughness formed during sliding, as shown in Table 1. Apart from the correlation, a marked change of the particle size is also of present concern.

Another important correlation was observed in some cases between the particle size and the thickness of surface layer. Tsuya considered that the depth at which cracks propagate should be correlated with the size of the particles.[29,32] She reasoned further that, as the cracks propagate at the base of the highly micronized crystal layer, the thickness of this layer should have a definite correlation with the amount of wear in a fixed sliding distance. This would be true under certain limited conditions, and she found a linear relationship between the amount of wear in volume and the thickness of the highly micronized crystal layer cubed, as shown in Fig. 18[29] for a series of unlubricated wear experiments with copper.

A more detailed observation was made by the present author and others.[18,21,36,37] In a study of unlubricated wear of nickel, the thickness of the wear particles having mean volume[37] was estimated for each sliding condition and compared with that of a total deformed layer under a corresponding condition. The result is shown in Fig. 19, where all the points seem to fall around a straight line suggesting a close correlation between these two thicknesses.

Although these correlations should not be generalized hastily, it implies some important features in the process of wear. The close correlation between the size of wear particles and the thickness of the deformed layer suggests that some factor determining the latter also determines the former, and, if the rate of particle formation does not vary appreciably, it will also govern the wear characteristics

Table 1. Correlation of Wear Particle Size to Surface Roughness Formed During Sliding Copper on Copper, With Various Lubricants (Ref 35)

Lubricant	Particle size, μm	Rank	Roughness, μm	Rank
None	240	1	71	1
Cetane	12	2	11.4	2
Dow Corning 200, 100 cSt	9.5	3½	9.15	3
Ucon LB-70X	9.5	3½	4.06	4
Shell 200 10 cSt	8	5	3.66	5

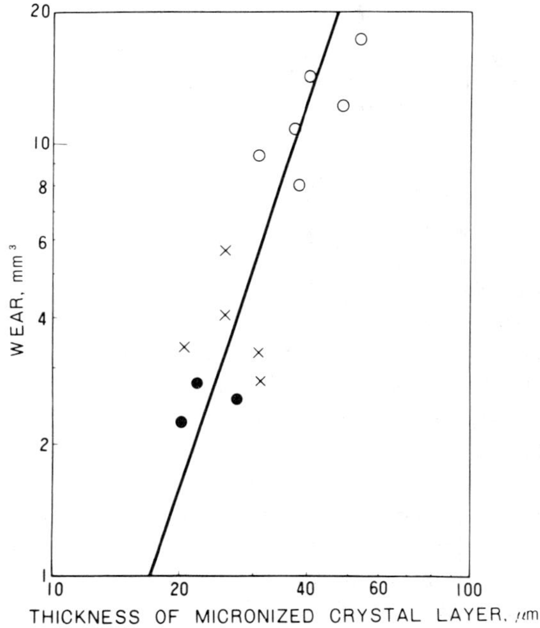

Fig. 18. Relationship between the amount of wear and the thickness of the highly micronized crystal layer, found with a polycrystalline copper block slid against a ring of similar material in a Timken-type machine. The loads are ○:194 N, ×:32 N and ●:22 N. The speed was varied between 0·08 and 5.6 m/s. The straight line represents a dependence of the worn volume on the thickness of the micronized crystal layer cubed. *(Ref 29)*

themselves, as perhaps is the case with Tsuya's experiment. Further, when it is recalled that a mechanical interpretation of the deformed layer thickness is possible, as stated in the preceding section, it is concluded that actual forces at the real contact points are responsible for this aspect of wear.

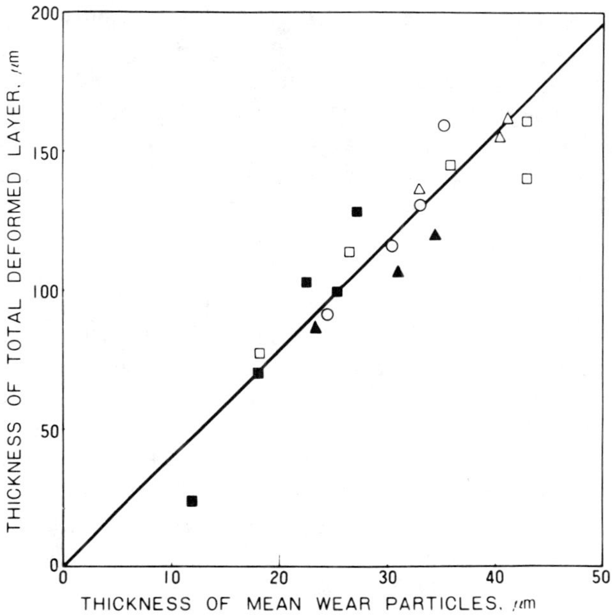

Fig. 19. Correlation between thickness of the wear particles having mean volume *(Ref 37)* and thickness of total deformed layer, observed in a series of experiments with nickel, as described in Fig. 11(a) and (b). ○:pressure is varied, □:speed is varied in vacuum; ■:speed is varied in air; △:load is varied in vacuum, ▲:load is varied in air. *(Ref 21)*

The results appearing in Table 1 can be interpreted in this connection as follows. First, the different surface roughnesses formed during sliding in the presence or absence of various lubricants would have caused difference in contact geometry, which in turn determined the magnitude of the actual forces at the contact points. Secondly, the removal of particles of a certain size would leave a certain roughness behind. Thus, these must constitute a feedback loop which has not yet been analyzed sufficiently.

The author has a view that initiation of cracks in the surface does not pose a critical problem in the process of wear, with the exception of the case of a single-pass experiment. As stated in an earlier section, some damage has usually accumulated in the surface which is newly exposed. Typical examples are given in Fig. 11, where it should be noted that some cracks are seen at the surfaces, which have been left when a particle was detached earlier. Consequently, microscopic cracks are considered as being available at the surfaces, and propagation of the existing cracks becomes the most important stage. A thorough quantitative analysis of this process was undertaken in the delamination theory, which is discussed elsewhere in this book, and no further mention seems necessary here. Instead, some recent results from close examination of worn surfaces will be described.

Despite the frequent reference to fatigue effects during sliding situation, evidence of fatigue-like striations has been surprisingly lacking, as Rigney wrote in a letter to the author. Some of the plausible reasons for it were listed above, but, when developments in electron microscopy or other observation techniques are considered, it may be reasonable to expect some definite evidence like the presence of striations.

Quite recently, Heilmann found an interesting example.[38] Figure 20(a) is a part of the worn surface of a Ni-40%Cu alloy slid against 440C stainless steel, in which features like striations are clearly seen. After closer examination of a stereo pair, Heilmann found that the striation-like pattern appeared in a deep pit, which would have caused protection from being erased by subsequent passes.

Inspired by this discovery, an effort was made in the author's laboratory to find similar evidence. A result is shown in Fig. 20(b), which shows a pattern, similar to that found by Heilmann, on a worn surface of a work-hardened copper disk which was slid against a pin of a copper-base sintered material used for current collectors without lubrication.

Though fractography is not in the author's line, some comment should be made on the nature of these characteristic features. First, the wavelengths of the striated patterns are rather long—typically, 4 to 5 μm in both examples. Further, the wavelength is not constant but shows rather marked change as is evident in Fig. 20. This suggests that the patterns appearing in Fig. 20(a) and (b) might not be fatigue striations, but like "tire tracks." Similar markings may be formed on a surface when it is beaten repeatedly by a ridge on the other surface which is changing its position gradually. This interpretation seems reasonable if we refer to Fig. 11. When contact takes place at the top of the potential wear particle, an indentation mark is formed on the lower surface of the crack, with a protrusion on the probably harder upper surface. Since the potential particle would move gradually in the direction of friction, "tire tracks" such as those shown in the micrographs might have been formed.

These markings may furnish further clues about the occurrence of fatigue fracture in wear. Of course, their relative importance in the whole process is open for further study.

Quantitative Correlations

In the introduction, a definition of wear was noted which requires the loss of substance from the operating surface to be progressive. This carries an important implication.

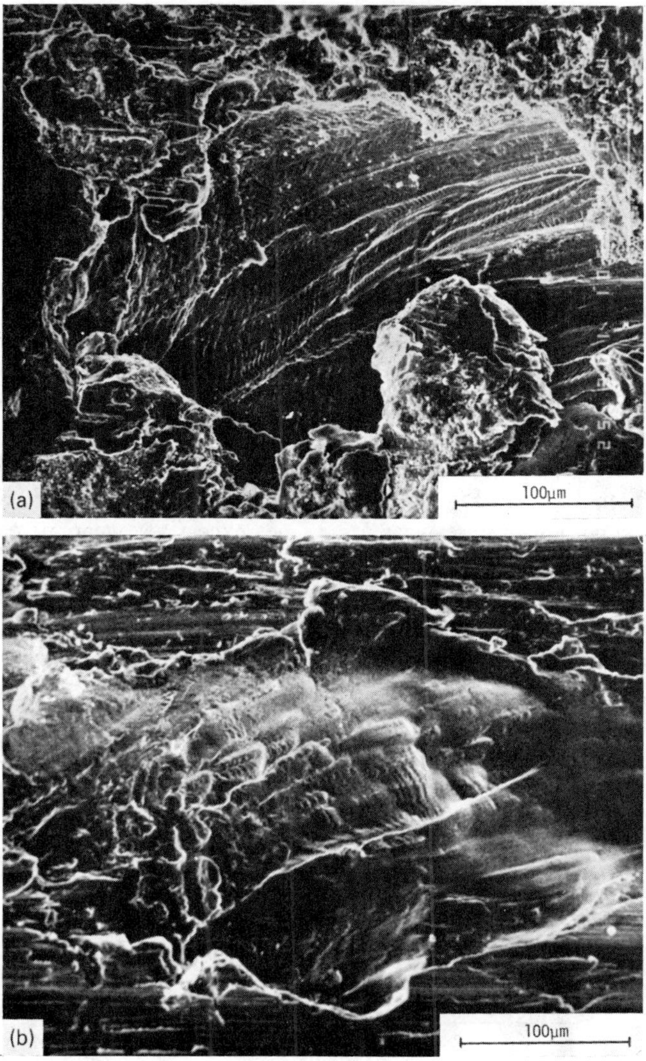

Fig. 20. Striations found on wear surfaces. **(a)** Surface of a Ni-40%Cu block slid against a 440C stainless steel ring in LFW-1 machine, at the speed 50 mm/s under sequentially increasing load up to 200 N in air. *(Courtesy of D. A. Rigney)* **(b)** Surface of a 99.99% pure copper disk slid against a pin of a copper-base sintered material at the speed 131 mm/s under a load of 2.55 N in air. Direction of motion of the mating surface is from left to right in both micrographs.

Loss of substance from surfaces is often observed accompanying such tribological failures as scuffing, scoring, seizure, pitting, flaking and spalling. Generally speaking, however, substance is lost rather rapidly in these failures, and further operation becomes difficult once they happen. Thus, what is important is the critical condition under which these failures initiate (and also the life for which the surfaces can operate without these failures). On the other hand, the progressiveness in the case of wear allows the operation to continue until a certain amount of substance is removed. Hence, it is always the amount of the lost substance, or the amount of wear, which is important, and the so-called wear laws relate the amount of wear to operating variables or material properties. Therefore, some quantitative aspects of wear should be considered in relation to fatigue.

Endo conducted a series of experiments on lubricated wear of steels. Basically, he has regarded wear experiments as part of materials testing, and he has tried to correlate the results with fatigue resistance of the materials.

In an earlier paper,[39] Endo and Fukuda analyzed progressive wear behavior as shown schematically in Fig. 21. When virgin steel surfaces were worn under lubricated conditions where wear loss is not excessive, wear increased rather rapidly at the outset and then gradually levelled off until a point of inflection, A, was reached. This behavior was explained as the initial stage of wear during a break-in period when abrasive mechanisms predominated. The successive decrease in wear rate was associated with improved conformity between the surfaces. Asperities on the surfaces were subjected to repeated stresses, and fatigue fracture started at point A and continued thereafter, resulting in a linear relationship between the amount of wear and the sliding distance. This feature appears frequently in papers by Endo and his colleagues. Further it should be pointed out that another explanation is possible, using the substrate fatigue theory described in an earlier section; typically the curves shown in Fig. 7 include analogous behavior.

Figure 22 is a plot of the results obtained with a 0.40%C steel when an annular end surface of a rotating cylindrical specimen was slid against a similar surface

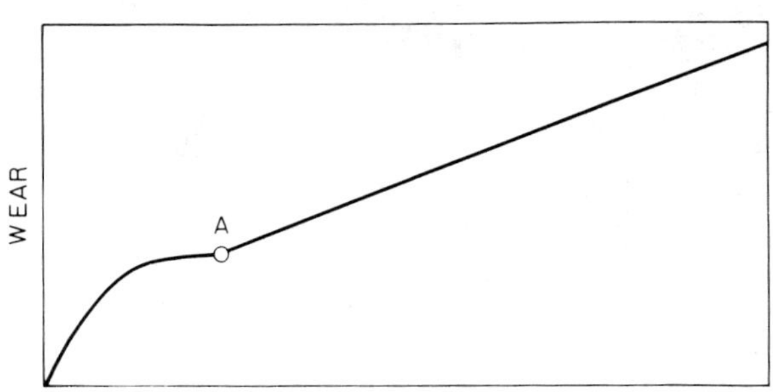

Fig. 21. Schematic representation of progressive wear behavior proposed by Endo et al., Ref 39. The point of inflection is indicated by 'A'.

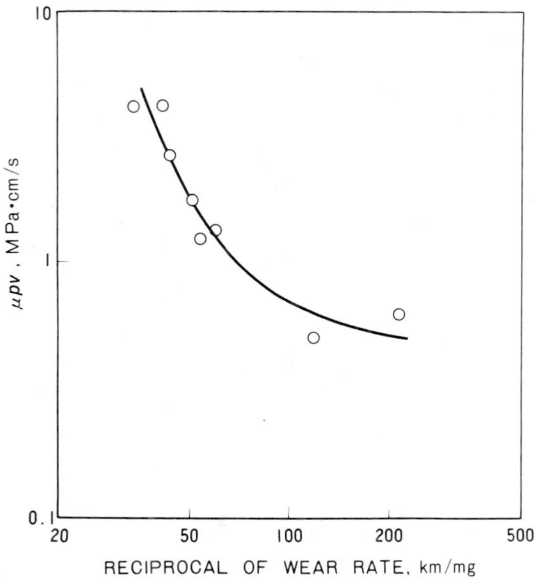

Fig. 22. Lubricated wear results showing similarity to S-N curve in fatigue testing. Sliding was made between the end surfaces of cylinders of 0.40%C steel at speeds of 0.13 to 0.32 m/s under apparent contact pressures of 4.1 to 6.6 MPa in machine oil. *(Ref 39)*

of a fixed specimen in a machine oil. The abscissa is the reciprocal of the wear rate in the linear region, and the ordinate is the product of the coefficient of friction, μ, average contact pressure, p, and the sliding speed, v. Endo and Fukuda regarded the abscissa as the number of shear stress cycles necessary to remove a unit volume of material from the surface, and concluded that Fig. 22 compared with an S-N curve in fatigue testing.

A similar result was obtained by Endo and others when effects of internal stresses on wear were determined.[40] A plate of a 0.07%C mild steel was worn against the periphery of a rotating disk of a hardened tool steel under lubricated and unlubricated conditions. Tensile or compressive loads were applied statically to the plate specimen so that known uniaxial stresses were induced in it. For sliding with lubrication by a machine oil, progressive wear behavior as shown in Fig. 21 was obtained except with the largest compressive stress. As an extension of their fatigue wear theory, they reasoned that the induced static stress would concentrate at valleys in surface microtopography, which could help or prevent the removal of protuberances by playing the role of a mean stress in fatigue testing. Therefore, wear with varying internal stresses should correspond to fatigue with different mean stress levels under a constant stress amplitude.

The result is shown in Fig. 23, in which the reciprocal of the wear rate in the linear region was plotted against the static stress, and shows characteristic

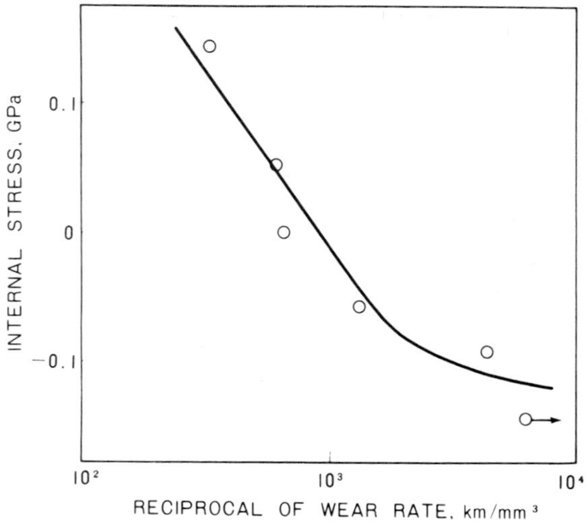

Fig. 23. Effects of internal stresses on wear of a 0.07%C mild steel plate slid against a disk of hardened tool steel at a speed of 0.31 m/s under a load of 161 N and lubricated by machine oil. *(Ref 40)*

features of an S-N curve. At the largest compressive stress, −0.147 GPa, the fatigue (linear) wear region did not appear during the experiment, as mentioned above. This value of the stress might be compared with that of a residual stress in the wearing surface determined separately, typically −0.2 GPa.

Only a few cases are known where we find positive correlation between wear resistance and fatigue resistance of materials. A change in the latter by some means is usually accompanied by changes in the contact mechanics discussed earlier, which can have marked effects more than sufficient to mask the correlation just discussed.

An example was presented by Endo and others.[41] They tried to determine effects of cold-working on the resistance to wear and fatigue of a 0.15%C mild steel. Fully annealed material was then cold-worked to different degrees and the specimens were machined from it. Wear experiments were conducted in a rotating cylinder machine as described above with a machine oil as the lubricant. Care had been taken to avoid the effects of work hardening during machining by electrolytically polishing the sliding surface. Reversed bending fatigue tests were carried out by using notched test pieces of the same material. The wear rate and the fatigue limit are compared with the degree of cold-working in Fig. 24, which shows a definite correlation.

Another example was afforded by the present author and others.[36] The dependence of wear on atmospheric pressure was studied with nickel and gold; gold was used as a reference material which is effectively free from chemisorbed

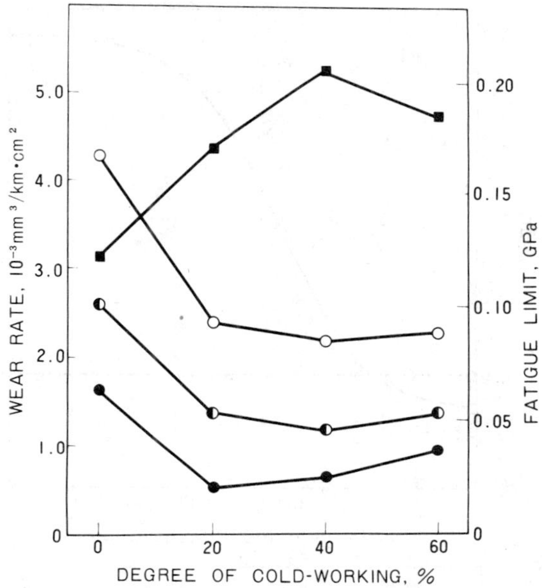

Fig. 24. Comparison of effects of cold-working on wear rate and fatigue limit of fully annealed 0.15%C mild steel. Wear was determined in sliding between the end surfaces of cylinders at a speed of 0.56 m/s under the loads ●:82 N, ◐:124 N and ○:147 N in machine oil. Fatigue limit (■) was determined by reversed bending fatigue tests of notched plate specimens 25 mm wide and 4 mm thick having a central hole 1.5 mm in diameter. *(Ref 41)*

films from the atmosphere. Sliding was made between similar materials in a pin/disk machine placed in a vacuum chamber, and the resultant wear was plotted in Fig. 25(a) against the atmospheric pressure. In this series of experiments, the contact mechanics were substantially unchanged by the change of atmospheric pressure with each material. Hence the mean size of the wear particles remained almost constant independent of the pressure, and the observed changes in the amount of wear were almost all due to changes in the number of particles. This happens to be a condition under which differences in fatigue resistance could become apparent.

The fatigue of the materials was studied in reversed bending in a modified Schenck machine. Figure 25(b) shows the results, in which the reciprocal of the number of cycles to failure was plotted against the atmospheric pressure.

As is expected, the amount of wear and the reciprocal of the fatigue life behave in very similar manners when the atmospheric pressure is varied. With nickel, both decrease fairly sharply with a reduction in the atmospheric pressure at intermediate pressures. Changing the pressure at low or at high pressure has a

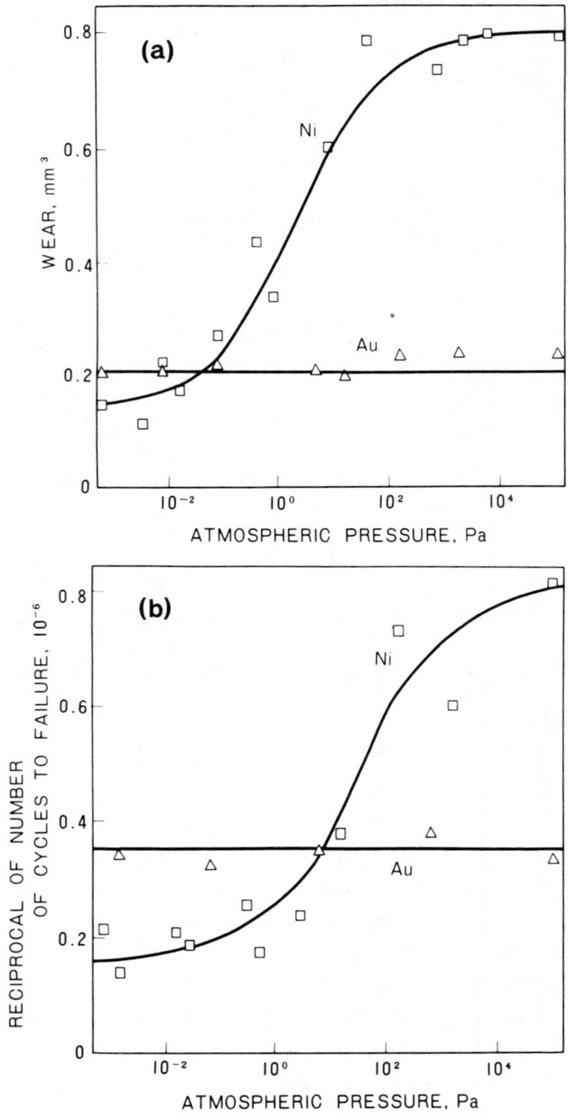

Fig. 25. Results showing similar effects of atmospheric pressure on wear resistance and fatigue resistance. Materials are 99.6% nickel of 210 VHN and 99.99% gold of 62 VHN. **(a)** Wear was determined in a pin/disk machine at a speed of 168 mm/s under a load of 2.55 N. **(b)** Reversed bending fatigue tests were made in a modified Schenck machine with plate specimens of 6 mm × 2 mm in the minimum section size, at a frequency 16.7 Hz and at a constant amplitude. The initial maximum bending stresses were 0.16 GPa for nickel and 0.08 GPa for gold.

smaller effect. In contrast, with gold, both the amount of wear and the reciprocal of the fatigue life are unaffected by the atmospheric pressure.

This result clearly shows that environments can have important effects on fatigue, and therefore wear. The problem becomes more serious in lubricated wear, because of complicated mechanical and chemical effects of lubricating oils, especially those containing chemically active additives. However, environmental effects are to be discussed in another part of this book, and no further mention of them will be made here.

Conclusions

Possible roles of fatigue in the process of wear have been discussed in several aspects.

The fatigue theories of wear originated in an attempt to give a more reasonable interpretation to what is commonly called "adhesive wear." Since wear is the loss of substance from the surface in response to mechanical action, microscopic fracture in the surface should form the essential part of its process. When we think about the repetitive nature of practical sliding systems, it would be a natural consequence that fatigue mechanisms have been considered to operate. The fatigue wear theories developed to date assume considerable complexity when compared with the adhesion theory. However, the present author thinks the complexity is a characteristic feature that a practical, meaningful wear theory should have. The apparent simplicity of the familiar adhesion theory forms one of its shortcomings, because it cannot be accommodated to the complicated wear behavior we frequently encounter.

Experimental evidence is being accumulated which supports proposed fatigue wear mechanisms directly or indirectly. Observations of the surfaces, subsurface layers and wear particles clearly indicate that fatigue mechanisms are operating during wear. However, positive evidence to show quantitative correlations between wear resistance and fatigue resistance of materials is sparse. Poor correlations are rather common except for cases where suitable conditions for the comparison are deliberately established. This is caused by irregular surface geometry, complicated contact mechanics, uncertain environmental effects, and other factors peculiar to tribological systems. The complexities in the formulation of fatigue wear theories also owe much to these factors. Further analysis of the details of wear processes is necessary, and this will certainly involve contributions from materials scientists.

Finally, it should be worthwhile to determine whether or not the mechanisms are common between lubricated and unlubricated wear, for the majority of wear experiments have been conducted under unlubricated conditions, whereas most practical wear problems occur in the presence of lubricating oils. There are some sound reasons why researchers prefer unlubricated conditions. For example, dry wear proceeds rather rapidly and it takes place on a larger scale, making observation far more convenient. Moreover, introduction of lubricating oils brings about a number of complicating factors.

Of course, some characteristic features, such as tenacious adhesion, may be peculiar to unlubricated wear processes. However, as far as fatigue mechanisms are concerned, essentially similar features have been observed with and without lubrication, though substantial differences are sometimes found. These differences should be taken into account in future developments of fatigue wear theories.

References

1. R. Holm, "Electric Contacts," H. Gebers Förlag, Stockholm, 1946, p. 214.
2. D. Dowson, ''History of Tribology,'' Longman, London, 1979, p. 12.
3. Japan Society of Lubrication Engineers, ed., "Lubrication Handbook," Yokendo Publishers, Tokyo, 1970, p. 72.
4. "Glossary of Terms and Definitions in the Field of Friction, Wear and Lubrication—Tribology," OECD, Paris, 1969, p. 64.
5. N. Soda, Y. Kimura and M. Sekizawa, Bull. Japan Soc. Mech. Engrs., vol. 15, 1972, p. 866.
6. "Glossary of Terms and Definitions in the Field of Friction, Wear and Lubrication—Tribology," OECD, Paris, 1969, p. 14.
7. I.-Ming Feng, J. Appl. Phys., vol. 23, 1952, p. 1011.
8. D. Scott, W. W. Seifert and V. C. Westcott, Sci. American, vol. 230, 1974, no. 5, p. 88.
9. J. F. Archard, J. Appl. Phys., vol. 24, 1953, p. 981.
10. L. Rozeanu, Wear, vol. 6, 1963, p. 337.
11. I. V. Kraghelsky, Trans. ASME, J. Basic Eng., vol. 87, 1965, p. 785.
12. J. Halling, Wear, vol. 34, 1975, p. 239.
13. E. F. Finkin, Wear, vol. 47, 1978, p. 107.
14. J. A. Greenwood and J. B. P. Williamson, Proc. Roy. Soc. Lond., ser. A, vol. 295, 1966, p. 300.
15. R. G. Bayer et al., Wear, vol. 5, 1962, p. 378.
16. G. Lundberg and A. Palmgren, "Dynamic Capacity of Rolling Bearings," Ingeniörsvetenskapsakademiens Handlingar, no. 196, 1947, p. 19.
17. Y. Kimura, J. Japan Soc. Lub. Engrs., vol. 16, 1971, p. 59.
18. N. Soda, Y. Kimura and A. Tanaka, Wear, vol. 33, 1975, p. 1.
19. Y. Kimura, H. Terada and K. Yamaguchi, Prep. Autumn Annual Meeting, Japan Soc. Lub. Engrs., 1973, p. 117.
20. L. E. Samuels, J. Inst. Met., vol. 85, 1956-57, p. 51.
21. N. Soda, Y.Kimura and A. Tanaka, Wear, vol. 40, 1976, p. 23.
22. A. W. Ruff, Wear of Materials 1977, ASME, 1977, p. 64.
23. Y. Tsuya, Bull. Japan Soc. Prec. Engrs., vol. 2, 1967, p. 214.
24. K. Nakajima and Y. Mizutani, Wear, vol. 13, 1969, p. 283.
25. N. Ohmae and T. Tsukizoe, Proc. 4th Intl. Conf. on Product Eng., Tokyo, 1980, p. 989.
26. J. P. Hirth and D. A. Rigney, Proc. Intl. Conf. on Strength of Metals and Alloys, V, Aachen, West Germany, Aug. 26-Sept. 1, 1979.

27. J. P. Hirth and D. A. Rigney, "The Application of Dislocation Concepts in Friction and Wear," in "Dislocations in Solids," ed. F. R. N. Nabarro, 1981.
28. D. A. Rigney and J. P. Hirth, Wear, vol. 53, 1979, p. 345.
29. Y. Tsuya, Tech. Rep. no. 81, Mechanical Engineering Laboratory, Tokyo, 1976.
30. P. Heilmann and D. A. Rigney, 8th Leeds-Lyon Symposium on Tribology, Leeds, England, Sept. 9-12, 1980.
31. D. A. Rigney, private communication.
32. Y. Tsuya, J. Mat. Sci. Soc. Japan, vol. 1, 1964, p. 35.
33. T. F. J. Quinn, Proc. Inst. Mech. Engrs., vol. 182, pt. 3N, 1967-68, p. 201.
34. N. P. Suh, Wear, vol. 25, 1973, p. 111.
35. E. F. Finkin, Wear, vol. 6, 1963, p. 293.
36. N. Soda, Y. Kimura and A. Tanaka, Wear, vol. 43, 1977, p. 165.
37. N. Soda, Y. Kimura and A. Tanaka, Wear, vol. 35, 1975, p. 331.
38. P. Heilmann, work described by D. A. Rigney, private communication.
39. K. Endo and Y. Fukuda, Proc. 8th Japan Cong. Test. Mat., 1965, p. 69.
40. K. Endo et al., Bull. Japan Soc. Prec. Engrs., vol. 4, 1970, p. 1.
41. K. Endo, T. Okada and Y. Iwai, Proc. JSLE-ASLE Intl. Lub. Conf., Elsevier, Amsterdam, 1976, p. 92.

Wear and Fracture Mechanics

ALAN R. ROSENFIELD
Battelle–Columbus Laboratories
Columbus, Ohio 43201

Abstract

Examination of the literature suggests that wear rates depend on fracture toughness only in the special case of hard, brittle alloys. Even in this case linear elastic models are deficient because they are inconsistent with the occurrence of flaking. The shortcomings of such models are discussed and areas for improvement identified.

Introduction

Fracture mechanics deals with flawed structures under load. Since it overlaps the traditional disciplines of metallurgy and mechanical engineering, it has two principal objectives:

(1) To determine failure loads
(2) To develop fracture-resistant materials.

To the metallurgist, the most important result of fracture mechanics has been the discovery of a material property called fracture toughness (denoted K_{IC}) that measures the resistance to crack growth and that, at best, is loosely related to conventional properties, such as strength and ductility.[1]

Wear is perhaps the most complex phenomenon to be examined by fracture mechanics. The analysis is complicated by rapidly varying complex stress systems, environmental effects, large plastic strains, and crack sizes on the order of microstructural dimensions. All of these effects have served to make analyses of wear both inexact and limited to idealized cases. Nevertheless, since wear can involve the creation of new surfaces via processes such as flaking and delamination, it should, in principle, be amenable to a fracture-mechanics treatment.

Perhaps the best way to begin under such circumstances is to determine empirically whether wear resistance is dependent on fracture toughness. Here the evidence is ambiguous. Figure 1 is a plot of data on abrasive-wear resistance of steels[2] that has been prominently cited[3] as evidence that improved fracture

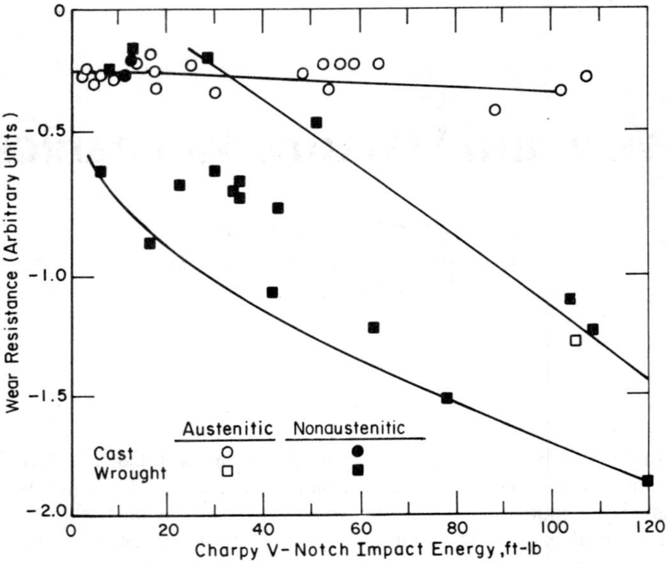

Fig. 1. Effect of impact energy on abrasive-wear resistance of several steels and cast irons. *(Data of Diesburg and Borik, Ref 2)*

toughness can be deleterious. For the wrought steels investigated, higher impact energy is associated with lower abrasive-wear resistance. However, a plot of the same data as a function of hardness reveals a different aspect of the problem. Figure 2 shows that the harder wrought steels are increasingly more resistant to abrasive wear. Since the scatter in Fig. 2 is less than that in Fig. 1, it is reasonable to conclude that hardness is the dominant property and that fracture toughness plays a lesser role. The apparent effect of toughness can be viewed as reflecting an approximate inverse relation between hardness and toughness. In contrast to wrought steels, the seeming independence of wear resistance of cast steels to both properties is puzzling and needs to be taken into account in any model of abrasive wear.

Bearing this in mind, this paper selects two areas of wear in which fracture-mechanics analyses have been presented: in abrasive wear, one of the surfaces is much harder than the other, while in sliding wear, the hardnesses of the two surfaces may be reasonably comparable.

Abrasive Wear

Hornbogen[4] and Zum Gahr[5-8] have examined the relation between fracture toughness and abrasive-wear resistance in some detail. Figure 3, taken from data reported recently by Zum Gahr and Doane,[7] illustrates the kind of observation that

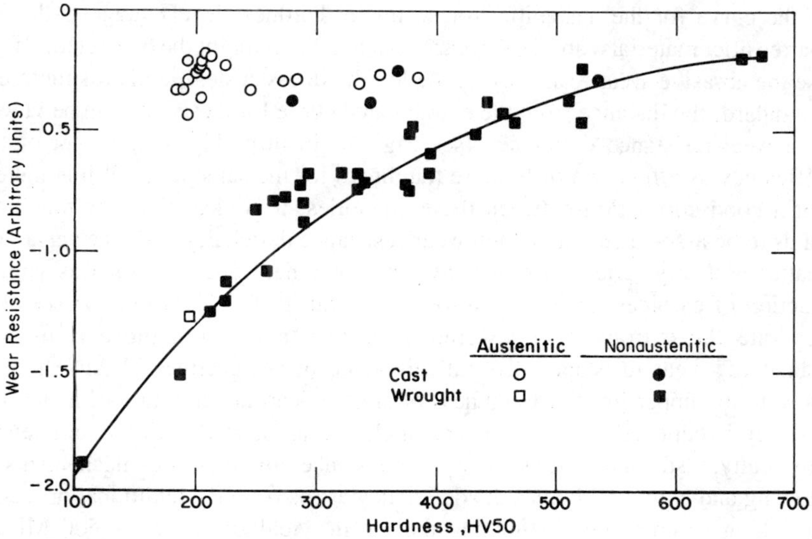

Fig. 2. Effect of hardness on abrasive-wear resistance of several steels and cast irons. *(Data of Diesburg and Borik, Ref 2)*

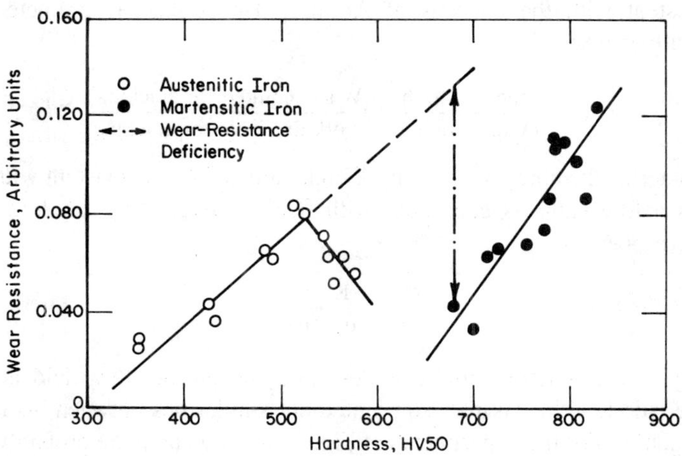

Fig. 3. Effect of hardness on wear resistance of cast irons. *(Data of Zum Gahr and Doane, Ref 7)*

gave rise to their approach. The data indicate that the major variables that affect abrasion resistance of white cast irons are hardness and microstructure. However, the hardest austenitic specimens (HV50 >525) are not as resistant as might be expected from an extrapolation of the rest of the data. Furthermore, extrapolation of the curve for the austenitic iron to higher hardness levels suggests that the martensitic materials are less wear resistant than might be expected. If we use the abrasive-wear resistance of the low-hardness austenitic microstructure as a standard, the distance from the extrapolated curve for these data can be viewed as a wear-resistance deficiency (see Fig. 3). In turn, Fig. 4 is a plot of this deficiency as a function of fracture toughness for the same data.[7] While there is not a good correlation between these quantities, it is clear that low toughness tends to be associated with a high wear-resistance deficiency and vice versa. The scatter probably reflects a direct influence of other variables, such as volume fraction of carbides. In Fig. 4, a toughness value of ~21 $MN \cdot m^{-3/2}$ seems to separate those irons with lowered wear resistance from those irons with "standard" wear resistance. To put this value in perspective, 21 $MN \cdot m^{-3/2}$ is close to the upper limit of toughness of sintered-carbide cutting-tool materials.[9]

There is believed to be a mechanism change associated with the deficiency.[5] Physically, "standard" abrasive-wear resistance involves the mechanisms of plowing and cutting, whereas the deficiency arises from the additional processes of flaking or chipping. Estimating the tensile yield stress, $\sigma_y \approx 500$ MPa at HV50 = 525 for white cast irons, the threshold occurs at a toughness-to-yield-strength ratio of $K_{IC}/\sigma_y \approx 0.04\ m^{1/2}$ (0.25 $in.^{1/2}$), which is an unusually small upper limit for brittle behavior. This idea of a threshold toughness/strength ratio is consistent with the analysis of Atkins.[10] He defines a parameter used in metal-cutting research

$$A = \frac{\text{Imposed strain}}{\text{Yield strain}} \times \frac{\text{Work required for cracking}}{\text{Work available for cracking}} \qquad [1]$$

to characterize the abrasion system. A high value of A is consistent with cutting, whereas a low value is consistent with flaking. Evaluating Eq 1 in terms of material properties:

$$A = \frac{K_{IC}g}{\sigma_y\sqrt{h}}, \qquad [2]$$

where g is a geometric term that describes the plastic-flow field and h is a characteristic length. Atkins also pointed out similarities between his result and Hornbogen's.[4] For the experiments in Fig. 3 and 4, g and h are probably material independent, and a transition K_{IC}/σ_y ratio is expected. However, currently it is not possible to evaluate A from independent estimates of g and h and make a rigorous test of Atkins' analysis.

Both Hornbogen[4] and Zum Gahr[5] have developed models that focus on the question of the threshold condition at which toughness begins to influence wear resistance. Figure 5[6] shows a crack pattern consistent with abrasive-wear observations. In Hornbogen's model, the crack cannot extend until the strain at

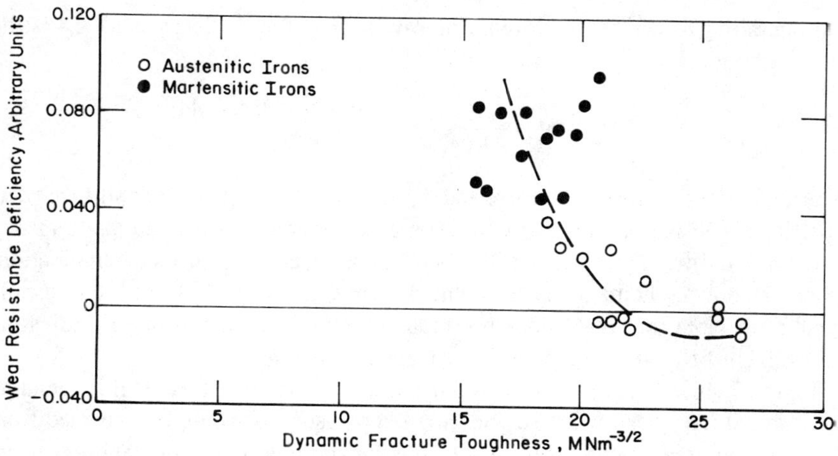

Fig. 4. Loss of wear resistance due to low fracture toughness. *(Data of Zum Gahr and Doane, Ref 7)*

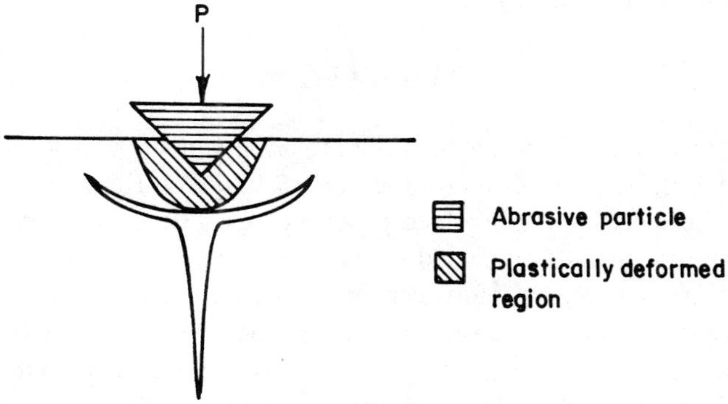

Fig. 5. Mechanism of flaking. *(After Zum Gahr, Ref 6)*

its tip exceeds a critical value. In Zum Gahr's model, fracture toughness becomes important when the normal force becomes large enough to link adjoining microcracks associated with second phases. Although these models differ in detail, both result in predictions that the pressure at which toughness begins to influence wear resistance will increase strongly with increasing fracture toughness and decrease with increasing hardness or yield strength. Evidence that high fracture toughness and low yield strength inhibit flaking can be derived from studies of crack formation around microhardness indentations.[11]

Zum Gahr[5,6] also has derived an equation for wear rate when both mechanisms are operating; if ΔV is the volume removed, H is the hardness, and K_{IC} is the fracture toughness,

$$\Delta V = \frac{\Psi}{H} + \frac{\varphi H}{K_{IC}^2}^{1/2}; \varphi = \begin{cases} 0 \text{ if } N \leq \text{threshold} \\ \varphi \text{ otherwise} \end{cases}, \qquad [3]$$

where N is the normal force exerted by the abrader on the wear surface. The coefficients Ψ and φ increase with normal force and also contain some microstructural and frictional characteristics of the abrader/wear-surface combination. Figure 6 shows schematically the combined predictions of Eq 1 and the threshold condition. Note that Eq 3 does not reduce to Eq 2 at the threshold, indicating unresolved differences between Zum Gahr and Atkins.

Figure 7 shows data provided by Zum Gahr[8] in support of Eq 3. It is seen that wear rate does indeed increase with applied pressure as would be expected from Eq 3. In the experiments using Al_2O_3, the threshold pressure for flaking was not reached and the softer, tougher cast iron exhibited higher wear rates (consistent with the solid lines in Fig. 6a). When SiC was used, the threshold was exceeded and the softer cast iron exhibited lower wear rates because of its higher toughness (consistent with the dashed lines in Fig. 6a).

Sliding Wear

The idea that wear can occur by a flaking or delamination process dates back at least to Styri's 1925 paper on rolling-contact fatigue.[12] He later suggested that such wear was caused by subsurface shear stresses.[13] Because of the differences between rolling-contact fatigue and sliding wear, these suggestions were largely overlooked for many years. However, both of Styri's suggestions are important in fracture-mechanics calculations of delamination, which involve a simple model of an asperity on the wear-causing surface passing over a delamination crack. The forces associated with the asperity are illustrated by the vector diagram of Fig. 8. If the normal force is N, there is an associated friction force, $N \cdot \tan \alpha$, where $\tan \alpha$ is the coefficient of sliding friction. Therefore, the resultant force is inclined at an angle α to the surface. In turn, both shear and normal forces are exerted on the crack. These can be estimated from simple elasticity theory.

Figure 8(b) illustrates the situation as the asperity approaches the crack. Note that the point force of Fig. 8(a) has been replaced by a distributed force to provide for a finite size of the asperity. In addition, the calculations assume that both the crack and the asperity are infinitely long in the plane normal to the drawing (i.e., plane strain). While this assumption is clearly incorrect, it is essential if the calculations are to remain relatively simple. In Fig. 8(b), the forces on the crack are seen to be compression, σ, and the forward shear, τ. Under these conditions the crack can only grow by shear. This requires the top face of the crack to slide over the bottom face and introduces a second coefficient of friction, $\tan \beta$,

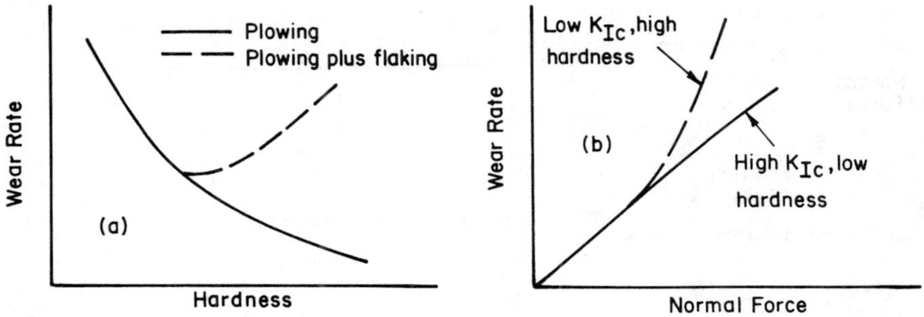

Fig. 6. Predictions of Zum Gahr's analysis of abrasive-wear resistance.

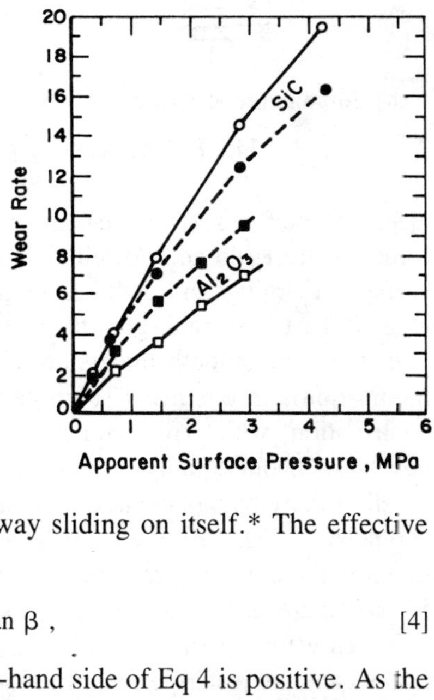

Fig. 7. Wear rate of fully supported specimens of two martensitic heat-treated states of lamellar cast irons with increasing load and different abrasive minerals. Open point symbols are for low-toughness/high-hardness; solid symbols, for high-toughness/low-hardness. *(Ref 8)*

associated with the material being worn away sliding on itself.* The effective shear stress on the crack then becomes

$$\tau_{eff} = \tau - \sigma \tan \beta , \tag{4}$$

and the crack can grow only when the right-hand side of Eq 4 is positive. As the force advances towards the crack, the shear stress decreases relative to the compressive stress. Eventually the right-hand side of Eq 4 becomes equal to zero and the applied shear stress is no longer large enough to overcome the crack-face friction (Fig. 8c). After the load has passed the crack, the situation changes and the compressive force becomes a tensile force. As is shown in Fig. 8(d), the crack can now open up and grow under combined tensile/shear loading.

All of the analyses in the literature use the geometry of Fig. 8, and all involve strictly elastic calculations. In my analysis,[14] a point force is considered and

*If the wear piece and the surface sliding on it are clean and self-mated, $\alpha = \beta$. In general, this relationship will not be true.

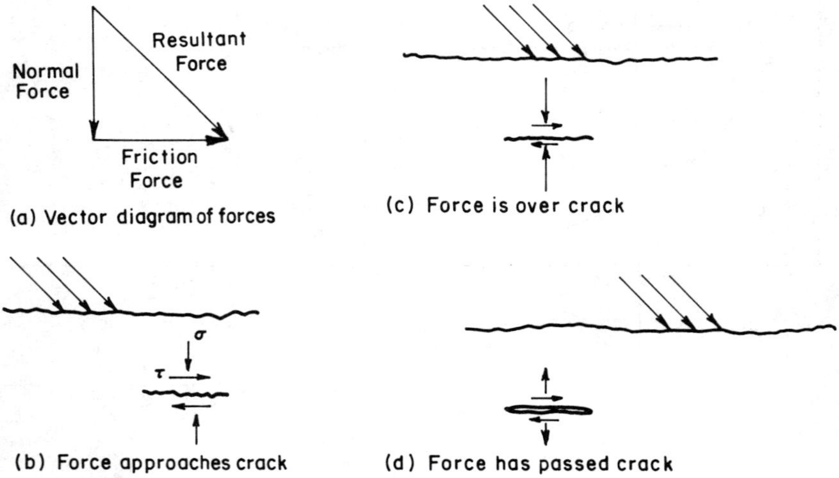

Fig. 8. Force causing growth of a delamination crack.

effects of the free surface are neglected. Fleming and Suh[15] provide for distributed force, while Hills and Ashelby[16,17] include both a distributed force and a free surface. Fleming and Suh[15] contend that only the tensile part of the cycle (Fig. 8d) causes crack growth by fatigue, but this is inconsistent with observations of the smooth underside of wear flakes (W. A. Glaeser, private communication). In contrast I suggested[14] that compression/shear makes the major contribution, while Hills and Ashelby[16,17] suggested that both components can contribute about equally. This question is unresolved.

All of these papers include estimates of the stress intensity associated with the geometry of Fig. 8. Initially, Hills and Ashelby[16] calculated values high enough to allow the crack to grow by a fatigue mechanism. They also found that about 50 passes are sufficient to extend the crack to the length of a typical wear-debris flake. However, their calculation required the entire bearing surface to be in the contact and not just asperities. In a later paper,[17] more realistic asperity sizes were used,and the resulting stress intensities became much too low, as was found by the other authors.[14,15]

The question of whether cracks can grow under compression/shear has been answered experimentally by Shieh,[18] who employed a highly constrained specimen. He found that stress levels almost twice the yield were required to form such cracks, a behavior that indicates that the phenomenon is indeed associated with plastic flow. Another indication that linear-elastic fracture mechanics may not be applicable was the observation that the fatigue-crack growth-rate equation for these cracks was far different from the usual one.[19] Shieh also derived an approximate expression for the stress intensity at the tip of a plastic crack. He pointed out that the stress intensity for an infinite stack of parallel cracks is larger than for a single isolated crack. His solution could serve as a first approximation to the case of a single delamination crack parallel to, and near, a surface. This effect could explain partly the very low stress intensities associated with the delamination-crack models discussed above.

In addition to estimating the driving force on a subsurface crack, it is also possible to predict the angle of crack growth. For example, Fig. 9(a) shows a kink formed under the influence of combined shear and compression. In a second paper, Shieh[20] pointed out that such a crack shape is expected theoretically. After reviewing the literature on rolling-contact fatigue, he concluded that the observed angles of deviation of the kinks from the original crack were 15° to 45°, in quantitative agreement with theory. Since these observations were made on hard and brittle alloys, it is not surprising that the elastic analysis was apparently verified.

However, there is a problem with this analysis in that it is inconsistent with the existence of cracks parallel to the wear surface; Fig. 10 shows an example.[21]

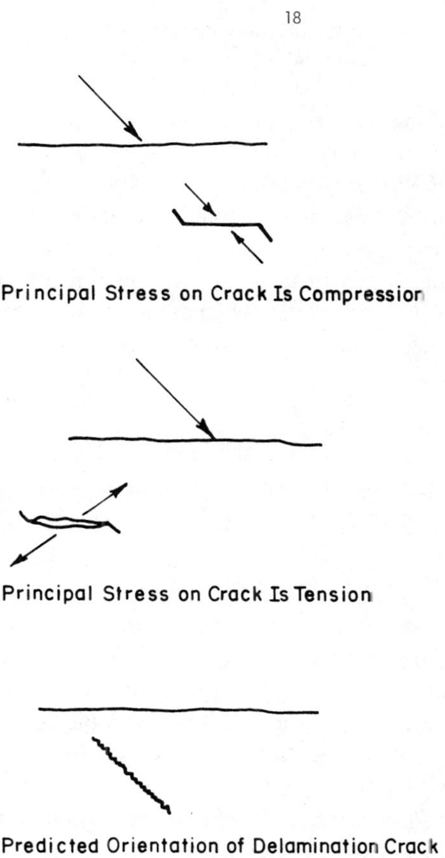

Fig. 9. Direction of crack growth according to the maximum-hoop-stress theory and the modified minimum-strain-energy theory.

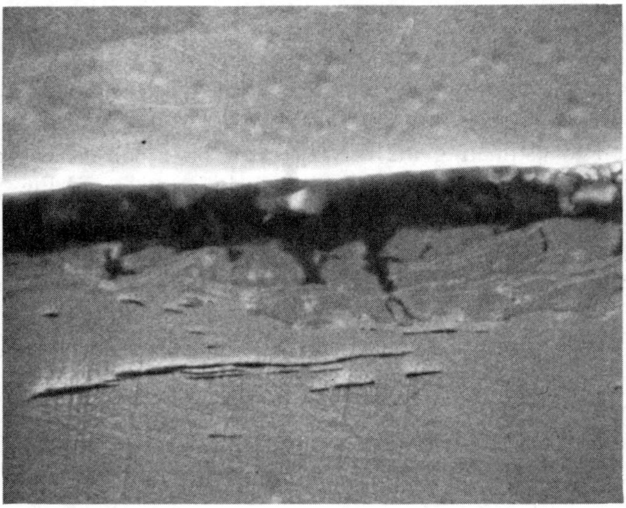

Fig. 10. Longitudinal section of Type 304 stainless steel which has experienced sliding wear. *(Micrograph by Hsu, Ref 21)*

Examination of the literature[22,23] shows that there are two theories that predict the direction in which this crack will grow: the maximum-hoop-stress theory and the minimum-strain-energy-density theory. Both these theories always agree closely for tensile/shear loading. The prediction for compressive/shear loading theory also agrees closely, provided that tan β (the coefficient of sliding friction between the opposing faces of the delamination crack) is less than unity.

Consider first the maximum-hoop-stress theory, recognizing that the stress field in the vicinity of the crack tip is given by

$$\sigma_y = f_y(\sigma, \tau, \tan \beta, a, r, \theta), \qquad [5]$$

where 2a is the crack length and r, θ are coordinates in a system centered at the crack tip. The crack will extend in the direction of maximum hoop stress, i.e., along a direction given by

$$\frac{\partial f_{\theta\theta}}{\partial \theta} = 0 .$$

Figure 9 shows the predicted directions of crack growth. Initially, as in Fig. 9(a), the principal stress on the crack is compressive. In this case, the front end of the crack will propagate deeper into the wear material and the back end will propagate towards the surface. When the contact point has passed by, as in Fig. 9(b), the principal stress on the crack becomes tensile. However, it turns out that the direction of crack propagation relative to the surface will be the same. If delamination wear is considered to be a fatigue process, with the crack growing a small bit each cycle, the net result would be that the crack will align itself at an angle to the surface. This is clearly at variance with observed delamination cracks, which tend to be parallel to the surface as was shown in Fig. 10.

A different result can arise from consideration of the minimum-strain-energy-density theory. In this theory, the crack will propagate in the direction in which the strain-energy density is a minimum (i.e., the potential energy is a maximum). If tan $\beta < 1$, its predictions agree with the maximum-hoop-stress theory. However, if tan $\beta > 1$, the compressive crack will propagate in the direction opposite to the prediction of the maximum-hoop-stress theory, as is shown in Fig. 11(a). For all values of tan β, the tensile prediction is the same for both theories. Therefore, Fig. 11(b) is identical to Fig. 9(b). However, unlike the previous case, the front of the crack will alternate going towards the surface and away from it. The net result will be a crack parallel to the surface, as is shown in Fig. 11(c). Of course, the crack will not be serrated, because the rubbing together of its faces during the compression part of the cycle will tend to make them featureless.

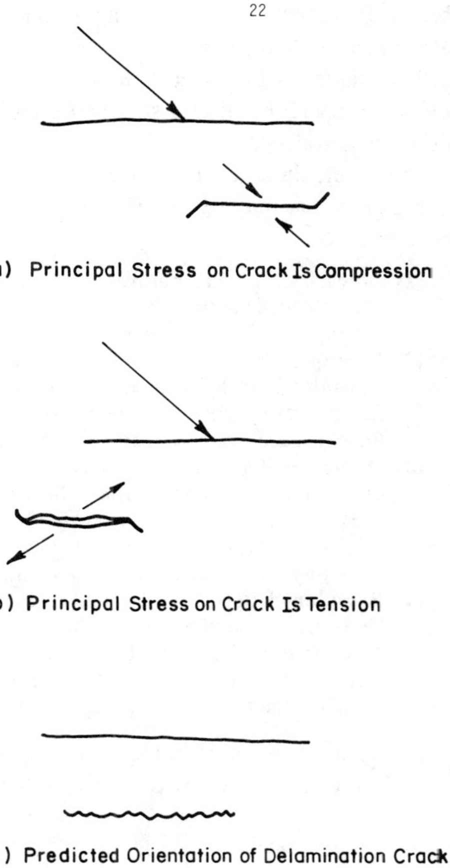

Fig. 11. Direction of crack growth according to the minimum-strain-energy-density theory.

While this approach overcomes the problem with the maximum-hoop-stress theory, it may be unsupportable. Swedlow[22] has developed a modified minimum-strain-energy theory that overcomes some objections to the original theory but results in predictions close to those of Fig. 11.

In summary, it appears that linear-elastic fracture mechanics is unlikely to provide a viable means of describing delamination wear. Thus, it is necessary to turn to approaches that involve plasticity. However, the elastic approach has been useful in providing insights into the problem. Among these are the recognition of fatigue as a significant factor in delamination wear and of the importance of the friction between the opposing faces in determining the crack driving force.

Discussion

Perhaps the most important point arising from this review is that linear-elastic fracture-mechanics analyses of wear are useful only when the wear piece is quite brittle. Brittleness, of course, implies that there is a limited amount of plastic flow associated with fracture. Conversely, in the case of tough materials, plasticity effects have to be taken into account. For very tough materials, such as OFHC copper, the plastic strains are enormous,[24] and elastic solutions may be quite inappropriate.

However, there are modifications to the usual rules of linear-elastic fracture mechanics that may be involved for the case of the material near the surface of a wear piece:

(1) **Linearity.** The material being worn away is heavily work hardened and is loaded and unloaded continually by the action of asperities. Thus, any given loading cycle can be considered as resulting in essentially elastic behavior.
(2) **Isotropy.** The wear process produces a highly textured material, and the governing equations have to be modified to take this effect into account. In addition, the fracture resistance may be anisotropic, and this effect may resolve the question of the existence of cracks parallel to the surface.
(3) **Homogeneity.** The wear piece is microstructurally inhomogeneous along the direction normal to the surface. In addition, it may contain inclusions that can be sites for subsurface crack formation. For example, Rice[25] observed much greater subsurface cracking and somewhat higher wear rates in a free-machining aluminum alloy than in a commercial high-strength aluminum alloy.
(4) **Boundary Effects.** Wear cracks form near free surfaces (in the mechanics sense). However, this problem has been partially resolved.[16,17]
(5) **Environmental Effects.** Lubrication will affect the local stress distribution at the surface and influence the stress intensity. In addition, the effects of stress-corrosion cracking and/or corrosion fatigue need to be considered.

Above all, it must be recognized that the application of fracture mechanics to wear is a relatively new approach and that much remains to be done in this area.

Addendum

Following the ASM seminar on wear, a preliminary elastic-plastic delamination model was developed.[26] The model consists of a subsurface shear crack with a strip of plastic deformation extending from each tip; it uses the solution of Bilby et al.[27] As in the elastic analysis discussed at the seminar, the growth of the crack is opposed by friction between its faces. However, the elastic-plastic model predicts crack growth when the fully elastic one does not. It also predicts step changes in wear rate with increasing load under special conditions, as has been occasionally observed.[28] In particular, step changes are expected for dry wear when the opposing materials are similar and when the normal load reaches a critical value which is dependent on the flow strength of the wearing material and on the friction coefficient for the system.

Acknowledgments

I am grateful to W. A. Glaeser, E. Hornbogen, and D. A. Rigney for their comments on the manuscript. The research is being sponsored by the U.S. Army Research Office, under Project No. DAAG29-78-G-0138.

References

1. J. P. Hirth and F. H. Froes: Met. Trans., *8A*, 1165 (1977).
2. D. A. Diesburg and F. Borik: *Materials for the Mining Industry*, R. Q. Barr, ed., Climax Molybdenum Co. (1974), 15; reprinted in D. A. Rigney and W. A. Glaeser, *Source Book on Wear Control Technology,* ASM (1978), 94.
3. American Society for Metals: *Metals Handbook*, 9th ed, Vol. 1 (1978), 607.
4. E. Hornbogen: Wear, *33*, 251 (1975).
5. K. H. Zum Gahr: Metal Progr., *116* [4], 46 (1979).
6. K. H. Zum Gahr: Z. Metallk., *69*, 312 (1978).
7. K. H. Zum Gahr and D. V. Doane: Metall. Trans., *11A,* 613 (1980).
8. K. H. Zum Gahr: Z. Metallk., *69*, 643 (1978).
9. M. Nakamura and J. Gurland: Met. Trans., *11A,* 141 (1980).
10. A. G. Atkins: Wear, *61,* 183 (1980).
11. R. K. Viswanadham and J. D. Venables: Metall. Trans., *8A,* 187 (1977).
12. H. Styri: Mech. Eng., *47,* 490 (1925).
13. H. Styri: Proc. ASTM, *51,* 682 (1951).
14. A. R. Rosenfield: Wear, *61,* 125 (1980).
15. J. R. Fleming and N. P. Suh: Wear, *33,* 251 (1977).
16. D. A. Hills and D. W. Ashelby: Wear, *54,* 321 (1979).
17. D. A. Hills and D. W. Ashelby: Eng. Fracture Mech., *13,* 69 (1980).
18. W. T. Shieh: Eng. Fracture Mech., *9,* 37 (1977).
19. J. S. Huang and R. M. Pelloux: Metall. Trans., *11A,* 899 (1980).

20. W. T. Shieh: Int. J. Fracture, *14,* 633 (1978).
21. K. L. Hsu: M.S. thesis, The Ohio State University (1978).
22. J. L. Swedlow: ASTM-STP *601,* 506 (1976).
23. G. C. Sih: Naval Research Rev., *32* [3], 30 (1980).
24. J. H. Dautzenberg and J. H. Zaat: Wear, *23,* 9 (1973).
25. S. L. Rice: Wear, *54,* 291 (1979).
26. A. R. Rosenfield: "A Dislocation Theory Approach to Wear," submitted to *Wear*.
27. B. B. Bilby, A. H. Cottrell, and K. H. Swinden: Proc. Roy. Soc., *A272,* 304 (1963).
28. N. C. Welsh: Proc. Roy. Soc., *A257,* 31, 51 (1965).

Characterization of Wear Surfaces and Wear Debris

A. W. Ruff and L. K. Ives
National Bureau of Standards
Washington, D.C. 20234

W. A. Glaeser
Battelle – Columbus Laboratories
Columbus, Ohio 43201

Abstract

Fundamental understanding of the wear process can be obtained and accurate analytical modeling can be carried out only to the extent that valid data and information are available. Since it is generally agreed that wear is a complex process involving many considerations, e.g., mechanics, material science, physics, and chemistry, it is essential that all sources of information be carefully and quantitatively explored. This paper will describe the type of information that can be obtained from the characterization of worn specimens and the debris particles produced during wear. There are actually three potential sources for data and information — the worn surface, the subsurface volume, and the wear debris particles. This paper will describe three different experiments carried out to develop an improved understanding of the wear of metals. In each case, the characterization methods applied were able to contribute necessary information to better understand the complete processes involved. Microscopic techniques of optical and electron microscopy were applied in these studies and will be described. The studies included direct observations of wear damage formation in copper, examination of the significance of plastic deformation in the wear of different steels, and microscopic studies of wear microstructure formation in copper alloys. It is hoped that these examples will indicate the potential value of surface, subsurface, and debris characterization techniques in wear research.

Introduction

When surfaces come into rubbing contact, they develop features or scars that can be interpreted in terms of wear mechanisms. Wear debris particles that are released from a surface during wear form an important constituent in the analysis regarding wear mechanisms. The interpretation of surface patterns caused by wear can be likened to fractography. Just as clear differences can be detected between the fracture patterns for ductile fracture and fatigue fracture, many wear patterns can be interpreted to identify erosive wear or abrasive wear.[1] Both systems require microscopy to resolve the surface features essential to the diagnosis. The scanning electron microscope (SEM), which has proven indispensable for fractographic analysis, is equally essential in the examination of wear scars and debris particles. The use of the SEM in wear analysis has been a relatively recent development in the continuing effort to understand the wear process. Prior to the advent of the SEM, some limited analysis was done by transmission electron microscopy using surface replication (as was done in fractographic analysis). The use of the SEM and other tools has advanced our understanding of the wear process, but there is still much work to be done to improve the interpretation of the considerable data which can be obtained from wear surfaces and wear debris.

Determination of the wear mode in failure analysis is one use of surface characterization methods. In the erosion of pump impellers, for instance, it is essential not only to determine whether erosion is present, but also to tell whether the erosion was caused by water drop impingement or solid particle impingement and whether corrosion is a factor. It is possible with current experience to do this. Cutting abrasion, a wear process that causes rapid removal of a surface, can sometimes be identified by the shape of abrasion scratches and the characteristic "micro chips" found in the wear debris.

A better understanding of the complex process of wear can be achieved by the use of other analytical techniques to supplement the SEM. These techniques are finding increased use in research directed at the fundamentals of wear. For example, electron microscopy is generally combined with conventional light microscopy because certain features seen by one technique are usually not seen by the other. For instance, the color of surface films or transferred material can be seen only by light microscopy. Particles on a wear surface can be removed with needles or tweezers under the light microscope. On the other hand, elemental analysis can be performed on surface features in an SEM equipped with energy dispersive X-ray analysis (EDAX), and three-dimensional features of wear microtopography at high magnification can be visualized by stereoscopic techniques not possible in light microscopy owing to depth of focus limitations.

Subsurface features are now known to be a very important aspect of wear analysis. Taper-angled microsections made both parallel and at right angles to the wear track can be subjected to both conventional light metallography and examination in the SEM.[2] Taper sectioning makes possible detailed examination of near-surface structure and the profiles of abrasion scratches. Microhardness measurements on regions very close to the surface can also be obtained through this technique. Metallographic sections parallel to the wear track provide information on the extent of near-surface plastic deformation. Grain boundaries can be seen bending over in the direction of flow. Total strain can be estimated from these features.[3] Sections made normal to the direction of sliding provide information on the extent of deformation under each microgroove developed by asperity plowing. It also provides information on the extent of lateral flow of surface material during the plowing process. When viewing a taper section by SEM, the specimen can be broken out of its plastic mount and positioned in the microscope so that the wear surface and the subsurface section can be observed simultaneously. Care must be taken in the way sections are taken and processed in order to preserve essential near-surface details.[4]

Surface profilometry can be used to measure the depth and profile of wear scars and the exact profiles of surface scratches, plateaus, etc. In fact, surface profilometry has been used to develop accurate topographic maps of worn surfaces.[5,6] A relatively quick method for estimating microgroove depths and profiles involves depositing micron-size polyethylene balls on the surface and observing them and the surface features by stereoscopic techniques in the SEM.

Other surface analytical tools, such as Auger spectroscopy, electron spectroscopy for chemical analysis (ESCA), and ion scanning spectroscopy, are also finding increasing application in the study of wear processes.[7] These techniques provide information on the composition and structure of the outermost few layers (not "visible" to electron diffraction, electron beam analysis, X-ray diffraction, etc.) of surfaces after wear. These techniques are revealing surface segregation, transfer of minute quantities from one surface to another, changes in surface composition in very thin layers and clues to the bonding mechanisms of wear resistant coatings. Infrared spectroscopy and ellipsometry have been used to identify organic compounds which form on surfaces as the result of sliding in organic fluids or vapors.

One of the most difficult techniques used in the analysis of worn surfaces involves transmission electron microscopy of foils made by chemical or ion-bombardment thinning of sections cut from wear specimens. Dislocation structures observed by this means have provided considerable information on the microstructural changes that occur in near-surface regions during sliding or rolling contact.[8]

Wear debris analysis holds great potential for diagnosis of wear processes. The premise of this approach is that during removal from the wear surface, debris particles should develop characteristic markings and shapes that relate directly to the mechanisms by which they were removed. Wear debris analysis is not yet at a point in development where it is totally reliable. Since collection of debris is usually done by remote means, the origin of the debris may not be known accurately. The debris may have been additionally deformed or altered after release from the surface. Some significant fractions of the debris may disappear either by combining with lubricant additives to form soaps or by deposition elsewhere in the system. Some debris observations can be misleading. For instance, spherical shapes can originate from airborne contamination by welding or high speed grinding processes during which micron-size molten metal droplets freeze into spherical shapes. As the understanding of debris removal processes improves, the interpretation of debris geometry, size and surface markings will become more reliable.

Debris collection is done by a number of methods. Ferrography is a popular means and involves separating debris from suspension in a fluid on a microscope slide by a magnetic process.[9,10] Among other methods, the simplest is to place a container under a wearing component (dry wear) and allow the particles to drop in. Wear particles can also be extracted from a component (wire rope, for instance) by flushing with a solvent and concentrating by centrifuging. The concentrate can then be filtered or allowed to dry on a microscope slide. Since drying on a slide can produce growth of organic crystals with debris as nucleation sites, care must be taken in analysis of the resulting dried product. Debris can be removed from bearing surfaces by stripping the surface with plastic replicating material or adhesive tape.

The above techniques represent a wide spectrum of possible methods for wear characterization. As experience is gained and the limitations and capabilities of these techniques are well understood, their suitability for given problems can be determined and their use optimized. In order to demonstrate wear characterization specifically, three examples will be discussed next in some detail. These examples show varied ways in which microscopy is aiding in research aimed at understanding wear processes.

In-SEM Wear Experiments

Scanning electron microscopy has provided us with a new conception of the microtopography produced by wear. The depth of focus available at high magnification has made this possible. The SEM has been used to characterize the wear process in a number of ways, including stereoscopic viewing of wear topography, elemental analysis of the surface by energy dispersive X-ray analysis, and measurement of strain by electron channeling.

These techniques involve observation of a worn surface after the wear has taken place. Since the surface character of the wear scar generally reflects events that occurred during the last pass of the contacting body, little can be learned of the progression of events that produce the final scar. To investigate this gap in information, the wear process can be observed in detail by performing wear experiments in the SEM. The formation of plate-like debris from cast iron has been observed in progress.[11] The transfer of material between two model asperities viewed in profile has been observed.[12] The deformation of the surface flow and formation of plate-like fragments has been observed in progress.[13]

An investigation in progress at Battelle is using the SEM to observe sequential changes in a wear path with each pass of a slider. The contact configuration consists of a steel ball pressed against the rotating flat face of a cylinder. The microscope is focused on the region just behind the ball and photographs are taken at 1000× magnification of a bench-marked area once per revolution. A schematic diagram of the system is shown in Fig. 1. Photographs of the specimen holder and actuator system are shown in Fig. 2 and 3. The design makes use of a rotating stage available in the SEM used for the experiments. The wear track is continually monitored in real time by using the TV mode of viewing while the cylinder is rotated. The operating conditions were as follows:

Ball (AISI 52100 steel): 2.54 mm (0.1 in.) diameter
Load: 0.060 kg (0.13 lb)
Estimated Hertz stress:
- Aluminum specimen: 613 MPa (89,000 psi), before plastic deformation
- 15Zn-Cu specimen: 690 MPa (100,000 psi)

Speed: 0.1 rpm or 0.05 mm/sec (0.12 in./min)

Single micrographs taken in sequence of one area of the wear scar located by reference marks outside the wear track have been taken for up to 20 sliding cycles and the cycle-by-cycle changes in microtopography analyzed. A sequence is shown in Fig. 4 to 7 (an aluminum surface). Aluminum transfer to the steel ball occurred with the first pass, causing formation of a prow similar to a built-up edge on a cutting tool. This changed the rake angle of the ball contact (steepened it),

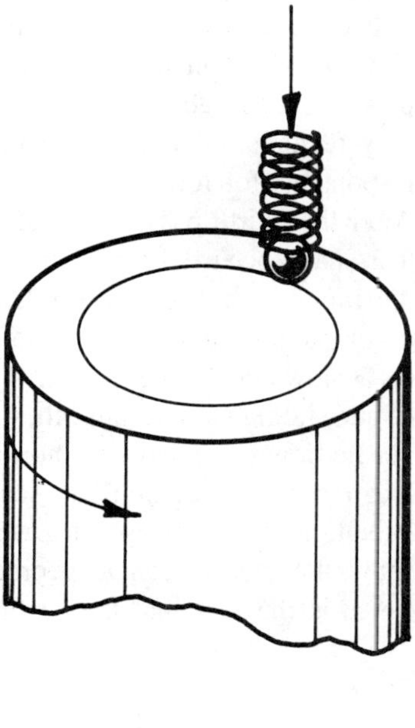

Fig. 1. Schematic diagram of ball and cylinder contact configuration used for in-SEM wear studies.

Fig. 2. Ball and cylinder specimens mounted on the SEM stage.

increased the depth of contact penetration, and narrowed the width of the wear scar. A conventional light micrograph of the aluminum "prow" on the steel ball is shown in Fig. 8.

It is quite obvious from a cursory observation of the sequence in Fig. 4 to 7 that surface changes occur rapidly. With each pass of the ball, most of the previous topography was altered. Note that the grooves marked 1 and 2 are the only features remaining after the first pass (Fig. 4). Presumably, the asperities responsible for forming grooves 1 and 2 remained intact for about four passes. After the fourth pass, a drastic change has occurred. A series of parallel lips or tears have occurred along the wear scar. The raised "lips" are not evidence of delamination but are more related to "chatter marks" or surface damage which accompanies the formation of a built-up edge on tools machining ductile materials at slow cutting speeds.[14] It is significant that these features appear as the result of one sliding pass because they indicate new surface formation and considerable plastic flow in a relatively short period of time. Note in Fig. 6 and 7 the apparent location of the beginning of separation (a) and the movement to (b).

Note also the release of a wear "sliver" at the side of the contact scar. This shows the effect of contact geometry on the shape of wear debris. Of equal significance is the fact that the surface damage shown in Fig. 6 is not changed in

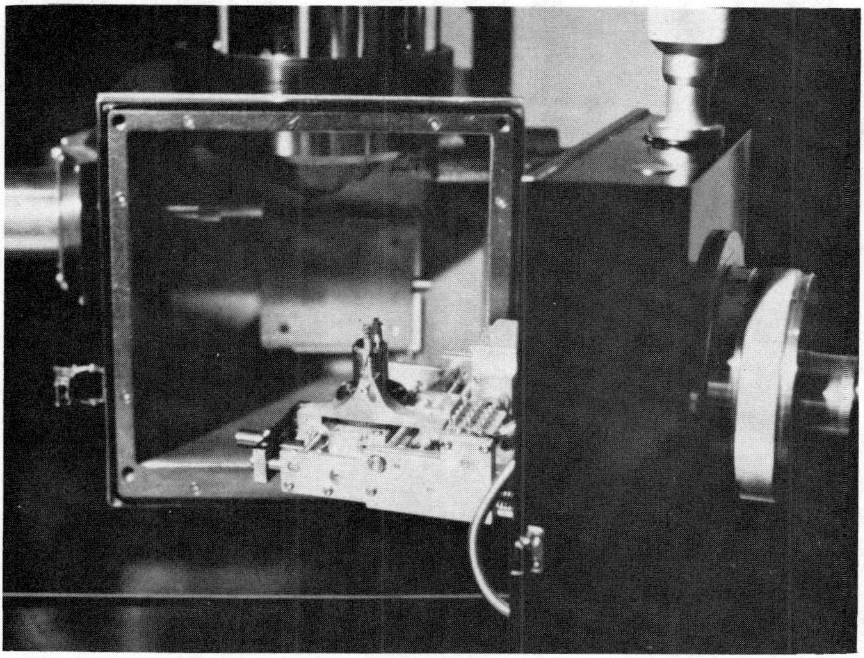

Fig. 3. SEM stage with mounted specimens and the SEM chamber.

Fig. 7. In fact, the basic configuration did not change over many ensuing passes of the contact. If anything, the lips tended to cold-weld to the surface and began to "heal."

The morphology of the early wear process in aluminum can be contrasted with that of a 15Zn-Cu binary alloy with approximately the same hardness and subjected to the same ball contact load and sliding velocity. The Zn-Cu alloy was coated with a thin organic film, the remnants of an alcohol cleaning process. The zone of interest in the sequence of wear track development was a radial scratch which the circumferential ball path intersected. The sequence shown in Fig. 9 to 11 reveals an entirely different sequence of microtopography. In this sliding contact, no gross adhesion or prow formation occurred. Contact was primarily by asperities on the ball surface. The thin organic layer on the copper alloy surface prevented adhesion and metal transfer during sliding contact. The contact area was wider than in the case of the aluminum experiment because the ball curvature dictated the size and shape of the elastic contact area. In the sequence example shown, the most obvious changes occurred at the cross-over zone of the radial scratch. A lip thrown up by the scratching process was deformed by the ball, and in the leveling process, the metal was extruded circum-

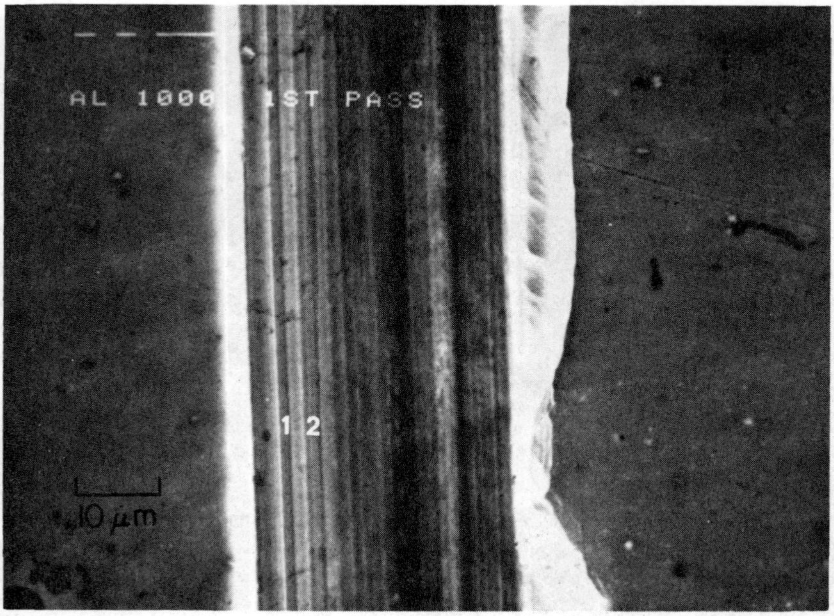

Fig. 4. First pass.

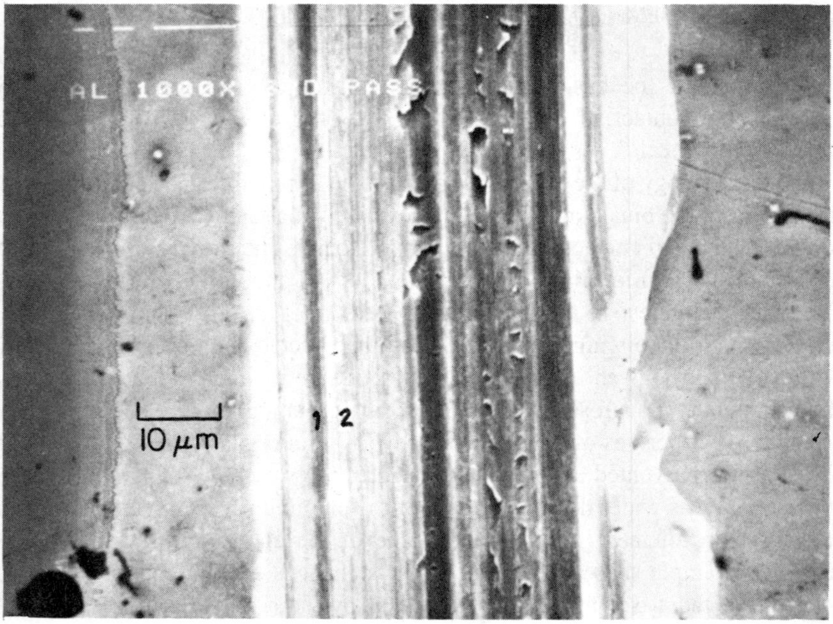

Fig. 5. Third pass.

Fig. 4 to 7. Wear from first, third, fourth and fifth sliding passes — aluminum

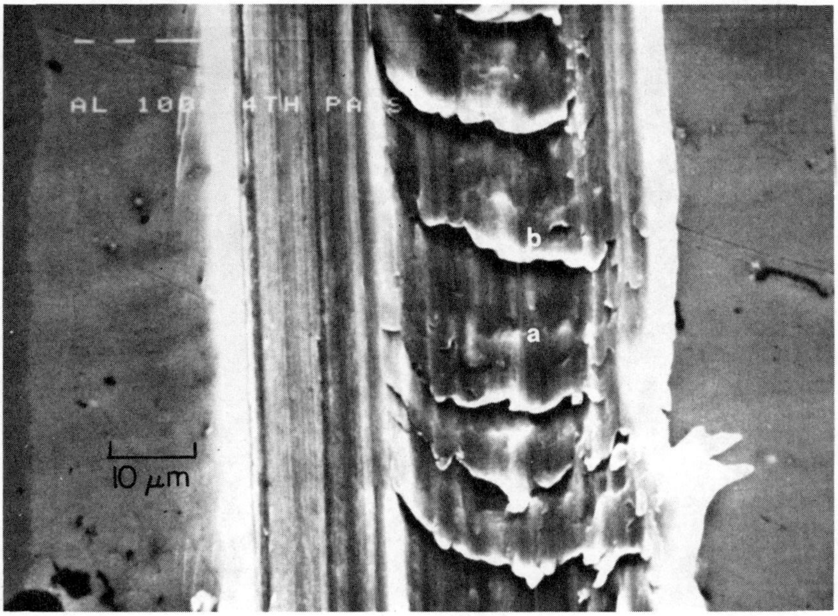

Fig. 6. Fourth pass.

Fig. 7. Fifth pass.

sequence. Direction of sliding is from bottom to top, all micrographs.

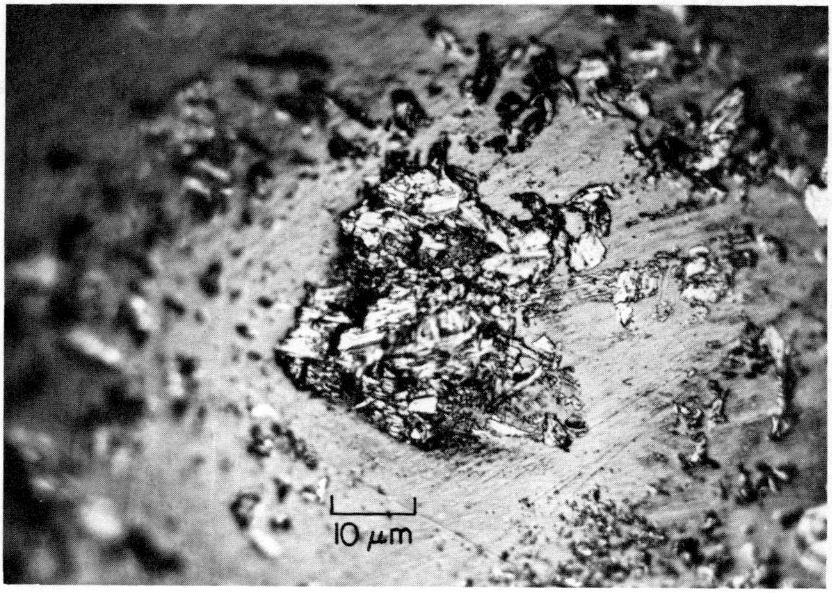

Fig. 8. Light micrograph of aluminum "prow" adhering to the steel ball surface. Direction of sliding is from bottom left to top right.

ferentially and radially. Several interesting changes in features can be noted. In the left center of the field, a deformed area of the ridge (A) is gradually extruding in the direction of sliding. Not only does this area grow in size, but an extruded lip of material is seen breaking loose (finger-like feature) and finally leaving the surface as a piece of debris.

A series of fine scratches in the upper right corner of the field (B) can be seen to change somewhat in precise shape, but their basic orientation and location remain fixed. The asperities causing these scratches, therefore, continue to track in the same grooves for several passes and the scratches seen in Fig. 11 can be interpreted as persistent features. Finally, considerable dark deposits can be seen surrounding the wear scar in this sequence. The deposit is part of the surface film scraped off by the ball and deposited at re-entrant ledges on the wear track. This deposition was a continuous process, causing a gradual increase in the size of the dark deposits. When this area was viewed by light microscopy, the deposit showed blue and orange interference colors. The deposit apparently was a lamination of thin transparent plates. These could be seen separately further down the wear path.

At another location, a "curled ribbon" type of wear particle was observed in the process of removal from the wear scar. The particle in the last stages of removal is shown in Fig. 12. The particle started as a sideward extrusion of a plowed ridge which subsequently tore in the direction of sliding.

The track widths in both sequences (the aluminum and the copper alloy) tended to increase with number of passes. When the track width was plotted as a function of number of passes, a linear plot resulted. This characteristic is shown in Fig. 13. The track growth rate clearly shows that plastic strain was occurring during each pass of the ball, but there was not necessarily any removal of material. The sequences represent a wear-in period of initial wear in which the size and shape of the wear scar are established rapidly. It is expected that if the experiments were continued for long times, a point would be reached where the rate of scar width increase would diminish considerably and steady-state conditions would be achieved.

Conclusions

The sequential *in situ* observation of changes in the microtopography of a wear scar reveals a number of significant characteristics of the mechanisms of wear. For ductile materials, the initial wear process involves considerable surface flow principally in the direction of sliding, but also normal to the direction of sliding. Some relatively major events such as surface tears and release of wear slivers can occur even during a single sliding pass. In general, most microscopic features seen in a wear scar represent what happened during the last sliding contact pass before making the observations. Some features such as fine scratch marks can persist for a number of passes, although they also change subtly. When the rake angle of a contacting asperity is high, the penetration increases and surface damage becomes more extreme.

These experiments have also demonstrated that the shape and size of debris produced during wear-in can depend on contact geometry. A point contact such as a ball-on-flat or pin-on-disk can produce fibrous debris originating from the sides of the wear scar. A line contact such as a journal in a bushing can produce flat plate-like debris broken off from extrusion at the exit zone of the contact rectangle. Interruptions in a surface which cross a wear path can generate flat debris as surface material is extruded into the groove and broken off.

The observation of wear in progress with the SEM is still being explored and new techniques are being developed. The experiments so far have shown phenomena occurring which had not been anticipated from the study of "after the fact" wear scars. The technique is useful because, as it has demonstrated, many transient events which occur on a microscopic scale in the wear scar lead in a step-by-step process to release of debris and the development of the final features seen on a worn surface after removal from contact. These transient events are lost to the investigator unless he can observe them as they occur.

Fig. 9. Wear scar after sixth sliding pass for 15Zn-Cu sequence. Direction of sliding here and in Fig. 10 and 11 is from bottom to top.

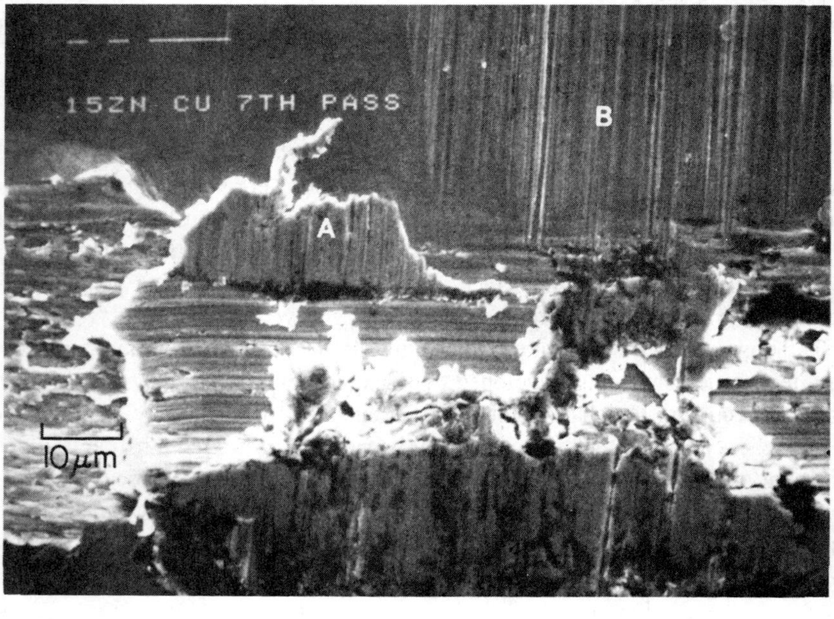

Fig. 10. Seventh pass for 15Zn-Cu sequence.

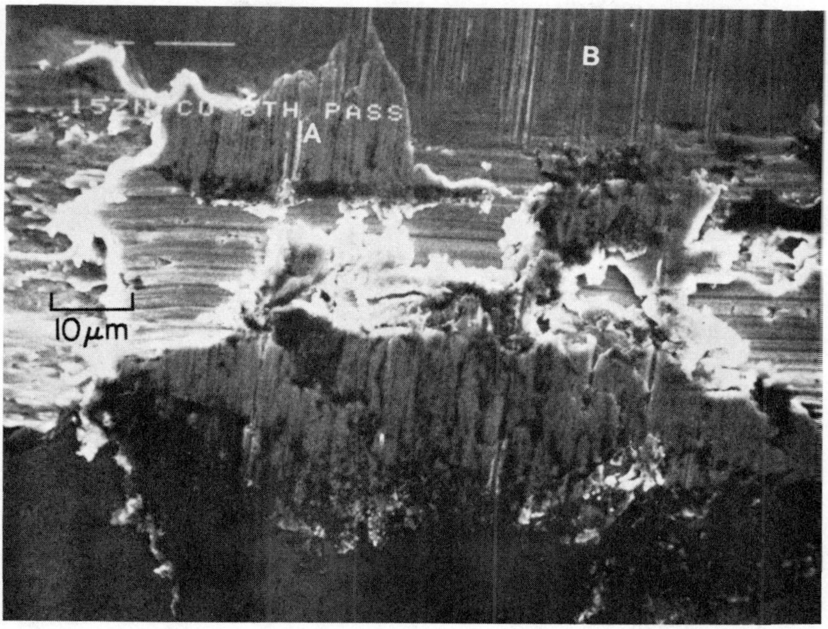

Fig. 11. Eighth pass for 15Zn-Cu sequence.

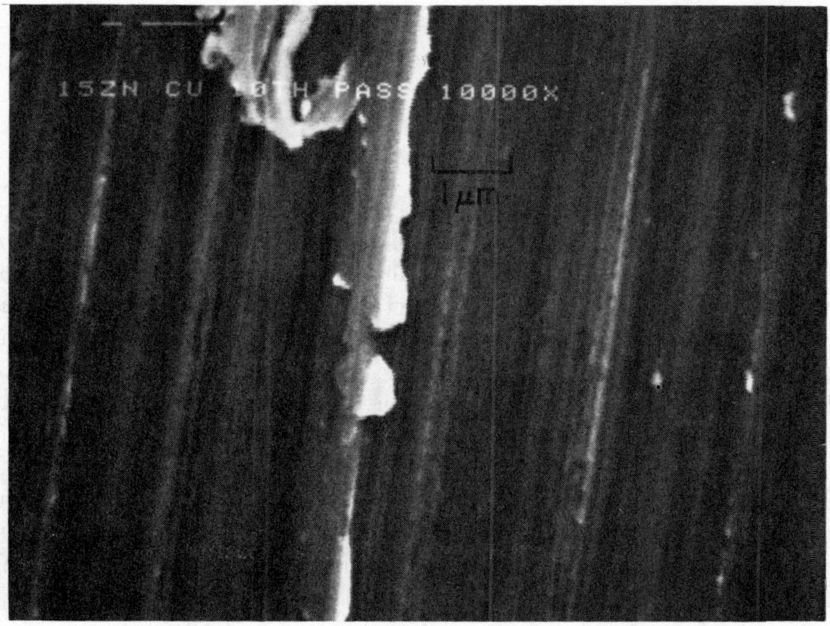

Fig. 12. Wear particle in the process of removal from the wear track in the 15Zn-Cu sequence.

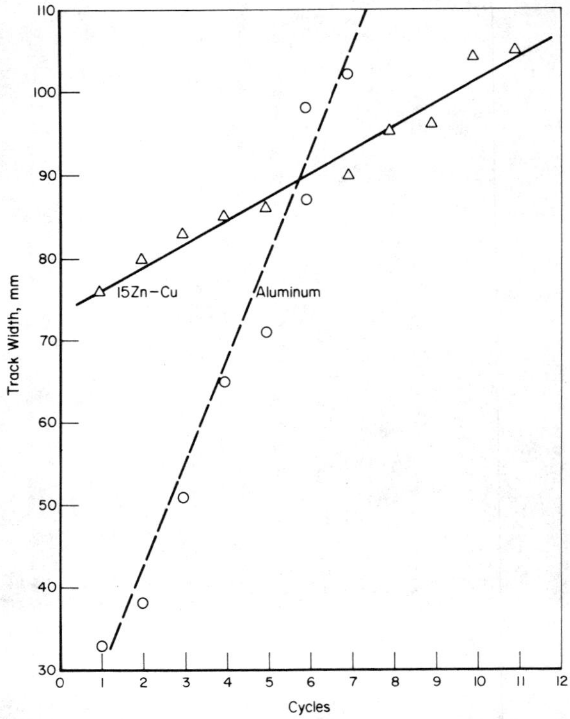

Fig. 13. Wear track width as a function of number of cycles of sliding from *in situ* SEM wear study.

Dry Sliding Wear Studies of Steels

This study was directed toward determining the effect of variation in the plastic work hardening and other properties of different steels on their wear characteristics under conditions of dry sliding wear. Particular emphasis has been placed on the structures of the worn surfaces and the type of wear debris formed using three different types of steel. The results indicate that in these experiments, plastic deformation played the major role in the wear process, although surface oxide effects due probably to water vapor in the argon atmosphere used are also a significant consideration. Several characteristic types of wear debris and surface structures were identified, and those aspects were consistent with the prominence of plastic deformation during wear. While previous studies of the wear of steels (see, for review, Ref 15) have identified several significant metallurgical factors, the question of plastic deformation characteristics and their role in wear has not received sufficient attention.

Experimental Methods

The wear test system used in this study involved a conventional block-on-ring geometry in a specially designed system utilizing a variable speed electric motor with a replaceable 52100 steel ring (34.9 mm in diameter, 8.7 mm wide) attached to the motor shaft to provide sliding motion. The test specimen was loaded on top of this ring and held in place on an instrumented arm which was free to pivot in a vertical plane. The holder used for the rectangular block specimens (6.3 × 12.7 × 25.4 mm) was designed so that weights could be placed over the contact in order to vary the load. The load used in all tests reported here was 10 N. A clear plastic enclosure around the wear contact region was used to maintain a flowing argon atmosphere. Humidity of about 30% RH was typical during the wear tests. A cam surface on the wear ring holder on the motor shaft was used to open and close a switch once per cycle. This switch signal was used to synchronize digital data sampling with the motor shaft position to reduce the effect of wear ring surface run-out. Surface run-out was typically 10 μm; however, this data sampling method reduced the effect by a factor of about 100. Wear debris was collected on a clean glass slide positioned below the wear contact. All the experiments carried out with this system used a sliding speed of 0.20 m/s (110 rpm).

The wear test system was interfaced to a minicomputer system for control and data acquisition purposes. Four variables were measured: friction force, contact displacement, enclosure or specimen temperature, and time, in a repeating cycle. Contact displacement was measured with a linear variable differential transformer (LVDT) which monitored the change in height of the specimen relative to the ring surface. Friction force was monitored through the use of two strain gauges attached to the pivoting arm; the resulting signal was directed to an amplifier. Temperature was measured by a thermocouple either placed within the enclosure or welded to the specimen. Running time was determined by the computer internal clock. The input signals were directed to an addressable switch scanner (100 Hz switching rate) and then to a programmable digital voltmeter (25 Hz reading rate). Program control enabled the computer to select a scanner channel, set the range and operating mode for the voltmeter and then read a value from the voltmeter. Data were printed during the wear test, usually the average of 10 consecutive readings at a 2 Hz rate. Data were also magnetically recorded for processing using the computer and a plotter. Processing software was developed to plot friction coefficient vs time, wear displacement vs time, and wear volume vs time. The wear displacement and wear volume plots could be made either with the raw data or with suitably averaged data.

The 52100 steel ring and metal specimen block were fastened in place to begin the wear test after cleaning in hexane and acetone. The specimen blocks were previously polished through 600 grit SiC paper. The block specimens were aligned using two set screws in the specimen holder. A small light was used to illuminate the contact area between the block and the ring and a magnifier was used to inspect the contact for parallel surfaces. When the specimen was properly aligned, argon was introduced into the chamber for wear tests. After about 10 minutes of argon flow the test was begun. When the test was completed, the specimen and the collected wear debris were removed and the ring was cleaned with hexane and acetone to remove loose wear debris so that the ring was available for another test. Test rings were changed whenever a new type of specimen was tested or when the rings began to show significant wear or transferred metal build-up. Typically two to four tests were conducted on one 52100 steel ring.

Materials Studied

Iron-based alloys of three types were selected for this investigation: a plain carbon steel, a microalloyed (dual phase) steel and an oil-hardenable tool steel. Plain carbon steels are utilized in many wearing situations since they offer a low-cost materials option. An AISI 1015 steel was selected for study here. The composition was 0.14%C, 0.4%Mn, 0.04%P, 0.05%S, the balance being iron. The material was obtained in hot-rolled strip form, 3.2 mm thick and 19 mm wide. Wear specimens were machined from this strip and annealed at 925°C for one hour in evacuated glass vials, and slowly cooled prior to testing. The microstructure prior to wear testing, shown in Fig. 14(a), consisted of ferrite and pearlite.

Microalloyed high strength, low alloy (HSLA) steels containing small amounts of alloying elements such as copper, chromium, columbium, silicon, and vanadium and a relatively low carbon content of about 0.2% have been developed in recent years.[16,17] Strengthening is achieved by a small grain size along with contributions of solid-solution and precipitation hardening. A related development of ferrite-martensite steels, so-called dual phase (DP) steels are a special class of microalloyed steel that offer improved formability at the same strength levels. They are obtained by reheating a HSLA steel in the ferrite-austenite region (intercritical annealing) to develop a structure of about 80% ferrite and 20% martensite with some retained austenite. Certain mechanical characteristics such as ultimate strength and ductility suggest the potential of these alloys for effective wear performance. One such dual phase alloy (DP80) will be reported on here. The indicated composition was 0.13%C, 1.56%Mn, 0.63%Si, 0.06%Al, 0.036%V, 0.01%P, 0.01%S, 0.014%Ce, 0.008%N. Wear specimens were cut from the sheet material after grinding about 0.1 mm from the surface but were otherwise used in the as-received (annealed) condition. The initial microstructure, shown in Fig. 14(b), consisted principally of ferrite with some martensite and retained austenite regions.

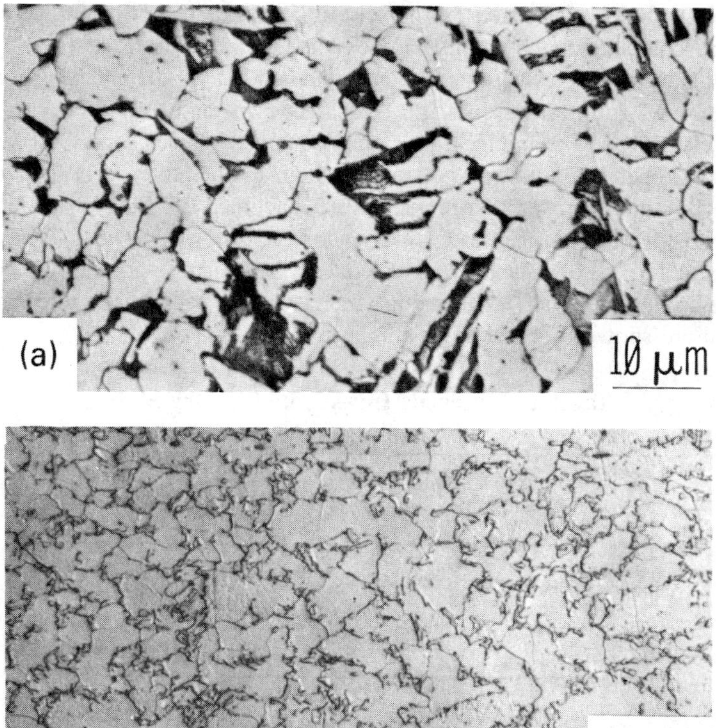

Fig. 14. (**a**) 1015 steel microstructure, etched in nital; ferrite and pearlite. (**b**) DP80 steel microstructure, etched in nital; ferrite and martensite. (**c**) O-2 tool steel microstructure, etched in nital; ferrite and carbides.

Tool steel alloys are used extensively in situations involving high wear rates and are characterized by high hardness and good wear resistance.[18] An oil-hardening tool steel, AISI type O-2, was studied here in the annealed condition. The nominal composition was 0.90%C, 1.6%Mn, and 0.25%Si. This steel is relatively inexpensive and the high carbon content produces good wear resistance in cutting operations that do not involve high temperatures. The material was obtained in bar form; wear specimens were machined from the bar and annealed at 815°C for one hour in sealed glass vials, and slowly cooled. The resulting microstructure, shown in Fig. 14(c), consisted of ferrite with carbide particles. The microhardness values and grain size values for each of the three materials studied are shown in Table 1 for comparison.

Table 1. Materials Investigated in Dry Sliding Studies of Steels

Material	Condition	Microhardness, KHN_{200}(kg/mm^2)	Average grain intercept length, μm
1015 steel	Annealed	178±8	49±7
DP80 steel.......	As-received (annealed)	235±8	8±3
O-2 tool steel	Annealed	233±5	30±7

Results and Discussion

Wear and Friction Measurements. Wear and friction measurements were carried out on the alloys using a 10 N load and 0.20 m/s sliding speed in an argon atmosphere at about 23°C. Sliding distances of either 363 m or 726 m were used. Examples of the friction curves and wear displacement curves obtained from several experiments on each material are shown in Fig. 15 to 17. These data show some of the variations encountered in the wear experiments. Figures 15(a) and 15(b) show different trends in the computed wear volume change* with time for two DP80 steel specimens. The steady-state wear rate on both examples is, however, in agreement as determined from the last 25% to 50% of the experiment. While the friction-distance curves in these cases are of a similar type, the beginning friction peak in Fig. 15(a) is more pronounced, possibly reflecting the nature of the break-in processes for this particular specimen-ring combination. The decreasing slope of the wear volume–distance curve at early times in Fig. 15(a) may also result from this same effect. In both curves the depth of the wear scar measured after the experiment was completed is indicated on the wear axis at the point P. A stylus profilometer was used to obtain this value. In this way

*Wear volume per unit length of wear scar is calculated (mm^3/mm).

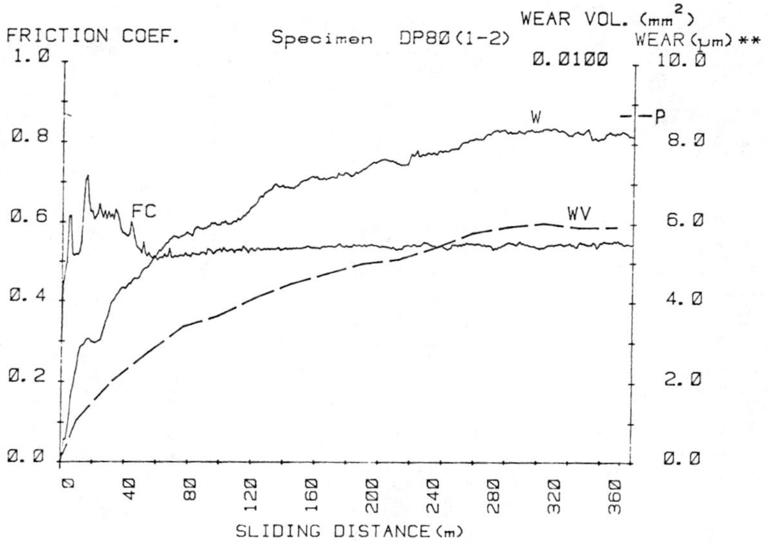

Fig. 15(a). Wear depth (W), wear volume (WV) and friction coefficient (FC) results for DP80 steel specimen.

an absolute measure of wear scar depth is obtained for comparison with the wear depth values obtained during the experiment using the LVDT system. The wear scar depth measured by profilometer was usually 10% to 20% greater than that measured by the LVDT system. This difference is attributed to the effect of wear debris in the contact zone and metal transferred to the ring surface, both cases leading to a reduced displacement value measured by the LVDT system.

The results of another experiment using a DP80 steel specimen and a new ring for a greater sliding distance are shown in Fig. 15(c). The wear curve here is not statistically smoothed before plotting and shows the extent of a measured variation in the wear contact relative position. The wear volume–distance curve in Fig. 15(c) shows a significant transition in slope at about 480 m sliding distance. The initial wear rate (wear volume per unit sliding distance) is linear but is considerably smaller than the final wear rate. This effect was reproduced in several additional experiments with the DP80 alloy and apparently reflects an actual change in wear process in the latter portion of these experiments. In view of this finding, two average wear rate values are reported for the DP80 steel, one for 363 m sliding distance experiments and one for 726 m sliding distance experiments. This effect was not found for either the 1015 steel or the O-2 tool steel.

Two experimental wear and friction curves for 1015 steel specimens are shown in Fig. 16. In Fig. 16(a) the wear data are averaged over 5-second intervals for a smoothing effect. Both experiments produced reasonably linear wear volume – sliding distance curves, although both showed a lower wear rate for the

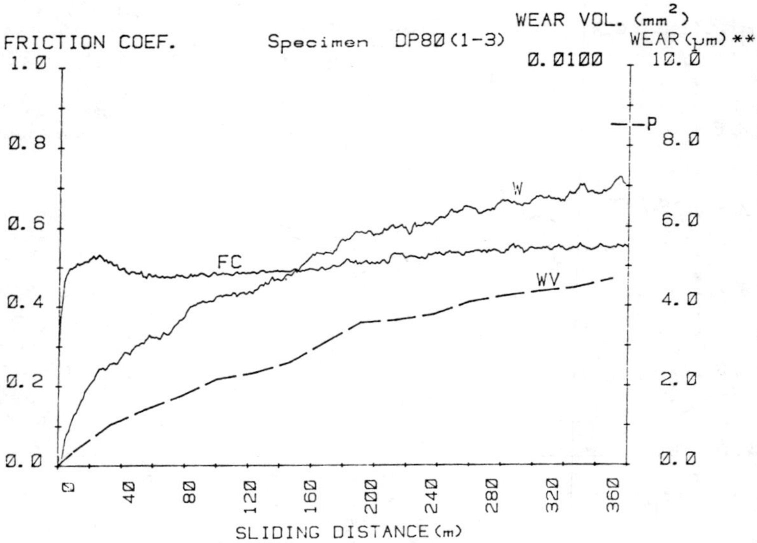

Fig. 15(b). Wear depth (W), wear volume (WV) and friction coefficient (FC) results for DP80 steel specimen.

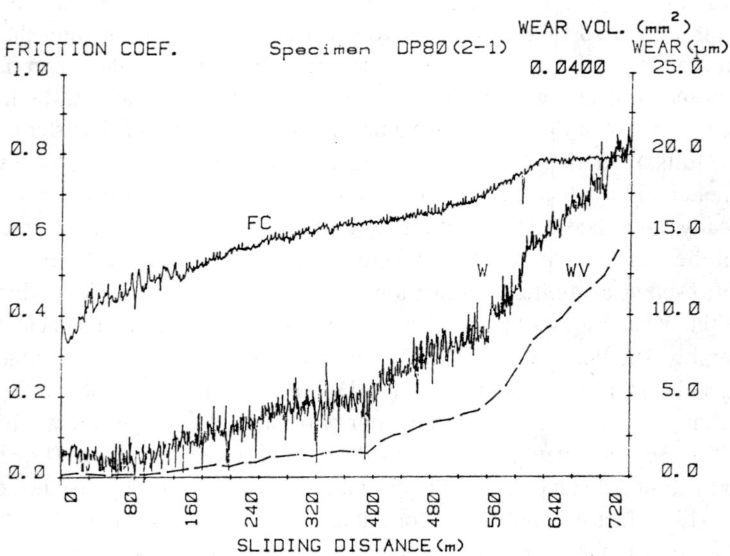

Fig. 15(c). Wear depth (W), wear volume (WV) and friction coefficient (FC) results for DP80 steel specimen.

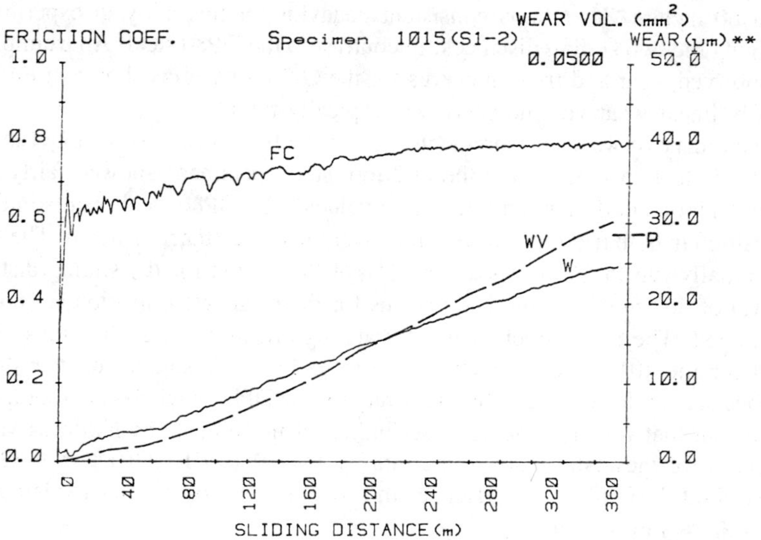

Fig. 16(a). Wear depth (W), wear volume (WV) and friction coefficient (FC) results for 1015 steel specimen.

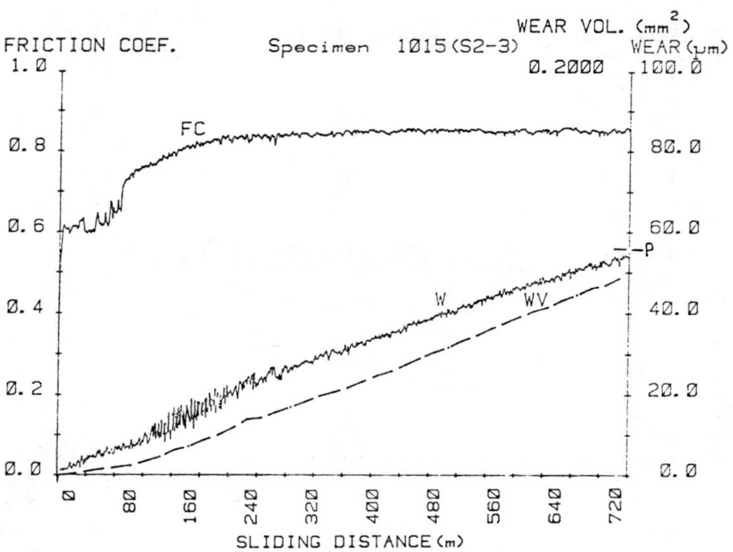

Fig. 16(b). Wear depth (W), wear volume (WV) and friction coefficient (FC) results for 1015 steel specimen.

initial 60 meters. There was consistent behavior for this alloy in experiments involving the two sliding distances, in contrast to the DP80 steel. An example of the observed wear and friction curves for the O-2 tool steel is shown in Fig. 17. A fairly linear wear volume curve was typically found.

A summary of wear measurements for the steels studied is shown in Table 2. The 1015 steel specimens and the O-2 tool steel specimens showed fairly consistent friction coefficient and wear rate values. The DP80 steel results indicate a transition in both friction and wear rate over the two distances studied. The wear rate initially (up to 363 m) is about 25% of that found for the sliding distance interval of 363 m to 726 m. The reasons for the wear rate transition are not yet understood. The DP80 steel wear rates are significantly lower than those measured for the 1015 steel, a factor of ⅛ to ½ lower depending on the sliding distance accumulated. The initial wear rate for the DP80 steel was also somewhat lower than that for the tool steel. As indicated in Table 1 the hardness values measured for these steels are in the ratio KHN(DP80)/KHN(1015) = 1.32 and KHN(DP80)/KHN(0-2) = 1. Hence, the wear rate ratios are not explained by the hardness ratios alone.

Wear Surface and Wear Debris Observations. Scanning electron microscope studies have been carried out on the worn surfaces and on debris particles recovered from the steel specimen wear tests. A schematic representation of the

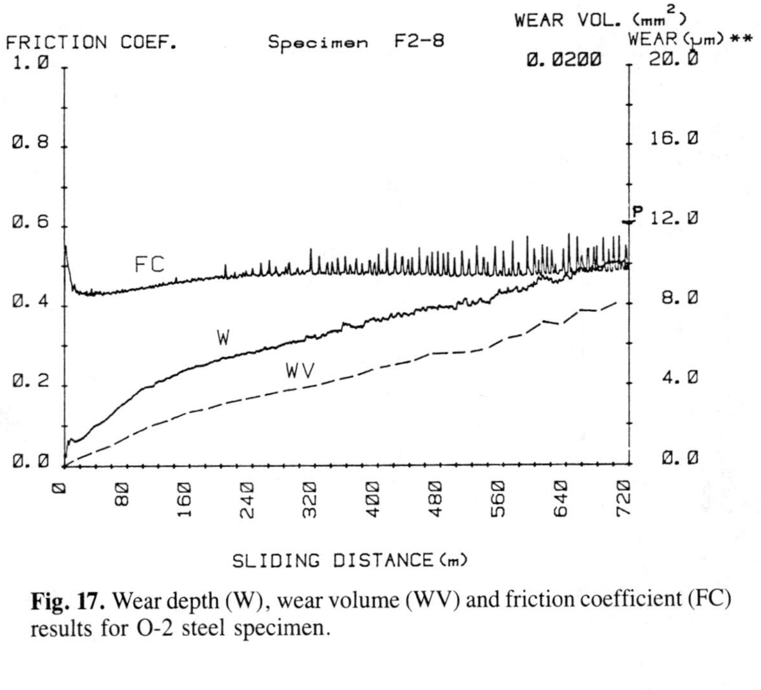

Fig. 17. Wear depth (W), wear volume (WV) and friction coefficient (FC) results for O-2 steel specimen.

Table 2. Wear and Friction Results on 1015 Steel, DP80 Alloy and O-2 Tool Steel

Sliding distance(a), m	Steady-state friction coef.	Final wear scar depth, μm	Final wear vol, mm^3	Wear rate(b), mm^3/m	Wear rate from LVDT data(c), mm^3/m
		Results on 1015 Steel			
363 (N)	0.65	20	0.19	5.3×10^{-4}	2.9×10^{-4}
363	0.80	27	0.31	8.4×10^{-4}	8.7×10^{-4}
363	0.72	30	0.36	9.8×10^{-4}	9.9×10^{-4}
363	0.80	22	0.22	6.2×10^{-4}	9.5×10^{-4}
Avg	0.74	25	0.27	7.5×10^{-4}	7.75×10^{-4}
726 (N)	0.76	33	0.41	5.7×10^{-4}	6.7×10^{-4}
726	0.83	56	0.91	12.5×10^{-4}	14.5×10^{-4}
Avg	0.80	45	0.65	9.0×10^{-4}	10.6×10^{-4}
		Results on DP80 Alloy			
363 (N)	0.50	5.4	0.028	0.77×10^{-4}	1.6×10^{-4}
363	0.55	8.7	0.056	1.5×10^{-4}	1.0×10^{-4}
363	0.55	8.6	0.055	1.5×10^{-4}	0.72×10^{-4}
363	0.55	7.1	0.041	1.1×10^{-4}	0.87×10^{-4}
363	0.55	6.5	0.036	0.99×10^{-4}	0.68×10^{-4}
Avg	0.55	7.4	0.044	1.2×10^{-4}	0.97×10^{-4}
726 (N)	0.80	27	0.30	4.1×10^{-4}	5.8×10^{-4}
726 (N)	0.80	29	0.34	4.7×10^{-4}	7.5×10^{-4}
726	0.75	28	0.32	4.4×10^{-4}	6.2×10^{-4}
Avg	0.78	28	0.32	4.4×10^{-4}	6.5×10^{-4}
		Results on O-2 Tool Steel			
363	0.48	9.9	0.068	1.9×10^{-4}	
363	0.53	11.6	0.085	2.3×10^{-4}	
363 (N)	0.48	7.3	0.043	1.2×10^{-4}	
363	0.52	8.2	0.051	1.4×10^{-4}	
363 (N)	0.45	6.9	0.039	1.1×10^{-4}	
Avg	0.49	8.8	0.056	1.6×10^{-4}	
726 (N)	0.53	13.5	0.11	1.5×10^{-4}	
726 (N)	0.52	12.7	0.098	1.3×10^{-4}	
726	0.52	14.9	0.12	1.7×10^{-4}	
726	0.50	12.1	0.091	1.2×10^{-4}	
Avg	0.52	13.3	0.11	1.5×10^{-4}	

(a) N signifies new ring. (b) Final wear volume/sliding distance. (c) Average over last one-third of experiment.

block-on-ring geometry is shown in Fig. 18 for use in this discussion. Examples of debris particles recovered as they fell from the 1015 steel block during wear experiments are shown in Fig. 19. Generally, two types of particles are found, large flat particles as at A and much smaller, equiaxed particles. Figure 20 shows other examples of this wear debris. In the lower micrograph one of the large, flat particles (A) appears to be partly fractured. The smaller particles (less than 1 μm in size) are frequently found clinging to the larger particles and also clumped themselves into a larger aggregate such as at B. Examination of the entrance zone of the wear scar on the block revealed a similar collection of debris particles. The exit zone of the wear scar on the block showed a different morphology as indicated in Fig. 21. The lower half of each micrograph is the wear scar; the upper half is unworn, original surface. There is some plastic flow of the steel out of the wear scar region causing the irregular edge, approximately perpendicular to the sliding direction (shown by the arrow). One of the plastically deformed protrusions is seen in more detail in Fig. 21. Clearly some evidence is present suggesting tearing of this material at the edges formed during the microdeformation process. Stereo scanning electron microscope studies of the morphology present provided convincing evidence for debris formation at many of these protrusions.

Studies of the wear scar surface within the contact zone also revealed evidence for considerable plastic deformation. Figure 22 shows two regions (A) within the wear scar where particles of size 3 μm to 10 μm appear to be forming by a partially ductile tearing process. Stereo viewing of micrographs taken from the wear scar region suggests that the normal load at any time is carried by only a small fraction of the apparent area and that plastic flow both in the direction of sliding (arrow) and transverse to that direction is probably significant in forming the ridges and lips of material, as well as forming debris particles.

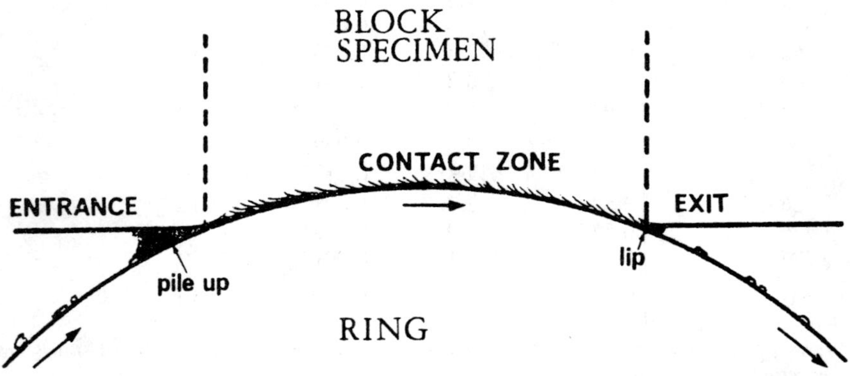

Fig. 18. Schematic representation of block-on-ring geometry.

Fig. 19. Scanning electron micrograph of wear debris particles from 1015 steel specimen.

Wear debris was also recovered from experiments involving the DP80 alloy. One collection of particles is shown in Fig. 23. Three different classes of particles were found; further details are presented in Fig. 24. Relatively large flat particles (A) were found that frequently show crack-like features (arrow). Comparably large particles of a distinctly different morphology (B) were also found. Further, much smaller (< 1 μm) wear particles of roughly equiaxed dimensions were always present. One area in the entry region of the wear scar is shown in Fig. 25(a). Wear debris particles found in that region (Fig. 25b) were similar to those recovered after falling from the wearing contact. Again score lines and crack-like features are seen (A). The morphology at the exit side of the wear scar is seen in Fig. 26. There is considerable evidence for plastic flow of material. In Fig. 26(b), the lip of metal shown has numerous crack-like features at the edges as well as transverse ripple markings that suggest constrained plastic deformation. Studies using stereo imaging of micrographs of the wear scar surface within the contact zone (such as Fig. 22) showed that the normal load was probably carried by a small proportion of the apparent contact area as evidenced by flat ridges of metal (A) extended in the sliding direction. Irregular edges of the ridges were commonly found. In most of the micrographs obtained, the regions within the contact zone that were depressed appear to have collected quantities

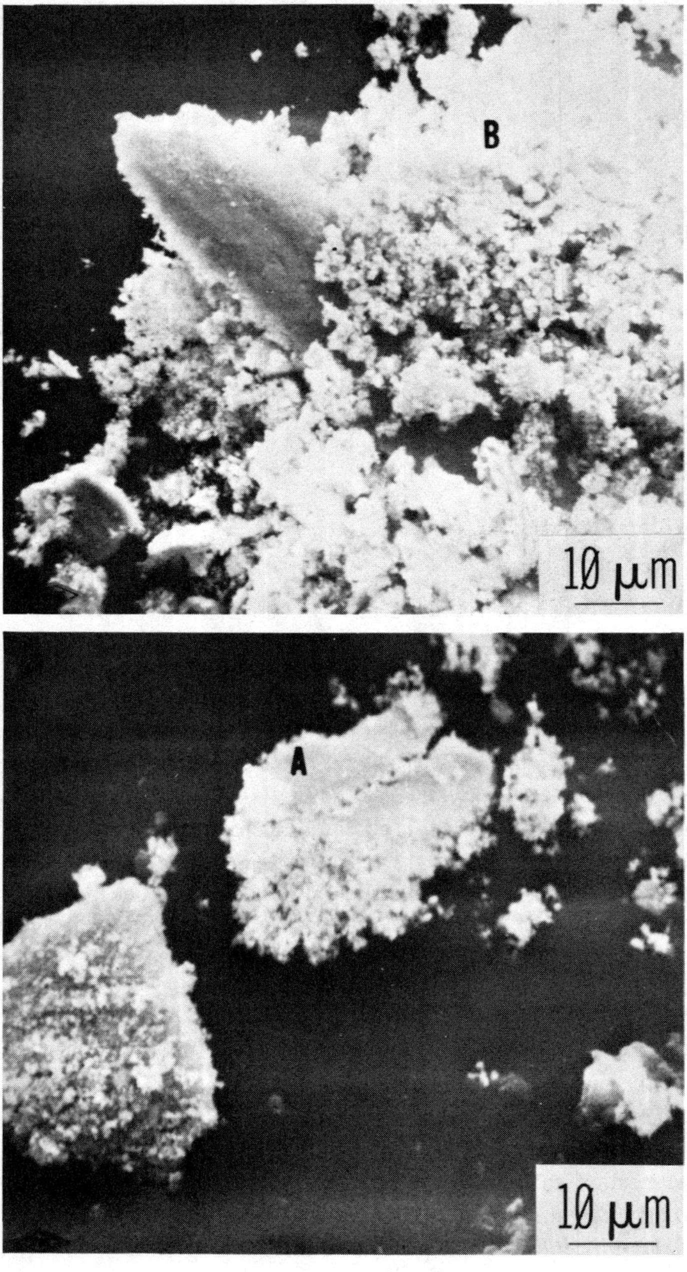

Fig. 20. Wear debris particles recovered from test of 1015 steel specimen. Note partial fracture of particles (A) in lower micrograph. An aggregate of much smaller particles is seen in upper micrograph at B.

Fig. 21. Scanning electron micrographs of exit edge of wear scar on 1015 steel specimen surface. Arrow in upper micrograph indicates sliding direction. Lower micrograph shows evidence of deformation and fracture at edges of displaced material.

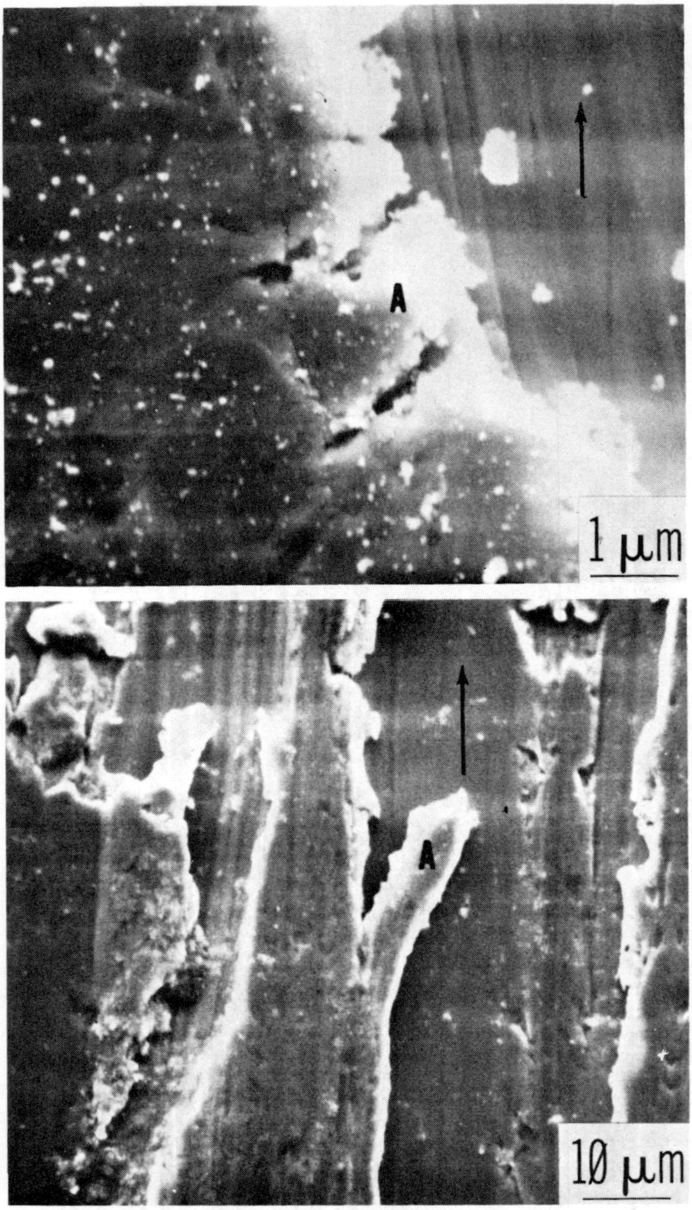

Fig. 22. Central region of wear scar on 1015 steel specimen. Wear particle formation appears under way at locations (A) where heavily deformed metal is present. Arrows indicate sliding direction.

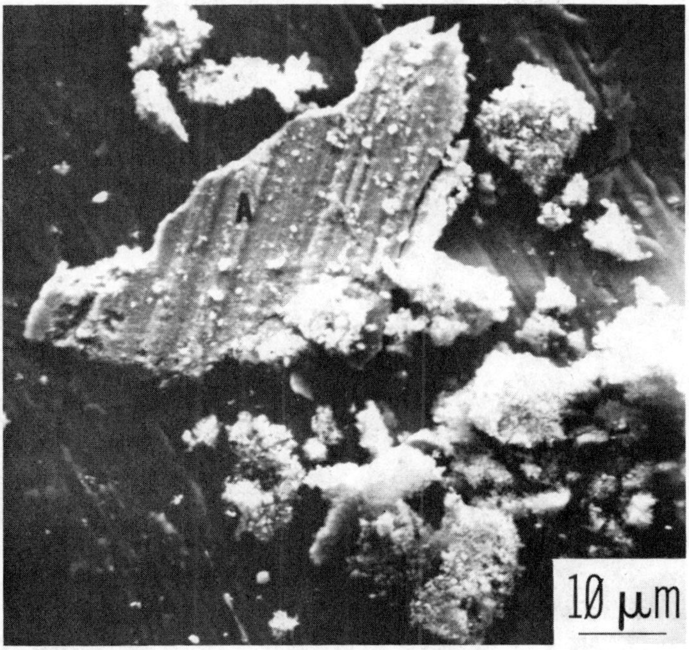

Fig. 23. Scanning electron micrograph of wear debris particles collected from test of DP80 steel specimen. Note scoring lines on surface of flat particle (A).

of the small debris particles, such as at B in Fig. 26(b) and 27. This process of fine debris collection and compaction within the wear gap may in fact produce the porous-appearing debris particles typified by the example at B in Fig. 24.

Mechanical Properties Measurements. Microhardness measurements were carried out on all three steels; the results are shown in Table 1. The values for the annealed specimens of DP80 steel and the O-2 tool steel showed comparable microhardness values. Both were about 32% greater than that measured for the 1015 carbon steel. Grain intercept lengths were measured for all three steels and are also shown in Table 1. The DP80 steel has a significantly smaller grain size than the others.

Mechanical properties measurements on the 1015 steel and the DP80 steel have been conducted using substandard-sized tensile specimens. The specimens were pulled until failure at an extension rate of 0.0083 mm/s at room temperature. The yield stress was determined at 0.2% plastic elongation and the tensile strength at failure. Ductility characteristics were determined from reduction-in-area measurements of the initial and final cross section of each tensile specimen. True

Fig. 24. Examples of large debris particles from DP80 wear tests having two different morphologies. Note crack-like feature in A at arrow.

stress – true strain data were also obtained. The comparison between DP80 steel and the 1015 carbon steel indicates that both the higher tensile strength values (682 MPa vs 500 MPa) and the greater potential ductility (49% R.A. vs 31% R.A.) of DP80 steel probably contribute to its greater wear resistance. Note that plastic deformation effects have been shown earlier to be a prominent feature of the wear process. Tensile data were not obtained for the O-2 tool steel; however, the microhardness measurements suggest comparable strength levels with DP80 steel. The somewhat greater initial wear resistance of the DP80 steel relative to the O-2 tool steel may be due principally to the greater ductility of DP80 steel. In addition, the oxidation characteristics of the two steels differ and would be important in their wear behavior, as noted by Budinski[19] for tool steels. Additional mechanical properties measurements on the steels are planned to further explore this matter. Multiaxial deformation data will be sought since such conditions are more closely related to the conditions in the wear contact.

Conclusions

Results have been presented on the wear of three different steels using a computer-interfaced block-on-ring wear system. Steady-state wear studies were carried out for total sliding distances of both 363 m and 726 m. Wear depth values were obtained continually through the course of each experiment. In this way the time periods required for steady-state wear could be determined and changes in wear rate during the experiment could be measured. In all experiments the generation of wear debris and the transfer of debris particles between the contacting surfaces were noted. Accumulation of debris at the entrance edge of the wear scar in the ring and block geometry used here was a major factor. Microscopic studies of the contacting surfaces after the experiments showed evidence of wear debris accumulation within the contact gap at locations where surface depressions existed. Previous studies, e.g., Landheer and Zaat,[20] have also emphasized the importance of material transfer in wear.

Three characteristic types of wear debris particles were identified. Flat metallic particles were prominent in the debris with major axis dimensions from about 1 μm to 40 μm and thicknesses up to about 1 μm. Examination of the worn surfaces revealed many features resulting from plastic deformation both within the contact gap and at the exit end of the wear scar. The size and features of these deformed ridges and lips suggested that production of flat debris particles could be explained by ductile failure processes on a microscopic scale. A second type of large particle was clearly the result of compaction of much smaller particles, presumably occurring within the contact gap or at the entrance edge. This type of process has been directly observed by Sasada et al.[21] using optical microscopy for several different metal combinations. The third type were small equiaxed particles about 0.1 to 1 μm in size that may have been formed by asperity rubbing

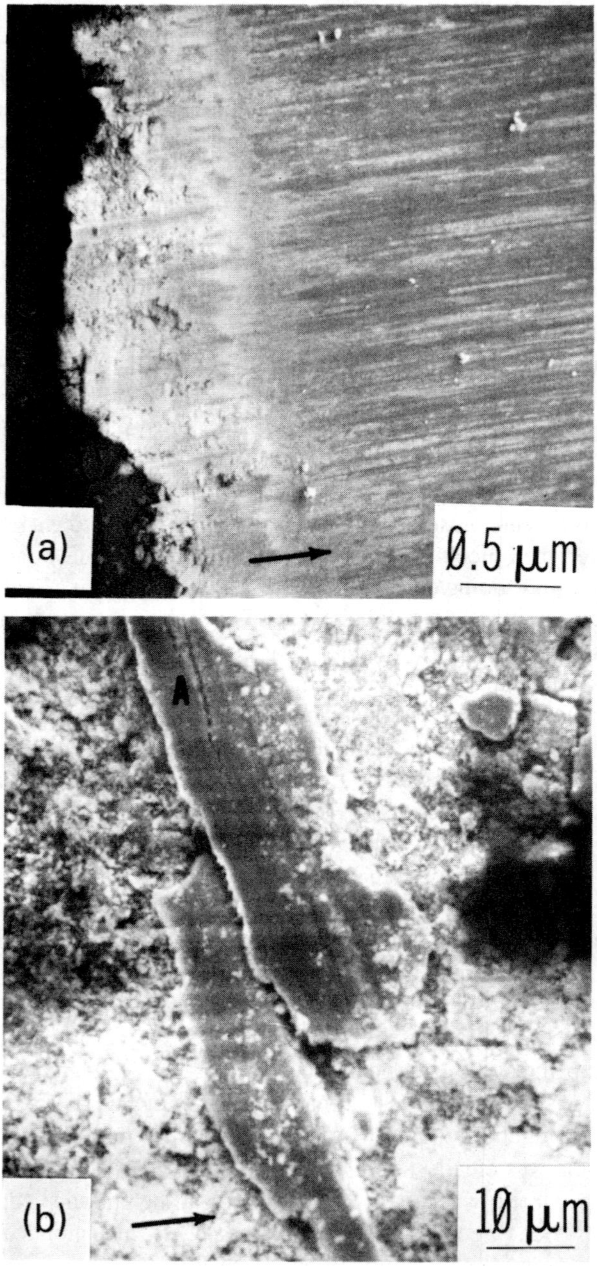

Fig. 25. (**a**) Entry region of wear scar on a DP80 specimen with sliding direction indicated by arrow. (**b**) Note crack in particle at A in one area of loose debris deposit.

Fig. 26. (**a**) Scanning electron micrograph of exit edge of wear scar on a DP80 specimen. (**b**) Details of deformation markings and crack-like features. Arrow indicates sliding direction.

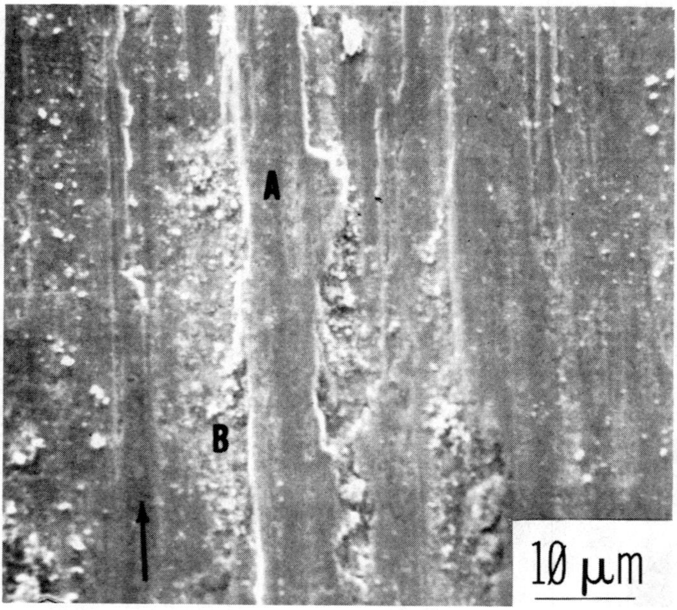

Fig. 27. Central region of wear scar on a DP80 specimen. Flat ridges of deformed metal (A) aligned in sliding direction (arrow) appear to carry imposed normal load.

and fracture within the contact. These smallest particles were compacted into one type of the larger particles as noted above. The presence of water vapor in the argon atmosphere within the test enclosure undoubtedly affected the oxidation rate of these steels. The asperity rubbing process just mentioned would probably be sensitive to the characteristics and thickness of this oxide layer at the contact surface.[22]

The wear rate measurements of the steels revealed some interesting differences. In particular, the DP80 steel wear rate, while initially about ⅛ that of the 1015 steel, increased for longer sliding distances until it was about ½ that of the 1015 steel. Clearly a change of this magnitude reflects a change in the dominant wear process. The measured coefficients of friction also showed a corresponding change for the DP80 alloy. Changes in the mode of wear and the magnitude of wear rate for different steels have been reported previously;[23,24] however, these changes involved changes in load, sliding speed, etc. The observations here seem to indicate another type of time-dependent process.

The mechanical properties measurements on the two steels revealed some significant differences that were generally expected. The DP80 steel alloy exhibited a slower work-hardening rate than that for the 1015 carbon steel specimens under uniaxial tensile deformation. The greater formability of the DP80 alloy is consistent with such behavior. In contrast, microhardnesss measurements of the 1015 steel in the deformed condition were only about 15% lower than for the DP80 alloy. Thus the difference in microhardness is probably only a small factor in the overall difference in mechanical behavior of the two types of steel and hence in the contrasting wear resistance.

Application of Transmission Electron Microscopy to the Study of Friction and Wear

Almost from the time that the transmission electron microscope (TEM) reached the threshold of becoming a commercially viable instrument and research tool, it has been applied to the study of wear. The use of TEM in this field, however, can best be described as exploratory rather than extensive. In this section, a brief review of TEM applications to wear together with a discussion of the techniques and methods employed will be presented. It is convenient to divide the presentation into three subsections: Replica Studies, Examination of Wear Debris, and Thin Foil TEM. For the most part, the discussion will be restricted to investigations concerned with the wear of metals, although the counterface material may be a non-metal. In the final subsection, an example of the application of TEM to investigate the wear of copper, taken from the authors' work, will be presented.

Replica Studies

The great advantage that early users perceived in the application of TEM to the study of wear lay in its significantly improved spacial resolution compared to optical microscopy. For early TEM's a resolution of <100 Å was easily obtained versus 2000 Å for optical microscopes. One of the first applications of TEM to the study of wear was that of E. F. Fullam and R. H. Savage in 1948.[25] The replica method was employed to investigate the wear of graphite commutator brushes after sliding on a copper ring. Replicas of the brush and copper-ring surfaces were prepared by stripping away a plastic film of nitrocellulose that was formed by the evaporation of amyl acetate solvent from a dilute solution that had been placed on the surface. A thin layer of silica was then deposited onto the plastic replicas in vacuum from a heated tungsten filament. The plastic was then dissolved, leaving the final silica film replica which could be examined in the TEM. An important aspect of the study and a main objective was the retention of debris particles on the final silica replica, and, in the case of the copper ring, a transferred layer of carbon. Thus, not only was the topography of the worn surfaces available for study at a relatively high resolution, but also particles or films that were removed by the replica could be examined with respect to their morphology and structure.

In 1949 Whitehead[26] reported that he had used two well-established replica techniques to study surfaces worn at very light loads. The techniques used were oxide replicas for aluminum and Formvar replicas for other metals. The oxide replica method is applicable only when the underlying metal forms a suitably strong and continuous oxide film that can be removed from the surface, generally by chemical attack of the metal. Aluminum is perhaps an outstanding example where this applies. Formvar replicas[27] are prepared by placing a dilute solution containing Formvar on the surface and, after the solvent has evaporated, mechanically stripping the plastic film from the surface. Surface topographic features are then revealed on both of these negative impressions of the surface by shadowing, i.e., evaporating a layer of gold, palladium, or other metal at an oblique angle to the replica surface. On examining the replica of a wear track produced by sliding steel on electropolished aluminum, Whitehead observed slip lines that were clear evidence of plastic flow. He also observed steel particles which he described as resulting from the formation and subsequent fracture of welded junctions. This was taken by Whitehead as support for the adhesion theory of friction advanced by Bowden and Tabor.

It appears that early investigators[28] held considerable hope that with the high resolution capability of the TEM direct evidence of welded asperity junctions might be obtained, and that an insight into the wear process that had not been forthcoming with optical methods might be brought about by TEM. Later studies, however, suggest that Whitehead's simple explanation of the presence of iron particles, for example, was premature. Barwell et al.[29] note that worn surfaces are quickly covered by numerous scratches and furrows rather than sites where discrete particles might have been torn away. The same authors also comment, in introductory remarks to a section describing the application of TEM to wear, "It is not to be expected that the complex mechanism of wear should easily be elucidated, for the very process itself tends to obliterate its own initial stages." Thus, as so often happens in science, a powerful new method, rather than confirming a currently held hypothesis, simply exposes another level of complexity.

A number of other TEM investigations[27,30-33] were carried out using the replica technique. Archard and Hirst[30] in particular made extensive use of replicas to study the mild wear of tool steels in air. In their experiments, it was found that the wear rate changed during the course of sliding, and it was possible to show that this was accompanied by a change in wear mode. Initially, welding and metal transfer occurred. With the TEM it was possible to discern that transferred lumps of material and debris particles were composed of an aggregation of smaller fragments. As sliding continued, the debris fragments became smaller and the wear rate increased. By using electron diffraction as well as other methods of analysis it was determined that the particles were metal oxides. Thus, the increased wear rate was associated with the presence of these oxide particles.

The replica method was also used to investigate lubricant performance. Templeman[31] prepared replicas of engine piston and ring surfaces to study the effectiveness of a phosphide engine oil additive. Oils containing the additive produced surfaces having a smoother and less distressed appearance. Young and Schwartz[32] reported success in employing a replica method to identify scuffing in diesel engine cylinder liners.

Although the major application of replicas was to the study of worn surface topography, application was also made to the investigation of subsurface damage. The most widely used method for studying subsurface changes was by optical examination of polished and etched cross sections taken through the worn surface. To obtain improved resolution of microstructural features, it is a straightforward step to prepare replicas of the same cross sections. Scott and Scott[27] indicate that subsurface cracks that were not observed by optical methods were detected by TEM examination of replicas. The same authors describe the application of replicas to the study of taper sections taken through scuffed surfaces. In a later paper, Tsuya[33] made extensive use of etched cross sections to study damage below sliding wear tracks in copper. The explanation of wear based on these observations has in many respects a marked similarity to the delamination process described some five years later by Suh.[34] On the basis of her studies, Tsuya deduced that under a layer of highly ''micronized'' crystals, microcracks are formed at inclusions and develop into ductile cracks.

The TEM replica method was employed extensively throughout the period 1950 to 1970, especially in the study of fracture surfaces. With improvements in the electron microscope itself and the development of high resolution carbon replica techniques by Bradley in 1954,[35,36] a resolution of 10 Å was claimed.[37] By 1970, however, despite a poorer resolution, the availability of commercial SEM's led to a substantial reduction of replica TEM in the study of surface topography. With the SEM, the actual surface can be examined directly with little or no preparation required. In most TEM's an area no larger than ~2.5 mm in diameter is available for examination at one time, while with the SEM several square centimeters, often constituting the entire worn surface, may be studied. Also, surface roughness is a limiting factor with the TEM replica method; there is almost no practical limit for the SEM. Artifacts associated with replica preparation and difficulties in the interpretation of images are other problems associated with the replica method. By utilizing different imaging modes in the SEM some distinction between different phases present can be obtained and, when equipped with an X-ray analysis system, elemental composition can be determined. Despite these advantages, however, the TEM replica method is still valuable when the study of features less than 50 Å in size is required, since this exceeds the capability of most commercial SEM's.

Examination of Wear Debris

Reference has already been made to the direct extraction of debris from worn surfaces by replicas. This so-called extraction replica method has the distinct advantage of permitting debris fragments to be studied in direct relationship to surface topographic features, which may provide important clues to the nature of the wear process and the origin of the fragment as well. For example a thin filament of metallic debris that lies adjacent to a groove in the surface might with some confidence be identified as a lip of displaced material from the groove. Of course, many, if not most, debris fragments may not have such a readily identifiable source.

Most early studies of debris were concerned with particle size and shape. Some advantage was also taken of electron diffraction to assist in the identification of the debris material.[38] Little effort, however, appears to have been expended on the study of the internal microstructure until fairly recently.[39,40] Such studies have the potential of leading to an understanding of the process by which the debris particles were formed. Glaeser[39] examined debris fragments extracted from worn bronze bearings. An example is shown in Fig. 28. Regions of twinning and a fine cell structure were observed. Striations that occur in chips formed by cutting were not found, and it was concluded that the fragments were probably not formed by a micromachining process. Dean and Doyle[40] studied particles from abraded tool steel and observed both striated and unstriated structures. The particles were formed by abrading the steel surface with a single point tool of known shape. Smooth points with a negative rake angle produced debris without a striated structure, while a rough point with effective cutting edges produced chips with the characteristic striated structure. Both types of chips produced continuous diffraction rings indicating a very fine grained structure.

TEM study of the internal structure of debris particles is limited to regions that are thin enough for electron transmission. For a conventional 100 kV instrument this results in a maximum thickness of 200-300 nm. Transmission may be obtained through edges of thicker particles; however, some caution must be exercised in evaluating such observations. With the exception of gold, metallic particles may be encased with an oxide layer. If diffraction is obtained only from this layer, the particle might be misidentified as consisting entirely of oxide.

In addition to the direct extraction of debris particles from worn surfaces, application has also been made to the collection of debris from unlubricated and lubricated systems. In the case of a dry system, the debris particles may be allowed to fall directly onto an electron microscope specimen support grid covered with a suitably transparent thin film, they may be brushed or blown onto the grid-support film, or they may be dispersed in a volatile solvent and recovered from that medium by a variety of means. If the debris particles are contained in a lubricant, it is necessary to eliminate the lubricant, for example, by repeated centrifuging and replacement with a suitable volatile solvent. Alternatively, it may be possible to drive off the lubricant at an elevated temperature under vacuum conditions. Several of these methods are described by Scott and Scott.[27] Ruff[41] has taken advantage of the fact that debris particles are dispersed according to size (for constant susceptibility) on ferrograms. An extraction replica technique

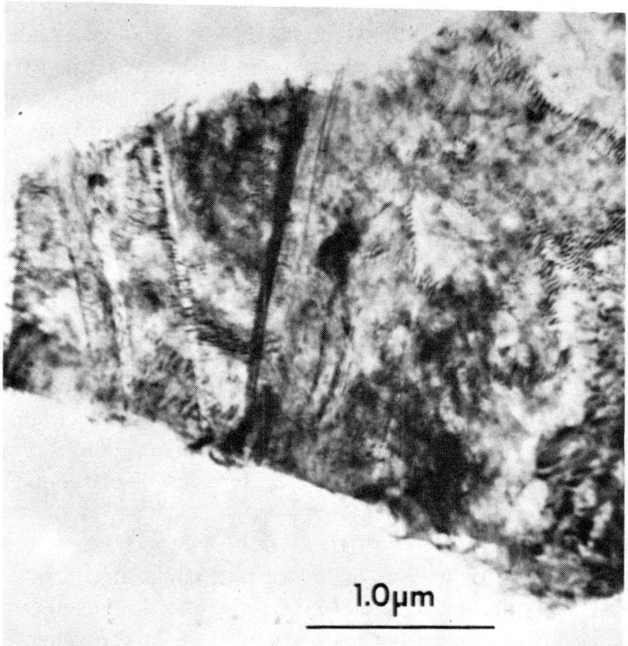

Fig. 28. TEM micrograph of debris fragment extracted from a worn bronze bearing. *(Ref 39)*

was used to obtain particles at selected regions along the ferrogram for study in the electron microscope.

Finally, reference should also be made to the fact that the capability now exists to obtain elemental analysis of particles directly in the TEM or STEM (scanning transmission electron microscope) on a fairly routine basis when these instruments are fitted with X-ray analysis systems (microscopes so equipped are sometimes referred to as analytical electron microscopes, or AEM's). Formerly this capability was only found with SEM's equipped with X-ray systems or with electron microprobes. With the AEM, one has the advantage of being able to obtain topographic, internal microstructural, diffraction and elemental composition analysis information from a single particle all in the same instrument.

In summary, although the application of TEM to the study of wear debris has been explored by several investigators and its value demonstrated, extensive use of the method has yet to be made.

Thin Foil TEM

Background. Until rather recently investigators employing TEM to study wear have neglected one of the most powerful and important methods developed in the last three decades to contribute to our understanding of materials behavior and properties. In that method thin sections of a material are examined directly in the electron microscope. Dislocations, grain boundaries, fine twins, stacking faults, microcracks, voids, precipitates and the interrelationship among these microstructural features that are fundamental to materials properties and behavior are revealed for direct study and evaluation. Methods for preparing suitably thin sections from the bulk for electron transmission and the principles guiding the application of TEM and the interpretation of the images obtained were well developed by the early 1960's.[42] Thus a means was available for examining the effects of friction and wear at the surface and throughout the affected volume. It appears, however, that virtually no such studies were carried out in the field of wear until nearly a decade after the method had been rather fully developed. This is somewhat surprising since optical examination of etched cross sections often showed a distinct layer of altered material below the contact surface. It was certainly well known that surface hardness was affected by sliding contact. Also, the extensive work by Wilman et al.,[43] using reflection electron diffraction, had provided a great deal of information regarding changes in microstructure at abraded surfaces. After abrading single crystals of copper and iron these authors reported the existence of a layer at the surface consisting of very fine polycrystalline grains exhibiting no preferred orientation. At greater depths, a zone of deformed grains having a fiber texture was present. The general appearance of this altered region under optical examination of etched sections has led to the description ''fragmented layer.'' Tsuya, as noted earlier, has used the term "micronized."[33]

Perhaps a greater expenditure of effort might have been directed towards the study of plastic flow and subsurface deformation effects had it not been for the predominance of the Bowden and Tabor adhesion theory to explain friction. As a direct extension of this theory to wear, it was assumed that the primary controlling factor was the formation of an adhesive junction. Although in most cases failure would occur at the junction, in a certain fraction of cases a path within the material would be favored, leading to transfer and eventual loss from the system of a fragment of material.[44] Deformation effects associated with junction shearing and indeed the root cause of weak failure paths within the material were largely ignored.

Preparation Methods. In general, two approaches may be taken in preparing TEM specimens from the near-surface region damaged by friction and wear processes. Perhaps the simplest and most direct way is to take sections parallel to the contact surface. In the second method sections are taken perpendicular to the contact surface. With the parallel section method, if the region immediately at the surface is to be examined, some means of surface protection must be employed during sectioning and subsequent thinning procedures. Most often, an inert, soluble protective lacquer is applied. This is removed at a stage prior to

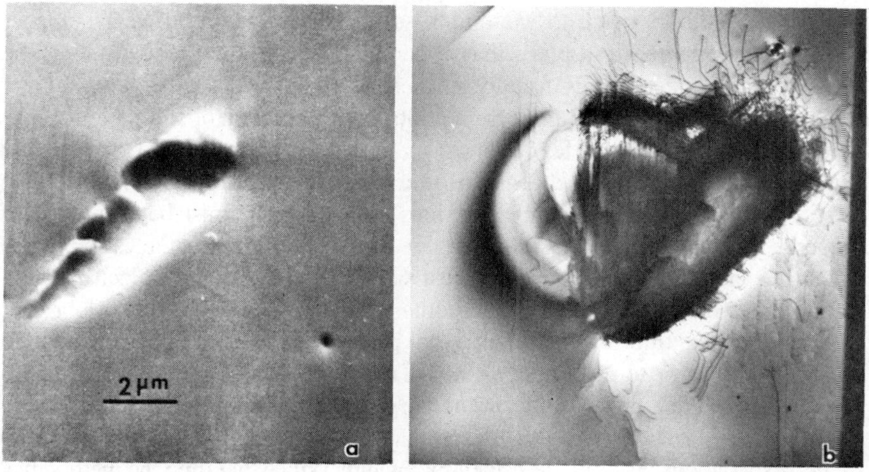

Fig. 29. Crater in 310 stainless steel produced by impact of 50 μm Al_2O_3 particle. The particle velocity was 59 m/s and attack angle was 90°. **(a)** SEM micrograph. **(b)** TEM micrograph at same magnification revealing high dislocation density in vicinity of crater. *(Ref 45)*

examination in the TEM, when no further risk of preparation damage exists. Thinning is, of course, carried out from the side opposite the worn surface. An example of a specimen prepared in this way is shown in Fig. 29(b). This illustration is taken from a study of the damage produced by erosive particle impacts.[45] The material is annealed 310 stainless steel. The surface had been exposed to impact at 60 m/s by an aluminum oxide particle ~50 μm in diameter. An SEM micrograph of the crater is shown in Fig. 29(a). Note that a region of very high dislocation density envelops the impact crater. The dislocation density falls sharply outside this envelope. Many dislocations have undoubtedly escaped from the surface or to distant grain boundaries outside the field of view.

To examine regions at greater depths below the surface, a known thickness of material may be removed before commencing thinning from the back side. When the depth of interest becomes large compared to the section thickness, conventional thinning from both sides of the section may be employed. Clearly, in order to obtain a representative picture of the changes that occur over a depth of many micrometers or even hundreds of micrometers, it may be necessary to prepare a large number of specimens. The typical, suitable TEM specimen thickness cannot exceed 200-300 nm when a conventional 100 kV instrument is employed. Examination of greater thickness at higher accelerating voltages may not be an advantage because of the overlap of closely spaced dislocations or other microstructural features, since features in three-dimensional space are projected onto a two-dimensional screen or photographic film for observation.

In the second approach to examining subsurface friction and wear damage, sections are prepared in a plane that is perpendicular to the worn surface. The orientation of the plane is usually chosen to be either parallel (longitudinal section) or perpendicular (transverse section) to the direction of sliding or rolling, since these orientations offer the most straightforward means of revealing the microstructural anisotropy that is so often associated with directional contact stresses. As in preparing parallel sections, some means must be employed to preserve the immediate worn surface region from damage during the various steps of cutting and thinning. Perhaps the most satisfactory method is to electrodeposit a thick layer of copper or other metal onto the worn surface. Residual lubricant or excessive quantities of loose debris that might prevent good adhesion must be washed away prior to electrodeposition. This method is, of course, one that is often used for edge preservation in the preparation of metallographic cross sections and taper sections for optical examination.[46] Application to the preparation of TEM specimens was first described by Swann;[47] since that time the method has been applied by a number of investigators to study near-surface microstructures. In a recent paper, Hogmark, Swahn and Vingsbo[73] present a detailed description of the method. The work of Turley[48] and Turley and Samuels[49] can be cited as excellent examples and as being among the first applications related to wear. Both the electrodeposited layer and the adjacent specimen are thinned simultaneously. The junction at the electrodeposited layer serves to mark precisely the location of the worn surface. Thinning may be accomplished by chemical, electrochemical, or ion beam methods, or a combination of these.

Although in some respects the parallel and perpendicular section methods are complementary, the perpendicular section method provides several advantages. First, with a single specimen it is possible to examine the change in microstructures from the original worn surface to a depth of tens of micrometers below the surface, or hundreds of micrometers when especially good success has been obtained in preparing a specimen. Second, the depth below the surface is accurately and precisely known. This is not always the case when parallel sections are prepared, because of uncertainties in measurement of section location and thinning rates. Third, the immediate subsurface region can be examined even when the original surface is exceptionally rough. It is difficult to prepare parallel sections of rough surfaces, and any thin regions available are usually very limited in extent. Fourth, it may not be possible to recognize the morphology of lamellar features that are parallel to the surface and comparable in thickness to the thin foil when parallel sections are employed. This was clearly illustrated by Turley in his investigation of the subsurface damage in α-brass associated with orthogonal machining.[48] In any case, sections from at least two orthogonal orientations may be necessary to display the structural morphology adequately. A disadvantage of the perpendicular section method lies in the more lengthy and complicated preparation procedures involved.

Applications. Despite the slow start in applying thin foil methods to the study of wear, an increasing number of investigators have employed the method over the past decade and a half. In some cases extensive use has been made of TEM (e.g., Ref 8, 48,54), while in others one or two micrographs were displayed to illustrate, often in a rather qualitative way, the nature of the microstructure at the worn surface. A variety of metals have been studied, ranging from commercial bearing steels and wear resistant steels[50,52,55-60] to pure materials, e.g., copper[8,54,61-64] or aluminum,[54,62,65-67] where the purpose has been to investigate basic microstructural changes in the simplest systems possible. The conditions of wear have been greatly varied: dry sliding,[8,51,53,58,61,64,68] lubricated sliding,[8,57,69,70] rolling,[50,55,56,59] abrasive sliding,[8,49] and solid particle erosion.[60,63] In some cases, a single, pointed slider or stylus has been employed to generate a wear track on a bulk specimen[54,66] or on a film[65,67,71,72] that is suitable for electron transmission without further preparation. Such experiments have been employed to determine the relationship between frictional work and the stored energy of generated dislocations[65,67,72] or to study the details of the slip systems and dislocation intersections that are associated with the resulting plastic deformation.[54]

The single cycle or single contact experiments generally represent only the initial stage of the development of an altered subsurface structure. For example, Ohmae[54] used a 2 million volt TEM to study the deformation produced in copper and aluminum by a single stroke of a slider. He observed a dislocation cell structure which extended to the immediate contact surface.

Studies employing multiple passes to obtain a steady-state condition of wear have demonstrated that a far more complex structure exists in the region immediately below the wear surface. In general, the structure that is developed depends on the load, sliding speed, initial microstructure, composition, ambient environment, etc. A fine-grained polycrystalline or subgrain structure has been reported to occur under many conditions of sliding contact when the stresses are sufficient to produce plastic deformation. In steels and cast iron this often gives rise to the so-called white etching layer. Under some conditions certain metals may undergo phase transformations.[52,57,58] A notable example is the investigation by Swahn, Becker and Vingsbo[52] describing the decay of martensite in ball bearings due to rolling contact fatigue. One of the simplest cases is copper, which exhibits a fine polycrystalline-like structure at the immediate surface under a variety of contact conditions.[8,61-64] A more detailed description of copper is given in the next section.

Subsurface Wear Damage in Copper. Many investigations have been carried out on the wear of copper, not because copper is an especially important bearing material (many of its alloys are important, however) or has unique wear resistant properties, but because its mechanical behavior and physical properties have been exhaustively studied. Specimen preparation procedures are also relatively simple and fully documented. Figure 30 shows the damage to copper produced by solid

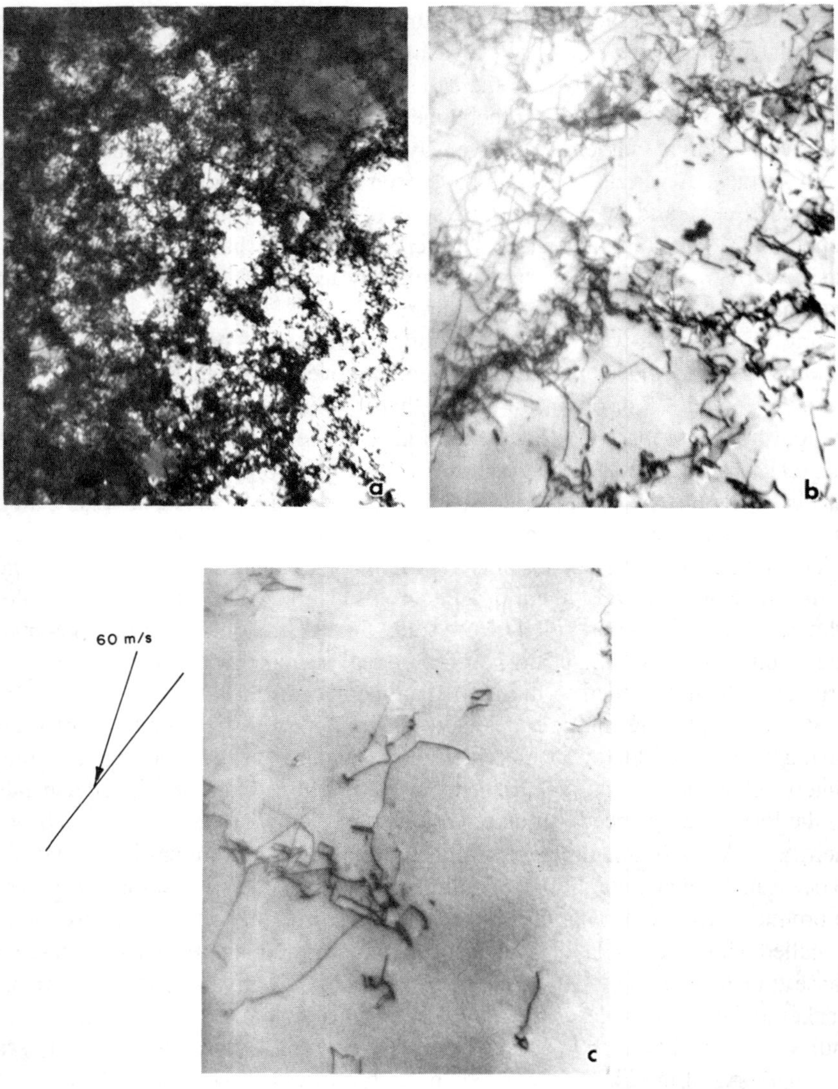

Fig. 30. Dislocation cell structure in copper at various depths below the surface. Specimen was eroded to a condition of steady-state wear. Erosion was carried out with 50 μm diameter Al_2O_3 particles at 60 m/s and an attack angle of 20°.

particle erosion at various depths below the surface. The TEM micrographs were obtained from perpendicular sections through the worn surface. As described earlier, the specimen preparation method consisted of first electrodepositing a thick layer of copper on the eroded surface, taking thin slices approximately 1 mm thick perpendicular to the eroded surface, and then electrochemically thinning these sections to produce foil specimens suitable for electron transmission. The specimen shown in Fig. 30 was eroded by a beam of aluminum oxide particles, nominally 50 μm in diameter, moving at a velocity of 60 m/s with an attack angle of 20° to the surface. At a depth of 52 μm below the surface, Fig. 30(c), the deformation is just large enough to initiate the development of a dislocation cell structure. A dislocation cell structure is typical of many materials that have been strained by a sufficient amount under conditions of multiple slip. At 36 μm below the surface, Fig. 30(b), a well-developed cell structure is present. As the surface is approached the average cell diameter decreases, reaching a minimum diameter of approximately 0.5 μm at a distance of a few micrometers from the surface, Fig. 30(a). However, closer to the surface a significant change in microstructure is observed. Figure 31 shows the region extending from the surface to a depth of about 5 μm. The curved acicular features are deformation twins. Twinning is apparently a result of the high strain rates associated with particle impact. Immediately below the surface, where the deformation is most intense, the twins are no longer distinct and a fine-grained polycrystalline

Fig. 31. Same specimen as for Fig. 30, region immediately below the surface. Dark acicular features are deformation twins. Surface location is indicated by dashed line. Arrow indicates the direction of the incident particles. Material at the surface is polycrystalline. *(Ref 63)*

layer is observed. A more detailed discussion of the erosion of copper can be found in Ref 63.

When copper was exposed to wear under other contact conditions the subsurface microstructures were similar in some respects and different in others. Figure 32 shows a region immediately below the surface of a copper specimen that had been abraded under three-body conditions utilizing a dry-sand/rubber wheel abrasive wear test machine.[8] Again, the immediate subsurface layer is polycrystalline. A diffraction pattern from the encircled region is shown in Fig. 33 and consists of fairly uniform rings. Another feature characteristic of surfaces that have been abraded or eroded with hard particles is the presence of a significant concentration of embedded particle fragments.[8,63] Embedded sand particles are identified in Fig. 32 and diffraction spots from these particles are present in Fig. 33. An important aspect of the embedding process and, indeed, of the wear process is the ductile shearing of metal across the surface. The fissure with trapped particles shown in Fig. 32 is a result of the ductile flow of copper along the surface.

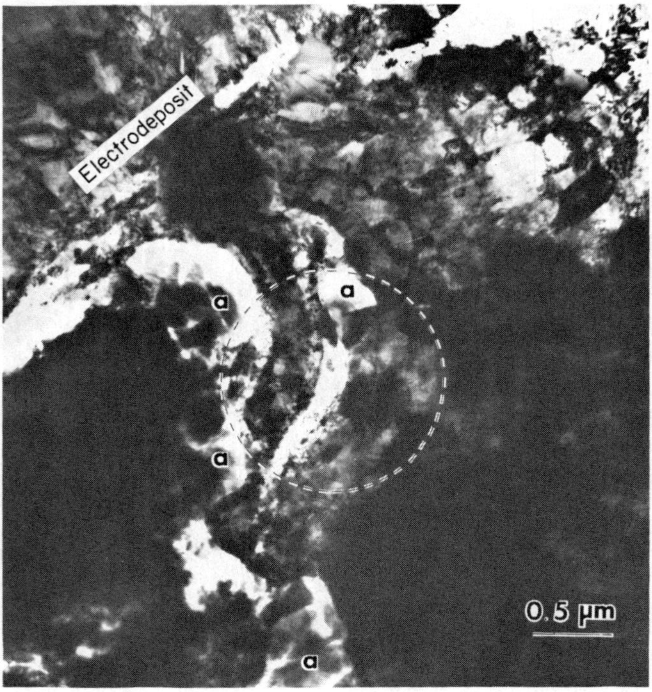

Fig. 32. Immediate subsurface region of abraded copper specimen. Fragments of sand abrasive particles are visible at (a). Arrow indicates direction of abrasive particle motion. *(Ref 8)*

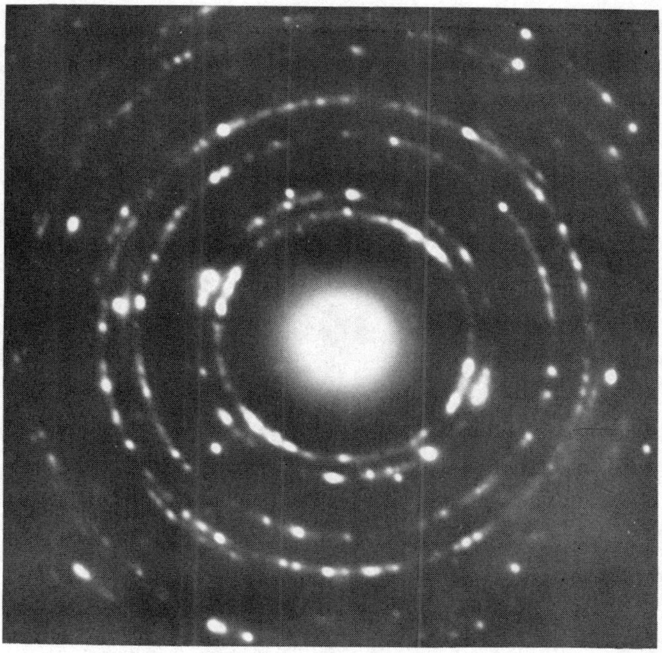

Fig. 33. Selected area electron diffraction pattern from encircled region in Fig. 32. Rings are from copper. A few spots from sand grains are also present.

At greater depths below the surface, abrasive sliding contact produced a dislocation cell structure quite similar to that shown in the case of solid particle erosion. A cell structure is also typical of other wear contact conditions, and simply reflects the fact that the material has undergone plastic strain by more than a few percent. Cell size and misorientation between adjacent cells (which is typically no greater than approximately 1° or 2°) reflect the amount of deformation and the direction of the applied stress, respectively.

The transition between the cell structure and the polycrystalline-like layer at the surface was studied under conditions of lubricated sliding.[8] An SEM micrograph of a longitudinal section, in which the grain structure is visible as a result of channeling contrast,[8] is shown in Fig. 34. A polycrystalline layer extends to a depth of about 8 μm below the surface. The original grains are too large to be seen at this magnification. However, an original annealing twin is present and is bent in the direction of sliding, indicating that shear of the lattice has occurred. The grains within the polycrystalline layer are also elongated in the direction of sliding. A TEM micrograph displaying the grain structure below the surface from the same specimen is shown in Fig. 35. Immediately below the surface the grains

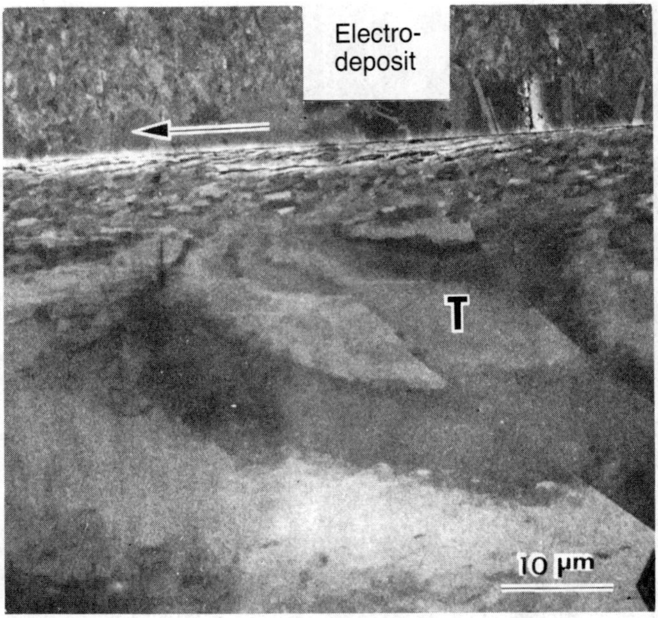

Fig. 34. SEM micrograph of longitudinal section of copper worn under a condition of lubricated sliding. Arrow indicates direction of motion of surface with respect to stationary slider. Layer of fine, elongated polycrystalline grains near the surface is visible as a result of channeling contrast. A large original annealing twin is also visible at T.

are relatively equiaxed and are crystallographically misoriented by a large amount. Diffraction patterns obtained from this region yielded spotty rings, perhaps justifying the description "polycrystalline." Van Dijck,[61] Bill and Wisander,[62] and, more recently, Dautzenberg[64] have observed a similar polycrystalline structure in copper near the surface. The elongated grains (or cells) at a slightly greater distance from the surface are only misoriented by a small amount. An example is shown in Fig. 36, where the misorientation between three adjacent grains is indicated. Selected area diffraction patterns from each grain were compared and found to differ by a rotation of 5° to 8° about a common [112] direction. The latter direction lay in the plane of sliding and was perpendicular to the sliding direction for this grain.

What appears to be the transition between the region of elongated grains and the dislocation cell structure is shown in Fig. 37. The cell boundaries are much less sharply defined and the misorientation between adjacent grains is less.

The process by which the polycrystalline structure is formed at the surface has been attributed to recrystallization,[62] to dynamic recrystallization,[64] or to grain boundary sliding.[3] The extent to which any or all of these processes may be

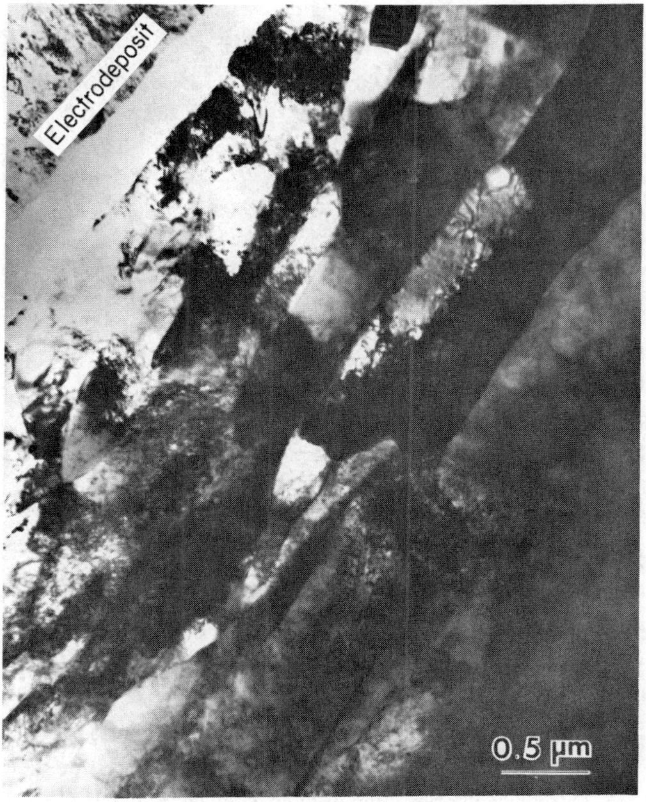

Fig. 35. TEM micrograph of specimen in Fig. 34, showing microstructure at the surface. *(Ref 8)*

involved under a given set of conditions is not entirely clear. The presence of elongated subgrains misoriented by only a few degrees with respect to each other suggests that recrystallization was not the process by which these grains developed. At the immediate surface, where diffraction patterns indicate grains of widely differing orientation, recrystallization may have occurred. Dynamic recrystallization refers to the recrystallization that occurs during the deformation process.[64]

An example where recrystallization of a thermal nature was clearly evident is shown in Fig. 38.[8] This is an SEM micrograph of a longitudinal section of a copper specimen. The test conditions were similar to those for the lubricated specimen described above with the exception that sliding took place without a lubricant in air (~50% relative humidity). During the course of sliding, oxidation occurred, leading to the accumulation of oxide and metal particles on the surface. Thick patches of this debris were distributed along the wear track, Fig. 39. These

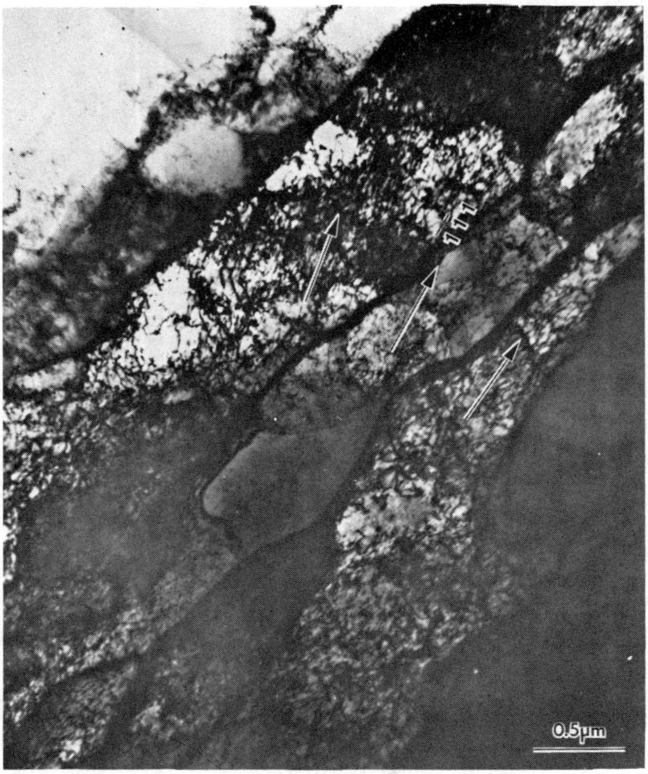

Fig. 36. Difference in orientation of three elongated subgrains or cells is indicated by arrows. Each arrow indicates the local $[\bar{1}\bar{1}1]$ direction. The difference in orientation corresponds to a rotation about a [112] direction perpendicular to the plane of the micrograph.

patches appeared to sustain the full load during sliding and possibly resulted in local heating. It is conjectured that this led to the recrystallized grains that are shown in Fig. 38.

Conclusions

The response of materials to friction and wear-related contacts is complex and varied. Transmission electron microscopy offers a powerful and in many respects unique means for investigating microstructural changes that are a function of both the contact conditions and materials properties. By using high resolution replica methods, one can reveal topographic details that may not be accessible with most SEM's. The shapes, sizes and internal structure of small debris particles can be studied. Electron diffraction can assist in particle identification, and when the TEM is equipped with the appropriate energy dispersive X-ray analysis and

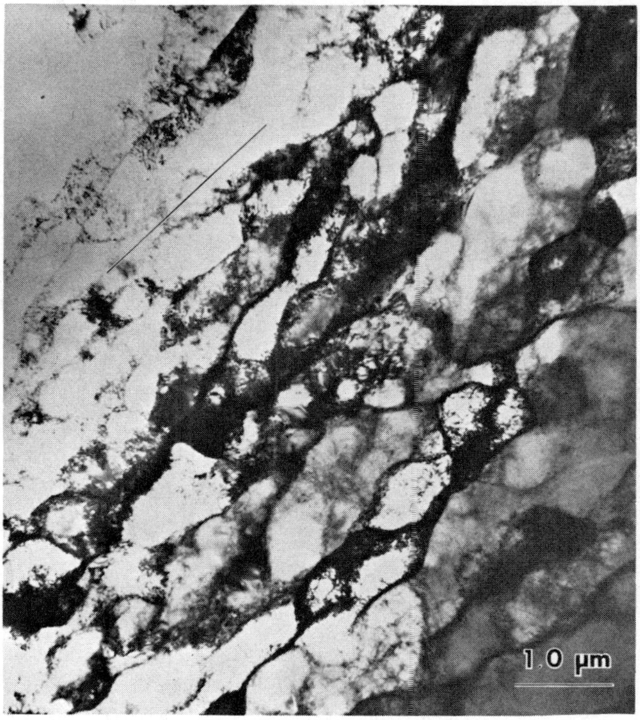

Fig. 37. Cell structure in region just below elongated subgrains. *(Ref 8)*

electron energy loss analysis systems, elemental chemical composition can be determined. TEM examination of thin foils prepared from the worn subsurface region permits direct investigation of changes brought about by wear. A high density of dislocations, dislocation cell structures, deformation twins, a near-surface region of extremely fine polycrystalline grains and changes in phase have been observed. Although much has been learned, further application of TEM to the study of wear should contribute to the fund of knowledge that must be expanded if the basic mechanisms of wear and the relationship between wear behavior and materials properties are to be understood.

Acknowledgments

Considerable assistance has been provided in the experimental phases of this work by Paul Boyer, Eric Whitenton, David Cooper, and Kirk Mettam. Numerous discussions with Peter Blau through the course of this work proved to be very helpful. Partial support from the Office of Naval Research is gratefully acknowledged.

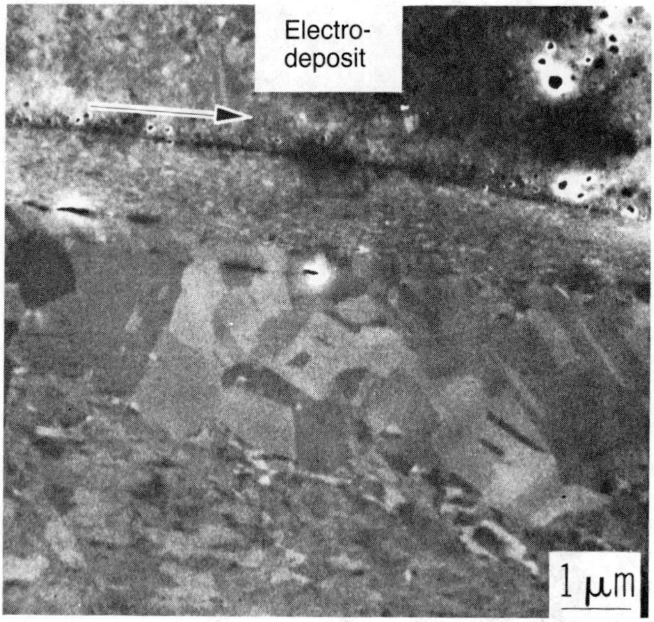

Fig. 38. SEM micrograph similar to Fig. 34. Specimen exposed to dry sliding wear. Note large recrystallized grains. Arrow indicates direction of motion of surface with respect to stationary slider. *(Ref 8)*

References

1. A. W. Ruff, "Debris Analysis of Erosive and Abrasive Wear," Proc. Intl. Conf. on Fundamentals of Tribology, M.I.T. Press, 1980, pp. 877-885.
2. W. A. Glaeser, "Wear Failure Analysis," Proc. Wear of Materials, 1979, ASME, April 1979, pp. 639-652.
3. J. H. Dautzenberg and J. H. Zaat, "Quantitative Determination of Deformation by Sliding Wear," Wear, *23*, 1973, pp. 9-19.
4. T. M. Ahn, P. J. Blau, K. L. Hsu, D. A. Rigney and J. D. Schell, "Metallographic Techniques for Wear Test Specimens," Wear, *56*, 1979, pp. 409-413.
5. J. A. Greenwood and J. B. P. Williamson, ''Contact of Nominally Flat Surfaces,'' Proc. Roy. Soc. London, *A295*, 1966, pp. 300-319.
6. H. T. McAdams, "Quantitative Microcartography, A Physico-Geometric Approach to Surface Integrity," Proc. Intl. Conf. on Surface Technology, SME, 1973, pp. 96-116.
7. D. H. Buckley, "Definition and Effect of Chemical Properties of Surfaces in Friction, Lubrication and Wear," Proc. Intl. Conf. on Fundamentals of Tribology, M.I.T. Press, 1980, pp. 175-199.
8. L. K. Ives, "Microstructural Changes in Copper Due to Abrasive, Dry and Lubricated Wear," Proc. Wear of Materials, 1979, ASME, 1979, pp. 246-256.
9. D. Scott, "Debris Examination, A Prognostic Approach to Failure Prevention," Wear, *34*, 1975, pp. 15-22.

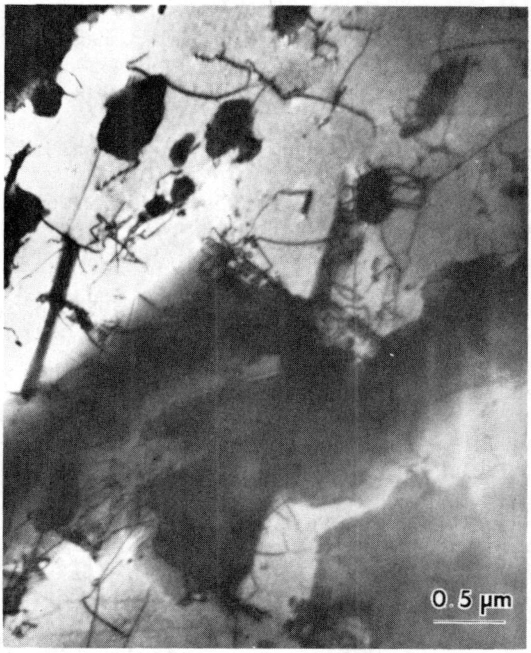

Fig. 39. TEM micrograph indicating low dislocation density in recrystallized grains. *(Ref 8)*

10. F. T. Barwell, "The Use of Temper Colors in Ferrography," Wear, *44,* 1977, pp. 163-171.
11. T. R. Bates, K. C. Ludema, and W. C. Brainard, " A Rheological Mechanism of Penetrative Wear," Wear, *30,* 1974, pp. 365-375.
12. T. Kayaba and K. Kato, "The Analysis of Adhesive Wear Mechanism by Successive Observation of the Wear Process in SEM," Proc. Wear of Materials, 1979, ASME, 1979, pp. 45-56.
13. Y. Tsuya, K. Saito, R. Takagi and J. A. Kaoba, "In-Situ Observation of Wear Process in a Scanning Electron Microscope," Proc. Wear of Materials, 1979, ASME, 1979, pp. 57-71.
14. K. Nakayama and M. C. Shaw, "Relationship Between Cutting Forces, Temperatures, Built-up Edge, and Surface Finish," Annals of the C.I.R.P., *14,* 1966, pp. 211-233.
15. P. L. Hurricks, Wear, *26,* 1973, pp. 285-304.
16. E. E. Fletcher, MCIC Report #79-39, Battelle-Columbus Laboratories, Columbus, Ohio, 1979.
17. T. Furukawa, Metal Progress, *116,* 1979, pp. 36-39.
18. E. M. Trent, in Wear—Treatise on Materials Science and Technology, ed. D. Scott, Academic Press, New York, 1979, pp. 443-489.
19. K. G. Budinski, in Proc. Wear of Materials, 1977, ASME, 1977, pp. 100-109.
20. D. Landheer and J. H. Zaat, Wear, *27,* 1974, pp. 129-145.
21. T. Sasada, S. Norose and H. Mishina, in Proc. Wear of Materials, 1979, ASME, 1979, pp. 72-80.

22. E. Tsuji and Y. Ando, in Proc. Wear of Materials, 1977, ASME, 1977, pp. 94-99.
23. T. S. Eyre, R. F. Iles and D. W. Glasson, Wear, *13,* 1969, pp. 229-245.
24. N. C. Welsh, Phil. Trans. Roy. Soc., *11,* 1965, pp. 435-439.
25. E. F. Fullam and R. H. Savage, J. Appl. Phys., *19,* 1948, pp. 654-661.
26. J. R. Whitehead, Research, *2,* 1949, pp. 145-147.
27. D. Scott and H. M. Scott, Proc. Inst. Mech. Eng. Conf. Lub. and Wear, 1957, pp. 609-612.
28. S. B. Twiss, C. R. Lewis and D. M. Teague, Lubrication Engineering, 1956, pp. 102-109.
29. F. T. Barwell, L. Grunberg, A. A. Milne and K. H. R. Wright, Proc. Fourth World Petroleum Cong., 1955, pp. 116-136.
30. J. F. Archard and W. Hirst, Proc. Roy. Soc., *A238,* 1957, pp. 515-527.
31. G. C. Templeman, Lubrication Engineering, *13,* 1957, pp. 141-145.
32. A. P. Young and C. M. Schwartz, Wear, *6,* 1960, pp. 235-240.
33. Y. Tsuya, Bull. Japan Soc. Prec. Eng., *2,* 1967, pp. 214-221.
34. N. P. Suh, Wear, *25,* 1973, pp. 111-124.
35. D. E. Bradley, Brit. J. Appl. Phys., *5*, 1954, pp. 96-100.
36. D. E. Bradley, J. Inst. Metals, *83*, 1954-55, pp. 35-38.
37. D. E. Bradley, J. Appl. Phys., *27,* 1956, pp. 1399-1412.
38. M. R. Piggott and H. Wilman, Proc. Inst. Mech. Eng. Conf. Lub. and Wear, 1957, pp. 613-619.
39. W. A. Glaeser, Wear, *43,* 1977, pp. 393-394.
40. S. K. Dean and E. D. Doyle, Wear, *35,* 1975, pp. 123-129.
41. A. W. Ruff, Wear, *42,* 1977, pp. 49-62.
42. G. Thomas, Transmission Electron Microscopy of Metals, New York, Wiley, 1962.
43. D. M. Evans, D. N. Layton and H. Wilman, Proc. Roy. Soc., *A205,* 1951, pp.17-30.
44. E. Rabinowicz, Friction and Wear of Materials, Wiley, New York, 1965, Chap. 6.
45. L. K. Ives and A. W. Ruff, Wear, *46,* 1978, pp. 149-162.
46. L. E. Samuels, Metallographic Polishing by Mechanical Methods, Elsevier, New York, 1971.
47. P. R. Swann, Acta Met., *14,* 1966, pp. 900-903.
48. D. M. Turley, J. Inst. Metals, *99,* 1971, pp. 271-276.
49. D. M. Turley and L. E. Samuels, J. Aust. Inst. Metals, *17,* 1972, pp. 114-119.
50. J. A. Martin, S. F. Borgese and A. D. Eberhardt, J. Basic Eng. Trans. ASME Series D, *88,* 1966, pp. 555-567.
51. K. Nakajima and Y. Mizutani, Wear, *13,* 1969, pp. 283-292.
52. H. Swahn, P. C. Becker and O. Vingsbo, Met. Trans., *7A,* 1976, pp. 1099-1110.
53. M. S. Beaton and C. R. Brooks, Wear, *41,* 1977, pp. 295-308.
54. N. Ohmae, Proc. Intl. Conf. on Fundamentals of Tribology, M.I.T. Press, 1980, pp. 201-222.
55. J. L. O'Brien and A. H. King, J. Basic Eng. Trans. ASME Series D, *88,* 1966, pp. 568-571.
56. J. A. Martin and A. D. Eberhardt, J. Basic Eng. Trans. ASME Series D, *89,* 1967, pp. 932-942.
57. A. A. Torrance and A. Cameron, Wear, *28,* 1974, pp. 299-311.
58. S. Hogmark and O. Vingsbo, Wear, *38,* 1976, pp. 341-359.
59. K. Fujita and A. Yoshida, Wear, *43,* 1977, pp. 301-313.
60. J. W. Edington and I. G. Wright, Wear, *48,* 1978, pp. 145-155.
61. J. A. B. Van Dijck, Wear, *42,* 1977, pp. 109-117.
62. R. C. Bill and D. Wisander, Wear, *41,* 1977, pp. 351-363.

63. L. K. Ives and A. W. Ruff, ASTM STP 644, 1979, pp. 5-35.
64. J. A. Dautzenberg, Wear, *60,* 1980, pp. 401-411.
65. G. Andarelli, D. Maugis and R. Courtel, Wear, *23,* 1973, pp. 21-31.
66. N. Takahashi and K. Okada, Wear, *28,* 1974, pp. 285-298.
67. D. Maugis, G. Desalos-Andarelli, A. Heurtel and R. Courtel, ASLE/ASME Lubrication Conf., Boston, 1976.
68. K. L. Hsu, T. M. Ahn and D. A. Rigney, Wear, *60,* 1980, pp. 13-37.
69. M. C. Inman and E. M. Kohn, Wear, *17,* 1971, pp. 33-49.
70. I. I. Garbar and J. V. Skorinin, Wear, *51,* 1978, pp. 327-336.
71. R. Feder and P. Chaudhari, Wear, *19,* 1972, pp. 109-112.
72. N. Gane and J. Skinner, Wear, *25,* 1973, pp. 381-384.
73. S. Hogmark, H. Swahn and O. Vingsbo, Ultramicroscopy, *1,* 1975, pp. 113-120.

Environmental Considerations in Wear Processes

R. M. BENTLEY AND D. J. DUQUETTE
Rensselaer Polytechnic Institute
Troy, New York 12181

Abstract

A review of research performed on the environmental effects on "dry" sliding wear processes during the last decade is presented. Studies involved primarily with gaseous media are considered, along with those concerned with the effects of temperature. A review of available literature shows that although some consideration has been given to aqueous media most effort has been directed at investigating the wear of common engineering materials in oxidizing atmospheres. Although oxide formation degrades the metal surface, the resulting film can produce a lubricating effect and in many cases reduces the coefficients of wear and friction. Several investigators have studied the thermodynamic and kinetic processes in friction activated oxidation as compared to static oxidation and have indicated that little difference exists. In nonoxidizing atmospheres friction, wear, and surface roughening are generally greater because of the absence of lubricating oxide films. Certain adsorbed species, such as chlorine, can provide lubrication, however. Water vapor in the environment is reported to have a marked effect in most studies, but there is little agreement as to the direction or severity of this effect. Some investigators have reported a single most damaging value of relative humidity at about 30-50%. Apparently humidity effects depend, to a large extent, on the details of the experiment. Sliding wear in aqueous media involves conjoint chemo-mechanical interactions paralleling those found in processes such as corrosion fatigue or stress corrosion cracking. The friction activated metal in the wearing interface is in general preferentially corroded, which in turn modifies the wear process. The exact nature of the process is very dependent on the specific metal/electrolyte combination. Changes in temperature affect both the chemical interactions occurring at the wearing interface and the mechanical properties of the bulk metal. In general, friction and wear increase with temperature as a result of an increase in ductility of contacting asperities, and the consequent growth of metal junctions. Finally, some special cases of hard wearing materials in extreme environments are presented.

Introduction

Although wear is usually thought of as a mechanical process, it is foremost a study of surfaces in contact — surfaces whose characteristics will be mitigated by the nature of the environment to which they are exposed. Accordingly, wear must be considered as a conjoint chemo-mechanical process, like corrosion fatigue or stress corrosion cracking. Indeed, some investigators discuss increased local corrosion due to "wear-activated" metal.[1-3] This parallels the corrosion fatigue phenomenon of preferential corrosion at persistent slip bands which emerge on the specimen surface. In wear the chemo-mechanical interactions at the surface are even more important, since the forces which produce wear must act across the contact interface. A rather unfortunate parallel to fatigue is that there is no general mechanism of wear, much less one that incorporates environmental effects. It seems, however, that most wear models are based on one or more of three broader mechanisms: adhesion, abrasion and plowing. The nature of these mechanisms will have received ample discussion elsewhere in this book, so it will suffice to state briefly that adhesion is concerned with local welding of contact points, abrasion is the damage produced when hard particles become embedded in one surface and gouge material on the opposing surface, and plowing occurs when asperities on the surface of a hard metal gouge material from the surface of a much softer metal. In recent years, considerable attention has been given to subsurface mechanisms which occur,[4] with the role of the surface being only to transmit forces.

Although a fair amount of work has been done in the area of environmental effects on wear processes, it is a small fraction of the total amount done on wear in general. Hopefully, this paper will serve as a reminder that environmental effects must be considered, and will provide impetus to control, and attempt to understand, the environment in the future.

Oxidizing Atmospheres

Most engineering metals readily combine with oxygen to form a stable oxide layer. Because of the broad scope of practical applications which are affected by oxidative wear it has received considerable attention in the literature. Oxidation may have opposing effects on the wear process: on the one hand it degrades the surface by removing metal atoms (a common measure of wear) and, on the other, oxide films may reduce friction and decrease the wear rate.[5-10] For example, Brown and Burton[11] found that the friction coefficient of mild steel sliding on a similar steel in air at a pressure of 1 atm was half that at an oxygen partial pressure of 10^{-6} Pa.

In 1969 Tao[1] undertook a study of the oxidation process in the wear of steel in air. The proposed models assumed oxidation to follow a parabolic rate law and, since oxide is continually scraped from the surface, a much higher rate constant and a lower activation energy for oxidation in corrosive wear than in static corrosion at the same "hot spot" temperature were calculated. The explanation was that the oxidation rate was greatly increased with wear-activated metal, i.e., metal which has an increased strain energy as a result of mechanical deformation from sliding. The two initially proposed models of oxide formation and attrition are shown in Fig. 1. In Fig. 2, the wear scar diameter for carbon steel sliding on carbon steel is plotted against time. It is seen that the data most closely agree with model 1, which proposes parabolic oxidation followed by rapid removal of the oxide layer. The author noted that severe wear in air could be eliminated merely by changing to an inert environment. He concluded that the rate of oxidation was more important than the mechanical wear process, because oxide attrition occurred very rapidly.

Molgaard and Srivastava[12] attempted to measure the activation energy of oxidation in wear as compared to that of the static oxidation process. The calculated oxidative wear activation energies were between 96 and 250 kJ/mole over the temperature range 550°-600°C. Based on these measurements, they concluded that it would be plausible to assume that the oxidation process is basically the same for oxidation in wear as for oxidation in the absence of wear. Other studies[13,14] reported activation energies for wear in various steels in the range 10-50 kJ/mole. However, these values are lower than the value of ~ 193 kJ/mole which has been assumed to be representative of the static oxidation process.[15]

As a subsequent step to determining the Arrhenius constant and activation energy in oxidative wear Aronov[16] has attempted to measure the kinetic parameters of the process. In dry friction experiments on steel in air at an unspecified

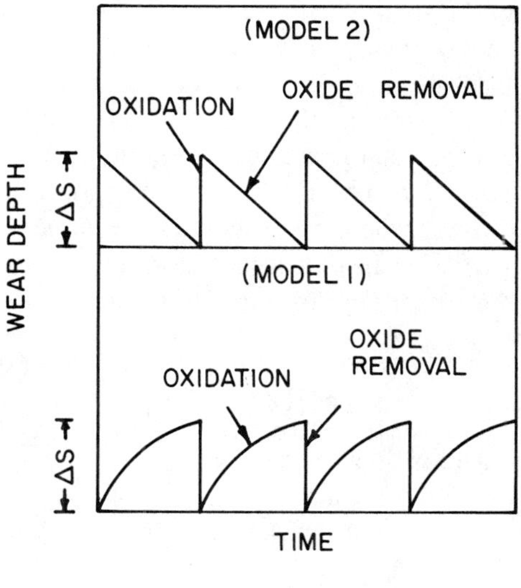

Fig. 1. Theoretical models for oxidational wear. Model 1 represents parabolic oxidation followed by rapid oxide attrition. Model 2 represents rapid oxidation followed by slow oxide removal. *(Ref 1)*

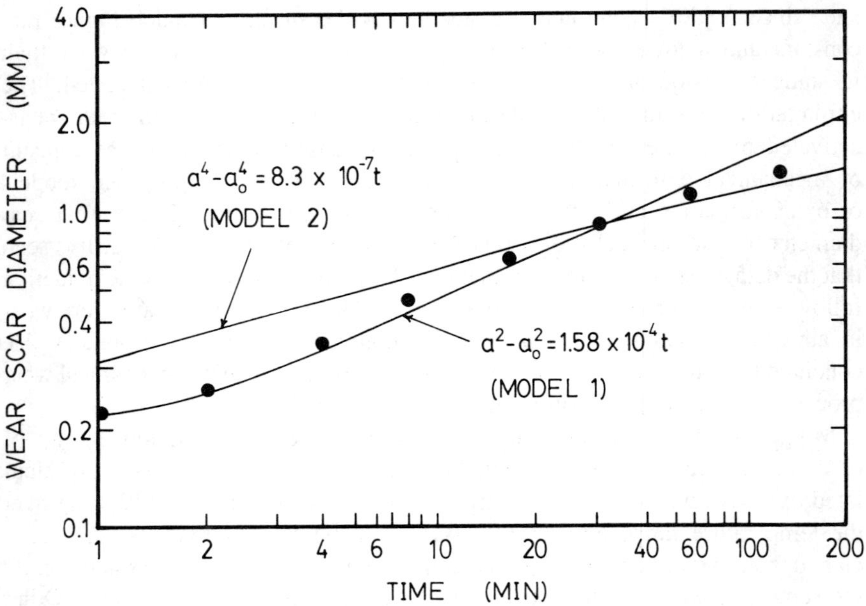

Fig. 2. Comparison of experimental data with the theoretical oxidational wear models. The wear scar diameter for a carbon steel pin sliding on a disk of the same material shows that the data most closely agree with Model 1—parabolic oxidation followed by rapid oxide removal. *(Ref 1)*

ambient temperature the normal oxidation process was well approximated by a parabolic relationship, with a rate constant close to that of the formation of Fe_3O_4 at 950°C. Two rate constants were in fact determined, that of the transformation of the friction activated metal to oxide and that of the destruction of the metal surface layers. It was assumed that this destruction occurred through a fatigue mechanism at the film-metal interface. As the activated layer became thicker (with increasing load) the diffusion path of the reactants became longer and the rate constant decreased, resulting in the increase in number of cycles to transformation and to failure, as shown in Fig. 3.

By far the most prolific author in the study of the mechanism of oxidation in wear over the past decade has been Quinn.[17-21] In his "oxidational theory of wear" the author developed the following equation to predict the wear rate:

$$w = \frac{L\ d\ A\ \exp\left({}^{-Q}/RT\right)}{V\ P\ f^2\ \rho^2\ \epsilon^2}$$

where w = wear rate
L = load
d = distance of a wear asperity contact

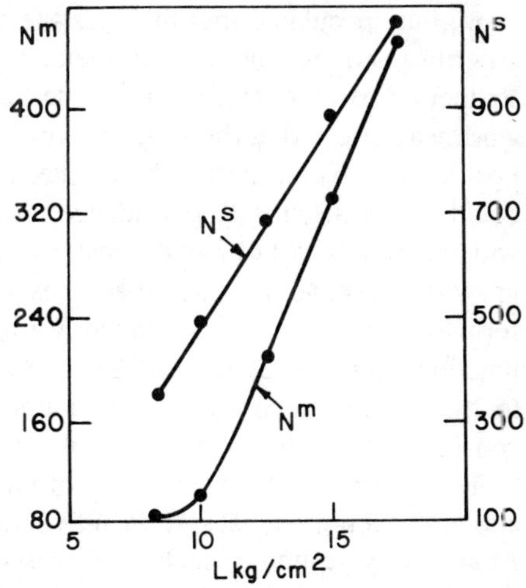

Fig. 3. Number of load cycles to transformation N^m vs the applied load L. N^m represents the time necessary for friction activated metal to transform to Fe_3O_4. N^s represents the time necessary for breaking up of the metal surface layer. *(Ref 16)*

A = Arrhenius constant of parabolic oxidation
Q = activation energy for parabolic oxidation
V = sliding speed
P = hardness of metal
f = fraction of oxide film which is oxygen
ρ = density of the oxide
ϵ = critical oxide film thickness.

Quinn states that "it is virtually impossible to obtain any information about the temperatures which actually occur at sliding interfaces by a direct method. . . . It is even more difficult to carry out oxidation experiments under conditions which approximate those occurring at the real areas of contact." In this oxidational theory Quinn states that information such as (*i*) the number of asperities, (*ii*) the size of the asperities, (*iii*) the thickness of the critical oxide film, (*iv*) the temperature at which the oxide is formed, and (*v*) oxidational constants can be obtained for many oxidizing metals. A major difficulty with Quinn's oxidational theory of wear is shared by many of the other theories of oxidational wear. It is intended to be a predictive theory, but many of the parameters (A, Q, ϵ, ρ, f) must still be inferred by indirect experimental observation—e.g., A and Q are based on contact temperatures indicated by the presence of a particular form of oxide. As will be discussed in a following section, this method of estimating the contact temperature is the subject of much controversy. It is questionable whether Quinn's oxidational theory of wear can easily be extended to many other wearing systems, as the author maintains.

Rabinowicz[22] found that for an oxide film to act as a solid lubricant it must be at least 10^{-5} mm thick. He had difficulty in attributing this to a physical phenomenon, but speculated that this was the thickness necessary to completely cover asperities and prevent metal-to-metal contact. Molgaard[23] addressed this issue in a paper on oxide thickness and oxide transfer in wear, although the thrust of this work was directed at the study of iron oxide layers greater than 10^{-4} mm thick. For thinner layers, the parabolic rate law adopted to describe the oxide growth could not be used with confidence. A model of oxide transfer was developed which parallels that of metal transfer in adhesion wear; e.g., after the oxide layer grows to a critical thickness it adheres to the oxide layer on the opposite surface, forms a junction, and subsequently ruptures along a plane below the original junction interface. Because of the hard and brittle properties of the oxide relative to the underlying metal, the traction, and thus the friction coefficient, was considered to be lower than that observed in metal-to-metal contact.

In other studies[24] the formation of an oxide "glaze" during the wear of Ni-Cr alloys was reported that significantly reduced friction coefficients and wear rates. Analyses by scanning electron microscopy and X-ray diffraction indicated that the glaze was composed of very fine compacted oxide particles. The particles were typically 10 to 50 μm in size and showed signs of having undergone significant plastic deformation. However, junction growth during sliding was reported to be relatively small because, as the junction grew, the contact pressures decreased, enabling the oxide to revert to its brittle nature. Thus the coefficient of friction was reported to be lower than that observed during metal-to-metal contact.

Barnes et al.[25] studied the wear behavior of Fe-5%Cr under unidirectional and reciprocating motion in which the oxygen pressure could be maintained between 10^{-6} Pa and atmospheric pressure. A summary of the friction and wear data correlated with the type of deformation observed is given in Fig. 4. The "islands" referred to took the form of protective raised areas of compacted debris which developed gradually but increased rapidly as the oxygen partial pressure was increased. These islands persisted for extended periods during subsequent evacuation of all gas, and were composed of a mixture of alloy and oxides in the form of oxide-covered particles. It was observed that the formation of oxide islands resulted in reduced friction and usually in reduced wear, with the amount of debris and extent of island coverage generally increasing with increasing oxygen partial pressure. The coefficient of friction and the wear rate for various contact geometries in the different oxygen partial pressures used are shown in Fig. 5.

Endo and Goto[26] reported that the initiation and propagation of fatigue cracks due to fretting wear were greatly accelerated in the presence of oxygen. However, a Vickers hardness measurement along the worn surface was lower in room air than in argon because of the lower tangential force (and thus friction coefficient) from the lubricating effect of the oxide debris.

A unique method for studying the interaction of oxygen and other gases with rubbing metal surfaces was developed by Momose and Namekawa[27] using a

Coefficient of friction	Type of deformation	Overall wear rate, $mm^3\ mm^{-1}$
1.8 ± 0.2	Metal-metal contact, with limited ductility due to work hardening	$(1 - 2) \times 10^{-6}$
1.0 ± 0.1	Some oxide and work-hardened metal debris causing abrasion, a few oxide islands, some metal-metal contact	$(1 - 4) \times 10^{-6}$
0.8 ± 0.1	Partial coverage on both upper and lower specimen with islands	5×10^{-7}
0.6 ± 0.1	Complete coverage of lower specimen and most of upper specimen with islands	2×10^{-7}
0.4 ± 0.1	Thick coverage with islands in air	$10^{-8} - 10^{-9}$

Fig. 4. Summary of friction and wear data correlated with the type of deformation observed in unidirectional wear of Fe-5% Cr surfaces under various oxygen partial pressures. The coefficient of friction and the overall wear rate decreased with higher oxygen partial pressures. *(Ref 25)*

Geiger counter to measure exoelectron emission. Two types of emission were measured, dark emission and optically stimulated exoelectron emission (OSEE). Dark emission was associated with a state of high excitement produced by the deformation of the metal surfaces. Dark emission ceased abruptly when friction was stopped. Optically stimulated emission was associated with the lower work function resulting from the adsorption of ambient reactive gases. When both oxygen and water vapor were present in the environment, optical emission did not increase immediately after friction began because it was prevented by the formation of oxide on newly exposed iron surfaces (Fig. 6). Eventually, however, water vapor was adsorbed, and emissions occurred at these sites. When friction was interrupted emission eventually decayed because of the cessation of new surface formation. Figures 7 and 8 show the change in optically stimulated emission from iron in various environments and after various amounts of wear.

Inert and Nonoxidizing Environments

Unlike fatigue damage, which is clearly reduced by protective atmospheres, the advantages or disadvantages of inert environments on wear are not straightforward. Although surface degradation by oxidation is absent, so is the oxide film which may tend to lubricate the opposing surfaces. In general, adhesion and metal transfer are more prevalent in inert environments, as indicated by the higher coefficients of friction. Although several reports indicated that damage, as measured by weight loss, was reduced in inert environments,[28-30] considerable roughening of the surface occurred.

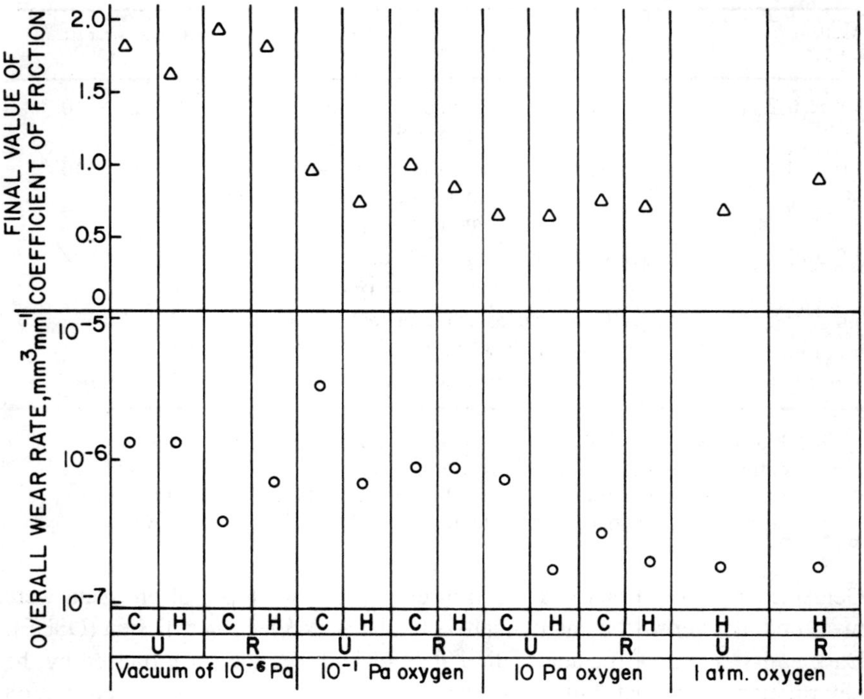

Fig. 5. The final value of the coefficient of friction and the overall wear rate for Fe-5% Cr during like-on-like sliding in different environments: C, conical specimens; H. hemispherical specimens; U, unidirectional motion; R, reciprocating motion. The oxygen partial pressure has a greater influence than the geometry or action of loading. *(Ref 25)*

In an early study[28] some of the phenomenological features of fretting wear of steels in carbon dioxide and helium as compared to dry air were described. Wear damage was greater in carbon dioxide and greater still in helium. Since no oxide formed in either environment, metal transfer was greater in the inert atmospheres. In CO_2 some metallic debris was observed. Helium, in contrast with CO_2, did not form an adsorbed film. Accordingly, metal transfer was massive and no loose particles developed.

Wheeler[31] measured the friction coefficients in the wear of copper, iron and steel in vacuum with trace amounts of chlorine or oxygen. The results are shown in Fig. 9 to 12. Adsorbed atomic oxygen and chlorine reduced the friction coefficient quite significantly, although oxygen reduced it more at saturation (monolayer coverage). In equal concentrations both species were identically effective, although at saturation oxygen formed a denser film which made it ultimately more effective. In contrast, Pepper[32] has studied the wear of copper, nickel and iron surfaces by a sapphire indenter in oxygen, chlorine, ethylene and

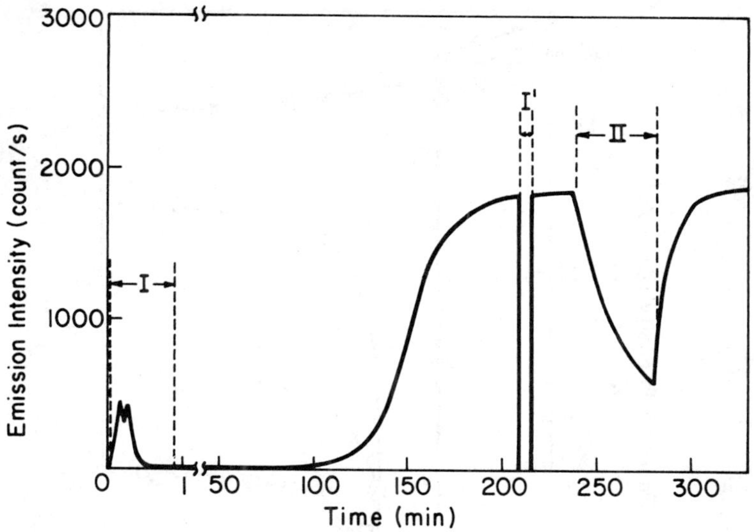

Fig. 6. Exoelectron emission from iron during sliding at a speed of 510 rpm: (I, I′) without optical stimulation; (II) rubbing interrupted. The incubation time for emission is due to suppression by oxide layers on newly exposed iron surfaces. *(Ref 27)*

vinyl chloride environments. These experiments were performed by first preparing atomically clean metal surfaces in an Auger spectrometer and then exposing them to given volumes of the various gases. The changes in friction coefficients are shown in Fig. 13. It was noted that oxygen increased friction in all cases, whereas chlorine decreased it. The lower coefficients of friction in chlorine were interpreted as being due to Van der Waals forces between the Al_2O_3 indenter and the metal chloride. The high coefficients in oxygen were thought to be due to strong interfacial bonds established by the reaction of metal oxide with Al_2O_3 to form the spinel $FeAl_2O_4$.

Endo and Goto[26] showed that the frictional force between steel surfaces was much higher in argon than in dry or room air (Fig. 14). However, the frictional force in the fretting of aluminum was lower in argon (Fig. 15). Metal particle debris was formed in both cases. The authors argued that metallic aluminum particles were soft and acted as good lubricants compared to aluminum oxide, whereas for steel, the iron oxide was a better lubricant than particulate iron. This argument could be plausible only if steel particles were work hardened to a much greater hardness than the surface and aluminum particles were not. Since both surfaces are also work hardened, this reasoning seems flawed.

Tsuya[33] compared the friction and wear of copper in hydrogen, vacuum and air. She found that the values of the coefficient of friction immediately after the

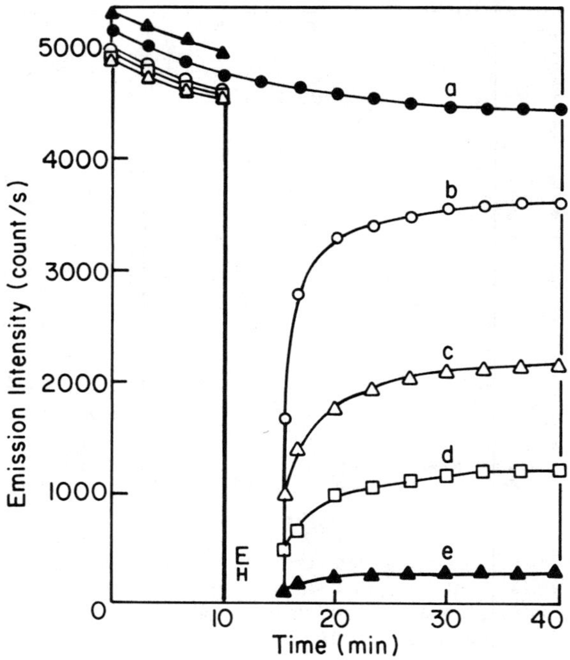

Fig. 7. Effect of exposure to argon-oxygen mixtures on the change of OSEE from iron after cessation of friction: **(a)** without exposure; **(b)** exposure under vacuum for 5 min; **(c)** 0.35 N/m^2 O_2; **(d)** 0.76 N/m^2 O_2; **(e)** 4.7 N/m^2 O_2. The activity of OSEE is suppressed by adsorption of oxygen molecules. *(Ref 27)*

oxides on the specimen surface were reduced by hydrogen were about 10 times higher than the values measured with electropolished surfaces. When contamination was prevented by surrounding the specimen with a titanium sponge, increases of 30 times were seen (Fig. 16). When surface oxidation was prevented, extensive welding and metal transfer occurred (Fig. 17). However, it was concluded that the wear tracks formed in air did not differ essentially from those formed in hydrogen. The major difference was that debris particles were smaller in air.

Soda and Sasada[34] studied the lubrication of sliding metal surfaces by adsorbed oxygen. The nature of adhesive wear of eleven pure metals was investigated in air under pressures of 10^{-6}–760 torr, and was divided into two types, mild wear and severe wear. To explain qualitatively the lubricating effect the authors introduced the concept of mean free time of a small contact point. They found that the critical transition point from mild to severe wear corresponded to a mean free time when a sheared junction was just covered by a chemisorbed monolayer of surrounding gas. Only transition metals, which chemisorbed oxygen, realized the mild wear state.

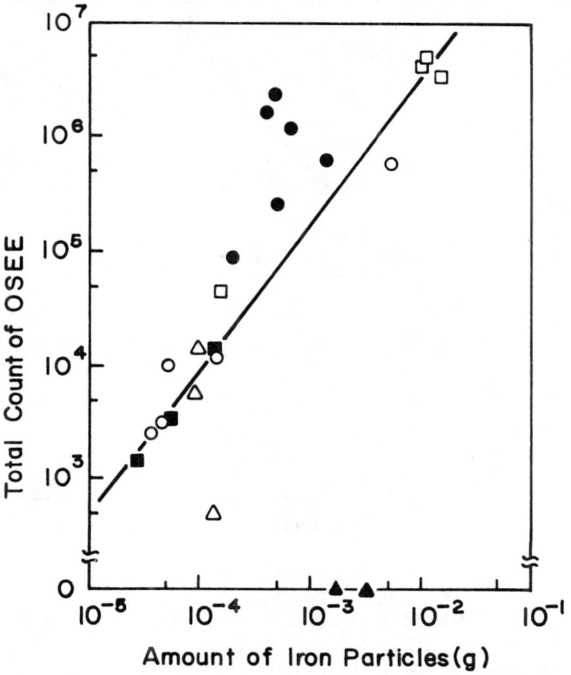

(□) Ar (10,700 N/m²); (▲) O_2 (1300 N/m²); (●) H_2O (2700 N/m²); (△) counter gas (C_2H_5OH) (2700 N/m²) + Ar (11,200 N/m²); (○) vacuum; (■) C_2H_5OH (2700 N/m²). The values in parentheses indicate the pressure of the environment. Exoelectron emission was shown to be more dependent on the amount of deformation than on the nature of the environment, although the latter determined the former.

Fig. 8. Relationship between OSEE from iron after friction in various environments and the amount of iron particles produced in this process. *(Ref27)*

Effects of Humidity on Wear

Moisture in the environment can have a pronounced effect on the fracture mechanisms of some steel and aluminum alloys with or without oxygen present. For example, it has been reported that fatigue lives for hardened steels were 10 times lower in moist argon than in dry argon.[35] Dahlberg[36] reported that moist air increased the crack propagation rate about 10 times over dry air for 4340 steel. The detrimental effect of water vapor on aluminum alloys is also well known and well documented.[37-43] It is not surprising, then, that water vapor also has a major effect on the wear of materials. What is surprising is that there are conflicting reports on whether the effect is beneficial or detrimental. For example, in fretting wear Feng and Uhlig[29] and Wright[44] found that humidity had little effect on steel in inert atmospheres but that damage decreased linearly with moisture content in air. Broom and Nicholson[45] reported, however, that exclusion of water vapor

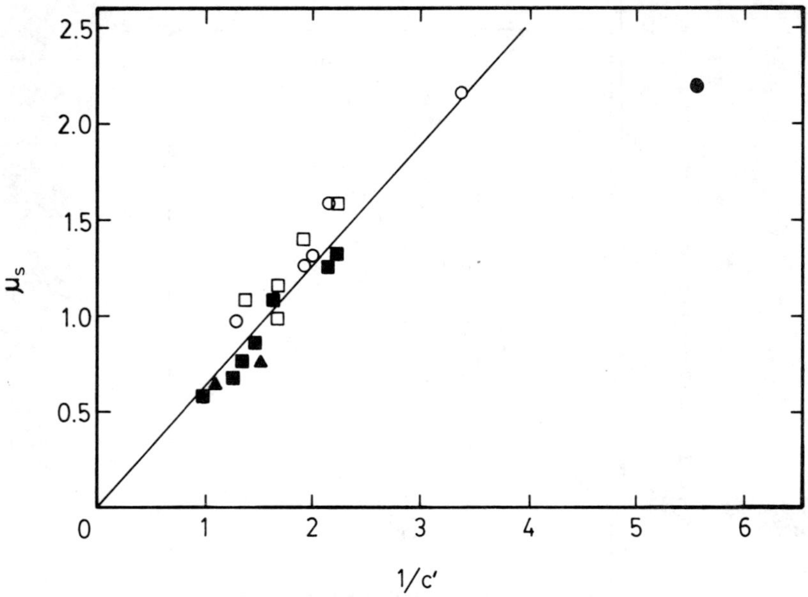

Fig. 9. Static coefficient of friction μ_s as a function of inverse of adsorbate concentration ($C' = \frac{C}{C_{monolayer}}$). ○, chlorine on copper. □, chlorine on iron. ■, oxygen on iron. ▲, oxygen on steel. ●, oxygen on copper. μ_s depended only on adsorbate concentration and not on the specific environment/metal couple. *(Ref 31)*

increased fretting fatigue life at a given stress level by a factor of eight. Endo and Goto[26] agreed that high humidities had a detrimental effect on the fretting of aluminum alloys but showed that the effect on carbon steels was negligible. Appledorn et al.[46] demonstrated that increasing humidity decreased the coefficient of friction in both air and argon. In fact, for both materials the friction for wet conditions was lower than for dry conditions (Fig. 18). Humidity had no consistent effect on the size of the wear scar, however, showing that wear rates and friction coefficients are not always related linearly.

The type of wear mechanism which is operating has a large effect on the measured value of tangential force. In general, adhesive conditions lead to larger friction coefficients than do abrasive conditions, in which much debris is produced. If the wear criterion used is weight loss, for example, the abrasive conditions will result in greater wear rates even though a lower friction coefficient is measured. It is important to interpret wear results with a knowledge of the criterion used.

In 1956 Godfrey[47] reported that variations in the relative humidity from a value approaching zero to a value near saturation had a marked effect on unlubricated fretting of steel. A relative humidity of about 30% produced the greatest amount of damage as measured by the area of the wear spot (Fig. 19 and 20). At relative

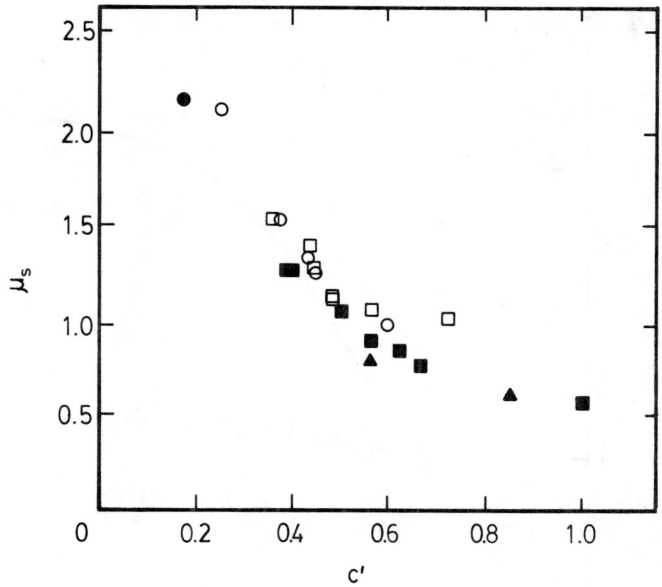

Fig. 10. Static coefficient of friction μ_s as a function of adsorbate concentration, C′. ○, chlorine on copper. □, chlorine on iron. ■, oxygen on iron, ▲, oxygen on steel. ●, oxygen on copper. *(Ref 31)*

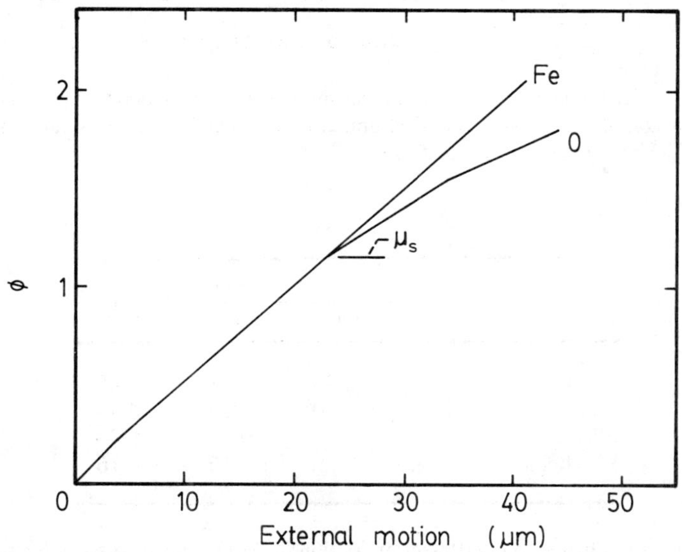

Fig. 11. Force coefficient φ (tangential force divided by normal force) of iron on iron as a function of external motion of rider holder arm. Load, 100/g. Fe, bare metal. O, with saturation (monolayers) oxygen concentration. *(Ref 31)*

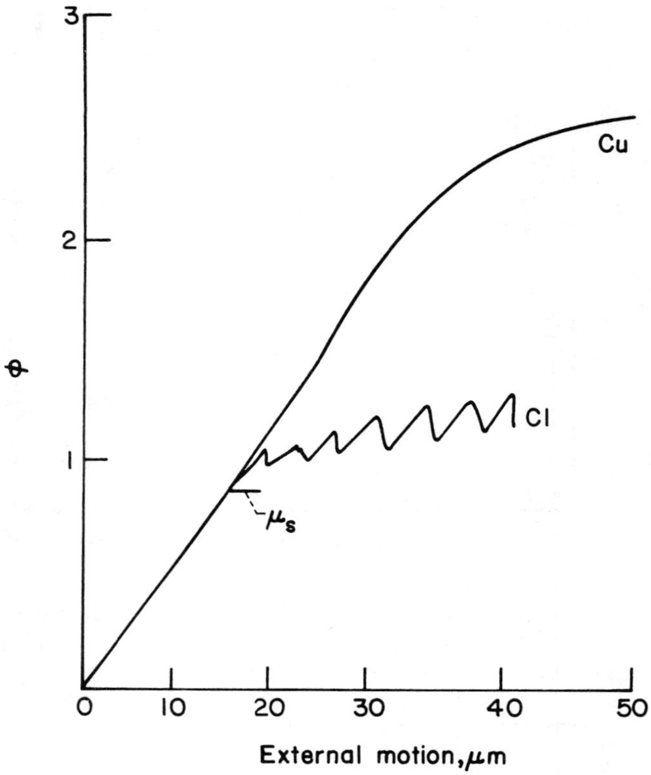

Fig. 12. Force coefficient of copper on copper as a function of external motion of rider arm. Load, 100 gm. Cu, bare metal. Cl, with saturation chlorine concentration. *(Ref 31)*

Gas	Metal Cu	Ni	Fe
Oxygen	+ 0.25	+ 0.41	+ 0.45
Chlorine	− 0.30	− 0.25	− 0.40
Ethylene	0	0	+ 0.30
Vinyl chloride	0	− 0.15	− 0.16

Fig. 13. Change in coefficient of friction of Al_2O_3 – metal system due to exposure to different gases. Load, 100 gm; sliding speed, 0.5 cm/sec. Oxygen increased the coefficient of friction for all metals and chlorine decreased it. *(Ref 32)*

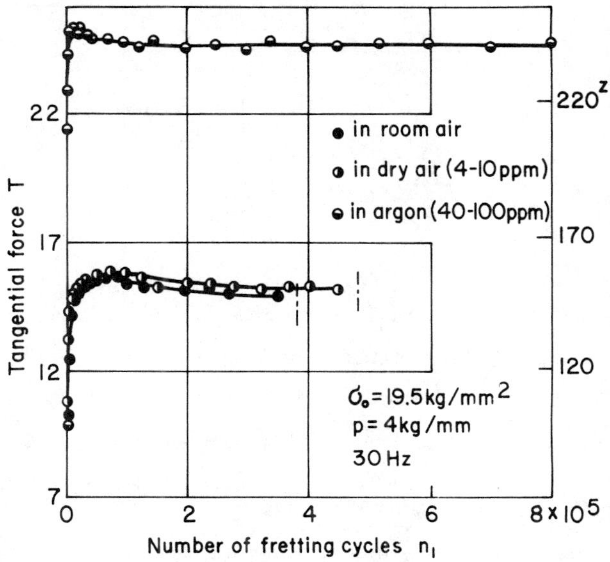

Fig. 14. Variations of tangential force in various environments for the fretting of carbon steel on carbon steel as a function of the number of fretting cycles. Moisture content did not have much effect, but oxygen greatly reduced the tangential force. *(Ref 26)*

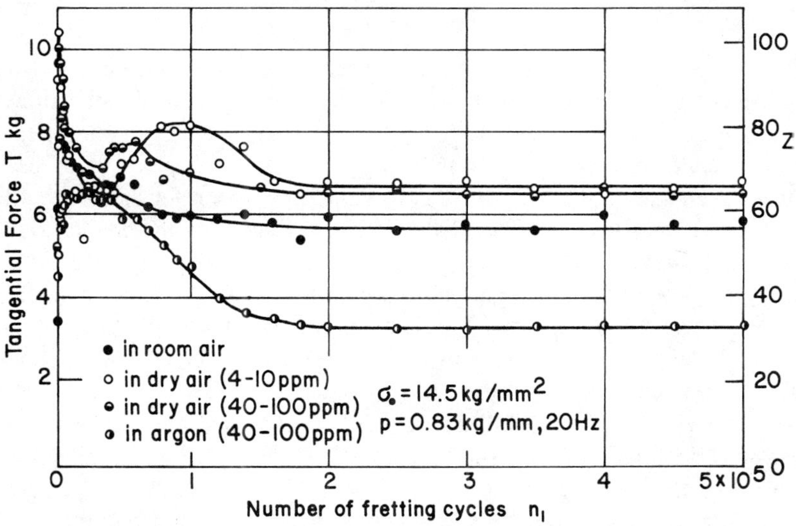

Fig. 15. Variations of tangential force in various environments for an aluminum-zinc-magnesium alloy fretted against itself as a function of the number of fretting cycles. Water vapor decreased the tangential force, but oxygen increased it. *(Ref 26)*

	Polycrystals
μ_0	0.68 ± 0.13
μ_1	7.05 ± 0.77
μ_2	8.54 ± 0.69
μ_3	5.68 ± 0.79
μ_4	20.5 ± 5.
μ_5	0.38 ± 0.12
μ_6	1.06 ± 0.42

μ_0: measured in air immediately after electropolishing in H_3PO_4 (d = 1.4).

μ_1: measured in H_2 after reduction at 500°C in H_2 purified by Cu.

μ_2: measured in H_2 after reduction at 500°C in H_2 purified by Cu and Pd.

μ_3: measured in H_2 after reduction at 500°C in H_2 purified by Cu, Pd, and Ti.

μ_4: measured in H_2 after reduction at 800°C in H_2 purified by Cu, Pd, and Ti, with Ti sponge surrounding specimens.

μ_5: measured in vacuum (oil diffusion pump) after heating to 500°C in vacuum.

μ_6: measured 16 h after the test for μ_4.

Fig. 16. The coefficient of static friction of pure copper sliding on copper after the application of various environments. With very clean surfaces (μ_4) coefficients of friction as high as 20 could be obtained. *(Ref 33)*

humidities of 10% or less no "stain" was observed and the surface was rough and matte gray. At humidities of 20 to 40% the scar was furrowed and reddish, while humidities greater than 50% produced black films. The author did not speculate on the chemical composition of these films.

In an even earlier study Wright[44] gave a possible reason for the above effect. He suggested that adsorbed water had a twofold effect: it promoted the oxidation of the metal surface but also lubricated the surface and broke up adherent oxide films. The 30% humidity singularity reported by Godfrey might then delineate that value where the corrosive mechanism was overtaken by the lubricative mechanism. Figures 21 and 22 show how fretting damage changes with humidity for various metal couples.

Larsen-Basse and Sokoloski[48] investigated the effects of humidity on abrasive wear of various materials by silicon carbide. As seen in Fig. 23, all materials showed an increase in wear rate between 0 and 65% relative humidity. Between 65 and 100% relative humidity the harder materials continued to show an increasing wear rate while the softer materials showed a decrease. It was suggested that atmospheric moisture decreased the fracture strength of the

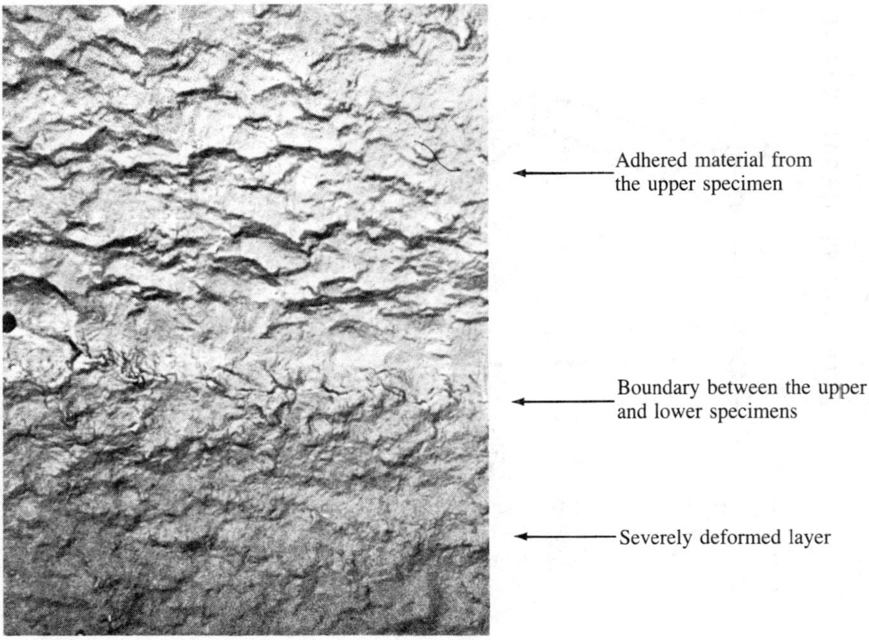

Fig. 17. TEM replica of a section of a copper specimen after sliding (copper rider) in H_2 (subsequent to reduction at 800°C in H_2 purified by Cu, Pd and Ti). Adhesion and metal transfer were prevalent. *(Ref 33)*

Ball-on-Cylinder Tests;
No Lubricant; Steel-on-Steel
(77°F, 240 g, 240 rpm, 32 min)

Environment	Wear scar diameter, mm	Coefficient of friction
Dry argon	0.95	1.32, 1.35
Dry air	0.91	0.75
Wet argon	0.25	0.57, 0.69
Wet air	1.07	0.29

Fig. 18. Wear scar diameter and coefficient of friction of steel sliding on steel in various environments. Water vapor decreased the coefficient of friction as did oxygen, but neither had a consistent effect on the extent of damage. *(Ref 46)*

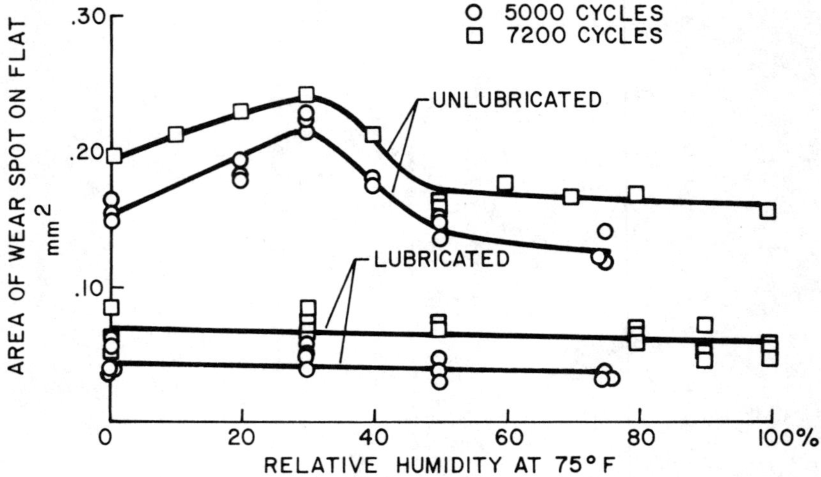

Fig. 19. Effect of relative humidity (in air) on the area of the wear spot on a carbon steel surface (carbon steel slider) after a specified number of fretting cycles. A relative humidity of 30% produced the greatest amount of damage. *(Ref 47)*

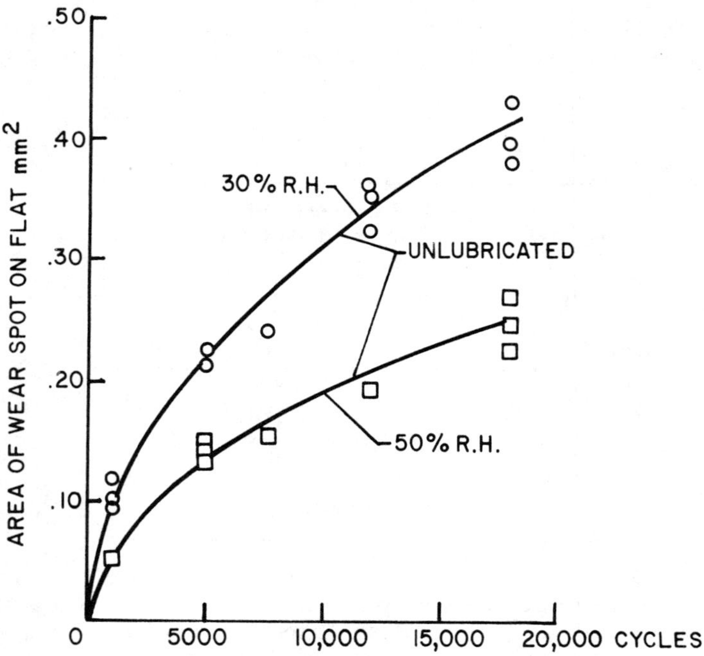

Fig. 20. Effect of relative humidity on the area of the wear spot on a carbon steel surface fretted with carbon steel. The damage at 30% R.H. was greater than that at 50% R.H. *(Ref 47)*

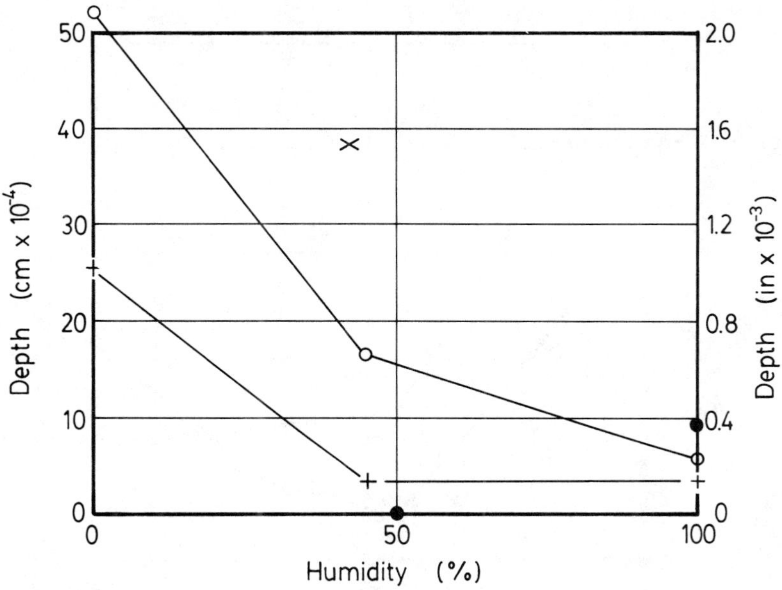

Fig. 21. Variation of the maximum depth of the fretting scar as a function of percent relative humidity: ×, nickel against steel, 50,000 oscillations; ○, chromium against steel, 50,000 oscillations; +, steel against steel, 50,000 oscillations; ●, gold against steel, 100,000 oscillations. *(Ref 44)*

SiC particles, enabling them to break up and expose more sharp edges. Above 60% relative humidity, however, so much breakdown occurred that the grit size decreased and wear rates dropped off. It was suggested that the breakdown of the SiC particles did not occur with the harder materials (sodium silicate glass and steel) because they would not permit extensive particle/metal contact by deformation. Such contact was necessary to transmit the forces which caused particle fracture.

Bill[49] has conducted an investigation into the role of humidity in fretting wear of iron, nickel and titanium. Iron and titanium showed maximum fretting wear at 10 and 30% relative humidity, respectively. Nickel showed a minimum at about 10% r.h. For iron and nickel increased humidity showed increased debris mobility with some evidence of increased plasticity of the metal surface. For titanium, more rapid fatigue of the fretting surfaces was seen to accompany increased r.h. levels. All three metals showed substantially reduced fretting in dry N_2 compared to dry air. Admitting moisture to the N_2 resulted in a proportionally greater increase in fretting than was observed when fretting was conducted in H_2O saturated air.

Tsuji and Ando[50] have looked at the effects of humidity on the sliding wear of high and low carbon steels and cast irons (nodular graphite and gray) at temperatures between 15°C and 30°C. They found that the effects of humidity were

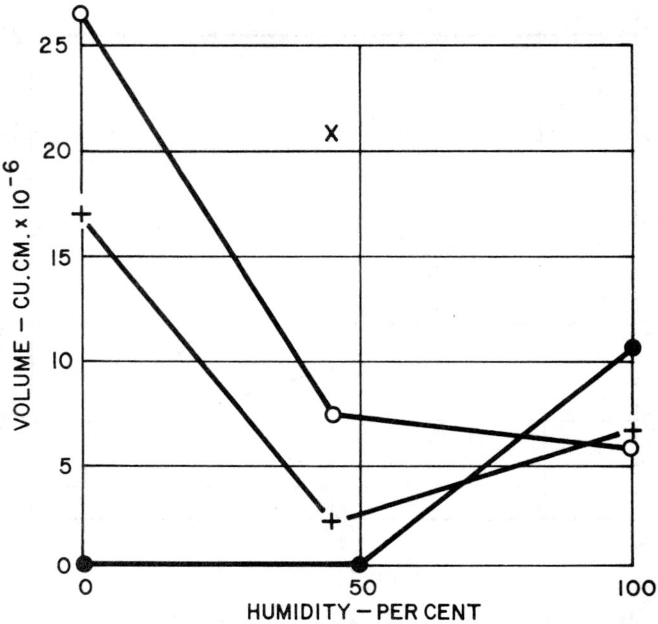

Fig. 22. Variation of the fretting scar volume as a function of percent relative humidity. The symbols indicate the same conditions as detailed in Fig. 21. Contrary to Fig. 19, the author showed that the minimum volume of damage occurred at about 45% r.h. for steel fretted on steel. *(Ref 44)*

more pronounced on cast iron than on steel, but that the exact effect was a complex interaction between the moisture content and the temperature (Fig. 24). It was speculated that the oxide film which forms on cast iron is very coarse and allows rapid additional corrosion when moisture is present in the environment.

Effect of Temperature on Wear

High temperature wear is a problem in many industrial and commercial applications. Chemical interactions with metal surfaces tend to speed up as the temperature is increased and may drastically alter wear properties. In addition, physical properties of the metals themselves are often quite different at elevated or depressed temperatures, and this may affect wear processes. By far the greatest amount of research has been conducted on how high temperature oxidation influences the wearing of metal surfaces.

Tenwick and Earles[14] have studied how the wear rate of carbon steel pins sliding on a Ni-Cr-Mo alloy (Nimonic 75) track is affected by the ambient temperature. Figure 25 shows that the wear rate is lowest at about 300°C. The

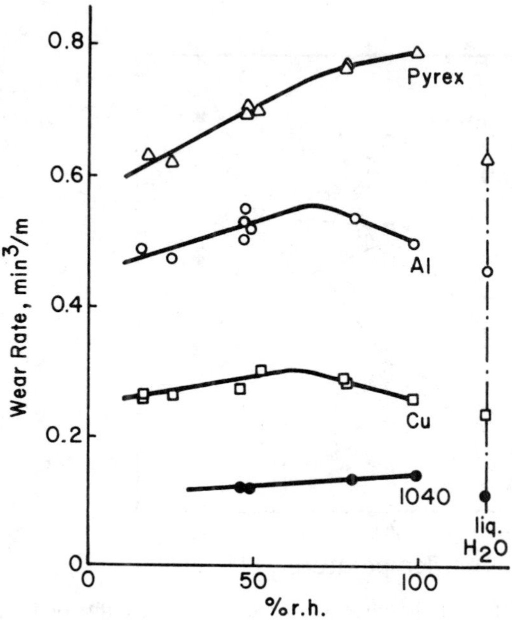

Fig. 23. Wear rate (volume loss per unit of sliding distance) in 2-body abrasion as a function of percent relative humidity (in air) for various materials abraded by silicon carbide. The harder materials (Pyrex and 1040 steel) showed increased wear above 65% R.H. but the softer materials showed decreased wear. *(Ref 48)*

authors stated that the formation of Fe_3O_4 was likely to predominate in the range 300°-500°C. Other investigators have also studied the effect of temperature on the wear of several nickel-chromium alloys. They showed that for nickel-base alloys a well-defined transition temperature exists above which significant reductions in friction and wear take place.[51-54] Below the transition temperature load-bearing areas were reported to be largely metallic, whereas at higher temperatures an oxide glaze composed of transient oxides with low shear strength was observed. The more dilute alloys exhibited lower friction and wear coefficients because NiO formation was favored over the formation of $NiCr_2O_3$, and NiO grew more rapidly. Data from a coordinated study[51] are shown in Fig. 26.

Several studies have been made on the friction of iron surfaces at high temperatures in vacuo.[55-57] For example, Bowden and Young[55] found that when samples were first outgassed at 1000°C, coefficients of friction of 3.5 were measured at room temperature. Seizure did not occur at room temperature but did in the range 300°-400°C. Buckley,[56] however, did find that seizure occurred at room temperature with low loads and sliding speeds. Barnes, Stott and Wood[58] carried out experiments on Fe-Cr alloys in ultrahigh vacuum between 20°C and

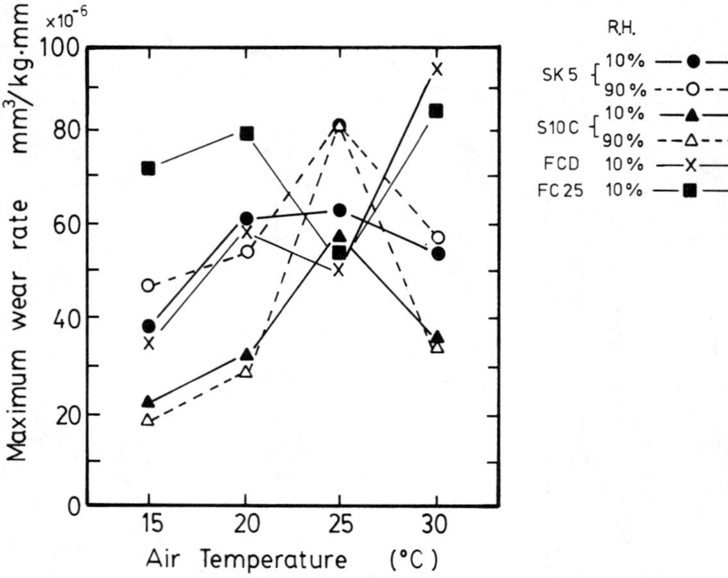

Fig. 24. Effect of air temperature and relative humidity on the maximum wear rate (volume per load-sliding distance) for materials: SK5, high carbon steel; S10C, low carbon steel; FC25, gray cast iron; FCD, nodular graphite cast iron. Wear rate was shown to be a complex interaction of R.H., temperature and material. *(Ref 50)*

850°C and found that, as in the nickel base alloys, a transition temperature seemed to occur at about 450°-500°C. Unlike the former alloys the friction coefficient rose with increasing temperature. It has been suggested that in general an increase in friction can be correlated directly with a decrease in hardness,[59] since adhesive junctions become larger as contact points become softer and flow more. However, hardness decreased markedly at 350°-400°C whereas the friction did not rise until about 500°C. Tabor[60] points out though that the hardness of metals has little direct effect on frictional behavior, since a decrease in hardness is usually accompanied by a decrease in shear strength. In a related note to this argument, Tegart[61] and Redfern and Sellers[62] have reported that iron and iron-chromium alloys showed limited ductility up to about 500°C, after which ductility increased considerably.

Bowden and Childs[63] have conducted a study on the effects of very low temperatures (25° to 300°K) on the friction and wear of twelve different pure metals (Au, Ag, Cu, Ni, Ta, Fe, Mo, W, Zr, Ti, Be, Co). The authors concluded that under conditions of strong adhesion (vacuum environment) the most important variable was the ductility of the metal. They pointed out, however, that in addition to adhesion, a plowing term must also be included in the analysis. For fcc metals, as the temperature was lowered a transition was observed, associated

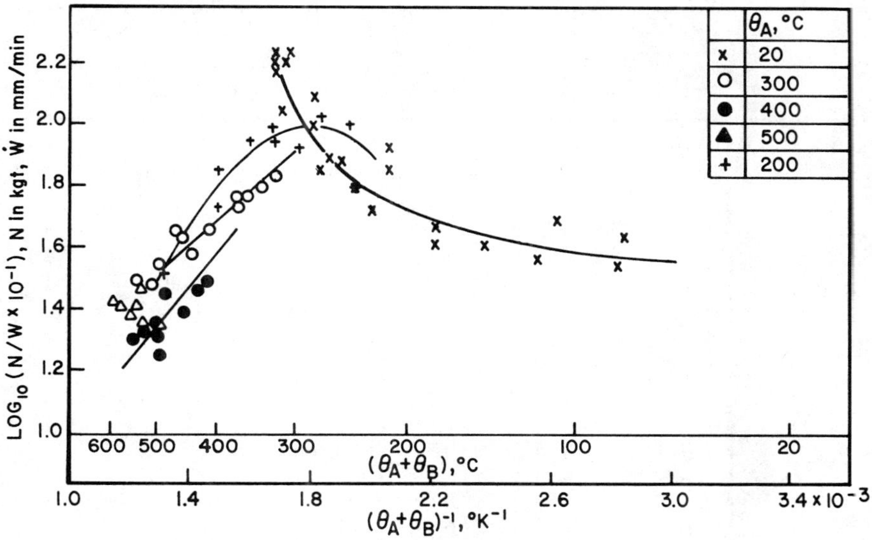

Fig. 25. Wear rate as normalized by the applied load (N is the normal load, ẇ is the wear rate as measured by depth of penetration per time) as a function of the ambient (θ_A) plus interface (θ_B) temperature, carbon steel sliding on Ni-Cr-Mo alloy. The lowest wear rate (highest N/ẇ) was measured at a temperature of about 300°C. *(Ref 14)*

Temperature, °C	N75	C263	N108	I901
20	Never	Never	Never	Never
100	Never	Never	Never	Never
150	0.9-1.8	Never	Never	–
200	0.9-1.8	1.8-7.2	1.8-7.2	Never
300	0.9-1.8	0.3-0.9	0.3-0.9	7.2-10.0
400	0.3-1.8	0.3-0.9	0.3-0.9	0.3-0.9
600	0-0.9	0-0.3	0-0.3	0-0.3
800	0-0.3	0-0.3	0-0.3	0-0.3

Fig. 26(a). Transition time from a high wear rate to a very low rate for a given temperature for several nickel chromium alloys sliding on themselves. A critical temperature of about 200°C was measured below which no transition occurred. *(Ref 51)*

with a marked rise in the work-hardening rate, and the friction coefficient dropped from 2.5 to 1.5. With bcc metals a similar transition was found but was associated with the ductile-brittle transition. Under brittle conditions, friction coefficients did not exceed 1. In hcp metals the behavior depended on the ductility of the metal, with cobalt recording the lowest value (friction coefficient = 0.5). Results are shown in Fig. 27. Wear rates increased with the

Alloy	Sliding time 10°/sec	Coefficient of friction						
		20°C	150°C	200°C	300°C	400°C	600°C	800°C
N75	0.01	1.2	1.2	1.0	1.0	1.0	1.0	1.8
	0.06	1.2	1.1	1.0	0.9	0.8	1.0	0.5
	0.3	1.0	1.0	1.0	0.9	0.8	1.0	0.5
	0.9	0.9	0.9	1.0	0.8	0.8	0.5	0.4
	1.8	0.9	0.6	0.6	0.5	0.5	0.4	0.4
	7.2	1.2	0.6	0.5	0.5	0.5	0.4	0.4
	18.0	1.2	0.6	0.5	0.5	0.5	0.4	0.3
	50.0	1.3	0.6	0.5	0.5	0.5	0.4	0.3
C263	0.01	1.2	1.2	1.2	1.1	1.0	0.9	1.3
	0.06	1.2	1.2	1.1	1.0	1.0	0.7	0.4
	0.3	1.2	1.1	1.0	0.9	0.9	0.5	0.4
	0.9	1.1	1.1	0.9	0.9	0.5	0.5	0.4
	1.8	1.0	1.1	0.8	0.5	0.5	0.5	0.4
	7.2	1.2	1.2	0.8	0.5	0.5	0.4	0.3
	18.0	1.3	1.2	0.6	0.5	0.5	0.4	0.3
	50.0	1.3	1.2	0.6	0.5	0.5	0.4	0.3
N108	0.01	1.0	1.1	1.0	1.0	1.0	0.8	1.0
	0.06	1.0	1.0	1.0	1.0	1.0	0.7	0.3
	0.3	1.0	1.0	0.8	0.8	1.0	0.4	0.3
	0.9	0.9	0.9	0.8	0.6	0.6	0.4	0.3
	1.8	0.9	0.9	0.7	0.6	0.6	0.4	0.3
	7.2	1.0	1.1	0.6	0.6	0.6	0.4	0.3
	18.0	1.1	1.1	0.6	0.6	0.6	0.4	0.3
	50.0	1.2	1.1	0.6	0.6	0.5	0.4	0.3
1901	0.01	1.0	...	1.8	1.7	1.2	0.8	0.8
	0.06	1.0	...	1.5	1.5	1.2	0.8	0.8
	0.3	1.0	...	1.3	1.4	1.1	0.6	0.6
	0.9	1.0	...	1.3	1.1	0.8	0.5	0.5
	1.8	1.0	...	1.2	1.1	0.8	0.5	0.4
	7.2	0.9	...	1.2	1.1	0.8	0.5	0.4
	18.0	1.1	...	1.3	0.9	0.8	0.5	0.4
	50.0	1.2	...	1.3	0.9	0.8	0.5	0.4

Fig. 26(b). Average wear volumes after various sliding times at temperatures between 20° and 800°C. *(Ref 51)*

coefficient of friction within a group of metals with a particular crystal structure.

A transition temperature of 120°-200°C was found in wear of mild steel in air,[64] and was attributed to the formation of a stable oxide film. In argon this film was absent and the wear rate increased as the temperature was raised to 500°C.

Thiery and Spinat[65] measured the wear rate of a Co-20Cr-10Ni alloy sliding on itself in the range of ambient temperatures between 0° and 800°C. As shown in Fig. 28 a transition to low wear rates occurred with increasing temperature after a maximum value of wear was reached at 250°C.

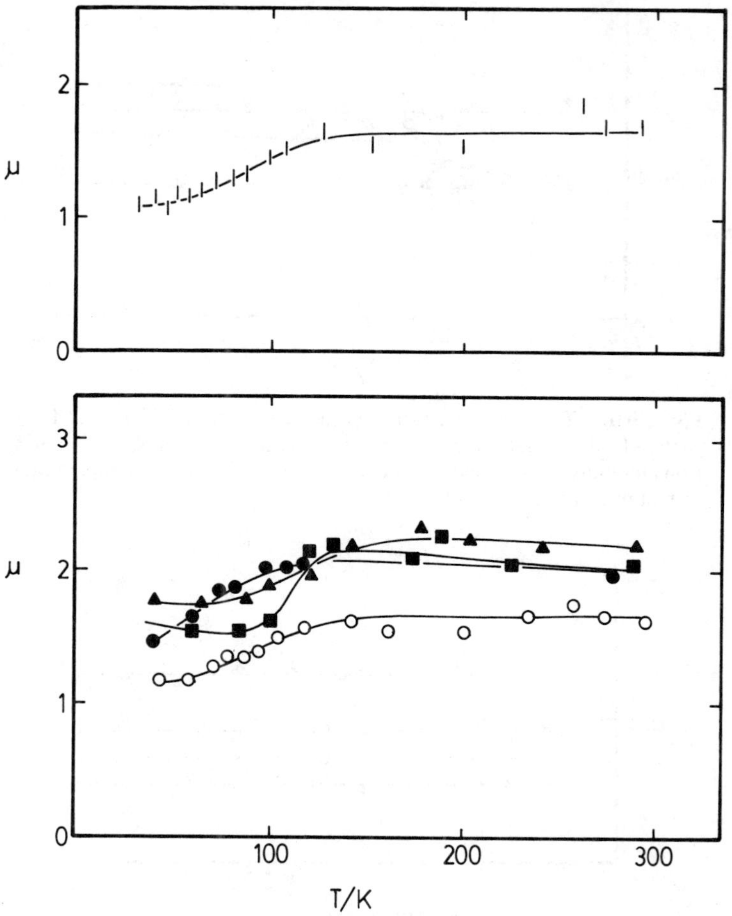

Fig. 27(a). Effect of temperature on the coefficient of friction of face-centered cubic metals. **(Upper graph)** The behavior of silver. **(Lower graph)** The behavior of gold ●, silver ○, copper ■, nickel ▲. A decrease in friction was seen at low temperature. *(Ref 63)*

On a Ti-6Al-4V alloy fretting fatigue damage decreased as the temperature was increased, but this was attributed to deteriorating mechanical properties rather than to a wear mechanism.[66]

Aqueous Environments

Wear behavior in aqueous environments parallels that of many other types of conjoint chemo-mechanical interacting systems such as corrosion fatigue or stress corrosion cracking. Therefore, the type of behavior exhibited depends largely on the specific metal/electrolyte combination. It may be stated in general, however,

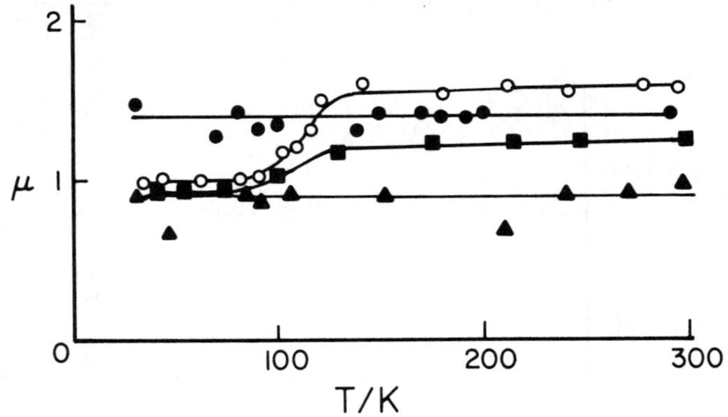

Fig. 27(b). Effect of temperature on the coefficient of friction of body-centered cubic metals. Tantalum ●, iron ○, molybdenum ■, tungsten ▲. Iron and molybdenum showed a decrease in friction at low temperatures. Tantalum and tungsten did not. *(Ref 63)*

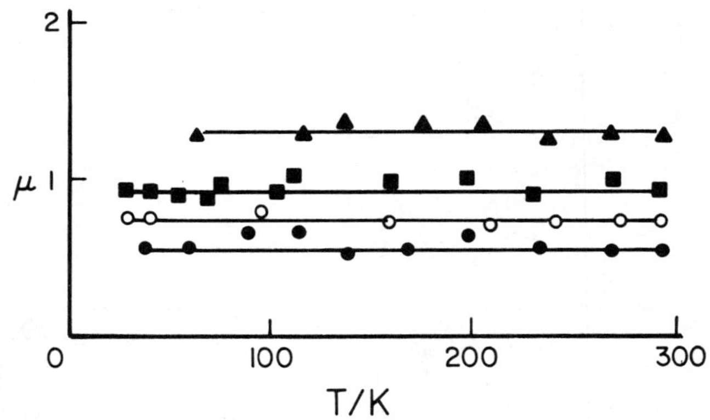

Fig. 27(c). Effect of temperature on the coefficient of friction of hexagonal close packed metals. Cobalt ●, beryllium ○, zirconium ■, titanium ▲. No temperature dependence was seen. *(Ref 63)*

that for a particular metal those environments which are the most detrimental in corrosion fatigue will also result in the greatest wear damage. For example, it has been demonstrated[46] that by making the environment more corrosive, wear is increased. Figure 29 shows the wear behavior of SAE 52100 steel in water and indicates that the wear scar is larger in aerated conditions than in de-aerated conditions.

Tse and Suh[2] have studied the role of chemical heterogeneities on chemical interactions between a sliding surface and its environment using 2024 aluminum alloy in sodium chloride solutions of various pH's and NaCl concentrations. They suggest that a metal surface should not be treated as homogeneous as it is usually

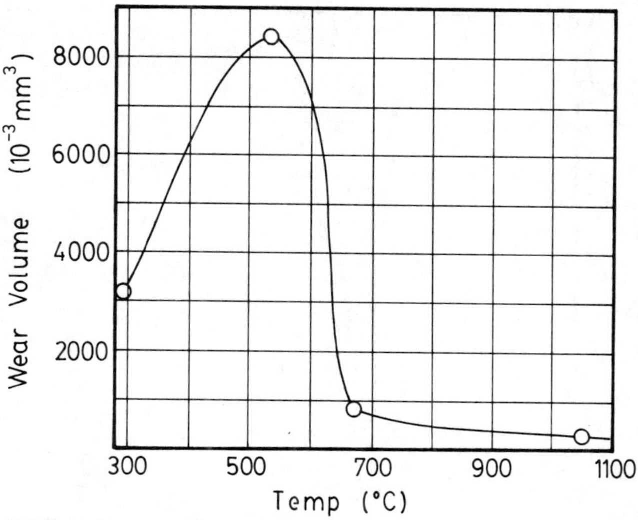

Fig. 28. Effect of temperature on wear volume produced by fretting on Co-20Cr-10Ni alloy after 180,000 cycles under a load of 80 N and amplitude of 2 mm. The damage produced increased up to about 270°C after which it declined to low values. *(Ref 65)*

Four-Ball Wear Tests Using Water as the Lubricant
(1200 rpm, 1 kg, 15 min, 74°F)

	Hard steel (61 R_c)		Non-hardened steel (25 R_c)	
	Wear scar diameter, mm	Coefficient of friction	Wear scar diameter, mm	Coefficient of friction
Argon	0.34	0.31	0.32	0.30
Air	0.58	0.31	0.57	0.40

Fig. 29. Effect of aqueous environment on the wear scar and coefficient of friction of hard and soft steels sliding on themselves. The more corrosive oxygenated solution produced greater amounts of wear but did not have a pronounced effect on the coefficient of friction. *(Ref 46)*

treated in wear but that localized corrosion may occur. Chemical reactions may reduce the surface tractions which cause mechanical deformation, but deformation may prompt localized corrosion. Corrosion products can act as lubricants, lower the coefficient of friction and reduce wear unless the corrosion itself is so severe that it wears the surface. Some of these results are shown in Fig. 30 to 33. Chloride ions in solution result in higher wear and friction coefficients up to a

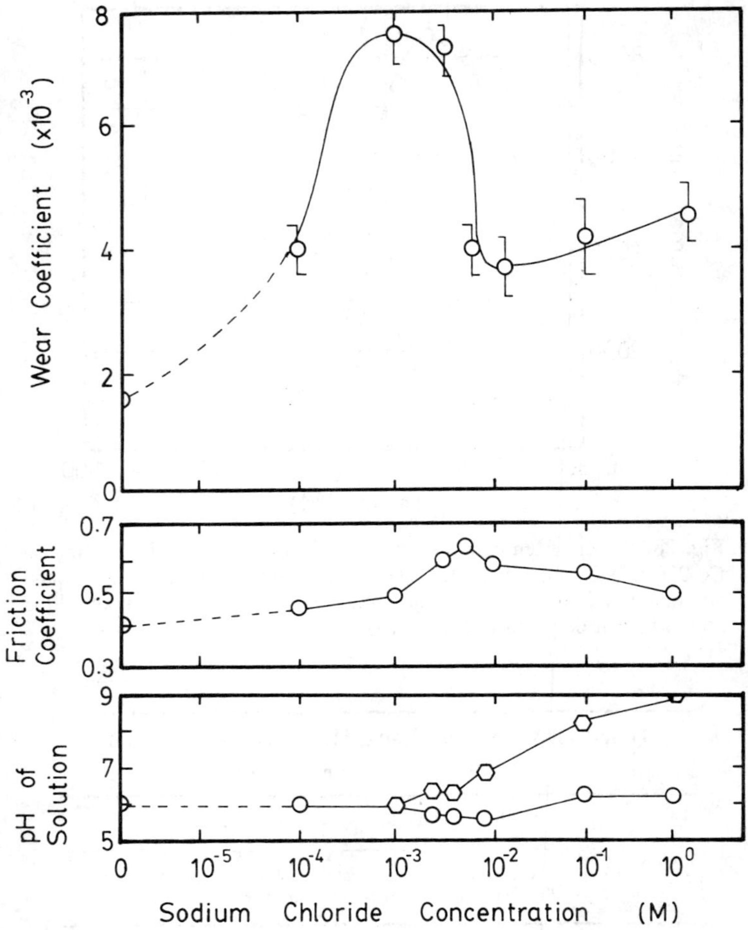

Fig. 30. Wear coefficient, friction coefficient and pH of solution after wear of 2024 Al alloy as a function of sodium chloride concentration: ○, before test; ⬡, after test. An intermediate chloride concentration produced the greatest wear. *(Ref 2)*

concentration of about 10^{-3} M, after which a gelatinous film formed and the coefficients dropped. The wear rate was shown to be high at both extremes in pH because of rapid chemical dissolution and thus weight loss. The black deposit was assumed to be composed of a blend of $AlCl_3$ and Al(OH).

Fretting wear increases the corrosion current of copper and aluminum in 1% NaCl solutions (Fig. 34).[3] The curves indicated in Fig. 34 were ragged as a result of film rupture and re-formation during the reversed sliding. Apparently fretting has an appreciable effect only when the solution is nearly neutral and the

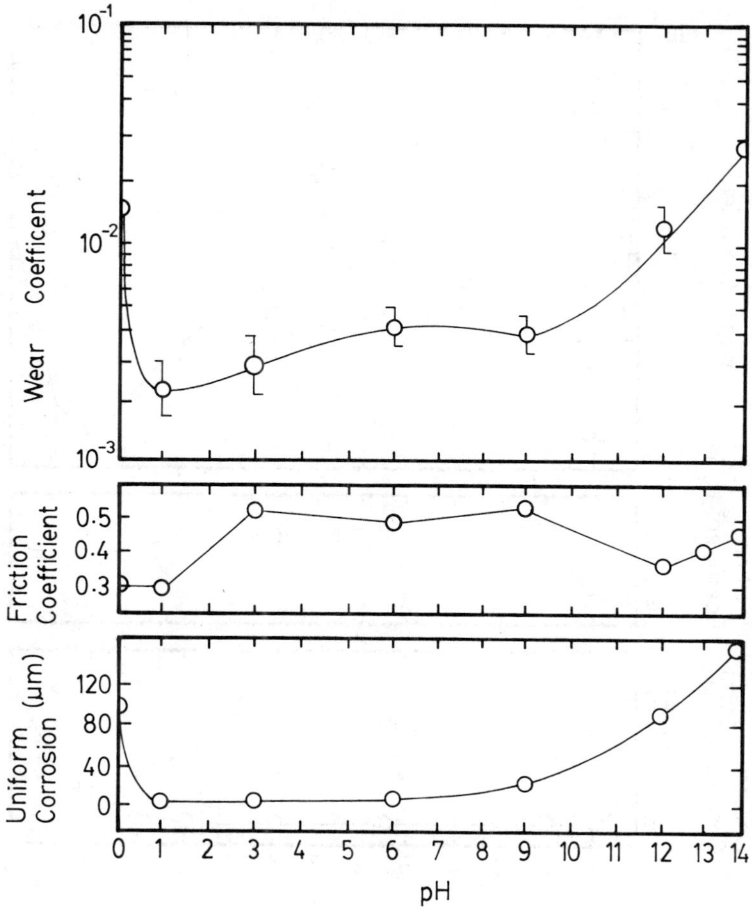

Fig. 31. Wear coefficient, friction coefficient and uniform corrosion of 2024 Al alloy as a function of pH in distilled water. The wear rate was greatest at very low and very high pH's. *(Ref 2)*

surface is covered with a protective oxide film (Fig. 35). At the extremes of pH the surface corrodes rapidly and the additional effect of fretting is negligible.

Special Applications for Wear-Resistant Alloys

Up to this point the wear of common engineering materials in common gaseous and aqueous environments has been discussed. Special alloys have been developed, however, for the purpose of being highly resistant to wear in certain specific environments. Frequently the environment is very severe and the system

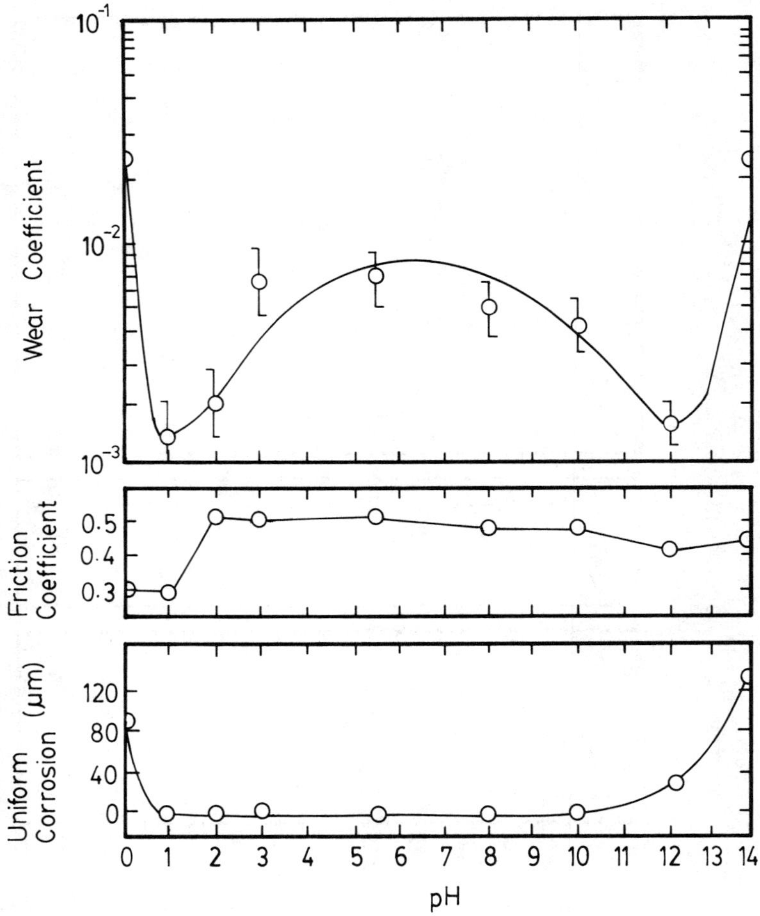

Fig. 32. Wear coefficient, friction coefficient and uniform corrosion of 2024 Al alloy as a function of pH of a 0.001 M NaCl solution. *(Ref 2)*

is such that the part must be able to function for long periods of time without maintenance. An additional requirement may be that the wear debris must not contaminate its environment. Some special alloys that meet these requirements are presented in the following section.

Tribaloy 700 in Sodium

The extreme conditions found in nuclear reactor service place special demands on wear-resistant low friction materials. These materials must exhibit long lifetimes in environments of neutron irradiation and high temperature liquid metal coolant. Tribaloy 700 (a high-Mo alloy) fulfills these requirements. Figure 36 shows the friction coefficient of this alloy as a function of amount of wear when

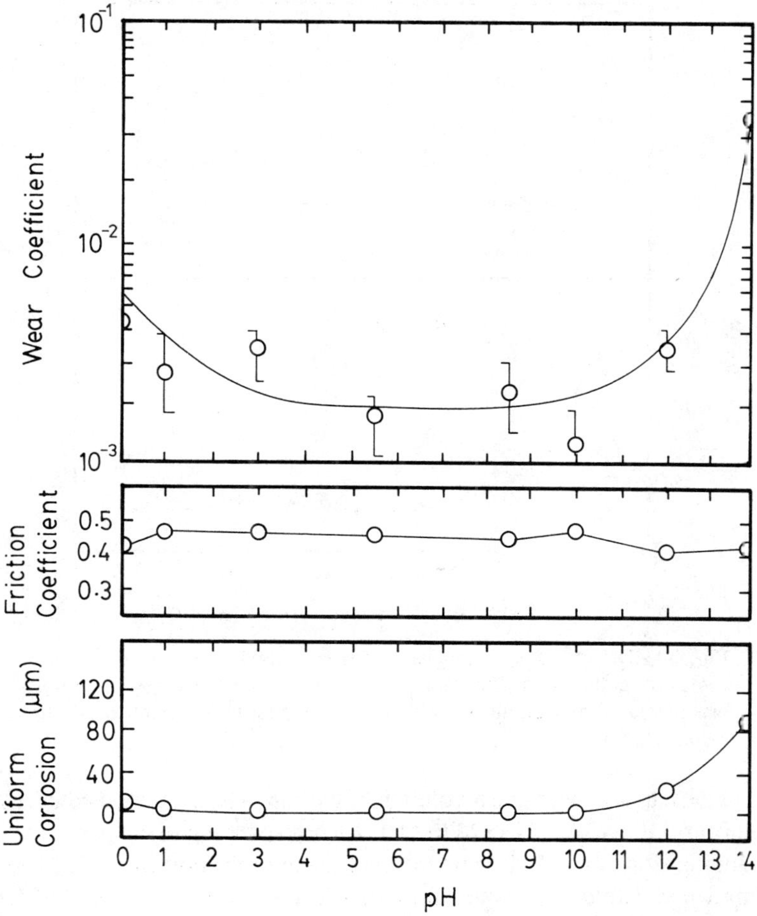

Fig. 33. Wear coefficient, friction coefficient and uniform corrosion of 2024 Al alloy as a function of pH of a 1.0 M NaCl solution. The increase in wear coefficient at intermediate pH values shown in Fig. 32 was diminished at this NaCl concentration. *(Ref 2)*

rubbed against itself in sodium.[67] The biggest advantage of this alloy is that the wear rate is so low that quantitative measurements of it have not yielded meaningful wear coefficients.

Wear of a Co-Cr-Mo Shoulder Prosthesis

Skorecki and Wheble[68] have designed an all-metal total replacement shoulder ball and cup joint with a mechanical life of about 20 to 30 years and a quantity of wear not exceeding that which can be accommodated by a human being without adverse effects. Figure 37 shows the amount of wear for the Co-Cr-Mo alloy as a function of the number of cycles. One cycle was defined as a ± 45°

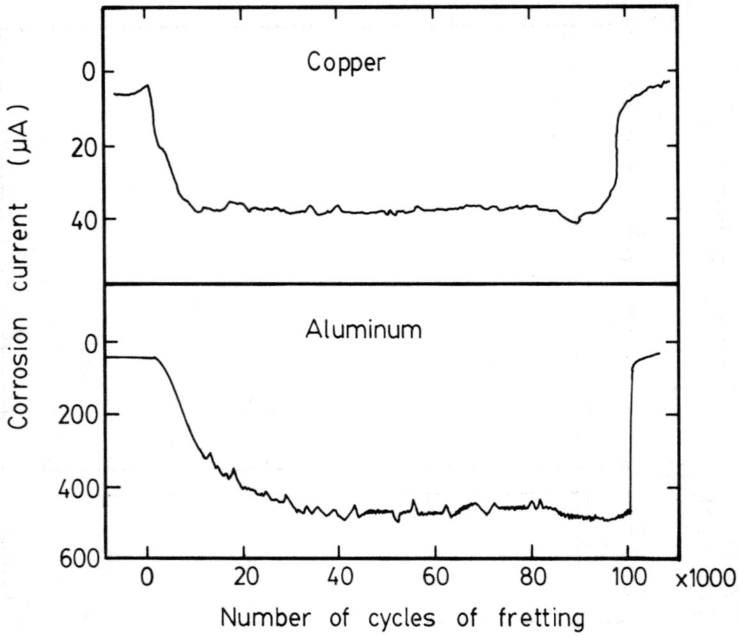

Fig. 34. The changes in corrosion current with the number of fretting cycles for copper and aluminum in 1% NaCl solution. Corrosion current was increased by fretting, with increases being larger the more active the metal. *(Ref 3)*

swing in the ball arm with one revolution of the cup. The maximum contact stress ranged from 140 to 380 MPa (20 to 55 ksi, corresponding to an axial separating load ranging from 10 to 70 lb) and the coefficient of friction was 0.15. The joint was lubricated with bovine synovial fluid. The amount of wear was considerably more for the ball than for the socket, but as the test proceeded there was a gradual improvement in the surface finish and the amount of wear per cycle steadily diminished. The temperature of the joint remained low and constant and the quality of the lubricant remained unaltered. Another joint pair composed of a ball of stainless steel and a cup of high density polyethylene were also tested. One failed by fracture of the polyethylene cup after 3000 cycles and the other after 8000 cycles, demonstrating the need for the strength of the all-metal joint. Several other investigators have also performed studies on the wear of artificial replacement joints.[69-74]

Discussion

Three years ago, in a critical synoptic review, Tabor[75] concluded that "due to the complex interaction of various wear mechanisms, the constant challenge to

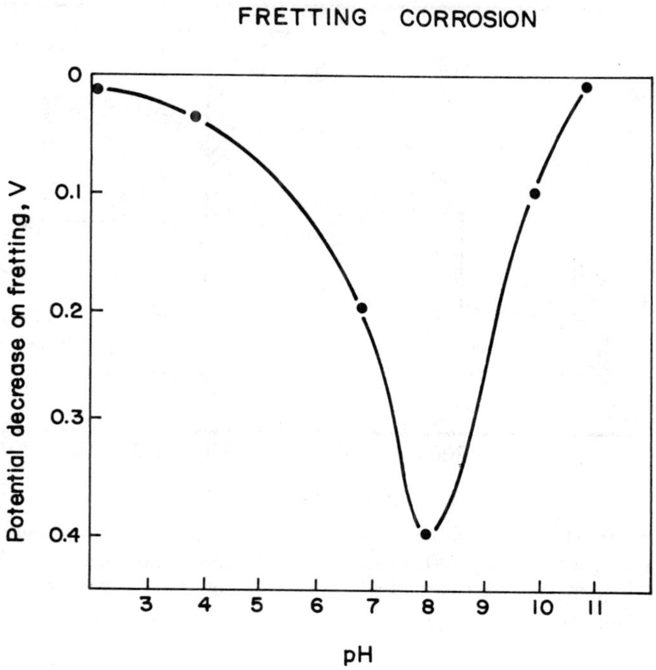

Fig. 35. The effect of pH on the potential decrease for Al-Zn-Mg alloy fretting in 0.1 N NaCl. The corrosion current (potential decrease) was greatest at neutral pH's. At the extremes of pH the metal corroded so rapidly that fretting had little influence. *(Ref 3)*

predict the wear of any given system, based on material properties, is unlikely to be surmounted for many years to come." Indeed, the most formidable challenge remains to directly observe, in some way, the dynamic interactions at the sliding interface. At present much of the information gathered in wear studies is inferred through indirect methods—e.g., the determination of hot spot temperature is often based on the presence of oxide debris which forms at a critical temperature, or mechanical properties of contacting asperities are deduced from bulk friction coefficient measurements. No one knows for certain what the actual environment of an asperity is during sliding. The focus on different aspects of the wear process and on different methods of studying it has inevitably led to a wide variety of proposed mechanisms.

One aspect which is very controversial is the debate over the contact, or hot spot, temperature. The temperature rise in fretting wear when a constantan pad was fretted against a 4130 steel specimen to form a thermocouple was measured to be only 18°C above the room temperature.[76] In a contrasting study,[3] the presence of $\alpha-Fe_2O_3$ in the debris from mild steel fretted on mild steel in air was

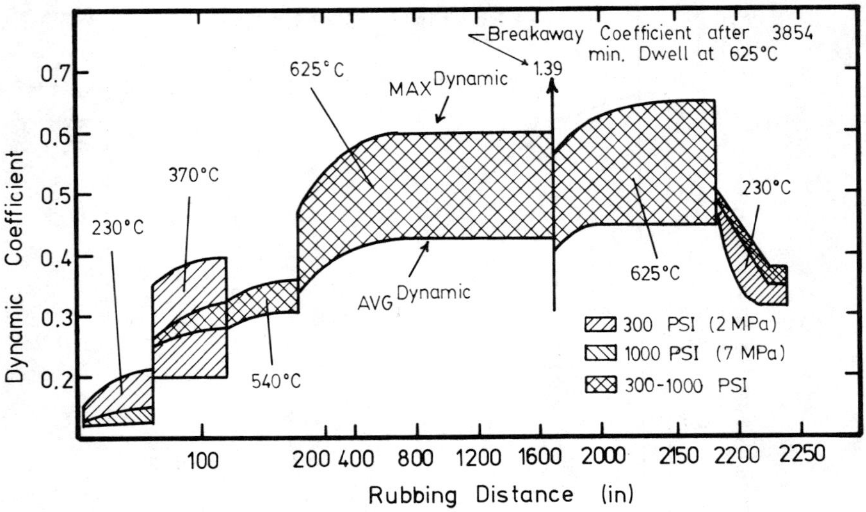

Fig. 36. The dynamic coefficient of friction as a function of rubbing distance for Tribaloy 700 at various temperatures and normal forces. The coefficient increased as sliding continued, but leveled off under most conditions. *(Ref 67)*

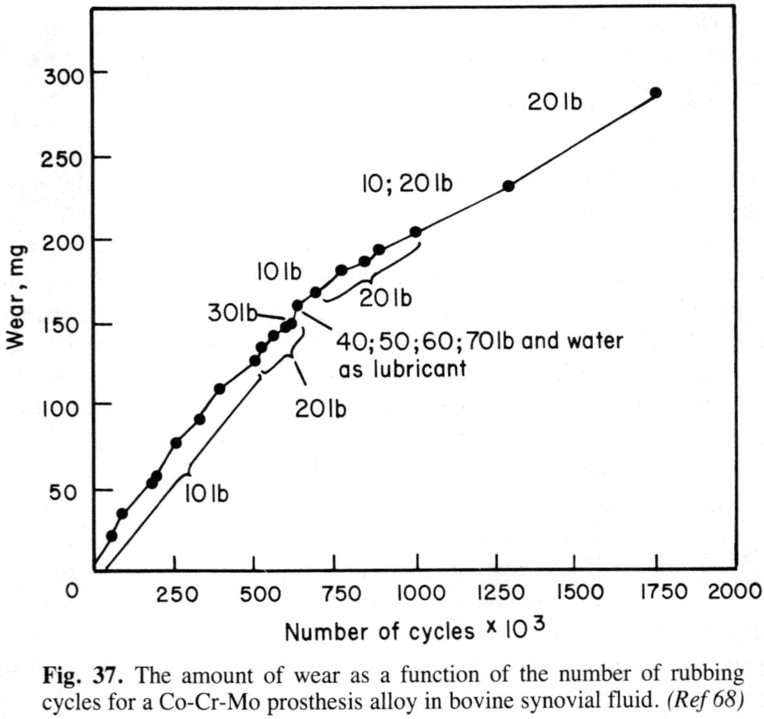

Fig. 37. The amount of wear as a function of the number of rubbing cycles for a Co-Cr-Mo prosthesis alloy in bovine synovial fluid. *(Ref 68)*

cited as evidence of a hot spot temperature of at least 290°C. Addressing the former investigation, it has been argued that any given asperity only occasionally contacts the opposing surface, and when it does, the contact may only last for a very short time.[77] A large temperature flash at the contact will probably be conducted away into the bulk metal before the thermocouple has a chance to respond. A lesser temperature rise will be measured, however, because of the accumulated "flashing" of the many asperities. An argument such as this would require that oxygen molecules be trapped within the contact interface in order to experience a high enough temperature to form the oxide. That oxygen does in fact reside within the contact is indicated by the fact that very small traces of gas significantly lower the coefficient of friction.[33,34] Oxygen trapped in this manner would be subjected to extremely large hydrostatic pressures, however,[78] and the critical temperature of oxidation may be lower than that at atmospheric pressure.

Further work needs to be done to sort out the cause and effect relationships among oxidation, temperature, and wear. On the one hand, high temperatures induced by wear are said to produce rapid oxidation, which may lubricate the surface and decrease the wear rate. On the other hand, high ambient temperatures commonly increase the wear rate. It seems unlikely that any simple relationship exists between sliding temperatures and oxidation rates. In some studies the dynamic oxidation activation energy is assumed to be the same as the static oxidation activation energy[15] and in others it is assumed to be significantly different.[13,14] A possible mitigating circumstance may be the geometry of the contacting interfaces. Little work has been done on investigating how asperity geometries, asperity interactions, and the real area of contact affect the contact temperature and oxidation.

When studies similar to that performed by Tsuji and Ando[50] are considered it is hard to reconcile the existence of large hot spot temperatures at all. These investigators found large differences in wear rates by varying the ambient temperature from 15°C to 30°C. If hot spot temperatures of several hundreds of degrees Celsius were produced by friction it seems unlikely that this small temperature change would have a noticeable effect.

Another point which causes much confusion is the use of the coefficient of friction as an indicator of wear resistance. The frictional force (tangential) is recorded in almost every wear study, and is divided by the applied normal force to give a coefficient of friction. The use of the friction coefficient to measure wear resistance is appealing because it is so easy to measure, but the practice may be a dangerous one since wear may[10,11,24,47,63] or may not[25,26,31,32,33,46] be directly related to the tangential forces. One problem in relating the two phenomena is that there is no single, accepted method of measuring the amount of wear. Typical methods include a measurement of the volume of removed metal, the width of the wear scar, the depth of the wear scar, or weight loss. Occasionally, however, other criteria, such as crack density or debris volume, may be used. As an example to demonstrate the anomaly which exists between a given wear measure and the coefficient of friction, consider the work of Barnes[25] as discussed in the section on oxidation. As the oxygen partial pressure was increased, a greater volume of oxide debris was generated and the coefficient of friction decreased. The reported wear rates (as measured by an unspecified volume per unit sliding distance) also decreased. Since oxidation definitely consumes metal atoms from the surface, which is also a wear process, it is unclear whether an increase in oxygen partial pressure produced more wear or less wear.

Another problem associated with using the coefficient of friction is that it is measured as a bulk quantity and may not be representative of any mechanism occurring at individual contact points. Assuming the model of adhesive wear proposed by Bowden and Tabor,[79] the coefficient of friction should be about 0.2—i.e., the tangential stress required to shear the metal ($\tau \simeq 0.5\ \sigma_{ys}$) divided by the maximum normal stress that the metal can supoprt ($\sigma_F \simeq 2.5\ \sigma_{ys}$). However, under conditions where the mating surfaces are very clean and adhesion may be expected, coefficients of friction can reach values greater than 10.[33] A possible explanation is that junctions shear at planes well below the original contact surface, where less plastic deformation has occurred, and that the normal stress is much less than the flow stress. Another explanation is that many asperities may protrude enough to block lateral movement of other asperities but may not be large enough to support any of the normal load. In this case the area on which tangential forces act would be larger than that on which normal forces act. Therefore, the ratio of measured tangential force to applied normal force (i.e., the measured coefficient of friction) will be larger than the ratio of tangential stress (shear strength) to normal stress (flow pressure).

The problems in associating the coefficient of friction with wear rates underscore the importance of considering environmental interactions in wear studies. The applied forces acting on the underlying metal must be transmitted across the wearing interface, which in turn may be greatly modified by environmental factors. To date, a vast number of studies on dry sliding wear completely ignore the nature of the media in which they are performed, allowing comparisons to be tentative at best. As more investigations are conducted on environmental effects on wear processes, it will become clear that the environment must be controlled in all wear studies. At the very least the ambient conditions must be recorded. It is doubtful whether a general, all-encompassing, model of wear will ever be developed, much less one that includes environmental interactions. Still, progress is being made in our understanding of tribochemical processes.

Acknowledgment

The support of the U.S. Army Research Office, Durham, North Carolina, under Contract No. 12554 MS, is gratefully acknowledged.

References

1. F. F. Tao, A study of oxidation phenomena in corrosive wear, ASLE Trans. *12* (1969), pp. 97-105.
2. M. K. Tse and N. P. Suh, Chemical effects in sliding wear of aluminum, Wear *44* (1977), pp. 145-162.
3. R. B. Waterhouse, *Fretting Corrosion*, Vol. 10, p. 186, Pergamon Press (1972).
4. N. P. Suh, The delamination theory of wear, Wear *25* (1973), p. 111.
5. D. H. Buckley and R. L. Johnson, ASCE Trans. *3* (1960), p. 93.
6. E. E. Bison, R. C. Johnson, M. A. Swikert and D. Godfrey, NACA Report 1254, 1956.
7. M. G. Hayler and S. W. E. Earles, Wear *18* (1971), p. 393.
8. N. C. Welsh, Phil. Trans. Roy. Soc. London *A257* (1965), p. 31.
9. F. H. Stott, D. S. Lin and G. C. Wood, Corr. Sci. *13* (1973), p. 449.
10. F. H. Stott, D. S. Lin, G. C. Wood and C. W. Stevenson, Wear *36* (1976), p. 147.
11. R. D. Brown and R. A. Burton, ASME WA/LUB 23, Paper 69 (1967), p. 1.
12. J. Molgaard and V. K. Srivastava, The activation energy of oxidation in wear, Wear *41* (1977), pp. 263-270.
13. T. F. J. Quinn, An experimental study of the thermal aspects of sliding contact and their relation to the unlubricated wear of steel, Proc. Inst. Mech. Eng. London, pt. 3, *183* (1969), p. 129.
14. N. Tenwick and S. W. E. Earles, A simplified theory for the oxidative wear of steels, Wear *18* (1969), pp. 381-391.
15. O. Kubaschewski and B. E. Hopkins, Oxidation of metals and alloys, Butterworths, London (1962).
16. V. Aronov, Kinetic characteristics of the transformation and failure of the surface layers of metals under dry friction, Wear *41* (1977), pp. 205-212.
17. T. J. F. Quinn, The effects of hot spot temperatures in the unlubricated wear of steels, ASCE Trans. *10* (1967), p. 158.
18. T. J. F. Quinn, The thermal aspects of sliding contacts and their relation to the unlubricated wear of steels, Proc. Instrn. Mech. Eng. *183*, pt. 3P (1969), p. 129.
19. T. J. F. Quinn, The division of heat and surface temperatures at sliding steel interfaces and their relation to oxidational wear, ASCE Trans. *21* (1978), p. 78.
20. T. J. F. Quinn, The dry wear of steels as revealed by electron microscopy and x-ray diffraction, Proc. Instrn. Mech. Eng. *182*, pt. 3P (1968), p. 201.
21. T. J. F. Quinn, J. C. Sullivan and D. M. Rowson, More developments in the oxidational theory of the mild wear of metals, Wear of Mat., Dearborn, Mich. (1979), p. 1.
22. E. Rabinowicz, ASLE Trans. *10* (1967), p. 400.
23. J. Molgaard, A discussion of oxidation, oxide thickness and oxide transfer in wear, Wear *40* (1976), pp. 277-291.
24. F. Stott and G. Wood, The influence of oxides on the friction and wear of alloys, Trib. Int. *11* (1978), p. 211.
25. D. J. Barnes, J. E. Wilson, F. H. Stott and G. C. Wood, The influence of oxide films on the friction and wear of Fe-5% Cr alloy in controlled environments, Wear *45* (1977), pp. 161-176.
26. K. Endo and H. Goto, Effects of environment on fretting fatigue, Wear *48* (1978), pp. 347-367.
27. Y. Momose and T. Namekawa, Exoelectron emission from metals subjected to

friction and wear, and its relationship to the absorption of oxygen, water vapor and some other gases, J. Phys. Chem. *82* (1978), pp. 1509-1515.
28. I. M. Feng and B. G. Rightmire, An experimental study of fretting, Inst. Mech. Eng. *170* (1956), p. 1055.
29. I. M. Feng and H. H. Uhlig, J. Appl. Mech. *21* (1954), p. 397.
30. A. J. Fenner, K. D. Wright and J. Y. Mann, Proc. Int. Conf. Fatigue of Met. (1956), p. 386.
31. D. R. Wheeler, Effect of absorbed chlorine and oxygen on the shear strength of iron and copper junctions, J. Appl. Phys. *47* (1976), p. 1123.
32. S. V. Pepper, Effect of absorbed films on friction of Al_2O_3-metal systems, J. Appl. Phys. *47* (1976), p. 2579.
33. Y. Tsuya, Microstructure of wear, friction and solid lubrication, Tech. Rep. of Mech. Eng. Lab. No. 81, Japan.
34. N. Soda and T. Sasada, Mechanism of lubrication by surrounding gas molecules in adhesive wear, Wear of Mat. Conf. Proc., St. Louis, Mo. (1977), p. 47.
35. E. R. Mantel, G. H. Robinson and R. F. Thompson, Influence of atmospheric moisture on fatigue of hardened steel, Met. Eng. Quart, ASM *1* (1961), p. 57.
36. E. P. Dahlberg, Fatigue crack propagation in high strength 4340 steel in humid air, Trans. ASM Quart. *58* (1956), p. 46.
37. F. J. Bradshaw and C. Wheeler, Appl. Met. Research *5* (1969), p. 112.
38. H. W. Liu and H. T. Corten, NASA Tech. Note D647 (1960).
39. J. A. Bennett, 10th Sagamore Army Mat. Res. Conf., Syracuse Univ. Press (1964).
40. H. A. Leybold, H. F. Hardrath and R. L. Moore, NASA Tech. Note 4331 (1958).
41. J. A. Dunsby and W. Wiebe, Mat. Res. and Standards *9* (1969), p. 15.
42. C. Laird and G. C. Smith, Phil. Mag. *8* (1963), p. 1945.
43. M. R. Archer, ASTM STP415 (1967), p. 181.
44. K. H. R. Wright, An investigation of fretting corrosion, Inst. of Mech. Eng. *1B* (1952), p. 556.
45. T. Broom and A. Nicholson, Atmospheric corrosion fatigue of age hardened aluminum, J. Inst. Met. *89* (1960), p. 183.
46. J. K. Appledorn, I. B. Goldman and F. F. Tao, Corrosion wear by atmospheric oxygen and moisture, ASCE Trans. *12* (1969), p. 140.
47. D. Godfrey, A study of fretting wear in mineral oil, Lub. Eng. *12* (1956), p. 37.
48. J. Larsen-Basse and S. S. Sokoloski, Influence of atmospheric humidity on abrasive wear—II: 2-body abrasion, Wear *32* (1975), pp. 9-14.
49. R. C. Bill, Fretting wear of iron, nickel and titanium under varied environmental conditions, Wear of Mat. Conf. Proc., Dearborn, Mich. (1979), p. 356.
50. E. Tsuji and Y. Ando, Effect of air temperature and relative humidity on wear of carbon steels and cast irons, Wear of Mat. Conf. Proc., St. Louis, Mo. (1977), p. 95.
51. D. S. Lin, F. H. Stott and G. C. Wood, The effects of elevated ambient temperatures on the friction and wear behavior of some commercial nickel base alloys, ASCE Trans. *17* (1974), p. 251.
52. A. C. Jesper, Proc. Inst. Mech. Eng. London *180*, 3k (1965), p. 265.
53. E. E. Bison, R. C. Johnson, M. A. Swikert and D. Godfrey, NACA Rep. 1254 (1956).
54. E. E. Bison, C. Lipson and L. V. Colsell, *Handbook of Mechanical Wear*, Univ. of Mich. Press, 1961.
55. F. P. Bowden and J. E. Young, Proc. Roy. Soc. London *A208* (1951), p. 311.
56. D. H. Buckley, NASA Tech. Note D-4775 (1968).
57. F. P. Bowden and T. H. C. Childs, Proc. Roy. Soc. London *A312* (1969), p. 451.

58. D. J. Barnes, F. H. Stott and G. C. Wood, The frictional behavior of iron and iron-chromium alloys at elevated temperatures, Wear *45* (1977), p. 199.
59. E. Rabinowicz, J. Appl. Phys. *32* (1961), p. 1440.
60. D. Tabor, Proc. Roy. Soc. London *A251* (1959), p. 378.
61. W. J. Tegart, Ductility, ASM Pub. (1968).
62. G. A. Redfern and C. M. Sellers, Proc. Conf. on Deformation under Hot Working Conditions, Univ. of Sheffield (1966).
63. F. D. Bowden and T. H. C. Childs, The friction and deformation of clean metals at very low temperatures, Proc. Roy. Soc. London *A312* (1969), p. 451.
64. P. C. Hurricks and K. S. Ashford, The effect of temperature on the fretting wear of mild steel, Proc. Instrn. Mech. Eng. *184*, 32 (1969), p. 165.
65. J. Thiery and R. Spinat, Comment reduire l'usure des pièces non lubrifiés dans les turbo-machines (Fretting in aircraft systems), AGARD Conf. Proc. No. 161 (1975).
66. M. M. Hamdy and R. B. Waterhouse, The fretting fatigue behavior of Ti-6Al-4V at temperatures up to 600°C, Wear *56* (1979), p. 1.
67. R. N. Johnson and D. G. Farwick, Friction, wear and corrosion of laves-hardened nickel alloy hardsurfacing in sodium, Thin Solid Films *53* (1978), pp. 365-373.
68. J. Skorecki and V. Wheble, Wear and fatigue tests on an all-metal Co-Cr-Mo prototype shoulder prosthesis, Trib. Int. *12* (1979), p. 20.
69. V. Wheble, J. Skorecki and G. Thompson, The design of a metal to metal shoulder prosthesis: Joint replacement in the upper limb., I. Mech. E. (April 1977).
70. L. Kessel and K. J. N. Wright, Prosthetic replacement of the shoulder joint: Joint replacement in the upper limb, I. Mech. E. (April 1977).
71. G. K. McGee, Developments in total hip joint replacement: Lubrication and wear in living and artificial joints, Proc. I. Mech. E. *181(3J)* (1967), pp. 85-89.
72. P. A. Ring, Total replacement of the hip joint: A review of a thousand operations, J. of B. and J. S. Brit. *56-B(1)* (1974), p. 44.
73. J. O. Galante, W. Rostoker and J. M. Doyle, Failed femoral stems in total hip prosthesis, J. of B. and J. S. Brit *57-A(2)* (1975), p. 230.
74. I. Duff-Barclay and D. T. Spillman, Total hip joint prosthesis: A laboratory study of friction and wear lubrication and wear in artificial and living human joints, Proc. I. Mech. E. *181(3J)* (1967), p. 90.
75. D. Tabor, Wear—a critical synoptic view, Wear of Mat. Conf. Proc., St. Louis, Mo. (1977), p. 1.
76. E. S. Sproles and D. J. Duquette, Interface temperature measurements in the fretting of a medium carbon steel, Wear *47* (1978), p. 387.
77. T. F. J. Quinn, A review of oxidational wear, Wear of Mat. Conf. Proc., St. Louis, Mo. (1977), p. 110.
78. A. S. W. Moore, Deformation of metals in static and sliding contact, Proc. Roy. Soc. London *A195* (1948), p. 231.
79. F. B. Bowden and D. Tabor, *The Friction and Lubrication of Solids,* Vol. 1, Oxford University Press, Oxford (1953).

Coatings for Tribological Applications

M. B. Peterson
Wear Sciences Corporation
S. Ramalingam
Georgia Institute of Technology

Abstract

Coatings are widely used in tribological applications, either being applied intentionally to reduce wear or formed *in situ* during the sliding process. In this paper the various types of coatings are considered. Detailed consideration is given to two areas: the controlled application of hard coatings and the tribological behavior of soft, thin lubricating films.

A review of recent developments with low shear strength films and the effect of critical variables is given. It is proposed that sliding takes place by three mechanisms: interface slip, interfilm shear, and interfilm slip. Which will take place with a given film is a function of the material, its substrate, and the energetics involved.

Processes for applying hard coats are reviewed. New techniques have been developed which are extremely beneficial. Coatings can be applied with controlled thickness and composition. Materials which can be applied include carbides, nitrides, borides, silicides, and oxides.

Introduction

Coatings are widely used in tribological applications, either being placed there intentionally to reduce wear, to avoid surface damage, or to modify frictional behavior or formed *in situ* during the sliding process. In either case a fundamental understanding of their behavior is essential to understanding friction and wear processes.

The use of coatings is growing and will continue to grow because they offer certain economic and technical advantages over the use of uncoated materials. Their main advantage is, of course, that coatings allow the base material to be optimized for strength purposes while the surface is optimized for wear and corrosion resistance. Since wear and corrosion resistant alloys are usually more expensive than low alloy structural alloys this approach is usually economically advantageous. Furthermore, replacing the coating during rework or repair is usually less expensive than part replacement.

It has also been shown that certain national goals can be achieved through the use of coatings.[1] In a study of materials conservation it was pointed out that a substantial percentage of certain critical metals is used in alloys or as coatings to promote corrosion and wear resistance (Table 1). The expanded use of metal coatings or, more important, non-metallic coatings could greatly reduce our dependence on metallic imports.

Coatings have found use in a wide variety of applications from simple, low cycle, sliding parts to complex, high cycle, long life components such as rolling contact bearings, gears, splines, and rings. They are used to provide resistance to both abrasive and adhesive wear. As our understanding of the wear and failure processes develops, special coatings can be developed for special sliding situations. Thus it is important that our understanding of the behavior and uses of coatings develops at the same pace.

In this paper the various types of coatings used in tribological applications are considered, recent research experience is reviewed, and areas for further development are explored. Detailed consideration is given to two areas: the controlled application (composition and thickness) of hard coatings and the tribological behavior of soft, thin coatings.

Classification

Coatings have been classified in a number of ways: by application technique, composition, or function, or by some other property. Examples of application techniques used for tribological applications are given in Table 2 and compositions in Table 3. Such classifications, though useful for purposes of illustration, are of little use in the selection of coatings for specific applications or

Table 1. Metals Used in Alloys for Corrosion and Wear Resistance

Metal	Usage, %	Metal	Usage, %
Aluminum	61	Nickel	24
Cadmium	60	Tin	35
Chromium	30	Titanium	30
Cobalt	20	Tungsten	50
Molybdenum	10	Zinc	50

Table 2. Coating Application Techniques

Electroplating	Painting
Electroless plating	Rubbing (transfer)
Hard facing	Melting
Flame spraying	Vapor deposition
Plasma spraying	Pyrolytic decomposition
Detonation spraying	Chemical vapor deposition
Diffusion	Ion plating
Surface reaction	Vapor deposition
	Sputtering

Table 3. Coating Compositions

Chromium plate	Carburizing
Electroless nickel	Boronizing
Cobalt and steel hard facing	Sulfiding
Al_2O_3 and other ceramics	Chromizing
Carbides and metal bonded carbides	MoS_2/graphite bonded coatings
	Teflon/nylon/polyethylene coatings
Oxide reaction films	Noble metal plating
Nitriding	Tin plating

problems. The most useful classification would be by "function"—that is, by identifying exactly what the coating is supposed to accomplish. In Table 4, some functions are listed along with some materials which have been shown to be useful for the listed purpose. Although a classification of this kind would be extremely useful in selecting materials, developing materials, and understanding friction and wear processes, our information is still too incomplete. Certain coatings are known to be beneficial, but we do not have a precise understanding of their functions. Hopefully, with continued research, a system will be developed with a priority table of useful compositions along with their most practical (minimum cost) application techniques.

For the present, a useful means to classify coatings is shown in Table 5. Although it does not uniquely classify all possibilities, it is informative in telling one what to expect from a friction and wear standpoint.

Hard surfacing essentially replaces one surface with another having, presumably, more desirable friction and wear properties. Hard coats, thick and thin, act as wear-resistant surface layers. They yield wear resistance by virtue of their hardness. The properties of the hard coating are essentially the same as those of a solid of the same composition. These properties are almost always known beforehand.

When the coatings are very hard ($H_v \simeq 2000$ to 3000 kg/mm^2), wear rate even under severe sliding conditions is small and often negligible. Coating thicknesses

Table 4. Coatings Used for Various Functions

Coating function	Coatings used
Reduce wear	Titanium carbide, nitride
Reduce friction	Teflon, MoS_2
Increase friction	Titanium, bonded abrasives
Improve lubrication	Copper, lead
Increase temperature or load capacity	Electroless nickel
Prevent adhesion	Silver/gold plate
Imbed particles	Indium, lead
Reduce corrosive wear	Chromium plate or diffusion
Retain fluid lubricants	Phosphating, nylon
Rebuild surface	Steel hard surfacing
Reduce surface roughness	Silver plate
Prevent fatigue wear	———
Prevent drop erosion	Polyurethane, neoprene
Prevent particle erosion	Cobalt alloy, molybdenum

Table 5. Coating Classification

Hard surfacing	Replace surface
Surface treatments	Modify surface
Low shear strength films	Lubricate surface

of 1 to 10 μm, which is well below the dimensional tolerance specified for most components, are sufficient for most applications. Thus the use of hard coatings offers a way of obtaining wear protection in finished components at a small added cost. However, such coatings are useful only if they have reproducible composition and thickness, and are well bonded to the base material.

Primary research efforts are therefore directed to getting good, consistent, strongly adherent coatings. The objective here is to ensure that the coating has the same physical properties as the original solid. In recent years a great many new approaches have been explored. The real advantage of these new technologies is that they allow close control of thickness and composition; they also permit combination of coatings. Some of these are described in a later section.

Surface treatments modify the surface either by hardening or by diffusion, or both. The modified surface has unique friction and wear properties which are often intermediate between the base metal (steel) and the resulting particle (carbide). Research efforts are usually devoted to the process variables and their control. Some of the most significant developments in the field of lubrication could result from some of the new techniques which allow placing specific ions in a surface. If successful, this would mean that lubricating additives could be placed within the surface rather than in the lubricant. Removal of the additives from the lubricant would allow concentration on such factors as thermal stability and viscosity.

Low shear strength films, if they are thick (>20 mil), act as replacement coating. However, since they are soft they usually wear or flow rapidly until they become extremely thin. When they are thin they are able to act as lubricant films with very unique friction and wear properties unrelated to the friction and wear of separate materials. Much of the fundamental effort in "solid lubricants" has been devoted to understanding their mechanism(s) of action. This is also discussed in a later section.

Low Shear Strength Films

Low shear strength or solid lubricant films are more prevalent in lubrication, friction, and wear processes than is usually thought. Solids such as Teflon, MoS_2, and graphite are applied to surfaces as lubricants. However, theories of boundary lubrication[2] propose that the mechanism of lubrication is a reaction between certain molecules in the oil and the surface to form low shear strength metal soap which provides the lubrication function. In addition to this reaction, EP (extreme pressure) additives (usually organic sulfides and disulfides) and antiwear additives (Tricresyl phosphate) are proposed to form surface reaction films which contribute additional lubrication functions. After long periods of operation under certain conditions, oxidation and polymerization of the lubricant can take place which produces a soft surface film. In addition to these *in situ* reaction films, a variety of lubricant adjunct films are put on the surface to accomplish one function or another, these being soft metal platings, polymer films, reaction films such as oxides, phosphates, chromates and various organic coatings. Another instance of lubrication with low shear strength solids is the preferential accumulation of certain atoms at the sliding interface of alloys (e.g., lead) which makes them free machining and better bearing alloys. In fact, there are those who propose that the high pressures associated with concentrated contacts solidify the fluid lubricant so that even so-called fluid film lubrication is really lubrication with a soft low shear strength solid. Although much of this is speculation it is clear that an understanding of the mechanism of solid lubrication will contribute a great deal to the understanding of friction and wear processes and to the development of improved lubricating materials. In the following paragraphs a brief review is given of lubrication with soft solids and a general model proposed for its behavior.

Types

Although we generally think of Teflon, molybdenum disulfide, and graphite as solid lubricants, a large number of compounds have been found which behave in a similar manner. The general types are listed in Table 6 along with some

Table 6. General Types of Solid Lubricants

Organics	Soaps, fats, waxes
Polymers	Teflon, polyethylene, methacrylates
Metals	Indium, tin, lead, silver
Inorganics	AX, AX_2, oxides, sulfides, chlorides
Glasses	B_2O_3, PbO-SiO_2, PO_4

examples. About the only thing that all these materials have in common is that they are soft. More specific compounds of each type, along with typical friction coefficients, are listed in Table 7 (stearate compounds), Table 8 (metals), Table 9 (AX_2 structure, chlorides, fluorides, bromides, iodides, etc.), Table 10 (oxides and mixed oxides), and Table 11 (glasses). These lists do not exhaust the possibilities; many more could be included based upon the temperatures being considered. The important point being made is that there are many soft materials which act as lubricating solids; they do not, however, perform in the same way. Certain solids have a much broader range of applicability and under a given set of circumstances have a longer wear life. These find general usage as solid lubricants.

Before considering possible mechanisms it is first necessary to review how different conditions affect the behavior of soft, thin lubricating films. Primary variables are the operating conditions, the film variables, the atmosphere, and the nature of the substrate. These will be discussed separately.

Frictional Behavior of Soft Films

The original hypothesis of Bowden and Tabor[2] was derived from the adhesion theory of friction:

$$f = \frac{n\,\bar{a}_r S}{L} + P^*$$

where n = number of contacts
$\bar{a}_r$ = average area/contact
S = junction shear strength
L = load
f = friction coefficient
P^* = plowing term.

When a soft low shear strength solid was placed on the surface, S for the substrate was replaced by the lower S for the film. However, the real area of contact was still determined by the substrate metal, so it remains the same. Thus, low shear strength films reduced friction. To this could be added a plowing term, P^*, which was proportional to track width. With thicker soft films a certain amount of

Table 7. Friction With Stearate Compounds

Stearate compound	Typical friction coefficient	Stearate compound	Typical friction coefficient
Iron	.05	Sodium	.08
Lead	.15	Copper	.10
Zinc	.09	Aluminum	.30

Table 8. Friction With Metal Films (100,000 psi steel/steel)

Metal	Friction coefficient	Metal	Friction coefficient
Indium	.06	Silver	.20
Tin	.09	Aluminum	.22
Cadmium	.10	Zinc	.34

Table 9. Frictional Behavior of AX_2 Compounds

	Fluoride		Chloride		Bromide		Iodide		Hydroxide		Sulfide	
Cu	—		$CdCl_2$		CdI_2	.06	CdI_2		CdI_2		FeS_2	.21
Fe	—		$CdCl_2$		CdI_2		CdI_2		CdI_2		FeS_2	
Ni	—		$CdCl_2$		$CdCl_2$		$CdCl_2$		$CdCl_2$		FeS_2	
Co	—		$CdCl_2$		$CdCl_2$		CdI_2		CdI_2		CdI_2	
Ti	TiO_2		—		—		—	.25			CdI_2	
Zn	TiO_2		$CdCl_2$		CdI_2		CdI_2		CdI_2		—	
Mo	—		—		—		—		—	.32	MoS_2	.05
W	—		—	.16	—		—		—		MoS_2	.08
Sn	—		—		—		—		—		CdI_2	
Cd	—		$CdCl_2$		CdI_2	.07	CdI_2	.06	CdI_2		—	
Pb	PbF_2	.60	$PbBr_2$	.45	$PbBr_2$		CdI_2	.28	—		CdI_2	.15
Hg	—		$HgCl_2$	.38	$HgBr_2$		$HgBr_2$	.18	—		—	
Ag	—		AgCl		—		—	.20	—		Ag_2S	
Ca	—		—		—		—		CdI_2	.18	—	

material had to be pushed out of the way, and this increased friction. The magnitude of the plowing term was determined by the geometry of the slider and the depth of penetration. Qualitatively this simple formula was valuable in that it could and still can be used to explain the effects of variables. All that was really needed was to realize that the real area of contact, $n\bar{a}_r$, could increase under certain circumstances and could even approach the apparent contact area.

Table 10. Frictional Behavior of Oxides and Compounds

	Oxide	Molybdate	Tungstate	Sulfate	Silicate	Chromate
Cu	.42	.41	—	—	—	—
Fe	.70	.42	.41	—	—	—
Ni	.70	.29	.51	—	—	—
Co	.29	—	—	—	—	—
Zn	.35	—	—	—	—	—
Mo	.21	—	—	—	—	—
W	.54	—	—	—	—	—
Pb	.11	.29	.35	—	.19	.33
Ag	—	.28	—	.14 (80°F)	—	—
K	—	.20	—	.17	—	—
Na	—	.17	.19	.70 (80°F)	.30	—
Ti	—	—	—	.20	—	—
B	.11	—	—	—	—	—

Table 11. Friction With Glass Films on Inconel (1100°F)

Glass	Friction coefficient	Glass	Friction coefficient
B_2O_3	.09	$PbO \cdot SiO_2$ glass	.18
Phosphate glass	.10	$PbO \cdot B_2O_3$ glass	.12
Lead borosilicate	.35	$CaO \cdot BaO \cdot B_2O_3$	.25

Experimental work by Peterson and Ling,[3] Kraghelskii[4] and Towle[5] showed that

$$f = \frac{S_0}{P} + \alpha$$

where P = pressure and α = constant. The constant α represents the increase in shear strength with pressure:

$$S = S_0 + \alpha P.$$

The difficulty with both of these expressions is that all the material properties are encompassed by a single term, S_0, and all variables must somehow affect S_0. Even more important, the exact meaning of S is not clear. Where is shear taking place—at the interface between the soft solid and the solid or within the soft solid? Some additional investigations have been conducted which attempt to predict frictional behavior using plasticity theory.

In the work of Peterson and Ling[3] friction is predicted for two conditions, interfacial slip and strong surface adhesion where all shear is within the layer. The following expression is derived:

$$f = B\frac{K}{P} + \frac{S_i}{2P}$$

$$B = 1/2\left[1 + \sqrt{1-\left(\frac{S_i}{K}\right)^2} + 4\beta - \frac{\pi}{2} - \cos^{-1}\left(\frac{S_i}{K}\right)\right] \cot \beta,$$

where P = pressure
K = yield strength
S_i = interfacial shear strength
β = 1/2 asperity cone angle of slider.

If one assumes an asperity cone angle of 170°, then two expressions for friction coefficient will result (choosing $S_i = 0$ for interface slip and $S_i = K$ for interfilm shear):

Interface slip $\quad f = .39\,\frac{K}{P}$

Interfilm shear $\quad f = .67\,\frac{K}{P}$

This says that it takes less energy to slip at the surface of the soft metal film than to shear the film; thus, surface slip is expected. However, this does not take into account crystal orientation, which might give an $S_i > K$. It also does not explain friction of MoS_2, where films are seen on both specimens after sliding.

Recently, Heilmann and Rigney[12] have developed an expression for the shear surface layers which they have applied to surface films. It takes into account a variety of material parameters such as stress-strain behavior and work hardening. Using the the authors' notations:

$$\mu = \left[\left(\frac{\bar{n}}{\bar{n}+1}\right)\left(\frac{NA}{L}\right)\bar{\tau}_o\bar{\gamma}_o^{\,1/\bar{n}}\right] \bullet \left[1 - e^{-(1+1/n)\bar{\alpha}T} + \frac{\tau_o\gamma_o^{\,1/n}n\,(\bar{n}+1)}{\bar{\tau}_o\bar{\gamma}_o^{\,1/\bar{n}}\bar{n}\,(n+1)}\right]e^{-\bar{\alpha}T}$$

where μ = friction coefficient
n = work-hardening coefficient
NA = real contact area
L = load
τ_o = shear stress factor
γ_o = shear strain factor
T = thickness
α = strain gradient factor.

The bars indicate parameters for the coating material.

This equation predicts certain trends correctly but still must be evaluated for general applicability. The important question, however, is not the accurate prediction of friction but the accurate understanding of significant material variables so that effective films can be predicted.

Effect of Time

The effect of time is shown in Fig. 1. If a powder is placed on the surface of a sand-blasted steel disk and tested, its friction-time curve will appear as in Fig. 1. Three regions are noted: film formation, film wear, and incipient failure where metal contacts through the film are influencing friction. In the film formation region the soft solid is being compressed and sintered into a continuous film. Friction is high because there is wear and metal contact between the mating metal specimens. After a short period of time (usually several minutes) a steady-state condition has been reached and friction is constant. Two different levels of friction can be observed. If there is only a film at the tips of the asperities the initial low friction is maintained. If sufficient solid was available to completely fill the spaces between the asperities a continuous film is formed and friction is higher. This is an area effect. The continuous film has a larger contact area and higher friction. After a certain period of time, metal contacts begin at the tips of the asperities, and friction will begin to increase. For the asperity film, failure is very sudden. For the continuous film this is a much slower process since there is a reservoir of material between the asperities. The region of interest is the film wear region, and solid lubricant effectiveness can be judged by two factors, the film life, L_f, and its friction coefficient, f.

If the experiment is attempted with a hard powder such as nickel oxide or iron oxide, no film will be formed and friction will be essentially steel vs steel. If the film is formed before the test, for example by plating, then the film formation regime will be avoided or reduced considerably. Obviously the failure region is

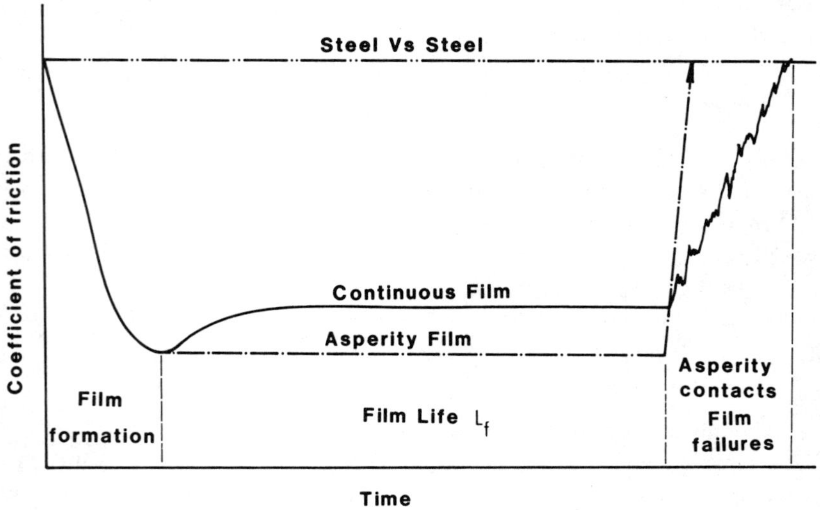

Fig. 1. Effect of sliding time on coefficient of friction when a low shear strength powder material is placed between a pair of sliding steel surfaces.

a function of the materials used. If the film is on a low friction, non-galling material, complete failure will be postponed for a long time with the continuous film.

In the following sections the effects of certain important variables on L_f and f will be illustrated.

Effect of Temperature

The effect of temperature on friction coefficient for a variety of soft solids is shown in Fig. 2 and 3. Figure 2 shows data for several conventional solids. The general trend with increasing temperature is illustrated by lead oxide, which has a gradual decrease in friction as temperature is increased. However, if a temperature is reached where the integrity of the film is destroyed, then friction will increase rapidly. Friction increases by melting (tin), by oxidation (MoS_2WS_2), by decomposition (Teflon and other polymers), and by poor film adhesion (graphite).

Figure 3 shows the frictional behavior of several other solids. Here friction was determined by decreasing the temperature from the melting point. The increasing trends with decreasing temperature are also apparent.

Obviously, changing the temperature changes the shear strength, S, of the film, which changes its frictional behavior. This correlation between friction and strength is shown for silver as a function of temperature in Fig. 4, and for a variety of materials at different temperatures in Fig. 5. From this experiment it is obvious that the strength of the film as determined by temperature controls the frictional behavior.

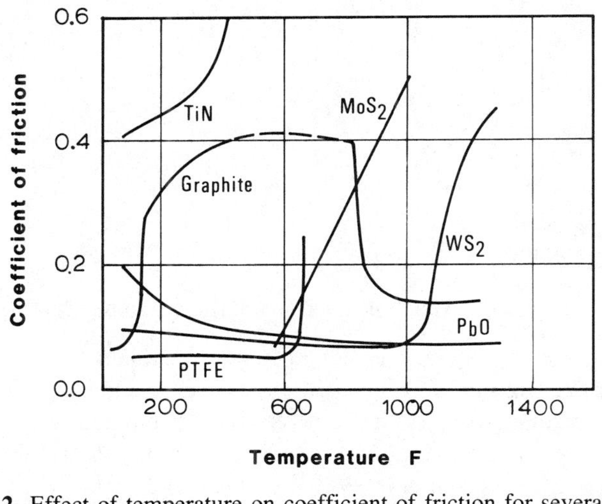

Fig. 2. Effect of temperature on coefficient of friction for several conventional soft solids.

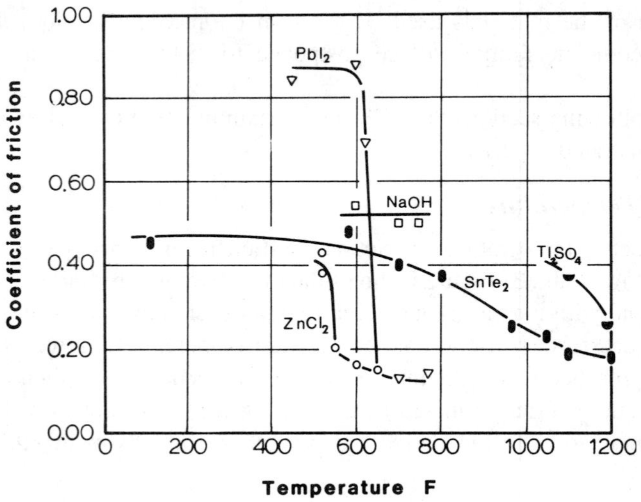

Fig. 3. Effect of temperature on coefficient of friction for a number of soft solids.

One may fit the data of Fig. 5 to the empirical expression

$$f = 2 \times 10^{-5}\sigma_T + 0.08$$

or if

$$S = .557\sigma_T$$

$$f = 3.59 \times 10^{-5}S + 0.08$$

or

$$f = 2\frac{S}{P} + 0.08$$

where f = friction coefficient
σ_T = tensile strength
S = shear strength
P = pressure = 56,000 psi.

Theoretically one would expect the friction coefficient to be proportional to $\frac{S}{P}$ not $2\frac{S}{P}$; however, the discrepancy is not difficult to explain. If the actual shear strength is higher (in this case $S = \sigma_T$), the correct relationship will result. It is well known that the shear strength increases with pressure: this, however, is accounted for by the 0.08. Finkin[6] has concluded in his studies that because of coupled stress effects, thin films will become much stiffer than thick films. Furthermore, the correct shear strength should be a dynamic value, not a static one. Work hardening is possible, especially with the harder materials.

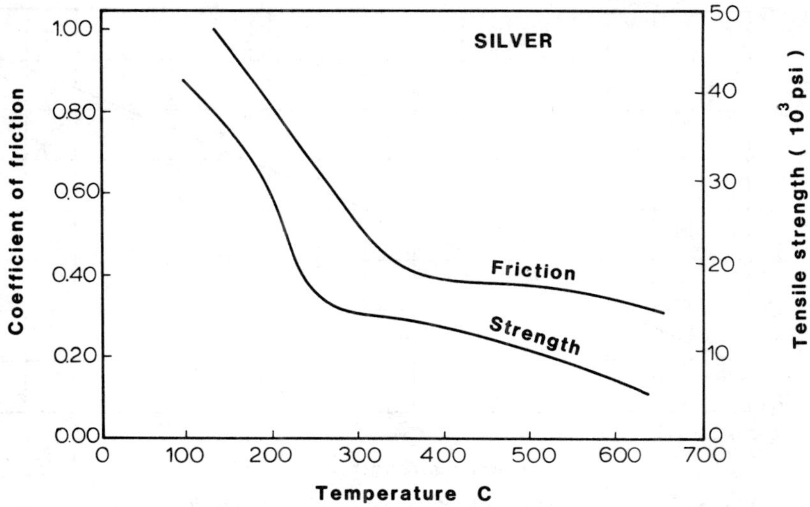

Fig. 4. Correlation between coefficient of friction and tensile strength shown for silver as a function of temperature.

Effect of Load

The effect of load on friction coefficient for lubricating films of indium and MoS_2 is shown in Fig. 6. It is seen that friction coefficient decreases significantly with increasing load and approaches an asymptotic value of 0.06. These curves correspond very closely to an equation of the form

$$f = \frac{S}{P} + 0.06\,,$$

the differences between the curves being accountable to the differences between S for indium and MoS_2. Thus, friction decreases with load because increases in load increase the pressure, P, while S remains constant or increases slightly. This effect of load is characteristic of almost all soft lubricating films.

Effect of Velocity

The effect of velocity is shown in Fig. 7. There, data by Rabinowicz and Burwell[7] taken over velocity changes of 10 orders of magnitude show that friction first increases with velocity and then decreases. The increase is attributed to shear rate effects, the decrease to increasing temperatures generated within the film. At high velocity the results of Bowden and Freitag[8] apply (Fig. 8). Extremely high shear rates exceed the capacity of the material to flow plastically and fracture occurs accompanied by high friction and wear. However, these velocities exceed normal usage and the primary effect of velocity is reduced friction due to reduced strength at higher generated interface temperatures.

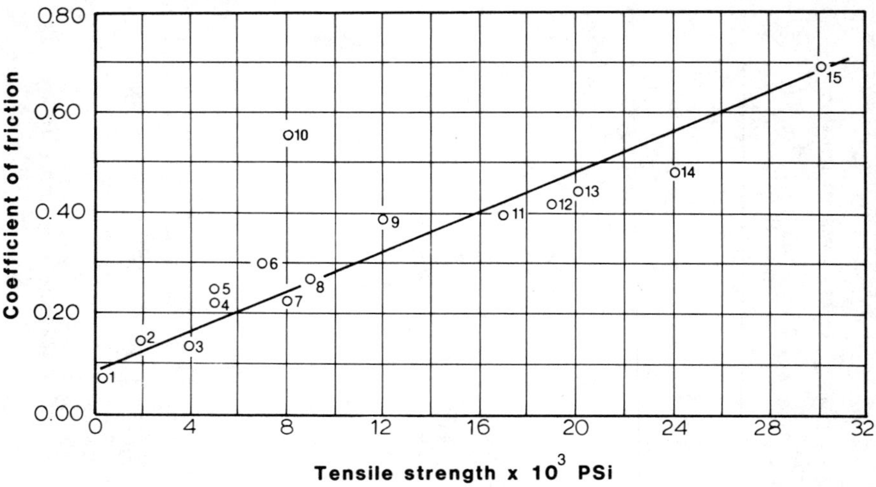

Fig. 5. Correlation between friction and tensile strength for a variety of materials at different temperatures. *(Ref 62)*

Effect of Contact Area

The effect of contact area can be understood using the simple expression

$$f = \frac{S}{P} + C = \frac{AS}{L} + C$$

where S = shear strength
A = apparent contact area
f = friction coefficient
P = pressure
C = constant
L = load.

It can be seen that friction coefficient should increase directly with the area, A. However, this cannot always be achieved. With thin films the load is supported by the base metal at the tips of the asperities. From a direct load standpoint conventional friction theory applies and the surfaces behave as if the film were not present. Thus, as originally discussed by Bowden and Tabor, the first term of the above expression can be:

$$f = \frac{A_m S_f}{L}$$

where A_m = real area of contact and
$A_m \alpha$ L = load.

If the film is only at the tips of the asperities, then this expression holds and friction will be independent of area. However, any surface geometry which traps

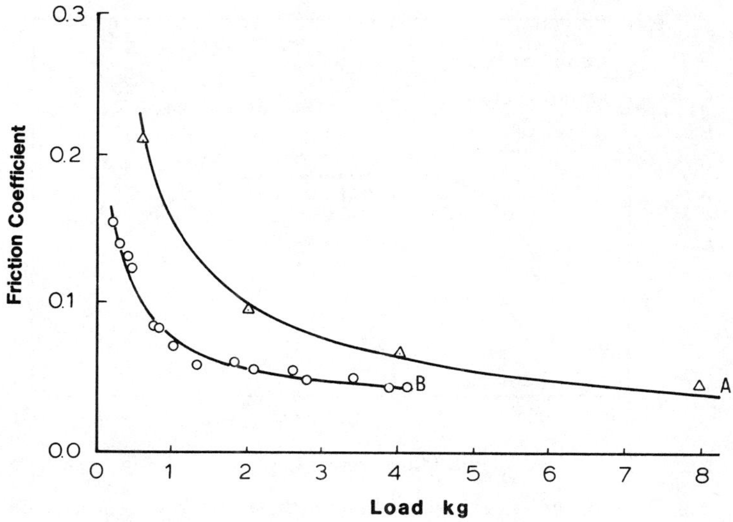

Fig. 6. Effect of load on friction coefficient for lubricating films of soft solids: (A) indium lubricating mild steel; (B) molybdenum lubricating chromium.

lubricant will behave according to the previous expression and friction will increase with increasing area. This is the reason that the continuous film of Fig. 1 gives higher friction than the asperity film.

There are three conditions which particularly favor large area effects:

(1) Larger film thickness
(2) Nonconforming geometries
(3) Higher loads.

The primary difficulty with any theoretical modeling or experiment verification is imprecise knowledge as to the exact shear area. Thus, all changes cannot be attributed to changes in materials properties.

Effect of Film Thickness

The effect of film thickness is extremely complicated. With very thick films only the properties of the film material and the adhesion of the film to the substrate need be considered. As the film thickness gets smaller the properties of both the film and the substrate influence the friction and wear life of the film. In addition to this, with thinner films one is more likely to obtain substrate metal contacts through the film, which influences the friction and wear behavior.

The generalized effect of film thickness is shown in Fig. 9. At large film thickness friction is constant, but it drops off to a minimum value and then increases again for smaller film thickness. The manner in which friction de-

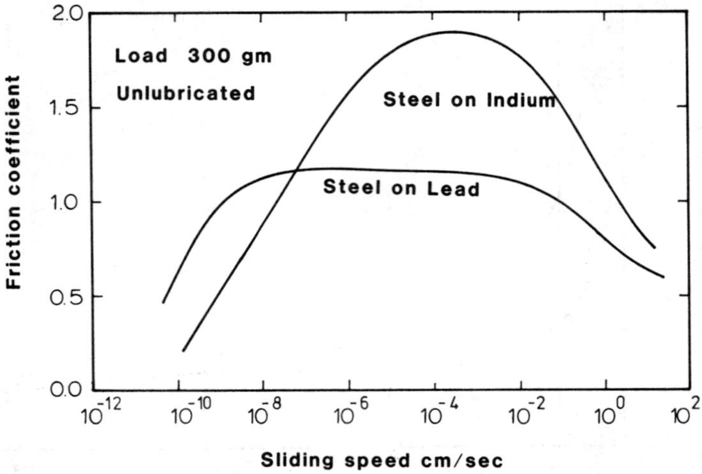

Fig. 7. Effect of velocity on friction coefficient of low shear strength materials. *(Ref 7)*

creases has been given careful consideration by Finkin.[9] He has concluded that the manner in which friction decreases is a function of the load/geometry/contact area relationship. For an elastic film:

$f \sim \sqrt{\frac{h}{L}}$	for spherical contacts
$f \sim \sqrt[3]{\frac{h}{L}}$	for cylindrical contacts
f independent of h	flats

The same analysis has not been carried out for other types of films; however, one expects the results to be similar. Various paths will be followed for different geometries depending upon the contact area.

The important point, however, is that soft films are easily squeezed out of the contact area so that very quickly the knee of the curve is reached and this is the normal operating range (10^{-3} to 10^{-5} mm); here friction changes very slowly with film thickness and one is operating with film thickness similar to that of the asperity heights.

Substrate Effects

In previous sections the role of the substrate in determining the shear area has been pointed out. There is another even more important effect of the substrate, and that is related to adhesion. Materials that have poor adhesion to the substrate will not form effective films if applied as powders and will be quickly removed if applied as films. MoS_2 will not lubricate titanium and has a much longer film

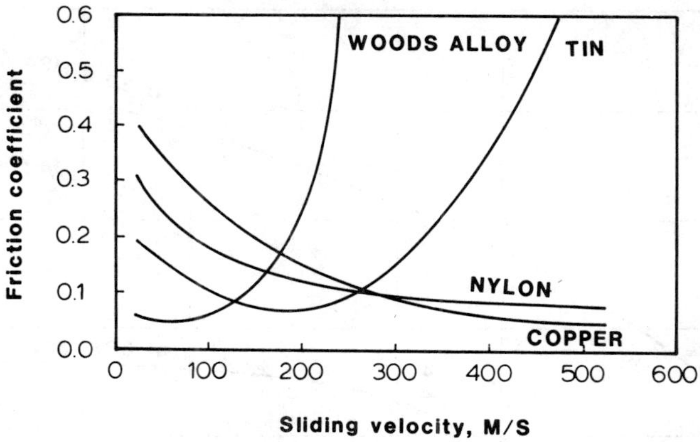

Fig. 8. Effect of velocity on friction coefficient at high sliding speeds. *(Ref 8)*

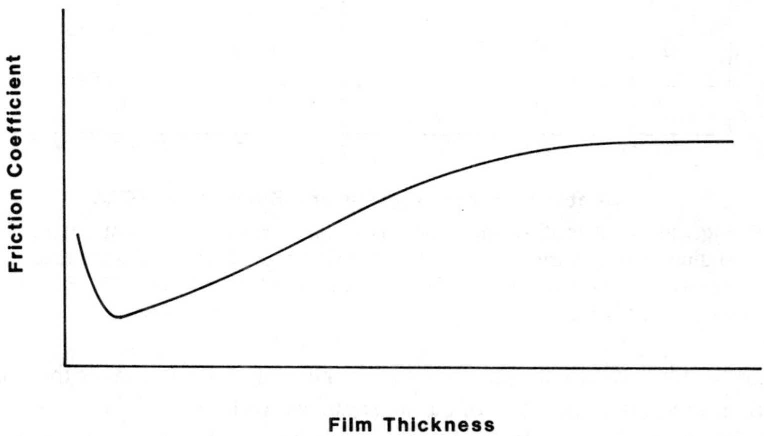

Fig. 9. Schematic illustration of the effect of film thickness on friction coefficient.

life on molybdenum.[10] Graphite is ineffective in a certain temperature region because of poor adhesion (Fig. 2). High shear strength compounds which are strongly bonded protect surfaces from wear even though they yield high friction. This high surface adhesion appears to be a necessary condition for good solid film lubricants. Where there is a complete lubricating film, the substrate has little effect, as shown in Fig. 10. Most material effects can usually be attributed to different surface roughness of the different materials.

Tsuya[11] has studied the endurance life of MoS_2 films on various substrates and under various operating conditions. The film life increases with substrate hardness and decreases with load. The significant parameter suggested is the "real

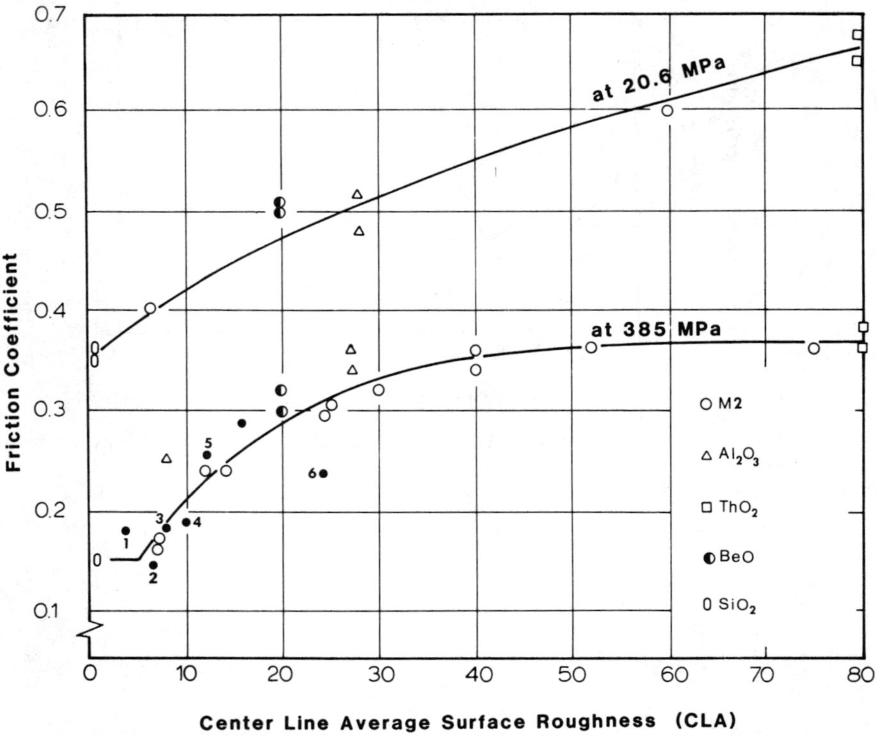

Fig. 10. Effect of surface roughness on friction coefficient of tin sliding against various materials at room temperature. The numbered materials are (1) S-Monel, (2) H12 steel, (3) SiC, (4) TiC, (5) B_4C, (6) ZrO_2, (7) Stellite-3.

contact length," which is the diameter of the real contact area in the sliding direction. In vacuum the effect of the substrate was different, with the longest life being found for silver, gold, and copper as shown in Table 12. This is understandable since the surfaces of both the base material and the MoS_2 are changed when in vacuum.

Atmosphere

The atmosphere is important in two ways. First, it influences the adhesion of solid to the substrate and reduces the life of the film. Sliding on any soft film results in rapid wear by plastic flow until the film thickness is equivalent to the thickness of the asperities. In the process the film is completely removed from some of the asperities. This must be resupplied by the residual film between the asperities. If certain molecules from the surrounding atmosphere adsorb and prevent film adhesion, then metal contact and short film life will result. This has been shown to be the case for MoS_2 in a humid atmosphere. Other solid films act in a similar manner.

Table 12. Endurance Life of MoS_2 Films (Sliding Cycles)

Material combination	Air	Vacuum
Titanium	—	40
Glass	100	100
Iron	—	300
Stainless steel	10,000	600
Molybdenum	200	8,000
Nickel	600	11,000
Tungsten	20,000	50,000
Copper	310	200,000
Aluminum	300	250,000
Gold	—	500,000
Silver	1000	700,000
Lead	2	—

A second effect of atmosphere is to influence the apparent shear strength of the soft solid. This was first noted by Savage in the 1940's and has since been confirmed by others. Some data for graphite are shown in Fig. 11. Nitrogen has little effect, but the other atmospheres lower friction substantially. It has been suggested that this effect is due to edge adsorption of vapor on the graphite crystals, which increases the bonding that holds the film together on the surface. For MoS_2 and other similar inorganic solid lubricants the opposite effect is found: friction is higher in air than it is in vacuum. The lowest friction is found in vacuum; different vapors increase friction by different amounts. Frictional behavior is related to the surface energy. Graphite has a high surface energy and

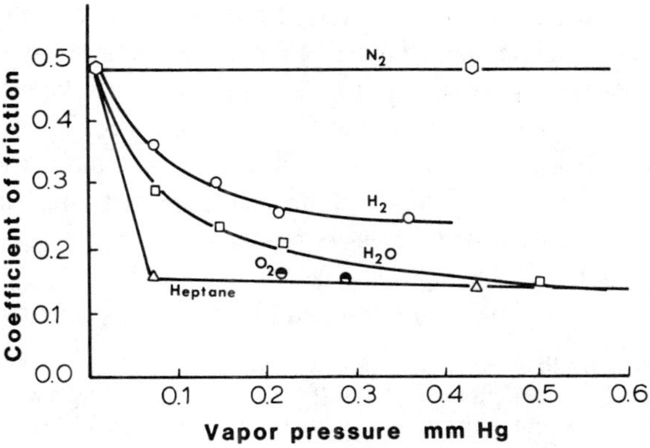

Fig. 11. Effect of atmosphere on coefficient of friction. The effect of nitrogen, hydrogen, water vapor, oxygen, and heptane on outgassed graphite is shown.

vapors reduce it. MoS_2 and other such materials have a low surface energy; vapors and impurities increase it. Teflon has a low surface energy and low friction.

Low Shear Strength Film Effects in Friction and Wear

In the previous sections a review has been given of the effect of the known variables on the frictional behavior of soft solids. From their behavior, three mechanisms of sliding can be proposed:

(1) Interface slip
(2) Interfilm shear
(3) Interfilm slip

Which of these will take place will depend upon which requires the lowest work.

In interface slip, a film is formed on one surface and adheres strongly. Slip takes place between the film surface and the slider where work is expended in both plastic deformation of the surface around the asperities and breaking the interface bonds. A silver film is an example of this type of behavior if the slider is iron or another oxidized surface. Wear of the film is caused by the gradual squeezing out of the soft solid.

Interfilm shear would be represented by strong adhesion to both surfaces and sliding taking place by plastic flow somewhere in between. A silver film with a nickel slider represents this type of behavior. Life under these conditions is very short. Work hardening and crystal rotation can remove large volumes of the film. Materials which have a dominant slip plane give much longer life. Friction can be closely related to flow behavior and can be characterized as plastic, viscous, visco-elastic, etc.

Interfilm slip takes place when there is strong adhesion to each surface and a low energy plane generated somewhere between the two surfaces. It is the same as if two surfaces were coated and gross sliding took place between them. The low energy slip plane may be inherent in the material (MoS_2, Teflon, etc.) or may be the result of adsorption (graphite). This mechanism may explain boundary lubrication where both the soft film and the adsorbing molecules are supplied by the lubricating oil. The task here is to properly match the soft film and the adsorbing molecule.

The question of what makes a good solid lubricant film can be summarized:

(1) Strong adhesion to at least one surface.
(2) A low energy plane either at the surface or in the coating material. The plane can be either inherent in the coating material, generated by adsorption, or generated by sliding.

Coatings for Wear Resistance—Hard Coats

Low shear strength films as discussed in the previous sections are used to lower friction in tribological pairs. Friction is lowered either by decreasing the forces needed to shear junctions produced at locations where real contact between materials occurs, or by confining the shear work within a film of low shear strength. The principles involved in the selection of coating materials to reduce friction are clear.

Films and coatings are also used to reduce the wear rate in friction pairs. The application techniques, coating compositions, and coating functions are listed in Tables 2 to 4. In general, when the wear rate in a friction pair is unacceptably high, one or more means of lowering the wear rate by surface modification, thin film deposition, or coating can be found. The surfaces prone to rapid wear are then simply replaced with those of a material which is more wear resistant.

What is a more wear-resistant material in the current context? A satisfactory answer provides guidance for the selection of appropriate surface coatings to decrease the wear rate. A brief examination of the elementary aspects of wear is useful in identifying the primary properties that a wear-resistant coating material must possess.

Wear is a complex process. When it is viewed as a process in which progressive loss of material occurs as a consequence of mechanical action, at least four principal mechanisms of wear can be identified: (*a*) adhesive wear, (*b*) abrasive wear, (*c*) corrosive wear and (*d*) surface (fatigue) wear. According to adhesive wear theory, the wear rate W, defined as worn volume V per unit sliding distance, is proportional to the load L and inversely proportional to the yield pressure P_o of the wearing material,[13] with

$$W = k\frac{L}{3P_o}.$$

Here k is a proportionality constant. A linear relationship is observed in wear experiments provided the nominal contact pressure does not exceed H/3, where H is the bulk hardness of the wearing material. Rowe's[14] modification of the elementary wear model takes the friction coefficient μ into account:

$$W = k\{1 + \alpha\mu^2\}^{1/2}\,\beta\frac{L}{P_o}.$$

Two additional constants, α to account for surface contamination and β to describe material behavior under complex stress loading, are used. However, wear volume is still an inverse function of the yield pressure.

Abrasive wear models also lead to wear equations where the wear volume is inversely proportional to the yield pressure. Conical indenters with a semi-apex angle of (90 − θ), for example, give an equation for wear rate of the form:

$$W = \frac{CL \tan \theta}{\pi P_o},$$

where C is a geometric factor. In both adhesive and abrasive wear situations, wear volume per unit sliding distance is therefore adequately described by a general equation of the form:

$$W = k^* \frac{L}{H}$$

The hardness H (or the yield pressure P_o) is seen to be a factor of prime importance in both cases. It is known that the hardness and yield pressure are related through a constant of the order of 3 for metallic materials. For non-metals with a much smaller elastic modulus, the hardness is a smaller multiple of the yield pressure.

With empirical observations of high wear resistance in materials with large H/E ratios, where E is Young's modulus, Oberle[15] has reported a correlation between wear resistance and H/E ratio. He has suggested that materials with low elastic moduli will accommodate the passage of abrasives across their surfaces through elastic deformation and thus avoid permanent damage. An example would be the action of well-rounded abrasive on surfaces of rubber-like materials. Wear rate is lower and the suggestion appears plausible. The elastic constant E may therefore have a role in wear resistance. Young's modulus E in fact turns out to be important when hard coats are used but the reasons are different, as will be shown later.

Fundamental aspects of adhesive and abrasive wear indicate that coatings of hard materials are necessary for wear protection. Hardness levels necessary to obtain wear resistance now need to be estimated.

By examining abrasive wear as a function of the hardness ratio H_a/H_m, where H_a is the hardness of the abrading species and H_m that of the abraded material, Khrushchov[16] reports little or no wear in the hardness ratio range of 0.72 to 1.12. Based on Richardson's data,[17] it has been suggested that hardness ratios of the order of 0.4 are desirable to obtain negligible wear.[18]

We note that to provide wear protection in sliding against the most commonly used constructional material, steel, in its fully hardened condition when it possesses Vickers hardnesses between 600 and 800 (depending on the alloy content, carbon content, tempering, etc.), counterface hardnesses in the Vickers range of 1500 to 2000 are sufficient. Wear-resistant coatings or films with hardnesses above $H_v \simeq 1500$ are thus seen to be sufficient for most applications.

Wear-Resistant Coating Material Selection

Wear-resistant coatings are used principally to guard against adhesive and abrasive wear. Wear considerations discussed in the last section indicate a need for coating materials with high intrinsic hardnesses. Many transition/refractory metal compounds (carbides, nitrides, borides, and silicides) and oxides of aluminum and magnesium, as well as carbides, nitrides, and oxides of silicon, are materials with the requisite hardness. A wide range of materials is available to afford wear protection.

Hard materials that can be and have been deposited as thin films or coats using one or another of the coating processes described in a subsequent section are listed in Table 13. Microhardnesses of these materials are tabulated in Table 14. Hard coatings of materials with even higher hardnesses, which are normally synthesized by high pressure techniques (cubic boron nitride, diamond), are possible with sputter coating.

More than one of the materials listed in Table 13 may be able to provide the wear resistance needed in a given application. The choice is then based on other relevant properties, ease of deposition, and other factors. First among the secondary considerations is the coefficient of friction. A coating material with an acceptable hardness and the lowest friction coefficient is preferred over one with a higher hardness and friction coefficient. This is so because a lower power dissipation in a friction pair is desirable.

Table 13. Hard Compounds for Use as Thin Coats in Tribological Applications(a)

Carbides	Nitrides	Oxides	Borides
TiC	TiN	TiO_x	TiB_2
HfC	HfN	HfO_2	HfB_2
ZrC	ZrN	ZrO_2	ZrB_2
TaC	TaN	Ta_2O_5	TaB_2
VC	VN	V_2O_3	VB
NbC	NbN	—	NbB_2
Cr_3C_2;Cr_7C_3; $Cr_{23}C_6$	CrN	Cr_2O_3	—
SiC	Si_3N_4	SiO_2	—
WC;W_2C	—	—	WB
Mo_2C	—	—	MoB
—	AlN	Al_2O_3	—
TiC-TiN(b)	TiN-TiC(b)	TiC-Ti_xO_y(b)	—
TiC-VC(b)	(Ti,V)N(b)	—	—
Ti-Si-C(b)	—	—	—
(Fe,Mn)$_3$C(b)	—	—	—
—	(Si,Al)N(b)	—	—
—	Fe_4N	—	—

(a) All materials listed have been deposited as thin coats with one or another of the coating technologies discussed in this chapter. For sources of primary data see Ref 38. (b) Complex or "alloy hard material" coatings.

Table 14. Microhardnesses of Typical Hard Coat Materials (For sources of primary data, see Ref 38.)

	Vickers hardness, kg/mm^2		
Element	Carbide	Nitride	Boride
Boron	3700	—	—
Chromium	1600 – Cr_7C_3 1300 – Cr_3C_2	2200 1083 – CrN	1800
Hafnium	2270 – 2650	1640	2250-2900
Molybdenum	1800 – MoC	—	2350
Niobium	2400-2850	1396 – NbN 1720 – Nb_2N	2100-2400
Silicon	3500	—	—
Tantalum	1800-2450	1220	2450-2910
Titanium	2000-3200	1200-2000	2200-3500
Tungsten	2100-2400 1450 – W_2C	—	2400-2660
Vanadium	2460-3150	1520-1900	2070-2800
Zirconium	2360-2600	1150	2250-2600

Literature microhardness values span a moderately wide range. A single specific value is usually not fully representative. Transition metal oxides, nitrides, and carbides can vary widely in stoichiometry and are mutually soluble. Variations in hardness reported are due to variations in stoichiometry and purity. Most borides, especially the hexagonal borides, are highly anisotropic.

The next important consideration is the coefficient of thermal expansion. Hard compounds with their stiffer lattices[19] exhibit smaller coefficients of thermal expansion than do metals. Since power dissipation and temperature rise occur in wear pairs, a coating material with the smallest coefficient of expansion mismatch with the base material is preferred. Otherwise, shear stresses, induced by temperature gradients at the coating-to-base material interface, tend to debond the hard coat. Thermal stresses in a material are proportional to the $E\alpha$ product, where E is Young's modulus and α is the linear thermal expansion coefficient. Elastic properties of hard coat materials are therefore also important in the choice of coating materials.

Thermodynamic properties of coating materials enter into consideration when coated parts are exposed to high temperatures during operation. Some hard compounds, such as tungsten monocarbide, tend to decompose at moderate temperatures. The choice of the hard coat material must hence consider thermodynamic stability. The standard free energy of formation of a compound is a measure of its thermodynamic stability and is sometimes used as an important criterion in the selection of coating materials. Selection of coating materials to hard coat cutting tool inserts is an example.

Coating Thickness

The surface treatment processes now used to improve wear resistance of parts (case hardening, induction hardening, flame hardening, laser glazing, etc.) rely on superficial heating and/or selective diffusion of alloying elements to obtain higher surface hardnesses. The highest surface hardness obtained is limited by the composition of the base material. In the case of iron and steel parts, surface hardnesses possible rarely exceed HV = 1000 kg/mm^2. The usual hardness range is some 200 to 400 points lower. Surface-hardened layers ranging in thickness from 0.4 mm to 1.0 mm or more are commonly used for long wear life and to permit refinishing of the surface-hardened parts by grinding or other processes.

In contrast, the surface hardness that can be obtained with hard coating is much greater (Table 14). With such hard coats, the H_a/H_m ratio relevant to abrasive wear[16] is lowered easily to values in the 0.2 to 0.6 range where negligible wear occurs. Coatings comparable in thickness to those produced by traditional surface hardening are then not necessary for abrasive wear protection.

The bonding in compounds used for hard coats is predominantly covalent in character. This is particularly true for transition metal carbides and nitrides with NaCl lattices.[20] It is therefore reasonable to expect that even under conditions of intimate contact, extensive cold welding of asperities will not occur. Even if it were to occur, the shear strength of the welded junctions can not be large. Loss of material from the hard-coated surface is hence expected to be small. Thin coatings should therefore be sufficient for protection against adhesive wear.

Rabinowicz[21] has suggested that the ratio of surface energy γ to hardness H is a good indicator for the choice of materials in sliding applications involving unlubricated surfaces. Materials possessing low γ/H ratios are recommended. Virtually all hard compounds possess γ/H ratios in the 0.02 to 0.1 mm^{-1} range while typical values for metals lie in the 0.5 to 10 mm^{-1} range. Hard coating is hence seen to offer a simple means of altering the surface energy/hardness ratio in order to lower wear.

Though surface energy considerations indicate the suitability of hard coats for lowering wear rates, they do not provide a basis for the choice of coating thickness. If hard coats that meet the surface energy criterion can be deposited, they will presumably suffer less wear, and will hence not need large thicknesses.

Experience with hard coats under severe unlubricated sliding conditions as in metal machining indicates that coatings as thin as 1 μm to 10 μm are sufficient to afford wear protection in many cases. Similar results are reported following 4-ball wear tests.[22] Empirically, therefore, thin coats a few micrometers in thickness are found to be satisfactory. Some recent experimental studies[23] indicate that there may be an optimal coating thickness for some applications. Unlike the case of low shear strength films,[24] analytical models have not yet been proposed that allow the computation of optimal hard-coat thickness.

The use of thin hard coats limits the possible temperature gradient and therefore the magnitude of thermal stresses at the coating-to-substrate interface. Thin layers of hard materials, visualized as Griffith solids, should also exhibit higher fracture stresses and elongations with lower coating thicknesses. They should therefore behave as compliant coatings when applied as thin coats. Wear studies involving titanium nitride coatings on elastically and plastically soft substrates at nominal contact stresses exceeding the yield stress of the substrate[25] lend support to this view. More direct evidence would be desirable.

No real basis exists at the present time to guide the choice of hard-coat film thickness. Based solely on experience, hard coats a few micrometers in thickness are used and found to be sufficient. Although several reasons can be advanced to support the choice of small coating thickness, theoretical justification does not yet exist.

Coating Bond Strength Requirements

A hard coating is used in a part with limited wear life to provide extended part life. Ideally, a coated part should fail by a mechanism other than wear, after a longer service life. To ensure this, the hard coat must remain on the wear surface throughout the extended part life without being worn away. Factors concerned with the selection of coating materials and thicknesses to assure longer life have been discussed in previous sections. It was tacitly assumed that coatings are well bonded to the substrate and hard coat debonding is not the life-limiting failure mechanism.

Well-bonded hard coats are often obtained when elevated temperature coating processes are used. Use of high temperature coating processes where graded interfaces can be produced does not assure high coating-to-substrate bond strength. Brittle interlayers may be produced and can limit the bond strength. Also, some materials cannot be hard coated at high temperatures since accompanying changes in substrate microstructure and properties are often unacceptable. Techniques of film deposition which permit hard coating at essentially room temperature are therefore also needed. Graded interfaces are difficult to produce in room temperature coating processes.

Several coating techniques are available for "room temperature" coating. They may be suitable for wear protection only if satisfactory substrate-to-coating bond strengths can be obtained with these processes. The suitability of a hard coating and a hard coating process can only be established after an estimate of the bond strength requirements is made.

Bond strength requirement estimates are made by examining the tribological environment in which the coated part is expected to work. The first tribological characteristic of importance is the wear environment (abrasive, adhesive, etc.). This has been dealt with earlier. The next set of important characteristics involves contact stress and conditions of contact.

In a typical contact with lubrication, the surfaces may be completely separated from each other by an intervening fluid film [hydrodynamic or full film elastohydrodynamic (EHD) contact] or the surfaces may be in intermittent contact with significant asperity interaction (boundary lubrication and partial film EHD). In hydrodynamic contacts, the maximum pressure rarely exceeds 75 MPa and the coefficient of friction lies in the 0.001 to 0.01 range. Maximum Hertz pressures of 0.5 to 3.0 GPa and traction coefficients of 0.02 to 0.12 are typical in full film EHD. In mixed film EHD, the Hertz pressure range is the same as in full film EHD, but the associated traction coefficients range between 0.02 and 0.3.

The shear stresses experienced at the coating-to-substrate interface are calculated for these cases with the Hertz contact analysis. The interface shear stress is a function of the Hertz pressure and the friction or traction coefficient. For thin films in the 1 μm to 10 μm range, interface shear stresses are of the order of 100 MPa to 300 MPa, larger values being obtained in mixed film EHD.

The use of "unlubricated" sliding contact is not common. A friction contact can be technically described as unlubricated when sliding contact assemblies are made with self-lubricating materials [grey cast irons, polytetrafluoroethylene (PTFE), impregnated porous solids, etc.] or when solid lubricants (graphite, MoS_2, WSe_2, etc.) are used. The contact pressures usually range between 3 MPa and 70 MPa in such cases. Typical values are of the order of 15MPa. Maximum coating-to-substrate bond strength requirements in such cases, when lubricant failure is total, will rarely exceed 70 MPa. The maximum shear stress that may occur is limited by flow within the softer body (Rigney-Hirth friction model, Ref 26). The 70 MPa bond strength requirement is hence realistic.

True unlubricated sliding contacts are encountered in metal working and metal machining tools. Plastic flow in the contacting material, as opposed to the tool, is the objective. The coated material surfaces are subjected to much higher normal and shear stresses than in the case of bearings; therefore very high bond strengths are necessary. The maximum shear stress at the coated surface can range up to 2 GPa or more in cold heading and forming tools, though in most cases, it is usually around 1 GPa (metal machining tools).

Bond strength requirements are thus seen to cover a broad range. Interface shear strengths of the order of 300 MPa are needed in bearings, and levels to 2 GPa may be necessary in tooling applications. The requirements in other unlubricated conditions are less stringent.

Summary of Requirements for Wear-Resistant Coatings

Wear-resistant coatings will necessarily have to be hard. Exceptional surface hardness levels are not necessary to assure extended wear life. Small wear rates obtained with hard coats indicate that thin coats are usually sufficient. When several materials that meet the hardness requirements are available, the material exhibiting the lowest friction coefficient is preferred. The coating material that matches the Eα product of the substrate most closely is preferred. All other conditions being comparable, in a choice among several materials, the most stable of the hard compounds is the preferred choice.

We close this section by noting that the coating-to-substrate bond strength requirements are critical in the choice of coating materials for tribological applications.

Selection of Hard Coating Processes

Many processes are available to coat parts with solid lubricant coatings and wear-resistant coatings. The choice, in the present context, is restricted to those capable of producing thin coats. The justification is as follows. The use of thin coatings eliminates the need for refinishing either to meet surface finish requirements or for dimension control (since coating thicknesses can be a fraction of the part tolerance). Thin coats require smaller coating times. They also lower the demand for scarce strategic materials.

The coating process selected must meet specific needs based on operational requirements. Coatings produced must consistently possess adequate coating-to-substrate bond strength. Since the coatings are compounds in many cases and compound properties (especially the hardness of compounds) are strong functions of stoichiometry, the process selected should be able to produce coatings with perfect or near-perfect stoichiometry. Coating thickness and uniformity control must be possible. Significant microstructure and property changes must not accompany coating. Coatings produced must be dense, and as far as possible, equiaxed. Process reproducibility, ease of coating control, and coating cost considerations are other standard requirements that any industrially viable process must meet.

Coating Processes

In earlier sections it has been shown that low shear strength films are useful in lowering the friction coefficient in tribological applications where conventional lubrication is either not possible or not desirable. Layers of hard materials have also been shown to be helpful in improving the wear life of parts. Techniques and processes that are used to produce soft and hard tribological coatings are described in this section. As noted earlier, coatings are useful in friction and wear applications only when satisfactory coating-to-substrate bond strengths are obtained. Therefore, coating processes that produce well-bonded thin coats are emphasized.

Pyrolytic decomposition, halide metallurgy, thermal evaporation, and sputtering are some of the physical phenomena that are used to produce coatings. Moderate and high temperatures are needed for the first two.

Diffusion rate is fairly high at the temperatures used and well-bonded coatings are produced easily.

Thermal evaporation and sputter coating are carried out over a wide range of temperatures. Thermally sensitive materials needing low temperature coatings are coated with the vacuum deposition processes. To produce coatings with good bond strength, some critical factors are recognized and the basic coating processes are modified accordingly.

Pyrolytic Decomposition

Many metallic and organo-metallic compounds are easily decomposed at elevated temperatures. Metallic and non-metallic coatings produced with these reactions are called pyrolytic coatings.

Soft coats of gold and silver are produced by decomposing tertiary dodecyl mercaptide of gold at 150°C[26] and silver carboxylates at low temperatures.[27] Hard coats of alumina and silica are produced by decomposing metal beta-diketonates.[28] Amorphous hard coats of alumina can be produced by decomposing aluminum isopropoxide on substrates maintained at 700° to 800°C.[29] By subsequent annealing, amorphous coats are converted into gamma alumina coats. Dicumene chromium and aluminum alkyls are decomposed[30,31] to produce metallic chromium and aluminum coats.

The pyrolytic coatings produced with the reactions cited are easily controlled and are carried out in relatively simple coating systems. Coating rates vary with reactions and reaction conditions. Rates ranging from a few microns per hour to a few microns per minute are obtained. Well-bonded coatings, soft and hard, suitable for friction and wear applications are produced fairly easily. Some of the compounds used are hazardous and require careful handling.

Chemical Vapor Deposition

Chemical vapor deposition (CVD) can also be regarded as a thermal decomposition process. Often, other reactions such as reduction of metal halides with hydrogen, displacement reactions, disproportionation reactions, etc., are used. Some typical reactions and temperatures used to produce metallic coatings are shown in Table 15.

The principal features of a typical CVD system are shown schematically in Fig. 12. Metal halides, transported in a mixture of gases, are reduced with hydrogen on the hot substrate. To produce coatings of compounds (carbides, nitrides, etc.), the chemical potential of the corresponding atomic species (carbon, nitrogen, etc.) in the coating chamber is raised and the corresponding hard compound coatings are produced. The acids produced by hydrogen reduction of halides are swept away as reaction products in an inert carrier gas (not shown in figure); argon is commonly used. Selected CVD gas mixtures and

Table 15. CVD Reactions Classified by Type

Pyrolytic Decomposition		
$(C_8H_{10})_2Cr$ (gas)	400-600°C $\longrightarrow$	Cr (solid) + $2C_8H_{10}$ (gas)
$Ni(CO)_4$ (gas)	150-200°C $\longrightarrow$	Ni (solid) + 4CO (gas)
Hydrogen Reduction of Halides		
$TiCl_4$ (gas) + $2H_2$ (gas)	700-1200°C $\longrightarrow$	Ti (solid) + 4HCl (gas)
WF_6 (gas) + $3H_2$ (gas)	350-1000°C $\longrightarrow$	W (solid) + 6HF (gas)
Displacement Reaction		
$SiCl_4$ (gas) + CH_4 (gas)	$\longrightarrow$	SiC (solid) + HCl (gas)
$CrCl_2$ (gas) + Fe (solid)	$\longrightarrow$	Cr–Fe (alloy) + $FeCl_2$ (gas)
Disproportionation		
$2GeI_2$ (gas)	$\longrightarrow$	Ge (solid) + GeI_4 (gas)

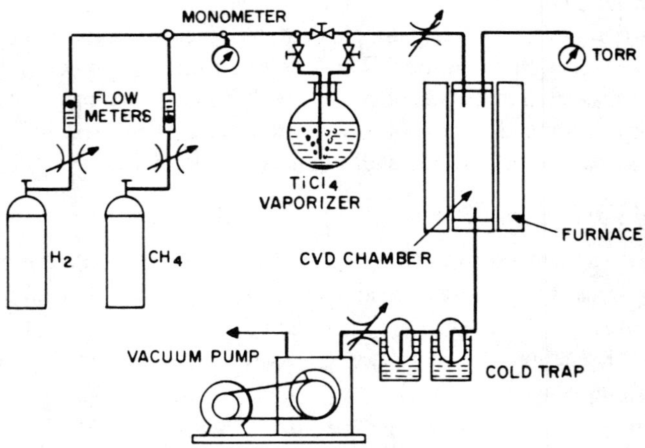

Fig. 12. Schematic illustration of a typical chemical vapor deposition system for TiC coating of materials.

temperatures to produce hard coats are tabulated in Table 16. Interested readers are referred to the Proceedings of the CVD Conferences sponsored by the American Nuclear Society and the Electrochemical Society for further details and specific data needed to produce different types of hard coats.

Table 16. Selected CVD Gas Mixtures and Temperatures for the Production of Hard Compound Coatings

Hard compound	Substrate material	Mixture of gases	Coating temperature
TiC	Cemented carbide steel	$TiCl_4$-CH_4-H_2	>800°C
TiC-N	17-4 P.H. stainless steel	$TiCl_4$-$N(CH_3)_3$-N_2-H_2	600-700°C
HfC	Graphite	Hf-Cl_2-Hf-Cl_4-CH_4-H_2	1300°C
ZrC	Graphite	$ZrCl_4$-CH_4-He-H_2	1850°C
TiN	Various	$Ti(NR_2)_4^*$-Ar-H_2 or N_2	250-600°C
ZrN	Various	$Zr(NR_2)_4^*$-Ar-H_2 or N_2	870°C
Al_2O_3	Cemented carbide	$AlCl_3$-CO_2-H_2	850-1100°C
TiB_2	Steel, graphite	$TiCl_4$-BCl_3-H_2	>800°C

*Radical R = methyl, ethyl.

Hard coating by CVD involves high temperatures (Table 16). Atomic mobility is high during coating, and "graded interfaces" are produced. If by graded interface it is meant that the composition changes across the coating-to-substrate interface are smooth and continuous, then truly graded interfaces are possible only in metallurgical systems with unlimited solubility. In all other cases intermediate phases are produced as in diffusion couples for the same systems. Bond strength of the hard coat produced may be limited by the properties of the intermediate phases. "Barrier coatings" are used to overcome the problem with unfavorable intermediate phases. Barrier coatings of chromium used in the TiC coating of AISI 52100 steel to produce hard-coated ball bearings is an example.[32]

In CVD, stoichiometry control is exercised through the control of the gas mixture, flow, and flow rate and through the choice of coating temperature. Coating rates of a few micrometers per hour are common. Coating rates obtained depend on the coating conditions. Thickness control is straightforward in parts with simple geometry. Coatings produced are usually columnar and exhibit preferred orientations, but these characteristics depend on coating parameters.

The difference in coefficient of expansion between the coating and the substrate can result in residual tensile stresses in coatings. When undesirable intermediate phases are present, debonding is possible and coatings are unsuited for wear applications. Interlayers of barrier coatings are used to overcome these problems also.

In many cases (coated cutting tools), the as-coated parts are used with no further finishing or treatment. For iron and steel parts subsequent heat treatment is necessary. This can be avoided in steels with adequate hardenability by providing a means for accelerated cooling following the coating run.

A large percentage of the carbide cutting tools now in industrial use are CVD-coated with carbides and nitrides of transition metals. Alumina-coated tools are also in use. Bond strengths obtained by CVD are quite satisfactory for use under severe sliding conditions as in metal machining and metal working.

Vacuum Coating Processes

Vacuum coating processes are particularly suited for the coating of thermally sensitive materials and polymers. A number of different processes are available and the principles involved will be briefly described. A more complete description of the processes and their utility for tribological coatings may be found in a recent edition of the *Wear Control Handbook*.[38] Interested readers are also referred to the Proceedings of the Metallurgical Coating Conferences sponsored by the American Vacuum Society.

A variety of metal coatings and coatings of congruently melting compounds can be produced by simple vacuum evaporation. Vapors produced by melting are deposited in vacuum on parts to be coated. While coatings with good bond strength are easily produced on non-metallic surfaces such as glasses, quartz, etc., evaporation coatings produced on metallic surfaces are usually not satisfactory for friction and wear applications. The difficulty is due to the presence of thin layers of oxides and other contaminants (adsorbed gases and water vapor) on part surfaces. Whenever part heating is permissible, this difficulty is easily overcome.

Methods have been devised to increase the kinetic energy of the vapor atoms condensing on part surfaces. In electrically assisted coating processes, vapor atoms are ionized and accelerated in an electric field to arrive at substrate surfaces with much greater energies (a few to several hundred electron volts) than their mean thermal energy (equal to 3/2kT, where k is the Boltzmann constant and T is the vapor source temperature). Coatings produced with energetic vapor fluxes are much better bonded than those produced by simple evaporation.

In ion plating (Fig. 13), ionization and acceleration requirements are met by requiring the coating flux to arrive at the substrate which is enveloped in a plasma.[33] A self-sustaining plasma requires a soft vacuum or higher voltages. Soft vacua are preferred. The vapor atoms experience multiple collisions in the plasma before they arrive at the substrate. The diffusion of the coating flux through the plasma is an advantage when complex shapes are to be coated. Ion plating under these conditions is not a line-of-sight process.

Well-bonded soft films of lead, ion-plated on ball bearing surfaces, are reported to perform satisfactorily in high vacuum bearings.[34] Gold and silver are ion-plated to produce soft coats. Ion plating of aluminum for corrosion protection is also common. Hard compounds such as silicon nitride are ion-plated by using a reactive gas mixture within the coating chamber.[35] The good adhesion of ion-plated films is thought to be due to concurrent sputter cleaning and coating. Net coating rate is the algebraic sum of the two and is only limited by the evaporation rate. Ion plating of thermally sensitive materials requires care to prevent radiant overheating of the substrate.

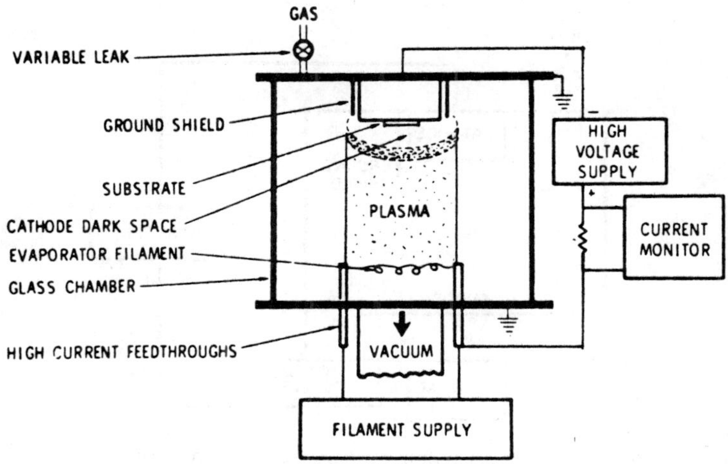

Fig. 13. Schematic illustration of an ion plating system.

In activated evaporation, the coating vapor atoms are ionized by "injecting" electrons with a "probe" electrode[36] interposed between the vapor source and the substrate (part surface). The substrate is maintained at ground potential or is biased as needed. Even with these modifications, activated evaporation is usually carried out with substrate heating to produce well-bonded coats.

In activated reactive evaporation,[36] a reactive gas mixture is used during coating to concurrently synthesize and deposit the needed compounds. Stoichiometry control requires close control of the gas mixture composition, gas flow rates, and evaporation rates. Hard coats produced by activated reactive evaporation on hot substrates exhibit good adhesion and excellent mechanical properties.[37]

Coating rates in activated reactive evaporation depend on melt power applied. Stoichiometry control requires gas flow rate and gas composition to be consistent with evaporation rate. Substrate temperature is an important variable in activated evaporation processes.[37] The structure and properties of films deposited are very sensitive to the coating process variables.

Sputtering

The vacuum coating processes that rely on melting to generate the coating flux are often classified as physical vapor deposition (PVD) processes. The sputtering processes rely on heavy ion bombardment to generate the coating flux. The sputtering process may be described, in simple terms, as follows.

A self-sustaining glow discharge is induced by applying a high voltage (D.C.) to a pair of electrodes in a soft vacuum as shown in Fig. 14. Electrons emitted by the cathode and accelerated by the applied electric field collide with and ionize the gas molecules in the vacuum chamber. The positively charged ions

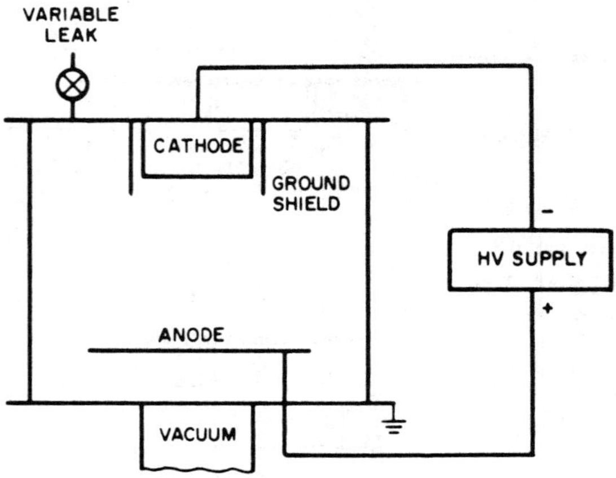

Fig. 14. Schematic illustration of a simple D.C. sputtering system.

produced are also accelerated by the field. They impinge on the cathode (target) and disintegrate it slowly at the atomic level. The atom flux generated is the coating flux. The process of generating an atom flux or the gradual removal of atomic layers by ion bombardment is known as sputtering. A detailed treatment of the physics of sputtering is found in Ref 39. Ritter[40] has given an introduction to sputter coating. A more extensive treatment of sputtering theory and practice may be found in Ref 41.

The atomic flux generated by sputtering diffuses through the low pressure gaseous environment and is deposited on the substrate to produce sputtered coatings. Parts to be coated may be placed on the anode to receive thin films of sputtered target material.

Target-to-substrate distances of 50 to 300 mm are used in normal sputtering. Operating pressure ranges between 1.33 and 3.0 Pa at cathode current densities of 1 to 100 A/m^2. Coating rates obtained vary with material sputtered. At the conditions cited, coating rates of 0.1 nm to a few tens per minute are common.

Well-bonded coats are obtained by sputter cleaning the substrate before target material deposition (done by reversing polarity, with a shutter used to cover the coating target to prevent target contamination during coating) and then depositing the film. Sputter cleaning conditions have to be chosen to prevent substrate overheating during the cleaning phase.

Quality of thin films (morphology, structure, texture, transport, and mechanical properties) produced and film adhesion are sensitive to sputtering conditions. The film quality problems are examined in two recent publications.[42,43]

Non-conducting materials cannot be sputtered in D.C. mode. Most hard compounds and soft compounds needed for tribological applications fall in this category. Reactive D.C. sputtering is a way of overcoming this difficulty.

Alternately, radio-frequency (R.F.) power supplies may be used. The charge accumulation on dielectric surfaces (targets) is neutralized by reversing the field for portions of each cycle. Customarily 13.52 MHz power supplies are used as required by the F.C.C. Soft films for solid lubrication, MoS_2 films, are routinely R.F. sputtered.[42,44]

Hard and soft films produced with D.C. and R.F. sputtering are suitable for friction and wear applications. Gas occlusion (argon) can occur in films produced at higher sputtering pressures (3 to 5 Pa). This and the slow coating rates are the principal limitations of D.C. and R.F. sputtering. These difficulties are avoided by using magnetron sputtering. Higher sputtering rates (one to two orders of magnitude greater) and lower sputtering pressures (an order of magnitude lower) are used.

An electron trap is used in magnetron sputtering to greatly increase the ionization efficiency at low pressures. A magnetic field at right angles to the electric field of the sputtering system is used. Electrons from the cathode acquire a drift velocity in the $\overline{\mathbf{E}} \times \overline{\mathbf{B}}$ direction ($\overline{\mathbf{E}}$ is electric field vector; $\overline{\mathbf{B}}$ is magnetic induction). The virtual distance between the cathode and anode is increased to improve the ionization probability. Much larger cathode current densities and sputtering rates are obtained. Clarke's[45] work is a useful reference. Use of reactive gas mixtures permits high rate reactive sputtering of hard compounds and soft coats.[25]

Following initial work by Clarke,[46] reactive sputtering has been successfully used to protect age-hardening aluminum alloys against wear with titanium nitride coatings. Hard coats produced at room temperature are satisfactory for friction and wear applications.[25]

Summary

Many different processes are available to produce hard and soft coats. Coatings suitable for friction and wear applications can be and have been produced. Bond strength requirements are foremost in these applications and have been met with CVD, PVD and vacuum sputtering processes.

Bond strength and coating quality (structure, properties, etc.) are strongly dependent on coating processes and coating conditions. A considerable amount of work is now in progress in characterizing coatings produced with various processes under a variety of coating conditions.

Previous work in the area has led to large-scale commercial use of CVD-coated carbide inserts. CVD-coated metal working tools are in use in Japan and the United States. All of the above are wear-resistant hard coats. RF sputtered soft coats of MoS_2[44] and ion-plated (Pb) ball bearings[34] are now in use in high vacuum applications. A survey of successful tribological applications of soft and hard coats is presented in the following section to acquaint the reader with the range of opportunities now available.

Tribological Applications of Soft and Hard Coats

Though the thin film coating techniques for friction and wear control have only recently been developed, a number of successful applications have been shown. They include sophisticated applications (aerospace) and industrial applications (metal working and cutting tools). Some examples are cited in the following paragraphs.

Rolling Element Bearings. While conventionally lubricated bearings can be built for high vacuum service, they require protection to prevent lubricant evaporation and surface migration of liquid films. Similar problems, as well as lubricant oxidation and decomposition, are faced in high temperature applications. For high vacuum and high temperature bearing applications, the lubricant problems are overcome by using appropriate soft coats (Pb, Au, Ag, or MoS_2).[34,42-44]

High vacuum bearings have also been produced by hard coating the tribo-elements of rolling bearings.[32,47] AISI 52100 balls have been coated with thin layers of TiC using CVD. The graded coating-to-substrate interface produced by the CVD process gives a well-adherent hard coat that is able to withstand high Hertzian contact stresses. The CVD-coated balls require post-processing by way of heat treatment (harden and temper).

Substrate structure changes accompanying CVD are avoided by sputter coating. This process also avoids mechanical stresses generated by thermal gradients that occur during CVD coating. Studies carried out at NASA-Lewis[48] show that even with thin layers (200 to 300 nm layers sputter deposited on 440-C stainless substrates) of hard materials (TiC, Mo_2C, Mo_2B_5 and $MoSi_2$) considerable wear life improvements are obtained.

Sliding Bearings. Compliant foil bearings are used in some aerospace applications. They are typically gas bearings and are prone to sliding contact under transient operating conditions (start and stop cycles). Hard coating of foil surfaces is useful to limit sliding wear. Thin hard coats (100 to 125 nm in thickness) magnetron sputtered from titanium carbide targets have been used to improve the life of spin axis gas bearings.[49]

Coatings for Erosion Resistance. Severe erosion of compressor blades occurs when sand and dust particles are ingested by aircraft turbine engines. Hard CVD coats of titanium carbo-nitride have been used to improve the erosion resistance of PH 17-4 stainless steel blades.[50] Erosion life (in seconds per mil) of CVD-coated blades is reported to be some 70 to 80 times greater than that of blades plasma sprayed with tungsten carbide.

Cannon and gun tubes are usually hard chromium plated to provide wear and erosion resistance. Small-bore gun barrels, CVD-coated with tungsten, exhibit markedly lower erosion rates.[51] Typically, gun steels (AISI 4150) are coated with up to 100 μm of tungsten by CVD. Tantalum – 10% tungsten coatings have also been produced in 7.62 mm rifled bores by CVD.[52]

Titanium diboride coatings applied by CVD have been found to afford erosion protection for Kovar, 446 stainless steel, and tantalum. Similar diboride coatings appear feasible for erosion protection of mild steels and tool steels.[53]

Hard Coats for Wear Resistance in High Stress Sliding Contact. In metal machining, the chips produced traverse across the tool face under high normal stresses. The tool-chip interface experiences high temperatures. During the machining of medium carbon steels with WC-TiC-TaC-Co tools, interface temperatures to 1100°C have been measured. The tribological conditions experienced during metal machining are perhaps the most severe that can be expected in dry sliding. A variety of coating techniques have been developed to produce thin coats (4 to 8 μm) of carbides, nitrides, carbo-nitrides, oxy-carbides, and oxides on cemented carbide substrates (usually WC-Co compacts or WC-TiC-TaC-Co compacts) for wear protection.

Commercial coatings available at present include titanium carbide, titanium nitride, titanium carbo-nitride, hafnium carbide, hafnium nitride, and aluminum oxide. Coated tools yield varying levels of improvement. By far the best performance is obtained with alumina coated tools.

During CVD, the basic tungsten monocarbide sometimes tends to produce brittle eta phases at the coating-substrate interfaces. Bond strength of the coating is then lower and tools perform inconsistently. Problems of this type have been overcome by the use of interlayers. Consistent tool performance also requires reproducibly high stoichiometry in the coated thin layers.

While CVD coatings are suited for carbide substrates, the typical temperatures involved in CVD processes are unacceptable for the hard coating of high speed steel tools. Under these conditions R.F. coating is used without the risk of overtempering.[54] Steel tool bits have also been hard coated with activated reactive evaporation.[55,56] Ion plating of titanium followed by induction hardening in selected atmospheres yields hard coats. Hard coats produced in this manner are now in use to produce circular and band saws[57] used in meat processing.

Coating for Metal Working Tools. Friction and wear problems are severe in metal working. High unit normal loads and boundary lubrication conditions are common. Surface modification techniques can be beneficial. As noted earlier, CVD-coated metal working tools are now in commercial use.

CVD hard coats have been used in injection molding dies. Durability improvements of two orders of magnitude have been obtained with CVD coats of TiB_2 on A-6 and H-11 die steels. Sputtered coatings of TiB_2, 5 μm in thickness, have also been shown to improve the durability of injection molding dies.[58]

Patterns used to produce sand molds are prone to severe wear by silica or zirconia sands. Pattern shapes are usually complex. Chemical vapor deposition techniques may be used to hard coat metal patterns or to build them up to shape on selected preforms. Using nickel carbonyls, low temperature CVD coats of hard nickel are produced. Patterns for automotive crank shaft sand molds have been fabricated by using this CVD process.[59]

Most metal forming tools now in use are alloy steels of one type or another that contain strategic materials. Hard coating of metal working tools is useful not only to improve wear resistance but also to conserve strategic materials.

Coatings for Non-ferrous Materials. Most non-ferrous materials in their bare form are prone to abrasive wear, adhesive wear, scuffing, and galling. Metallic soft coats and solid lubricant coatings (MoS_2, WSe_2, etc.) are appropriate for friction and wear protection.

Ion platings of soft metals (gold, silver, cadmium) as well as coatings of chromium and zirconium are useful to protect against different types of wear.[60] Titanium alloys are boronized to eliminate wear-in problems and to reduce wear rate in the oxidative wear regime.[61] Ion plating with thin coats of lead is also useful.[34]

Hard coats on soft metallic substrates can provide excellent protection against wear. Because of the high temperatures needed, it is not possible to produce hard coats using CVD on many non-ferrous metals. Sputter coating and reactive evaporation are feasible. The use of sputtered TiC coatings to protect soft beryllium components has been cited.[49] Aluminum alloys, including age-hardening grades, have been hard coated with sputter deposition. It has been shown that titanium nitride films approximately 5 μm in thickness are sufficient to protect 6061 aluminum alloys.[25] The same process is useful for 70-30 brass, titanium, and magnesium alloys.

Closing Remarks

The justification for the use of soft and hard coats in friction and wear applications has a sound theoretical basis. Many laboratory scale applications of soft and hard coats have been demonstrated. Coatings of both types are also in industrial use. But the use is not as widespread as it could be. Coating durability and uncertainty regarding coating-to-substrate bond strength, i.e., reliability of coatings, are among the reasons for the slow growth in the use of thin coats in tribological applications.

Soft films possess a finite endurance life. In open tribological systems, the film may be deformed and displaced from locations where it is needed. Film loss may also occur by "wear," i.e., physically lost from the bearing environment. In both cases, periodic replenishment of the soft coating material is essential. This can be accomplished by using coated or impregnated cages in ball and roller bearings. Cages made of sintered materials containing soft coat compounds are an alternate solution. These remedies are responses to the finite endurance life of low shear strength films. In the case of wear-resistant coatings, the needed life is obtained with ease. Unacceptable wear life is remedied by a harder, more stable, or thicker coat. Film endurance life is rarely the problem.

With both hard and soft coats, improvement in tribological performance is possible only if the coating is reliably bonded to the substrate. To assure this, satisfactory coating-to-substrate bond strength tests are essential. A number of empirical test methods are used to assess the bond strength. They include the peel test, the scratch test, and so on. None of these are satisfactory quantitative indicators of the coating-to-substrate bond strength. The development of sound bond strength tests which truly measure the coating-to-substrate bond strength will go a long way in easing the acceptance of coatings in tribological applications. The thinness of the coatings, internal stresses and textures within them, and the wide range of physical and mechanical properties exhibited by the coatings suggest that it may not be possible to devise a single test method meeting the requirements of all coating materials and substrates.

Many coating processes are available to deposit a wide variety of hard and soft coats. Coating structure and properties, as would be expected, vary with processes and coating conditions. While some work is in progress in establishing structure-property relations, there is a need for a great deal of additional work.

In some cases, as for example when coatings are produced with atomic or molecular species with mean energies some 2 to 3 orders of magnitude larger than their thermal energy in the vapor state, non-equilibrium and amorphous structures are possible. Electrically assisted vacuum coating processes employ coating fluxes with high mean energies. It may then be possible to produce new structures with properties that are especially suited for tribological applications. Deposition of cubic boron nitride is an example. A comprehensive program of study of the structure and properties (including film residual structures) as a function of coating processes and conditions is essential to identify other similar opportunities. Fundamental studies of the coating-to-substrate interface and of the problems limiting the coating-to-substrate bond strength are also needed if thin coating technology is to make a significant impact in friction and wear applications.

In recent years, several other powerful surface modification techniques have been developed. They include the beam processes (laser, electron, and ion) and they hold great promise. These techniques must be considered together with the coating processes discussed in this chapter when surface processing is viewed as a possible means of overcoming friction and wear problems.

Acknowledgments

The authors thank Mrs. Edith Wilkerson for careful manuscript preparation. One of the authors (S. Ramalingam) acknowledges partial support of this work by the Department of Transport (DOT Contract No. DTRS 5680-C-00015, Mr. Russel Zub, Project Monitor, TSC, Cambridge, Mass.).

References

1. M. B. Peterson, "Technical Options for the Conservation of Metals," 1979, Congress of the United States, Office of Technology Assessment, Washington, D.C. 20510.
2. F. P. Bowden and D. Tabor, *Friction and Lubrication of Solids*, Oxford Press, London, 1950.
3. M. B. Peterson and F. F. Ling, "Friction and Lubrication at Extreme Pressures," Paper No. MF69-103, American Society of Tool and Manufacturing Engineers, Detroit, Michigan, 1969.
4. I. V. Kraghelskii, *Friction and Wear*, Butterworth, Washington, D.C., 1965.
5. L. C. Towle, "The Failure of Amonton's Laws in the Friction of Polymer Films Under High Pressures," ASLE International Conference on Solid Lubricants, 1971.
6. E. F. Finkin, "Mechanical Aspects of Solid Lubricant and Elastomeric Coatings," Science and Technology of Surface Coatings, Academic Press, New York, 1974.
7. E. Rabinowicz and J. T. Burwell, "The Nature of the Coefficient of Friction," J. Appl. Phys., *24*, 1953, 136.
8. F. B. Bowden and E. H. Freitag, "Friction of Solids at Very High Speeds," Proc. Roy. Soc., *A248*, No. 1254, 1958, 350.
9. E. F. Finkin, "A Theory for the Effects of Film Thickness and Normal Load in the Friction of Thin Films," J. Lub. Tech., *91*, 1969, 551.
10. M. J. Devine, E. R. Lamson and J. H. Bowen, "The Effect of the Chemical Composition of Metals in Solid Lubrication," Proc. USAF Conference on Aerospace Fluids and Lubricants, 1963, 317.
11. Yuko Tsuya, "Lubrication with Molybdenum Disulfide Film under Various Conditions," ASLE Proceedings — International Conference on Solid Lubricants, 1971, p. 39.
12. P. Heilmann and D. A. Rigney, "Sliding Friction of Metals," Proc. Leeds-Lyon Conference, Sept. 1980.
13. J. F. Archard, "Contact and Rubbing of Flat Surfaces," J. Appl. Phys., *24*, 1953, 981.
14. C. N. Rowe, "Some Aspects of Heat of Absorption in the Function of a Boundary Lubricant," Trans. Am. Soc. Lub. Engrs., *9*, 1966, 100.
15. T. L. Oberle, "Properties Influencing Wear of Metals," J. Metals, *3*, 1951, 438.
16. M. M. Khrushchov, Wear, *28*, 1974, 69.
17. R. C. D. Richardson, "The Wear of Metals by Relatively Soft Abrasives," Wear, *11*, 1968, 245.
18. J. Halling, ed., *Principles of Tribology,* Macmillan and Co., London, 1975.
19. J. J. Gilman, *Micromechanics of Flow in Solids,* McGraw-Hill, New York, 1975. (See also J. J. Gilman and B. W. Roberts, Elastic Constants of TiC and TiB_2, J. Appl. Phys., *32*, 1961, 1405.)
20. K. L. Yip, A. B. Kunz and W. S. Williams, "Electronic Structure of TiC and VC from Iterative Extended Hucker Theory," Phys. Stat. Sol., *75*, 1976, 533. (See also J. J. Gilman, "Hardness of Carbides and Other Refractory Hard Metals," J. Appl. Phys., *41*, 1970, 1664.)
21. E. Rabinowicz, *Friction and Wear of Materials*, Wiley, New York, 1965.
22. D. Scott, Wear, *48*, 1978, 283.
23. K. Nakamura et al., "Application of Wear-Resistant Thick Films Formed by Physical Vapor Deposition Processes," Thin Solid Films, *40*, 1977, 155.
24. J. Halling, "The Role of Surface Topography in the Friction of Soft Metallic Films," Tribology 1978, Inst. of Mech. Eng. London, Pub. No. 1978-6, 1978, 131a. (See also Ref 9.)

25. S. Bair, S. Ramalingam and W. O. Winer, "Tribological Experience with Hard Coats on Soft Metallic Substrates," Wear, *60*, No. 2, 1980, 413-419.
26. D. A. Rigney and J. P. Hirth, "Plastic Deformation and Sliding Friction of Metals," Wear, *53*, 1979, 345. (See also Ref 12.)
27. H. M. Fitch, U.S. Patent No. 2,984,575, 1961.
28. H. M. Fitch, U.S. Patent No. 3,262,790, 1966.
29. R. L. Hough, "Chemical Vapor Deposition of Metal Oxides from Organometallics," Proc. 3rd Int. Conf. Chemical Vapor Deposition, American Nuclear Society, Hinsdale, Illinois, 1972, 232-241.
30. N. G. Anantha, V. Y. Doo, D. K. Seto and G. Pecenco, "Chromium Deposition from Dicumene-Chromium to Form Metal-Semiconductor Devices," Proc. 2nd Int. Conf. Chemical Vapor Deposition, The Electrochemical Society, New York, 1970, 649-654.
31. J. C. Withers and L. C. McCandless, "Aluminum Coatings by Pyrolytic Spray™ Chemical Vapor Deposition Process," Proc. 2nd Int. Conf. Chemical Vapor Deposition, The Electrochemical Society, New York, 1970, 393-398.
32. H. Gass and H. E. Hintermann, "Chemical Vapor Deposited TiC for Bearings," Proc. 4th Int. Conf. Chemical Vapor Deposition, The Electrochemical Society, Princeton, New Jersey, 1973, 563-576.
33. D. M. Mattox, "Fundamentals of Ion Plating," J. Vac. Sci. Technol., *10*, 1973, 47-52.
34. M. J. Todd and R. H. Bentall, "Lead Film Lubrication in Vacuum," Proc. 2nd Int. Conf. on Solid Lubrication, ASLE, Park Ridge, Illinois, 1978, 148-157.
35. M. Okada, M. Fukutomi, M. Kitajima and R. Watanabe, "Silicon Nitride Coatings on Molybdenum by RF Reactive Ion Plating," Proc. 6th Int. Conf. Chemical Vapor Deposition, The Electrochemical Society, Princeton, New Jersey, 1977, 235-245.
36. R. F. Bunshah and A. C. Raghuram, "Activated Reactive Evaporation Process for High Rate Deposition of Compounds," J. Vac. Sci. Technol., *9*(6), 1972, 1385-1388.
37. A. C. Raghuram and R. F. Bunshah, "The Effect of Substrate Temperature on the Structure of Titanium Carbide Deposited by Activated Reactive Evaporation," J. Vac. Sci. Technol., *9*(6), 1972, 1389-1394.
38. S. Ramalingam, in *Wear Control Handbook*, ed. M. B. Peterson and W. O. Winer, ASME, New York, 1980.
39. G. Carter and J. S. Colligon, *Ion Bombardment of Solids,* American Elsevier, New York, 1968.
40. E. Ritter, *Vacuum Metallurgy*, ed. O. Winkler and R. Bakish, Elsevier Publishing Co., Amsterdam, 1971, 803-818.
41. J. L. Vossen and W. Kern, eds., *Thin Film Processes*, Academic Press, New York, 1978.
42. T. Spalvins, "Sputtering Technology in Solid Film Lubrication," Proc. Int. Conf. on Solid Lubrication, ASLE, Park Ridge, Illinois, 1978, 109-117.
43. M. N. Gardos, "Quality Control of Sputtered MoS_2 Films," Lubrication Engineering, *32*, 1976, 463-480.
44. R. L. Christy and G. C. Barnet, "Development of a Sputtered MoS_2 Lubrication System for Spacecraft Gimbal Bearings," Lubrication Engineering, *34*, 1978, 437-443.
45. P. J. Clarke, "The S-Gun, a Direct Replacement for the Electron Beam Evaporator," Solid State Technology, *19*, 1976, 77-79.
46. P. J. Clarke, "Magnetron dc Reactive Sputtering of Titanium Nitride and Indium-tin Oxide," J. Vac. Sci. Technol., *14*, 1977, 141-142.
47. H. E. Hintermann, H. Vobing and N. Hanni, "Wear-Resistance Coatings for Bearing Applications," Wear, *48*, 1978, 226-236.

48. W. A. Brainard, "The Friction and Wear Properties of Sputtered Hard Compounds," Proc. 2nd Int. Conf. on Solid Lubrication, ASLE, Park Ridge, Illinois, 1978, 139-147.
49. R. P. Riegert, "Sputter Coatings for Gas Bearing Applications," AFML Technical Report No. AFML-TR-78-192, Air Force Materials Laboratory, WPAFB, Ohio, 1978.
50. C. L. Yaws and G. F. Wakefield, "Scale-up Process for Erosion Resistant Titanium Carbonitride," Proc. 4th Int. Conf. Chemical Vapor Deposition, The Electrochemical Society, Princeton, New Jersey, 1973, 577-588.
51. K. H. Meyer, "Chemical Vapor Deposited Tungsten as a Means to Reduce High Temperature Erosion in Gun Barrel Bore," Proc. 3rd Int. Conf. Chemical Vapor Deposition, American Nuclear Society, Hinsdale, Illinois, 1972, 625-640.
52. F. A. Galski and A. Crowson, "Tantalum Alloy Chemical Plating of Gun Barrels," Proc. 6th Int. Conf. Chemical Vapor Deposition, The Electrochemical Society, Princeton, New Jersey, 1977, 542-554.
53. H. O. Pierson and E. Randich, "The Coating of Metals with Titanium Diboride by Chemical Vapor Deposition," Proc. 6th Int. Conf. Chemical Vapor Deposition, The Electrochemical Society, Princeton, New Jersey, 1977, 304-317.
54. K. Y. Su and N. H. Cook, "Enhancement of High Speed Tool Life by Titanium Nitride Sputter Coating," Proc. 6th North American Metal Working Research Conf., SME, Detroit, 1977, 297-303.
55. R. F. Bunshah, A. H. Shahbaik, R. Nimmagadda and J. Covy, "Machining Studies of Coated High Speed Steel Tools," Thin Solid Films, *45*, 1977, 453-462.
56. K. Nakamura, K. Inagawa, K. Tsuruoka and S. Komiya, "Applications of Wear-Resistant Thick Films Formed by Physical Vapor Deposition Processes," Thin Solid Films, *40*, 1977, 155-167.
57. N. N. Engels, U.S. Patent No. 3,915,757, 1977.
58. R. S. Plumb and W. A. Glaeser, "Wear Prevention in Molds Used to Mold Boron-filled Elastomers," *Wear of Materials 1977*, ASME, New York, 1977, 337-343.
59. G. S. Vasilash, "Using Vapor to Make Nickel Foundry Patterns," Manufacturing Engineering, *84*, 1980, 25.
60. N. Ohmae, T. Nakai and T. Tsukizoe, "Ion-plated Thin Films for Anti-wear Applications," *Wear of Materials 1977*, ASME, New York, 1977, 350-357.
61. T. S. Eyre and H. Al Salim, "The Effect of Boronizing on Adhesive Wear of Titanium Alloys," *Wear of Materials 1977*, ASME, New York, 1977, 344-349.
62. M. B. Peterson and F. F. Ling, "Frictional Behavior in Metalworking Processes," J. Lubrication Technol. (ASME), *92*, 1970, 535.

Wear of Steels

Olof Vingsbo and Sture Hogmark
Uppsala University Institute of Technology
Uppsala, Sweden

Abstract

Wear resistance is discussed in terms of mechanical properties and internal structure for different classes of materials, with special emphasis on steels. A simplified treatment of wear mechanisms, based on fundamental tribo-fracture processes, is presented and applied to the case of wear of steels. The topography of worn steel surfaces is studied with respect to operating wear mechanisms with the aid of scanning electron microscopy. The formation of friction layers on worn steel surfaces is described in some detail with the aid of structure images and microhardness profiles for a number of different types of steel alloys, and the influence of the developed layers on wear properties is discussed. Five groups of steel alloys of high commercial and scientific interest are treated: austenitic stainless steels, structural steels, tool steels, abrasion resistant steels, and ball bearing steels.

1. Introduction

When studying the wear properties of a special class of materials like polymers or metals or, as in the present case, steel, it is essential to bear in mind the complexity of a general tribological system. A resulting lack of stringency — or even knowledge — implies that in tribology practically any statement can be verified or disproved with the aid of experiments, if only a suitable choice of parameters and variables is made. However, comparisons between different materials in similar or different tribological situations eventually focus on ranking with respect to wear resistance, or at least on distinguishing different wear behavior. The present survey of the particular case of steel will be based on a background of fundamental wear phenomena, described in a systematic way with the aid of a few basic tribological concepts. Some of these are already incorporated in textbook terminology, and some have recently been suggested by the Uppsala team.[1] A short review of these fundamentals is given below.

2. Background Fundamentals

2.1. Tribo-Systems and the Definition of Wear

A *tribo-system* is characterized by surfaces in mechanical contact under relative motion. Depending on the nature of the acting applied load components, a *tribo-surface* is subjected to gliding or rolling contact (wandering loads), or impact [transient (un)loading]. Wear can now be defined as *the removal of material from tribo-surfaces.*

The general tribo-system is affected by a multitude of parameters, the majority of which may be difficult to control, to measure or even to identify. It is practical to divide these parameters into three groups: design, material and environment. Examples of principal characteristics are given in Table 1.

Table 1. Tribo-System Parameter Groups

Design
Shape of the tribo-surfaces
Area of nominal contact
Mode of mechanical contact
Spectrum of applied forces
Spectrum of relative velocities of contacting surfaces
Material
Material properties: mechanical, chemical, physical, etc.
State of phase: gas, liquid, solid
Material combinations
Friction and wear induced alterations
Environment
Chemical properties: chemical potential, aggressiveness, etc.
Physical properties: temperature, pressure, etc.
Mechanical properties: friction coefficient, viscosity, etc.

2.2. Wear Mechanisms

Wear mechanism is the collective name for the micro-events by which wear takes place. The basic mechanism of material removal from a surface is *fracture*.

In principle, there are two general types of fracture, depending on the dominance of shear or normal force components with respect to the future fracture surface. In the case of tribo-surfaces, it is possible to distinguish four modes of *tribo-fracture* as described in Table 2.

In practice, cooperation of two or more of the tribo-fracture modes is common. In addition, tribo-fracture is generally associated with *interactive phenomena* such as plastic deformation, diffusion or corrosion. The effect of the interactive phenomena on the tribo-system and on the resulting wear is often decisive.

Table 2. Tribo-Fracture Mechanisms

Force component	General fracture type	Tribo-fracture mode
Shear	Shear, ductile	Shearing of adhesive junctions Microchipping
Normal	Cleavage, brittle	Impact cracking Fatigue cracking

Considering the great variety of possible parameters, the number of combinations of tribo-fracture mechanisms and interactive phenomena is practically infinite. This is reflected by the rich flora of wear mechanisms defined in current literature. A general trend in this development is that authors try to define more and more complex and sophisticated wear mechanisms, which are, in fact, characteristic only of the particular tribo-system used in the reported testing. In comparison to this, the present approach focuses on a small number of fundamental tribo-typical mechanisms together with their interactive phenomena. The definition of complex, specific mechanisms is no longer necessary, but can be replaced by a relevant description of the tribo-system in question (whether a service case or a laboratory testing machine). Irrespective of the approach chosen, the aim of research on mechanisms is to contribute to organizing wear science in a systematic or hierarchic structure, although today most investigators agree that it will never be possible to work out a unified "theory of wear," or to find "the equation of state" for wear.

Valuable information on wear mechanisms is often easily available with the aid of scanning electron microscope (SEM) images of the resulting surface topography. In Fig. 1 a few micrographs are reproduced, showing representative examples of the four suggested tribo-fracture modes.

2.3. Wear—a Surface Phenomenon

Wear is essentially a surface phenomenon. By definition, the tribo-fracture mechanisms refer to micro-events in a thin surface layer, resulting in a slow, gradual destruction, whereas macro-fracture gives rise to sudden catastrophic failure.

In a tribo-system, frictional energy is injected through the tribo-surfaces. Most of the energy is dissipated as heat in the bulk. The volume of the work piece is generally many orders of magnitude greater than that of only the superficial layer, and in steady state a steep, well-defined temperature gradient develops from the surface. (Dynamic temperature measurements in tribo-systems represent a major physical and technical problem, which is being continually discussed. It will not be treated here.) Some of the energy is injected into a thin superficial layer, resulting in plastic deformation, phase transformations, diffusion, material losses

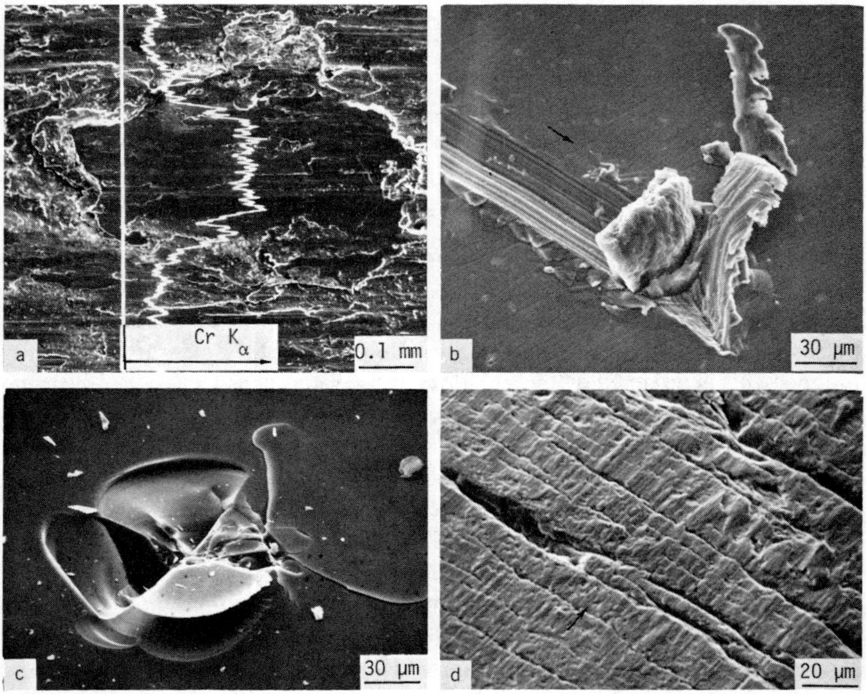

(a) Shear fracture of an adhesive Cr tool steel–carbon steel joint. Cr K_α scan along the vertical line proves that Cr steel has been transferred to the carbon steel. **(b)** Microchip formation during cutting by hard particle in copper surface. **(c)** Impact cracking in glass surface. **(d)** Fatigue cracking in the surface of a punching tool.

Fig. 1. SEM images of the surface topography after tribo-fracture.

by tribo-fracture, and a strain gradient from the surface. As a result, the topography, microstructure and mechanical and physical properties generally change during a short, initial (running-in) period, until, at steady state, a surface layer is developed. The properties of this layer may deviate drastically from those of the bulk material. It is, however, the *actual properties* of the *active layer* that determine the wear response of the material, rather than the *nominal properties* of the bulk. An instructive example is the indentation hardness, which is often considered a decisive parameter for abrasion resistance. Often bulk hardness values measured at room temperature are used, although the relevant parameter would be the microhardness measured at a representative temperature.

Thus, the main interest in wear research should be devoted to the active layer. The bulk material of a tribo-system only serves to (*a*) support and maintain the surface in a given functional shape, (*b*) transfer load, and (*c*) conduct heat.

Unfortunately, dynamic studies of the active layer are generally impossible, for the following reasons:

Wear phenomena occur on a microscale
The mating tribo-surfaces are screening each other
A measuring probe would critically affect the local wear process to be measured
Structure studies often require destructive specimen preparation.

Therefore, most investigations of the surface layer must be conducted *after* the test has been interrupted, and thus refer to a "passivated" or *resulting layer*, the structure of which does not in general coincide with that of the "active" layer. Knowledge of the actual parameters of the active layer can then be achieved only by deduction from the resulting layer, based on knowledge of the operating tribo-system. From an engineering point of view it may even be desirable to make a second deduction, in order to relate the active layer parameters to the nominal bulk material parameters *before* testing, the "catalogue" parameters of the material produced. Thus, when characterizing the wear response of tribo-surfaces with respect to, e.g., material properties, the interrelationships visualized in the block diagram of Fig. 2 have to be considered.

The above considerations of surface layers refer to layers *modified* during wear. A similar approach may be followed also concerning *applied* layers, i.e., surface coatings other than lubrication films, applied to a tribo-surface before service or testing, with the purpose of improving the tribological performance. A dramatic break-through in physical and chemical vapor deposition (PVD and CVD) techniques has taken place during the last few years, and today the field of applied surface coatings seems to be one of the most promising sectors in the

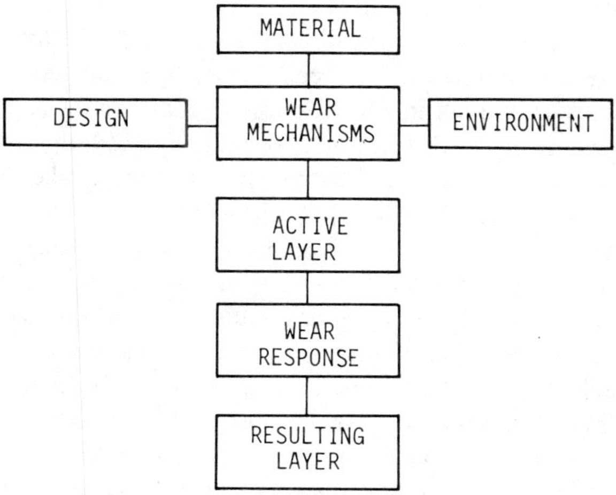

Fig. 2. Block diagram of tribo-system.

development of wear resistant materials. The effect of applied layers is treated separately at this conference, and will not be considered here.

3. Wear Resistance and Mechanical Properties

3.1. General

In general, fracture is primarily controlled by the mechanical properties of a material for a given environment. With respect to wear, the present choice of material loss and tribo-fracture as fundamental concepts is based on the dominance of the mechanical properties of the active layer over a variety of other parameters affecting a tribo-system. In order to compare different classes of material, the principal mechanical properties can be related to deformation or to fracture according to Table 3.

Table 3. Relationship of Mechanical Properties of Solids to Deformation or Fracture

Deformation	Fracture
Yield stress, σ_y	Fracture stress, σ_f
Young's modulus, Y	Fracture elongation, ϵ_f
Hardness	Cohesive energy
Toughness	Brittleness
Ductility	

It is, in principle, possible to relate the basic parameters σ_y, σ_f and ϵ_f to Young's modulus, if only relative values within each class of material are considered. Similarly, other deformation characteristics, like toughness, brittleness, etc., as well as material classes, can be roughly compared. The diagram of Fig. 3 represents an attempt to visualize this type of relationship. The diagram refers to room temperature and "normal" strain rate conditions, under which the zone of lowest Y corresponds to ideally plastic materials, like wet clays, with very low σ_y and an extremely high ϵ_f. The fracture stress is poorly defined, but approaches the yield stress when work hardening vanishes. The fact that $\sigma_f > \sigma_y$ implies an area of plasticity in this region of the diagram. The next zone is characterized by rubber-elastic properties, with low Y in combination with high ϵ_f and no plastic region. Typical materials in this region are the elastomers. With increasing modulus there is an elastic/plastic transition into an interval of increasing toughness, and with a plastic region but with rather low ϵ_f. These are typical metal properties, but also many polymers belong to this group. A region of brittleness is reached via a plastic/elastic transition which is characterized by high Y and σ_f and vanishing ϵ_f. Typical representatives are most ceramic materials. The steels constitute a special group within the metals, extending on both sides of the tough/brittle, plastic/elastic transition line. This is one of the reasons for the extremely wide field of applications of steels, from mild steels to martensitic steels.

Physically, "toughness" and "hardness" as well as "strength" are complex concepts, and a material may be better described by its ability to withstand plastic deformation and fracture. For this purpose σ_y, σ_f and ϵ_f are essential parameters, although they generally refer to the special case of uniaxial tensile testing. Basically, wear resistance is determined by the actual combination of parameters with respect to resistance to tribo-fracture under the influence of the interactive phenomena of the tribo-system. Physical models in terms of dislocation theory or fracture mechanics are well established, and can be more or less directly applied to wear to describe tribo-fracture as well as relevant interactive phenomena. It is important to bear in mind, however, that the nominal mechanical properties, as visualized in Fig. 3, are modified in the active layer of a tribo-system. An increase in temperature due to friction and deformation will generally lower the σ_y and σ_f curves, while the ϵ_f curve will rise, at the same time that the elastic/plastic transition lines will move to the right. This corresponds to moving to the left in the diagram. An increased deformation rate will have the opposite effects.

3.2. Application to Steels

Historically, the iron alloys have had a unique position as structural materials within a wide range of applications, from the simple binary Fe-C alloys of the

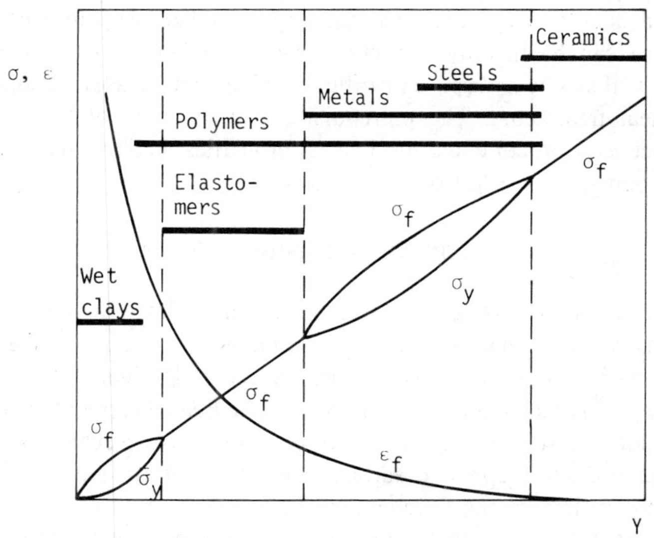

Fig. 3. Ranking of mechanical properties for various classes of materials. σ_y = yield stress; σ_f = fracture stress; Y = Young's modulus.

Iron Age to sophisticated stainless steels or powder-technology high-speed steels of the industrialized society of the 1980's. Man's knowledge of steel making has improved from the Viking swordsmith's trial-and-error methods to today's detailed structure tailoring via complex alloying, heat treatment and forming. Explicit development of wear resistant steels, however, has only been conducted during recent years. Instead, the wear properties used to be an inherent—and generally not considered—consequence of the mechanical properties in a wider sense. Yet wear resistance was often successfully achieved merely by choosing among existing steels with the aid of empirical rules of thumb.

A unique property of steel, compared with many other materials, is the wide range within which toughness and hardness can be matched. As schematically described in Fig. 3, this corresponds to the favorable position in the high-σ_y/high-σ_f range, covering both tough and brittle regions.

The actual performance of a steel is ultimately controlled by its *internal structure*. The structure, in turn, is mainly a result of the choice of chemical composition, heat treatment, and forming. Examination of the binary Fe-C phase diagram indicates the possibilities of structural variations. In addition to hardening by martensitic transformation, the mechanical properties of steel are due to the strong interaction between dislocations and

Interstitial elements (carbon or nitrogen)
Substitutional elements (metals other than Fe)
Precipitates (hard particles like carbides).

Considering the rich supply of more complicated alloy steel phase diagrams which are available today, it is clear that very specialized requirements for wear performance can be met by "structure engineering" (mainly controlling the content as well as size and space distribution of separate phases via alloying and relevant heat treatment). The prerequisite is knowledge of the appropriate relations between structure and mechanical properties, and this has been a field of rapid progress for the last two decades.

4. Wear Mechanisms in Steels

In Table 4 the tribo-fracture mechanisms are related to the most basic conventional wear mechanisms in metals commonly referred to in the literature today. Corrosive wear, diffusive wear, etc., are dominated by interactive phenomena, as pointed out earlier, and will not be treated explicitly here. Shearing of junctions and microchipping may be competitive mechanisms in tribo-systems. In addition, repetitive surface contact, as in erosion, can cause the superposition of impact and fatigue cracking.

It is important to realize that the actual combination of operating wear mechanisms depends not only on the material but on the total tribo-system, and consequently there are no wear mechanisms specific for a particular material.

Table 4. Tribo-Fracture Mechanisms and Corresponding Conventional Wear Mechanisms

Tribo-fracture mechanisms	Wear mechanisms
Shearing of junctions	Adhesive wear, seizure
Microchipping	Abrasive wear, erosion
Impact cracking	Impact wear, erosion
Fatigue cracking	Erosion, contact fatigue

Thus, a systematic characterization of materials, e.g., different steels, with respect to general wear resistance is virtually impossible. Instead, a limited number of instructive, carefully chosen applications will be discussed below, with the aim of covering a reasonably wide range of steels in representative tribo-systems of scientific and commercial interest. Examples are tool steels for cutting applications, abrasion resistant steels for mining, ball bearing steels, etc. Trends and tendencies will be emphasized at the expense of handbook-type details of all steels involved.

4.1. Abrasive Wear

The abrasive wear mechanism is usually defined as the plowing action of a hard particle or asperity sliding on a softer surface. The effect is demonstrated on a microscale in Fig. 1(b). Comparison with macrogrooving, as shown in Fig. 4, indicates that there is essentially no size effect, with respect either to chipping or to plastic deformation and crest formation. Depending on cutting geometry, abrasive chipping can also strongly remind one of the action of a tool edge in bulk cutting operations. The groove in Fig. 4 was cut in a plain carbon steel during a laboratory experiment with a pyramidal (negative rake angle) cemented carbide tip. Comparison of Fig. 1(b) and 4 shows that, because of the negative rake angle, metal is extruded by shear deformation in front of the abrasive tip in both cases.

4.1.1. Metal Cutting Tools. The abrasive mechanism is largely responsible for tool wear in metal cutting when the work material contains hard phase particles or has a hard surface layer. As an example, the edge region of a high-speed steel (HSS) twist drill is shown in Fig. 5. The work material was a tough, hardened, martensitic steel, in which the carbide precipitates have grooved the flank as well as the rake face of the drill. The direct action of an abrasive element is shown in the "snapshot" of a hack saw tooth in Fig. 6, where an oxide particle has been caught in its grooving operation in the developing flank of the edge. Similarly, single carbide particles from the tool steel can abrade the tool surface when they are torn out and dragged along by the mating work piece (see Fig. 7).

4.1.2. Other Applications. Abrasive wear is, by far, the dominating surface-deteriorating mechanism in a variety of tools and machine parts used in mines and

Fig. 4. Quick-stop grooving experiment with negative rake angle. Plain carbon steel. SEM.

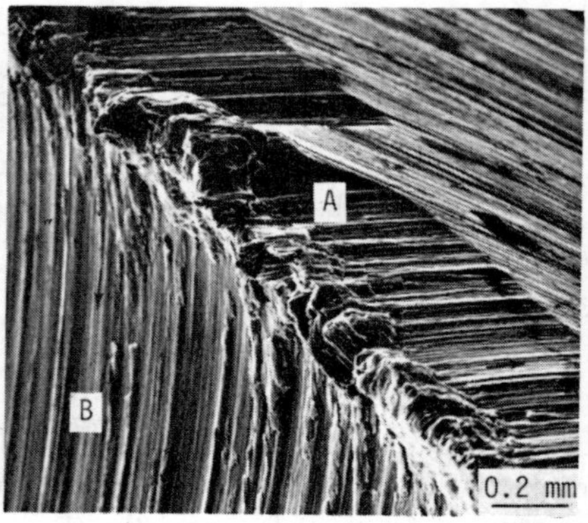

Fig. 5. Abrasive grooving of flank (A) and rake face (B) of high-speed steel drill. SEM.

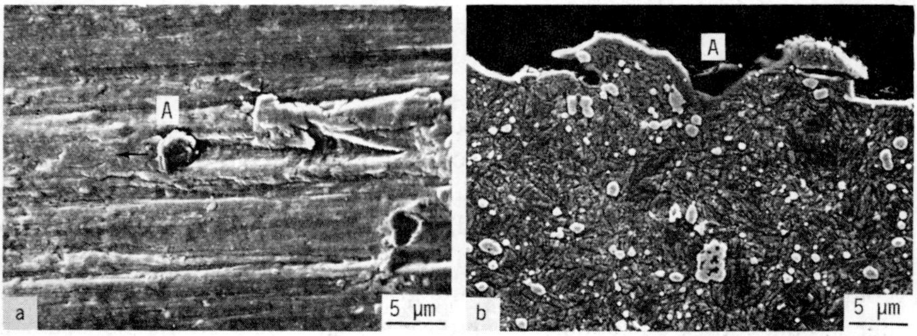

Fig. 6. (a) Oxide particle (A) acting as abrasive element in the flank surface of a high-speed steel hack saw tooth. SEM. **(b)** The situation of Fig. 6(a) viewed in transverse section. The particle (A) has cut a groove with high crests on both sides. SEM.

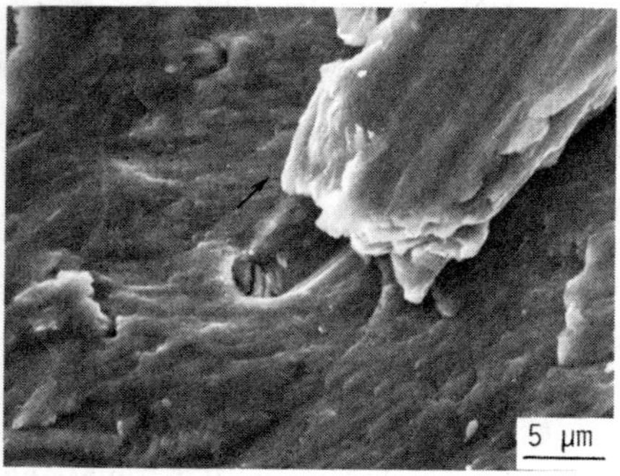

Fig. 7. Carbide particle torn out from the surface layer of a tool steel and acting as an abrasive element. SEM.

quarries. One example is given in Fig. 8, which shows the abraded surface of a tool steel excavator tooth after operating in sandstone (Fig. 8a) and in a granite quarry (Fig. 8b). Very similar effects, independent of groove size, can be found after sandblasting at low angles of incidence (see Fig. 9). This is a method of deliberate surface treatment by abrasion, another example of which is grinding. Figure 10 shows the surface topography resulting from the cumulative micro-grooving of a chromium steel during wet grinding.

4.1.3. Resistance to Abrasive Wear. In most abrasive applications there is a correlation between abrasion resistance and indentation hardness,[1] as demonstrated in Fig. 11. This is, in fact, one of very few rules of reasonably general validity in wear science, and it has been applied for decades. Recently, however, it has been suggested that information about resistance to abrasive wear can also be obtained by measuring the specific grooving energy, i.e., the grooving energy consumption per unit material loss.[2-4] Figure 12 shows a plot of specific grooving energy versus groove size for laboratory single-pass grooving experiments in some commercial steels.[4] This type of diagram provides the possibility of ranking different materials directly and with respect to different groove size intervals. The obtained ranking may differ from one groove size interval to another, thereby simulating wear cases of different severity. Agreement with results from service studies has been found by the authors in an investigation not yet completed.

A steel alloy can often be "reinforced" against abrasion by its content of hard phase particles (generally metal carbides) by the following mechanism. If the abrasive elements are smaller than the particles of the steel, a selective wear of matrix material occurs. Figure 13 shows the edge region of a steel punch used for

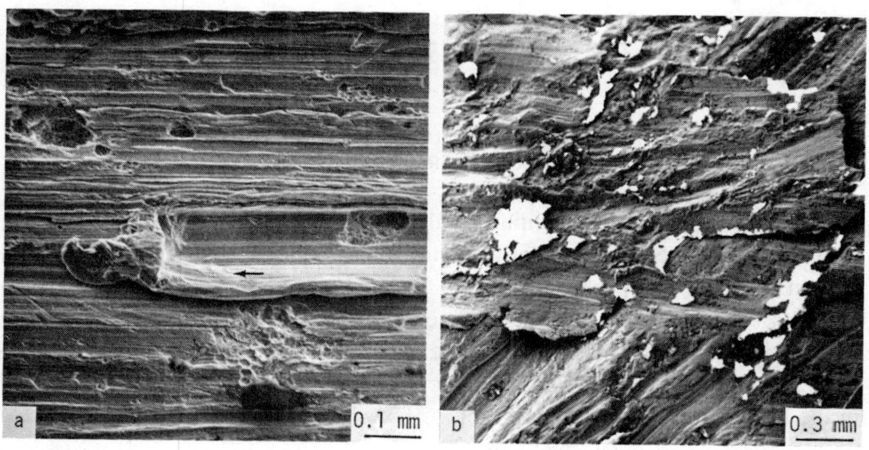

Fig. 8. Abraded surface of excavator tooth after operation in **(a)** sandstone, **(b)** granite. SEM.

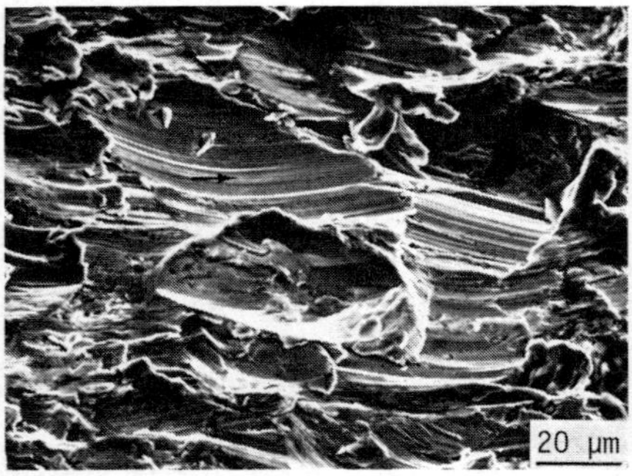

Fig. 9. Austenitic stainless steel surface after sandblasting at low angle of impingement. SEM.

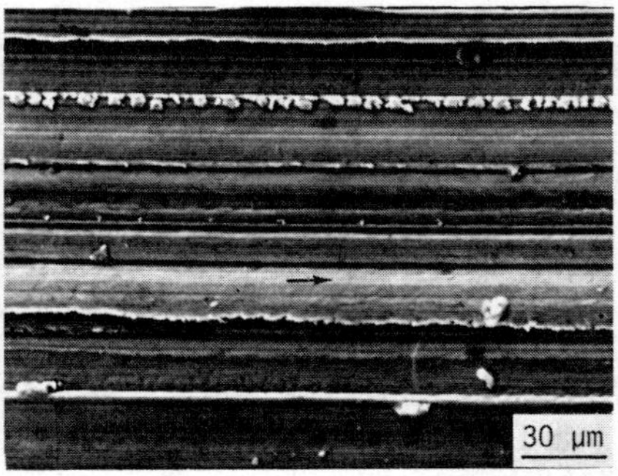

Fig. 10. Chromium tool steel surface after wet grinding. SEM.

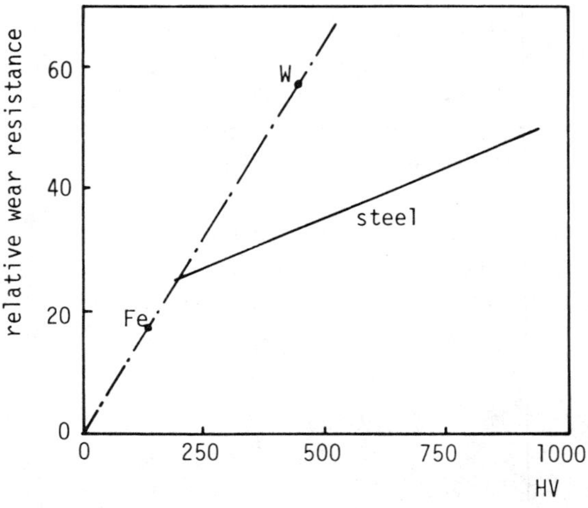

Fig. 11. Abrasion resistance versus indentation (Vickers) hardness according to Khrushchov and Babichev. *(Ref 1)*

cutting a work material containing small SiO_2 particles. The preferential matrix abrasion leaves steel carbides protruding from the tool surface. Matrix material is then screened and protected against further abrasion on the "lee" side of the carbides. The resulting ridges, each emerging from a protrusion, constitute so-called snail's eyes. In the particular case of SiO_2 abrasive particles, which are harder than the carbides, even the latter are slowly abraded, which can be seen in the high magnification micrograph of Fig. 14.

4.2. Adhesive Wear

4.2.1. Asperity Junctions. The most basic type of adhesive junction is made up of interacting asperities, as described in the literature. The phenomenon is

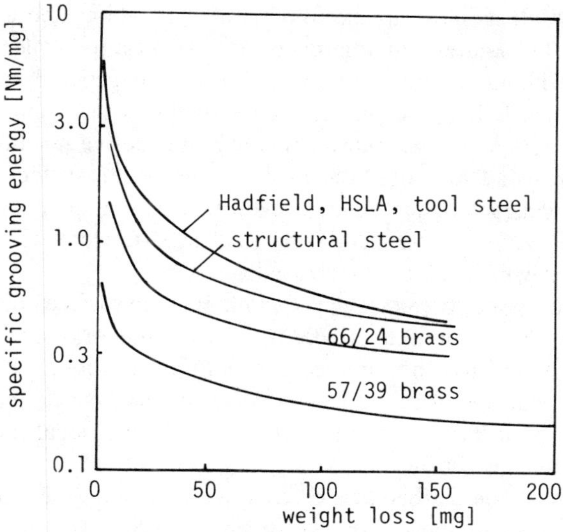

Fig. 12. Experimental plot of specific grooving energy versus groove size for laboratory testing of a number of alloys. *(Ref 4)*

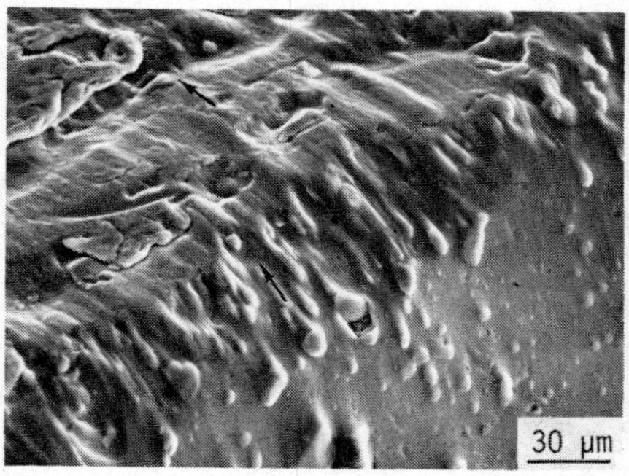

Fig. 13. Formation of "snail's eyes" by selective matrix wear of steel punching tool edge. SEM.

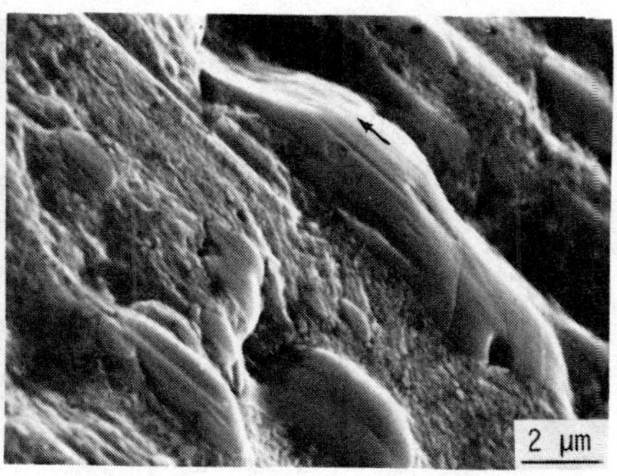

Fig. 14. Carbide particles, abraded by small SiO_2 particles. SEM.

typical of metal-to-metal contact, established under high local shear strain. Contamination layers break up, and direct metallic contact is formed, whereby former external surfaces are transformed to interphase or grain boundaries.

The total cross-sectional area of the developed junctions—the real contact area—is far less than the nominal contact area, and during sliding the junctions act as stress concentrators until fracture occurs. The material response is determined by the mechanical properties of the metal in the small volumes under momentary deformation. Referring to Fig. 3, the σ_y/σ_f interrelation is the critical condition, and as far as wear resistance is concerned, the favorable region is in the tough interval, as close to the tough/brittle transition line as possible. In developing steel structures with improved adhesive wear resistance, the ambition is to raise σ_y without passing the transition line. This generally implies that σ_f must also be raised. However, a common situation is that, under the extreme conditions of stress and temperature prevailing in adhesive junctions, the actual properties correspond to a position displaced to the left in the diagram, i.e., towards ductile characteristics. Also, steels which are nominally brittle may well be in the tough zone. Consequently the final tribo-fracture will be of ductile character, and it will involve an increased ϵ_f.

By definition, a wear prerequisite is that the fracture surface does not coincide with the initial adhesive joint interface. If it does, however, the locally absorbed frictional energy still has the same components (mainly heat, deformation and fracture work), but the process does not contribute to wear defined as material transport. This is, in fact, a common situation, which is ideal for high-friction applications (clutches, brakes, etc.).

4.2.2. Prows. The probability of asperity collision (the "collision cross section") increases with the size of a considered asperity. Adhesive wear implies the growth of one of the partners, and a resulting increase in the chance of a future collision, junction formation, etc. If a growing asperity in this way passes a critical size, it will come in more or less continuous sliding contact with the mating surface, and a so-called prow is formed.

Adhesion, rotation of the contact interface, and successive shear are the consecutive three steps by which prow formation occurs[5] (see Fig. 15). Depending on tribo-system parameters, the result will be either localized patches of transferred material (discontinuous layer), Fig. 16(a), or a continuous built-up layer (BUL), Fig. 16(b). The fresh fracture surfaces of prows and layers show a characteristic scale-like topography, which is indicative of a shear fracture mechanism (see Fig. 17).

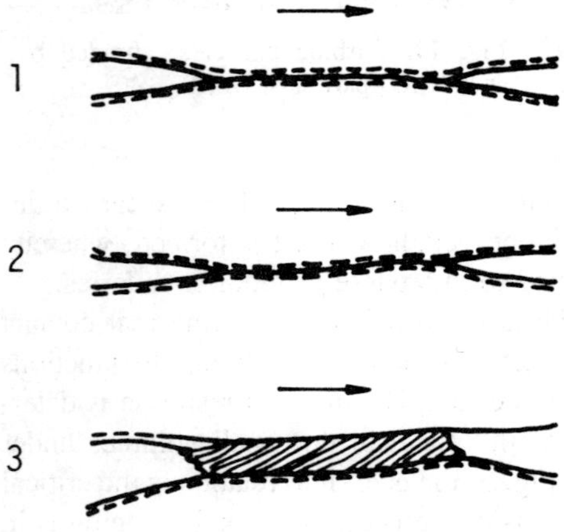

Fig. 15. Three stages of prow formation, according to Cocks. *(Ref 5)*

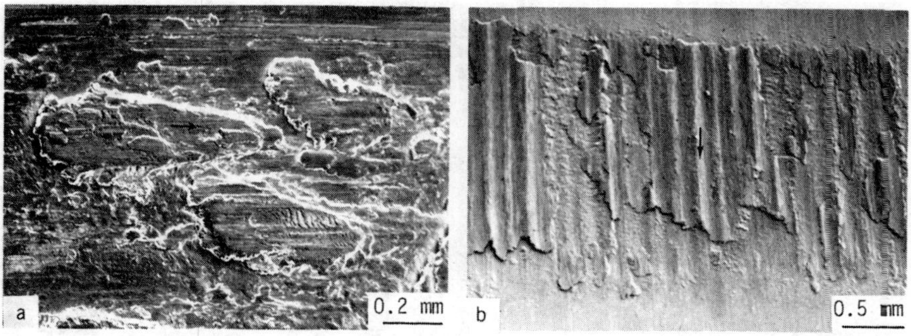

Fig. 16. Material transfer by adhesive wear in pin-on-ring tests: **(a)** discontinuous tool steel layer on tool steel ring; **(b)** continuous stainless steel layer on high-speed steel pin. SEM.

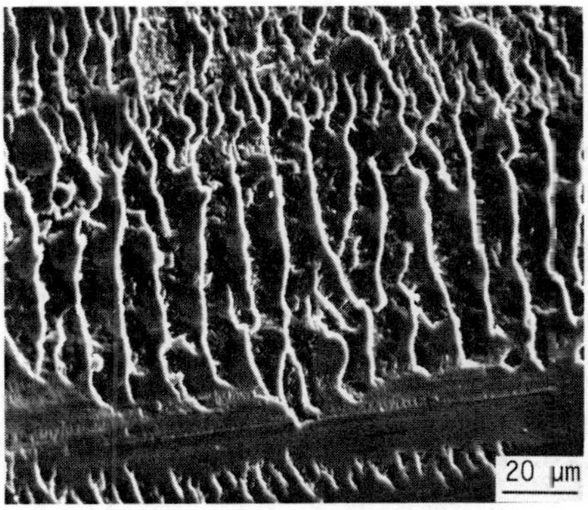

Fig. 17. Scales in fresh shear fracture surface in tool steel. SEM.

The prow formation mechanism can also be compared with metal cutting from the point of view of material removal by successive shear. The morphology and internal structure of a BUL or prow in steel reminds one strongly of that of a machined chip, consisting of thin layers separated by regions subjected to heavy localized shear (see Fig. 18). Also, the formation of built-up edges (BUE) in cutting operations, Fig. 19, involves adhesion of work material and the formation of a stagnation zone; these stages are largely similar to those of prow or layer formation.

4.2.3. Resistance to Adhesive Wear. A necessary condition for adhesive wear to occur is that a strong enough junction can be formed during sliding. Considering the tribo-system parameters according to Table 1, adhesion is promoted by increasing the parameters given in Table 5. Naturally, the dominating parameter in suppressing adhesion is lubrication, which reduces the real contact area, and weakens the metallic adhesion across the junction interface.

For the case of steel the interactive phenomenon of corrosion/oxidation is important, and can be affected via environmental and material parameters. As an

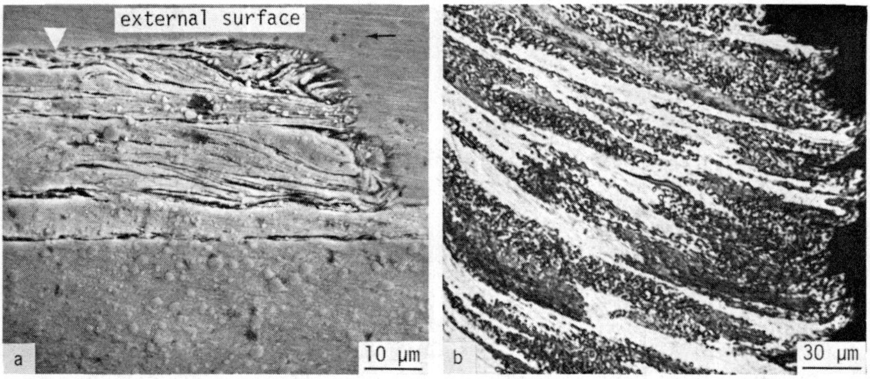

Fig. 18. Longitudinal sections of **(a)** prow (SEM), **(b)** lamellar macrochip (light micrograph).

Fig. 19. Built-up edge on high-speed steel edge after cutting a carbon steel. SEM.

Table 5. Parameters Promoting Adhesive Wear

Design	Material	Environment
Contact pressure	Ductility	Chemical inactivity
Sliding velocity	Corrosion resistance	Temperature
Sliding distance	Heat resistivity	

example of environmental effects, it has frequently been confirmed by dry sliding experiments that a transition from mild oxidative to severe adhesive wear occurs when the oxygen partial pressure of the surrounding atmosphere is reduced. As for material parameters, austenitic stainless steel is a typical example of the well-known fact that high inherent corrosion resistance promotes adhesion. In addition, this class of steels displays an instructive overall constellation of material parameters which promote adhesion and cladding: high ductility, low heat conductivity, low corrosivity.

4.3. Combined Effects of Abrasion and Adhesion

Abrasive wear often occurs as a secondary effect of adhesion. Figure 20 shows part of the worn surface of a pin from a pin-on-ring test with tool steel used for both friction partners. It is seen how a prow, strengthened during its formation by adhesive wear, can abrade material of the same composition. The large groove in the upper part of the image has been formed in the same way. The scale pattern in the worn bottom surface of the groove indicates that prow formation has occurred in the ring surface by successive adhesion and shear in the pin material. The small grooves within the upper large groove were probably cut by carbide particles in the mating surface. Protruding carbides can also be seen with protected ridges of matrix material, illustrating the action of the carbide phase against abrasive wear, discussed in section 4.1.

A similar case is seen in Fig. 21, which shows the edge region of a punch tool used to pierce austenitic stainless steel sheets. Fragments, broken out from the edge line, have stuck to the work material and grooved the tool flank during successive slip. This is an example of "self-abrasion." In fact, self-abrasion by

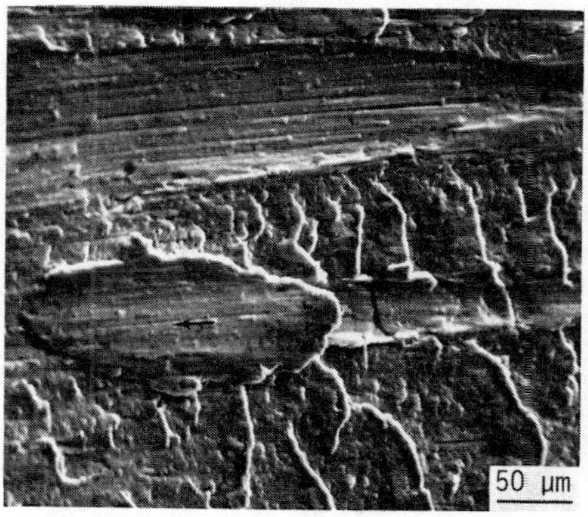

Fig. 20. Combined abrasive/adhesive wear of a tool steel surface. SEM.

Fig. 21. Self-abrasion of tool edge. SEM.

torn-off prows, adhered to the mating tribo-surface, is the main reason behind the devastating effects usually connected with "severe" adhesive wear.

Adhesion between the mating surfaces of a tribo-system can also result in material loss by cleavage fracture. This mechanism has sometimes been named "attrition," and it may be responsible for microspalling in the edge region of cutting tools (see Fig. 22a-b). The phenomenon is attributed to the repetitive exchange of the BUL or BUE.[6]

4.4. *Contact Fatigue*

Contact fatigue is defined as observable *cumulative effects of repetitive contact stress variations*. The effects of contact stresses, the magnitudes of which are harmless during one load cycle, gradually accumulate during load cycling and eventually give rise to nucleation and propagation of fatigue cracks. Particularly during cyclic metal-to-metal sliding, temperature cycling may be superimposed on the stress variation. In such a case, contact fatigue is a phenomenon of thermo-mechanical character.

4.4.1. Sliding Conditions. In terms of wear, fatigue cracking must lead to the loss of material from the tribo-surface, which requires crack propagation with a geometry such that fragments are created and loosened. During the cutting of ductile steels, metal cutting tools often exhibit stress distributions which give rise to the nucleation of series of fatigue cracks normal to the cutting direction. Material loss occurs by microspalling associated with cracks of high stress concentration during passage of the work material. Figure 23 shows the edge region of a punch tool after cutting in austenitic stainless steel. The transverse cracks and microspallings observed in the tool flank reveal contact fatigue effects superimposed on the self-abrasion mentioned in the previous section. The diagram of Fig. 24 is a plot of tool wear versus number of cuts from the same experiment. A sharp increase in slope of the wear curve indicates the onset of measurable fatigue effects.

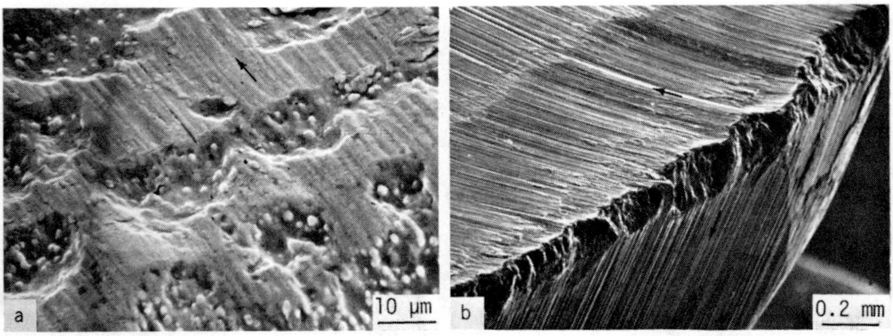

Fig. 22. (a) Near-edge flank zone of a tool steel after cutting austenitic stainless steel. SEM. **(b)** High-speed steel edge after cutting carbon steel. SEM.

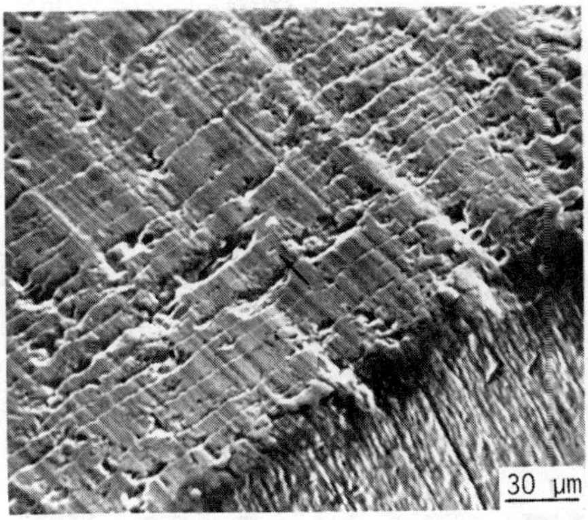

Fig. 23. Contact fatigue cracks with occasional microspallings in a worn tool steel punch. SEM.

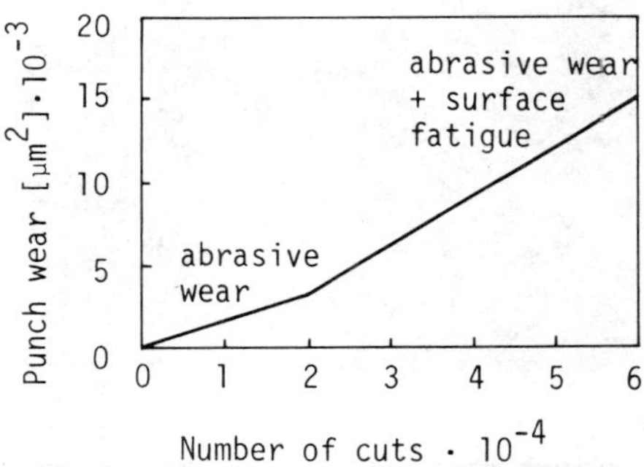

Fig. 24. Tool wear curve from punching in austenitic stainless steel.

Figure 25 shows the propagation morphology of thermo-mechanical fatigue cracks in a saw tooth, also used for cutting austenitic stainless steel. An obvious consequence of this type of cracking is that large scale spalling near the tool edge can lead to catastrophic failure by fatigue.

Another characteristic feature of contact fatigue under wandering (sliding or rolling) loads is that fatigue cracks of two main orientations may develop. The surface is fragmented by one set of cracks roughly normal to the surface, and the fragments are eventually released by a second set of cracks in a subsurface plane parallel to the surface. This phenomenon is one type of *delamination process*. In sliding applications delamination is often a wear mechanism competing with abrasion. As an example, Fig. 26 shows a cross section of a heavily abraded excavator tooth (compare with Fig. 8), exhibiting delamination in the friction layer. Also, prows and built-up layers are often fragmented and torn off by delamination type mechanisms, as demonstrated by Fig. 27 and 28. The subsurface crack propagation plane may be situated within the prow/layer as well as in the underlying base material (Fig. 27), or it may coincide with the base/layer interface (Fig. 28).

4.4.2. Rolling Contact Fatigue, Ball Bearings. As mentioned above, fatigue effects generally overlap other wear phenomena in most tribo-systems. Studies of

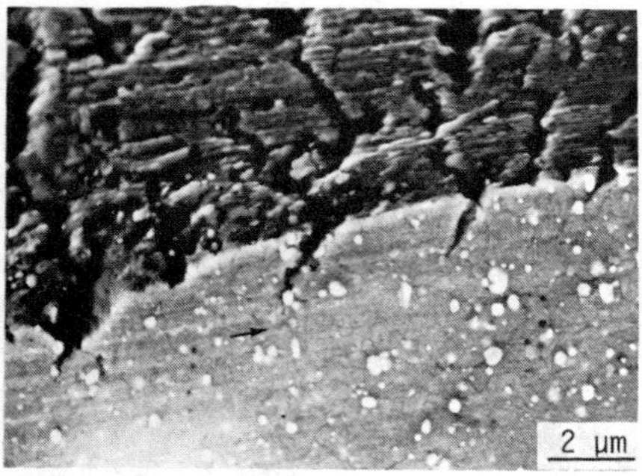

Fig. 25. Thermomechanical fatigue cracks in sectioned high-speed steel saw tooth. SEM.

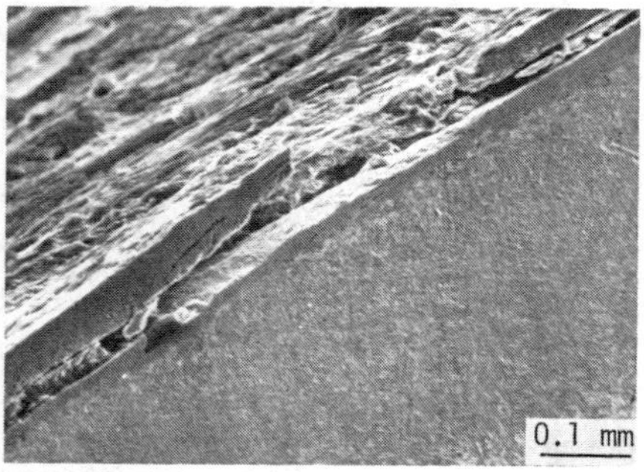

Fig. 26. Delamination wear in friction layer of abraded high strength low alloy (HSLA) steel surface. SEM.

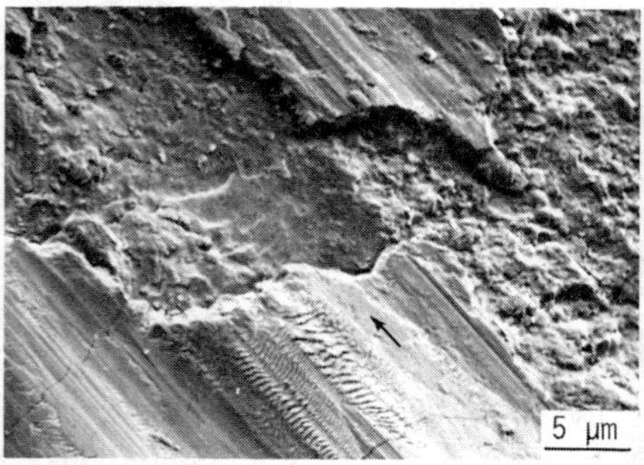

Fig. 27. Delamination of austenitic stainless steel built-up layer on high-speed steel. SEM.

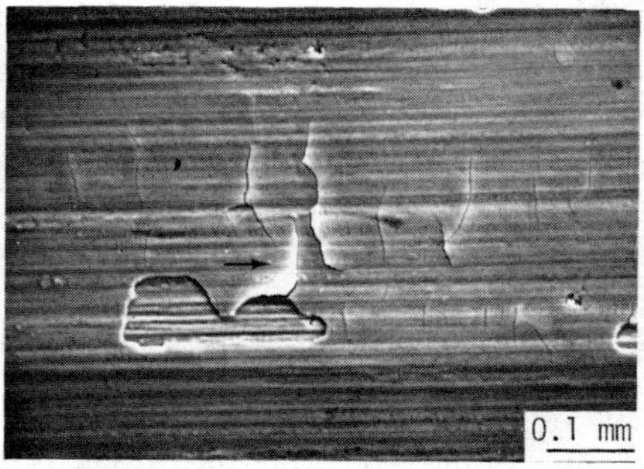

Fig. 28. Delamination of built-up layer formed by extensive flow of surface material during abrasion (tool steel). SEM.

isolated surface fatigue mechanisms are best performed by experiments with rolling contact systems, in which sliding effects are minimized. In such a system, a considered point on the studied tribo-surface is repeatedly passed by a rolling body—a cylinder, an ellipsoid or a sphere. If the surface is plane, the contact region is rectangular, elliptic or circular, respectively, and in the case of steel the nominal contact area is small. Rolling contact fatigue is defined as *observable cumulative effects from repetitive rolling* over a given surface. Typical features of a rolling contact tribo-system are low friction and low effect per rolling pass.

The case of ball bearings has been chosen to illustrate pure contact fatigue in steel. Ball bearings constitute an example of extreme tribological performance, and martensite or lower bainite ball bearing steels are frequently used in other applications also.

In radial ball bearings, which dominate most applications, the inner ring fails by fatigue cracking of more or less typical delamination geometry ("spalling"). The cracks mainly propagate at a depth under the ball track where the Hertzian shear stresses are near maximum. An SEM micrograph of a spalled ring is shown in Fig. 29. Spalling leads to vibrations, and macroscopic failure is generally defined by vibrations of a certain minimum amplitude.

5. Friction Layers of Worn Steel Surfaces

As frequently emphasized in the present paper, the wear response of a tribo-surface is given by the structure and properties of the active surface layer during the wear process, while investigations are generally performed on the resulting, passivated layer after testing. Well-established techniques for studies of the resulting so-called friction layer are therefore essential for successful deduction of the conditions of the active layer. The internal structure of the friction layer is documented as a function of depth with the aid of microscope methods (SEM, TEM, light microscope). The only simple mechanical investigation possible with respect to the required space resolution is testing for microhardness to provide hardness profiles (hardness as a function of depth). In the following, a systematic procedure is outlined for the interpretation of hardness profiles of friction layers formed in a few important types of steels in different types of tribo-systems.

5.1. Hardness Profiles

During a microevent of wear, the primary effect of the frictional energy, injected through a tribo-surface, is to generate plastic deformation and heat.

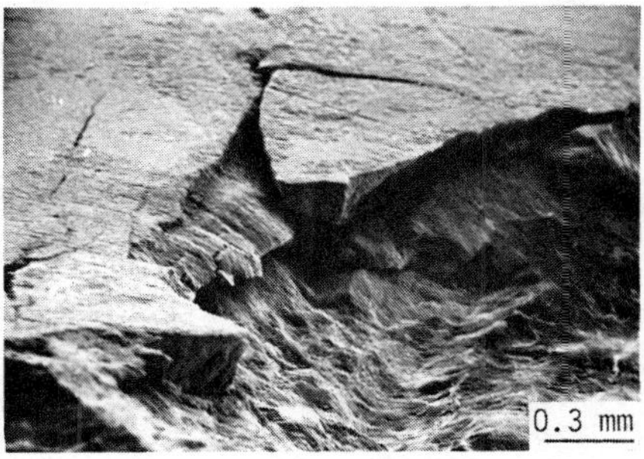

Fig. 29. Spalling in ball bearing inner ring. SEM.

Averaging over a number of events (in time and space), strain, $\epsilon(z)$, and temperature, T(z), depth profiles develop, generally with steep gradients near the surface, as schematically shown in Fig. 30. The $\epsilon(z)$ and T(z) profiles affect the resulting microhardness profile H(z), which is made up of thermal and structural (athermal) components. All changes, except in dislocation mobility, are due to changes in structure which can be strain induced or temperature induced, according to Table 6. Some components, like strain hardening, are found in all types of steels, while others, like precipitation hardening, are typical only of particular steels.

The dislocation mobility component (b in Table 6 and Fig. 30) is purely thermal and is due primarily to the temperature dependence of the Peierls-Nabarro stress via the shear modulus. In pure metals it can, in principle, be obtained from hot hardness data for relevant temperatures. During temperature changes, it is reversible. The strain hardening component (c), on the contrary,

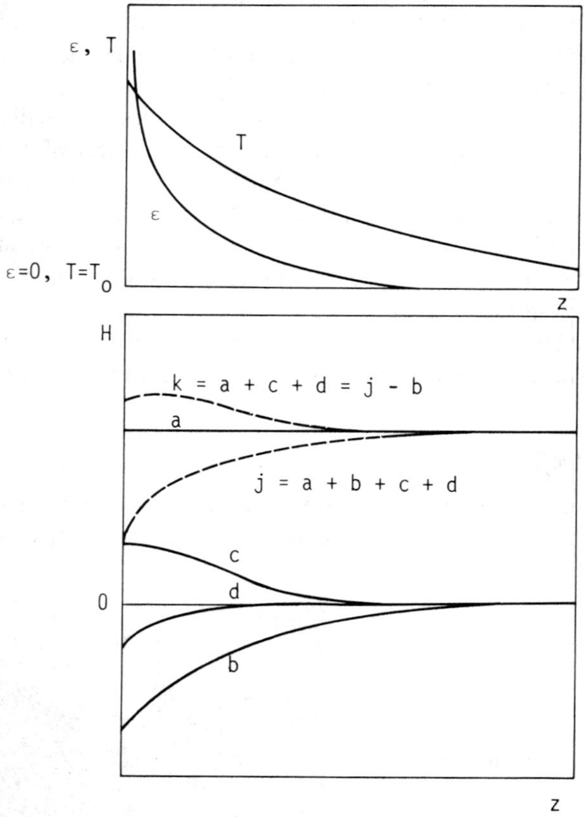

Fig. 30. Wear-induced ϵ, T and H depth profiles. ϵ = strain; T = equivalent temperature (defined in text); H = microhardness.

Table 6. Microhardness Components in Friction Layers

Microhardness component (a)	Type of steel	Strain induced	Temperature induced
b–Dislocation mobility	All	–	X
c–Strain hardening	All	X	–
d–Recovery/recrystallization	All	X	X
e–Austenitization	α-phase	–	X
f–Strain-induced martensitic transformation	Hadfield; some austenitic stainless	X	–
g–Thermal martensitic transformation	Quench hardenable	–	X
h–Precipitation hardening	High alloy, high carbon	–	X
i–Tempering	Martensitic	–	X

(a) Letters b to i correlate with Fig. 30 and discussion in text.

refers to completely athermal effects of changes in dislocation structure during plastic deformation. At constant temperature, it is irreversible. At elevated temperatures, it can be reduced by recovery or recrystallization softening mechanisms (d), which are thermal, dependent on pre-strain and thermally irreversible.

The remaining components of Table 6 involve phase transformations. The martensitic transformations (f, g) imply an increase in hardness, whereas tempering (i) results in softening. As for austenitization (e), it leads to softening compared with the martensitic phase, but not necessarily compared with ferritic phases. (Note that in this respect only structural effects are considered, while the simultaneous temperature effects are included in the b component.)

A particular advantage of this approach is that temperature effects can be separated from structure effects. Thereby it is possible to describe the hardness profile in the active layer as well as in the resulting layer by choosing relevant temperature, phase and structure for each depth level.

During frictional heating, however, the temperature concept has to be considered with some care. Under non-seizure conditions, the volume momentarily undergoing deformation is very small. Typical values of the ratio of real to nominal contact area are of the order of 10^{-4} to 10^{-2}. In addition, the time for a single contact event is very short ($\sim 10^{-3}$ to 10^{-2} s). Although the energy density in the affected volume may be very high, it is doubtful whether a bulk high temperature in terms of a complete phonon spectrum can develop during the formation and shear of an adhesive junction. Instead, the temperature deduced from the resulting structure in a friction layer should be interpreted as an "equivalent temperature," i.e., that bulk temperature which would be required for achieving the observed structural changes in thermal equilibrium (stable or metastable). In the following, all temperature/depth distribution curves, T(z), refer to this equivalent temperature.

The simplest types of hardness profiles are composed of the components common for all steels (b-d), when no phase transformations occur within the range of T(z), as demonstrated in Fig. 30. The bulk hardness of the specimen, represented by the straight line "a" in the diagram, corresponds to bulk pre-treatments (rolling, forging, etc.). (In a real case, difficulties may be met in terms of unknown surface treatments, like grinding or polishing.) The hardness profile j of the active layer is generated by adding all three curves b, c and d to the line a, and corresponds to

$$j = a + b + c + d.$$

Cooling to room temperature, when the surface contact has ceased, implies an increase in hardness corresponding to the subtraction of the thermal component b, and the hardness profile k of the resulting friction layer is described by

$$k = a + c + d = j - b.$$

The j and k curves will serve as a base for developing more complicated profiles. Figure 31 describes the case of a Hadfield steel, the well-known abrasion resistance of which is due to strain-induced martensitic transformation (MT) under tribo-conditions. The MT component f resembles the work hardening component c. If, however, the effective A_3 and A_{cm} temperatures are reached in the outermost surface layer during the frictional heating, as indicated in Fig. 31, the stable phase is austenite, and the f component drops to zero above A_1. The active layer hardness profile is then obtained as

$$l = j + f.$$

After cooling to room temperature, the austenite in the surface layer is again metastable and susceptible to strain-induced MT. However, as the tribo-action has ceased, no straining occurs, and the profile of the resulting friction layer is given by

$$m = l - b = k + f.$$

As a last example, consider the effects of thermal MT and precipitation of a second phase, according to Fig. 32. During frictional heating, austenitization (e) may occur in the outermost layer with temperatures higher than A_1, followed by thermal MT (g) during cooling. Similarly, if the conditions of precipitation are fulfilled in a temperature interval $T_1 < T < T_2$ for a sufficient total time, the corresponding depth interval will undergo precipitation hardening (h). The n and p hardness profiles will then exhibit local maxima and minima according to Fig. 32.

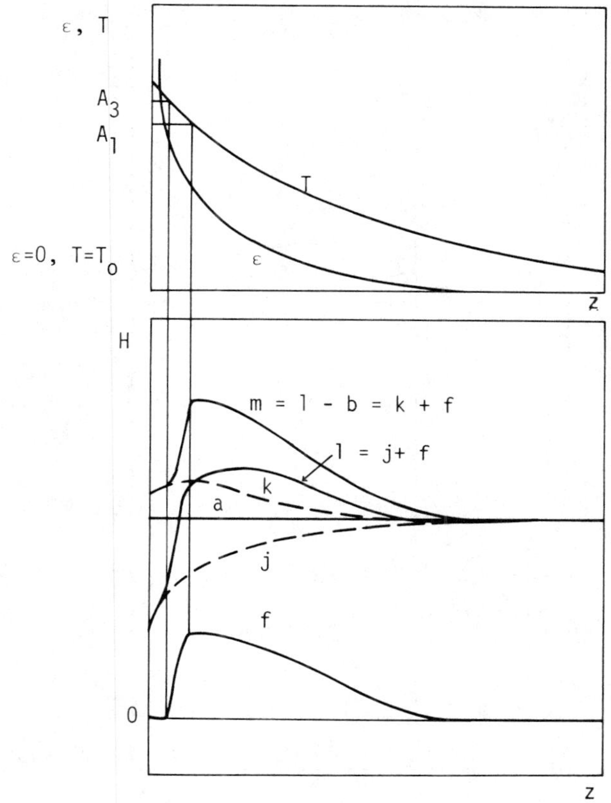

Fig. 31. Wear-induced ϵ, T and H profiles.

The active layer hardness profile n is found by adding the e or h branch (or both) to the j curve. Analogously, the resulting layer profile p is made up of the k curve plus either the g or h branch (or both).

This demonstration of how different hardness components add up to hardness profiles is based on one and the same set of $\epsilon(z)$ and T(z) distribution curves, and is thus typical of one given tribo-system only. In section 5.3 below, friction layers obtained in selected experiments and service cases are demonstrated, and it will be seen that changes in the magnitude and depth of extension of $\epsilon(z)$ and T(z) will influence both the shape and depth distribution of the hardness profiles. Complex profiles are to be expected, particularly when the initial surface is already constituted by a friction layer as a result of pre-treatments such as grinding, polishing, blasting, etc.

The preceding discussion was based on simple linear addition of independent contributions to generate such profiles as j, k, l, m, etc. This is a rough

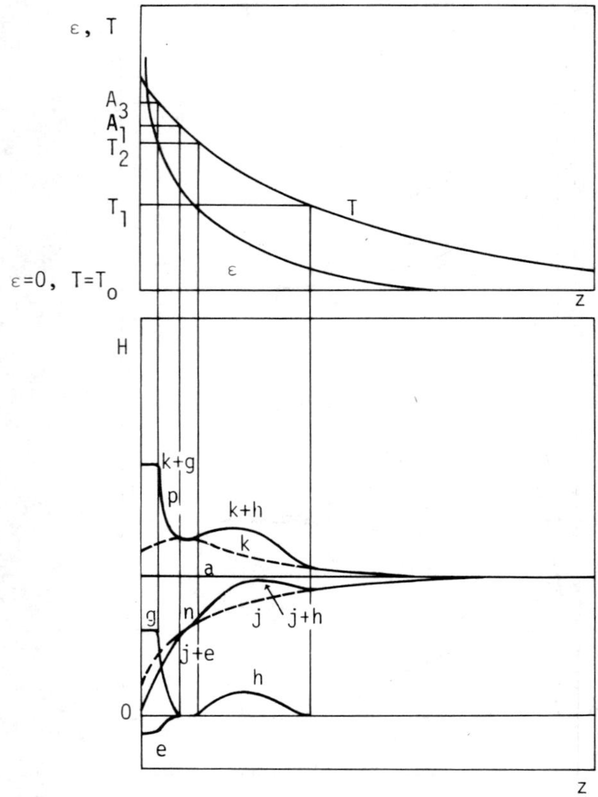

Fig. 32. Wear-induced ε, T and H profiles.

approximation used to provide qualitative predictions and explanations of strain, temperature, and hardness profiles. It is not intended as an accurate quantitative method.

5.2. The Internal Structure of Friction Layers

From a structural point of view, the generated friction layers can be schematically subdivided into zones. In the case of steels the ε(z) and T(z) distributions are often such in many tribo-systems that a four-zone model is appropriate (A-D in Fig. 33).

For high enough surface temperatures, typical of the outermost zone A, dynamic recrystallization or recovery has been reported.[7,8] The resulting microstructure is characterized by extremely fine grains or dislocation cells. In carbon steels the dynamic structural changes take place in the austenitic phase, and the resulting structure after cooling and MT is generally referred to as *friction martensite*. Zone B contains a fine, heavily textured structure, partly transformed

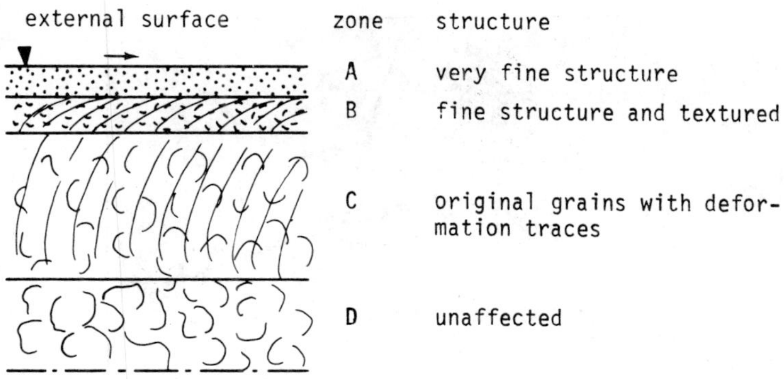

Fig. 33. Schematic zone structure of friction layers.

to martensite in carbon steels. The structure of zone C is typical of work hardening. The original grains are preserved, but the dislocation density has increased and a textured structure is being developed. Thermodynamically stable austenitic steels may show mechanical twinning, while metastable austenitic steels may begin to undergo strain-induced MT in this zone. The unaffected zone D, finally, constitutes the supporting material which will be frictionally affected when the $\epsilon(z)$ and $T(z)$ distributions move inwards as a result of wear of the friction layers.

5.3. *Selected Examples of Friction Layers*

Figures 34(a) and 34(b) are LM structure images of cross sections from a saw slot bottom and from the ring in a pin-on-ring test. In both examples the test material is a thermodynamically stable austenitic (18/8) stainless steel. The measured microhardness depth profiles of the friction layers are of the simple type demonstrated in Fig. 30, because no phase transformation has occurred.

A longitudinal section of a grooved Hadfield steel together with the corresponding microhardness profile is shown in Fig. 35. This material exhibits a deep zone of strain-induced MT below a superficial soft layer of dynamically recrystallized or recovered austenite. The hardness profile reveals the actual state of strain and temperature during grooving from which, in principle, the $\epsilon(z)$ and $T(z)$ profiles may be estimated (compare Fig. 31). The internal structure is similar to that of the austenitic stainless steel. However, martensite twins are found in the deformed zone C.

A typical structure image of a worn pearlitic carbon steel is shown in Fig. 36, collected from a laboratory grooving experiment. The A zone appears white after nital etching. Microhardness measurements reveal a hard structure, which at high magnification proved to consist of extremely fine grains or dislocation cells. This is an example of friction martensite, the high hardness of which indicates that little tempering has occurred.

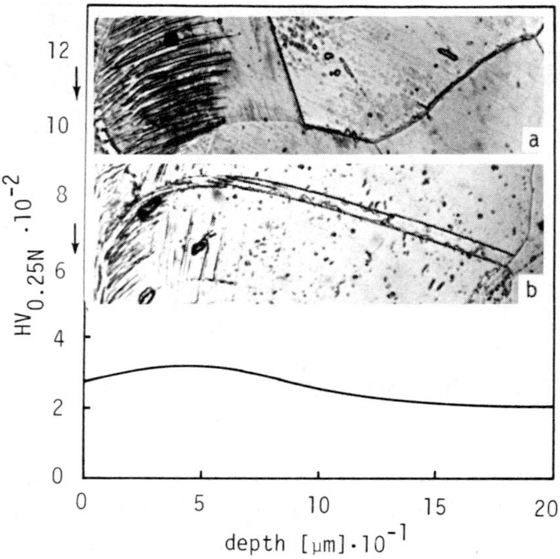

Fig. 34. Structure and microhardness profiles of friction layers observed in longitudinal sections of 18/8 stainless steel: **(a)** saw slot bottom; **(b)** ring from pin-on-ring test. Arrows indicate sliding direction.

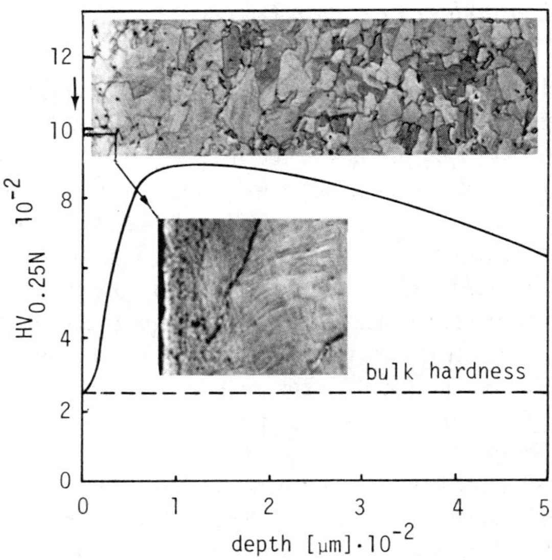

Fig. 35. Structure and microhardness profile of friction layer observed in longitudinal section of a Hadfield Mn steel. Arrow indicates sliding direction.

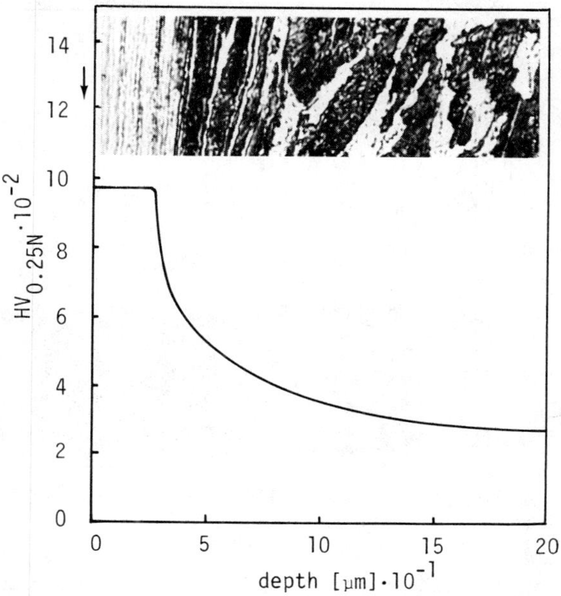

Fig. 36. Structure and microhardness profile of friction layer observed in longitudinal section of a pearlitic carbon steel. Arrow indicates sliding direction.

A laboratory-grooved, quenched and tempered (QT) carbon steel is seen in Fig. 37 together with its corresponding hardness profile. Considering estimated $\epsilon(z)$ and $T(z)$ distributions, the hardness drop in zone B and the hardness peak in zone C can be explained as the result of the matching of strain hardening and softening by tempering in this depth interval. A TEM micrograph of the transition between the fine recovered structure of zone A and the coarser textured structure of zone B is shown in Fig. 38. The inserted selected area diffraction patterns are consistent with the fine grained and textured structures, respectively.

When the same type of QT steel is subjected to intense wet grinding, one can obtain a friction layer like the one shown in Fig. 39. Compared with the case of grooving, shown in Fig. 37, the temperature distribution has deepened, and a thick layer of hard untempered friction martensite remains after the test. Strain hardening effects are small, localized to a thin surface region, and not resolvable by microhardness measurements. Below the friction martensite ($T < A_1$) the virgin martensite has softened considerably by tempering, and the hardness profile of the friction layer is equivalent to that of an induction or flame hardened layer.

A grooving test performed on a QT tool steel gave the structure and hardness profile of Fig. 40. The outermost hardness peak corresponds to the white-etching

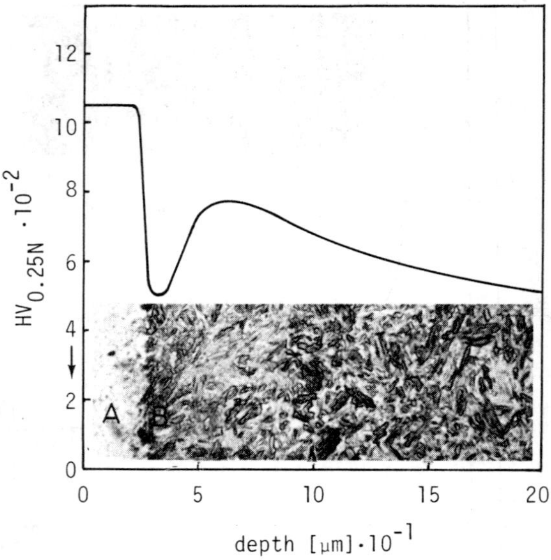

Fig. 37. Structure and microhardness profile of friction layer in quenched and tempered carbon steel. Arrow indicates sliding direction.

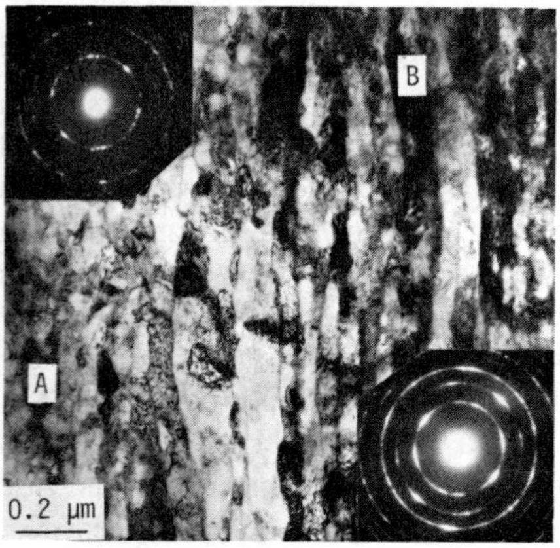

Fig. 38. Internal structure of zones A and B of Fig. 37. Selected area diffraction patterns from a specimen area of about 0.5 μm diameter.

friction martensite. No plastic deformation can be detected, and thus a second hardness peak cannot be due to work hardening. Instead, it corresponds to secondary hardening as described in Fig. 32 (curve h). The adjacent hardness dip to values below virgin hardness is clearly a result of overtempering.

Two examples of friction layers in a high-speed steel are shown in Fig. 41. The LM micrographs show the internal structure of a hack saw tooth flank and a pin from a pin-on-ring laboratory test. Conventional quenching and tempering of

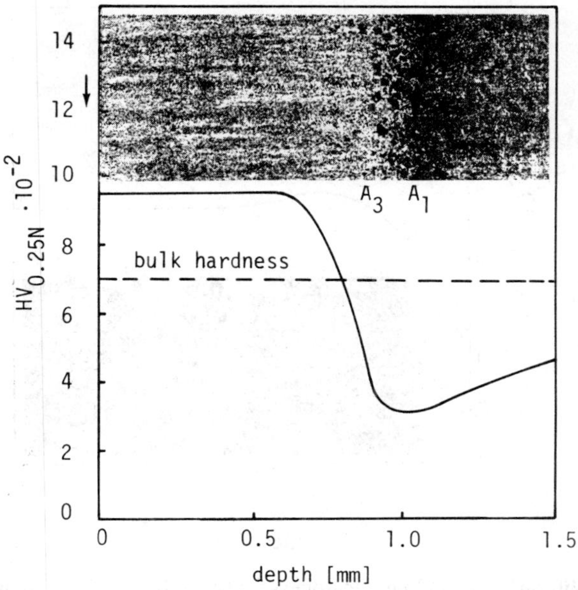

Fig. 39. Structure and microhardness profile of friction layer in wet ground quenched and tempered carbon steel.

high-speed steels has the purpose of adjusting the structure to give a secondary hardness peak, which implies resistance to deformation even at high temperatures, and extensive plastic deformation can only take place within the austenitized layer. At the same time, the structure just beneath will suffer over-tempering and soften. A quench to room temperature will force the austenitic layer to transform partly to martensite, while a large fraction is retained as austenite. The hardness of this surface structure is not likely to exceed that of the bulk structure, which was carefully prepared by secondary hardening during double tempering. All these considerations are consistent with the measured microhardness profile of the friction layer. Particularly, it can be noted that the hardness of the friction martensite is lower than that of the virgin bulk hardness.

In addition to the examples above, which refer to the basic phenomena described in Table 6, other effects may have to be considered from case to case, particularly in zone A. Redistribution and dissolution of carbides can occur in tool steels together with embedded oxide particles and transfer from the mating surfaces. The etched longitudinal section shown in Fig. 42, taken from a pin-on-ring test, shows an example of both effects. The oxide distribution was obtained by a K_{α} X-ray line scan displaying the relative variation in oxygen concentration.

In ball bearings tested under high stress conditions, a sequence of tribologically interesting phase transformations leads to the formation of a friction layer. In a

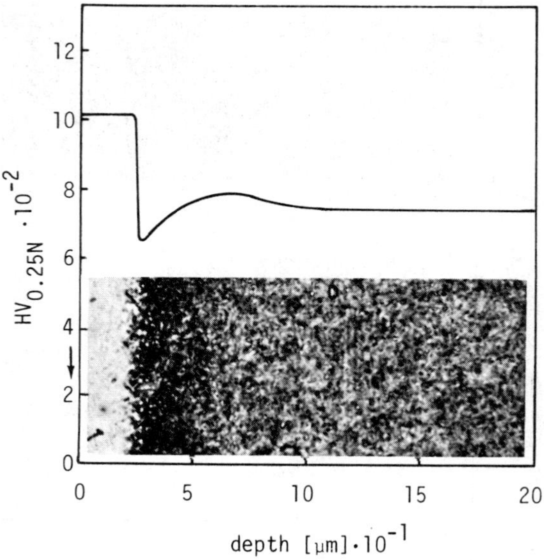

Fig. 40. Structure and microhardness profile of friction layer in quenched and tempered tool steel.

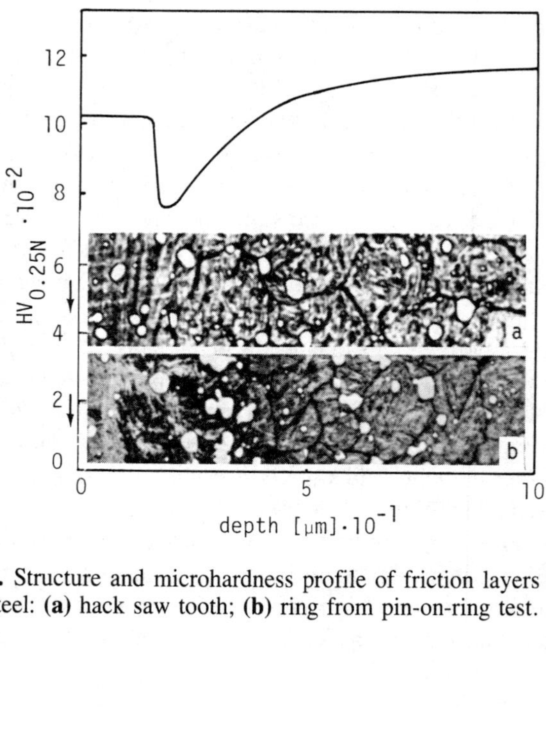

Fig. 41. Structure and microhardness profile of friction layers in high-speed steel: **(a)** hack saw tooth; **(b)** ring from pin-on-ring test.

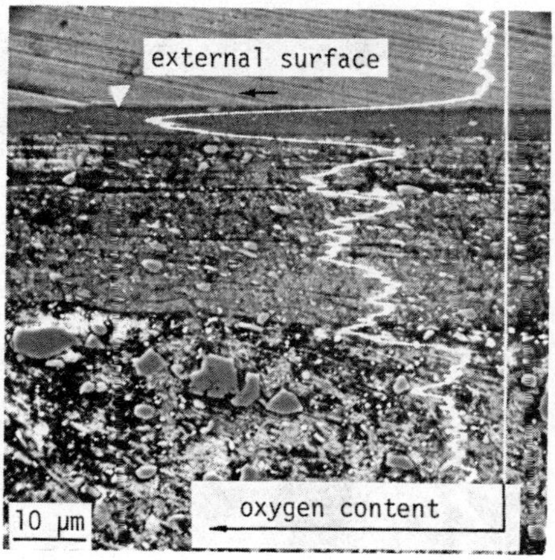

Fig. 42. Structure and oxygen distribution of friction layer in tool steel. SEM.

depth interval of 0.1-0.5 mm, surrounding the level of maximum Hertzian shear stresses, a zone of a ferrite-martensite phase mixture develops by stress-induced martensite decay. In bearings surviving 10^8-10^9 revolutions under extreme loads, short-range carbon diffusion leads to the formation of a sandwich structure of ferrite and carbide discs. Eventually, extensive martensite decay results in a set of large ferrite discs. The structure and hardness profile of the friction layer is reproduced in Fig. 43. The low rolling friction does not imply high enough temperatures for austenitization or dynamic recrystallization, but strong plastic deformation takes place in the ferritic phase, which generates a very fine dislocation cell structure (Fig. 44) This phase, which has been called "cell ferrite,"[9] resembles the friction martensite in sliding friction layers, and also appears white-etching in high magnification LM.

5.4. Wear Properties of Friction Layers

In sections 5.1-5.3 the friction layers have been treated with respect to microstructure and hardness depth distribution. It is obvious that the layers represent great changes compared to the bulk material, and that steels often possess the favorable property of self-strengthening during wear.

Hitherto, the research emphasis has been put on layer formation. However, the next step will have to be experimental studies of the performance of friction layers during continued wear. A first question will deal with the adhesion between layer and matrix (compare applied coatings), which is generally found to be good. Particularly for the case of friction martensite on carbon steel, this property can be explained by the softer, tempered interface layer (see Fig. 37, 39-41), which serves as a plastic work absorbent, inhibiting crack initiation and delamination.

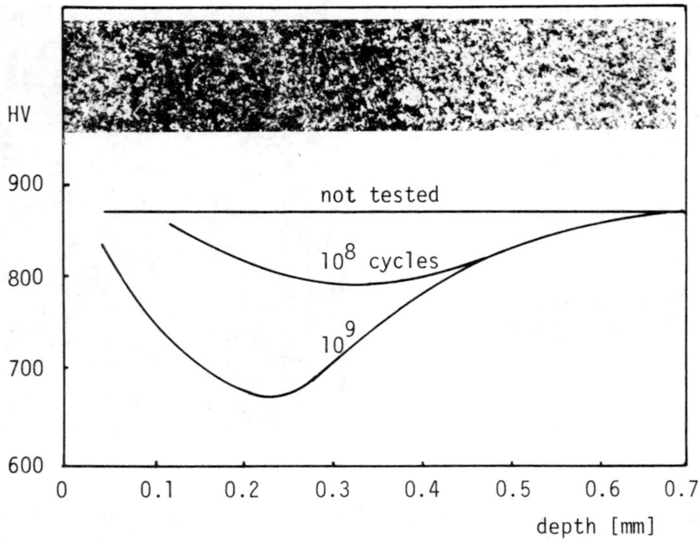

Fig. 43. Structure and hardness profile of surface layer in ball bearing steel.

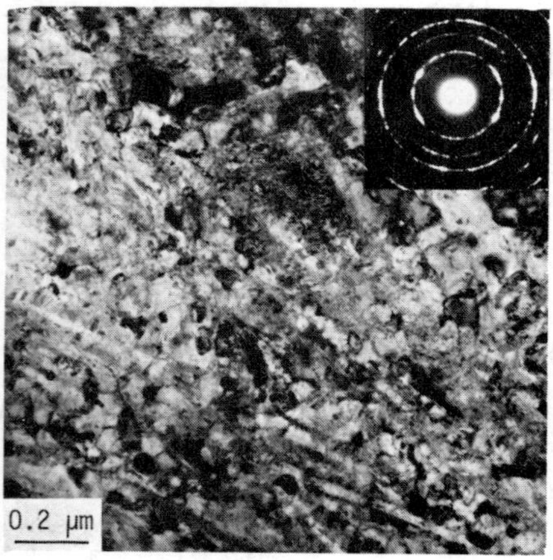

Fig. 44. Cellular structure in fatigued ball bearing steel. TEM.

The second question is concerned with the action of wear mechanisms within a friction layer. Experimental work on this subject has just been started, and a few findings will be briefly discussed.

Repetitive grooving in the same track has been performed in a laboratory test.[3] By adjusting the feed (chip thickness) the abrasive action can be chosen to coincide with a desired depth level in the friction layer. Figure 45 shows experimental results from repeated grooving in a high strength low alloy (HSLA) steel,

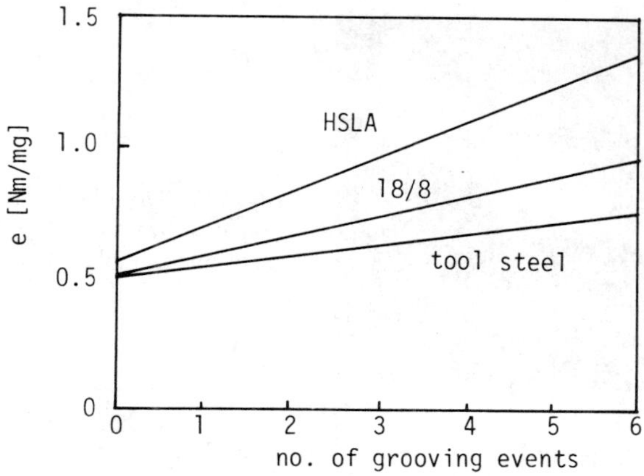

Fig. 45. Increase in specific grooving energy e during repeated grooving, according to Ref 3.

an austenitic stainless steel, and a soft annealed, low chromium tool steel. The feed was adjusted to within the outermost part of the friction layer. Ranking of the steels was made with the aid of the specific grooving energy, which was found to increase with the number of grooving events, as shown in Fig. 45. The HSLA steel proved to be superior, as expected. The stainless steel reached the second position because of its high ductility and corresponding capability of absorbing plastic work, while internal delamination in the friction martensite layer was responsible for the lower abrasion resistance of the tool steel.

With the aid of quick-stop equipment it has been possible to show that active layers are formed during grooving even ahead of an abrasive tip.[4] This phenomenon is expected to be particularly advantageous for the Hadfield type steels, in which the hardness in the layer to be worn will highly exceed that of the bulk material before the abrasion has even occurred (compare Fig. 31 and 35).

This type of investigation is very difficult to pursue in service cases. Occasionally, however, it has been possible to observe local effects of repeated wear action. Figure 46 shows an example of a local double friction layer on an excavator tooth (compare Fig. 8b).

6. Conclusions

The philosophy of the present way of collecting and presenting wear data has been to show that it is possible to apply to wear phenomena concepts from the fields of mechanical properties, dislocation theory, and fracture mechanics. These fields are today much better established and represent a more mature level of knowledge than the young science of wear. To a large extent it is also possible to apply knowledge from processes such as extrusion, hot forming, and cutting operations to wear problems.

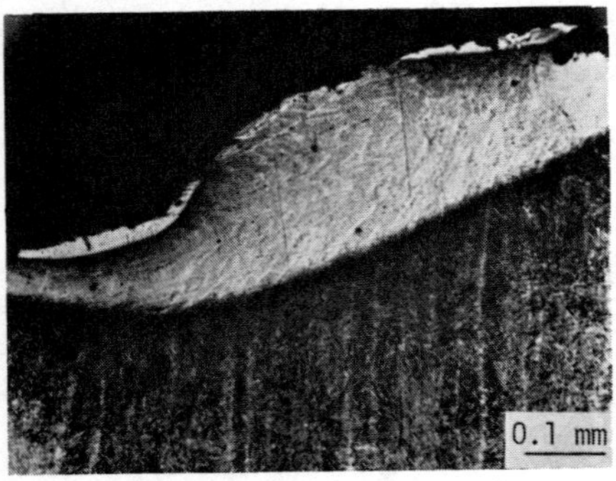

Fig. 46. Double friction layer after service abrasion of excavator tooth. Light micrograph.

The critical threshold in this transfer of knowledge is the concept of friction layers. The present work shows that it is possible to modify studies of the mechanical properties and parameters to represent the friction layers which are found in post-mortem studies of worn surfaces. In addition, the separation of pure thermal effects facilitates a necessary deduction to the properties of the layer which actually existed during the wear process. Because of the flexibility of the internal structure of this class of metals, the application of this approach to steels is a very rewarding task. It seems extremely promising to direct structure engineering—which is practiced with great skill in other applications today—towards an explicit interest in wear and tribo-systems.

References

1. M. M. Khrushchov and M. A. Babichev, "Abrasive wear resistance and modulus of elasticity of heat treated steels," in Friction and Wear in Machining, ASME Trans. 17(1962)9.
2. U. Bryggman, S. Hogmark and O. Vingsbo, "Abrasive wear studied in a modified impact testing machine," in Wear of Materials 1979, ASME, New York 1979, p. 292.
3. U. Bryggman, S. Hogmark and O. Vingsbo, "Evaluation of abrasion resistance by means of controlled grooving," Proc. 4th International Abrasion Colloquium, Grenoble, May 9-11, 1979.
4. U. Bryggman, S. Hogmark and O. Vingsbo, "Force and energy measurements during controlled grooving—a basic study of abrasive wear," ASLE preprint, No. 80-LC-8B-2.
5. M. Cocks, "Interaction of sliding metal," J. Appl. Phys. 33(1962)2152.
6. P. K. Wright and E. M. Trent, "Metallurgical appraisal of wear mechanisms and processes on high-speed-steel cutting tools," Metals Technology, January, 1974, p. 13.
7. S. Hogmark, S. Söderberg and O. Vingsbo, "Surface deformation and frictional heating of steel during machining and laboratory testing," in Mechanical Behaviour of Materials, ICM3, Pergamon, Oxford 1979, p. 621.
8. J. H. Dautzenberg, "The role of dynamic recrystallization in dry sliding wear," Wear 60(1980)401.
9. R. Österlund and O. Vingsbo, "Phase changes in fatigued ball bearings," Met. Trans. A 11(1980)701.

Friction and Wear of Thermoplastic Polymers

E. HORNBOGEN and K. SCHÄFER*
Institut für Werkstoffe
Ruhr Universität Bochum
West Germany

Abstract

The origin of unlubricated friction and wear of polymers is discussed, with emphasis on thermoplastic materials. Most of the experimental results reported were obtained with polytetrafluoroethylene, polyethylene, polypropylene, and polyamide.

The systematic discussion is based on the factors that affect the coefficient of friction: surface energy and plastic deformation, including hardness. For the understanding of wear additional mechanical properties are required which characterize separation processes: elongation at fracture and fracture toughness. The differences between the friction and wear of metals and polymers are due to the low surface energy, high oxidation resistance, low hardness, low heat conductivity, and strong temperature dependence of all mechanical properties of polymers.

Macroscopic properties are explained from molecular structure (type of polymer) and morphological structure (arrangement of molecules). A structural basis has been used to discuss the limits of friction and wear properties. Finding materials which combine a low friction coefficient and high wear resistance is the most important concern for research in this field.

Introduction

Polymers have some friction and wear properties that cannot be obtained by any other group of materials. For example, the materials which can produce the minimum friction coefficient are polymers. In addition, the high chemical stability of many polymer molecules leads to a surface which is not considerably changed by reactions with the environment, such as oxidation. There are also

*Now at barmag Barmer Maschinenfabrik AG, D-5630 Remscheid–Lennep.

secondary properties of polymers which favor their application in our field of interest — for example, their high capacity for damping vibrations (internal friction) and their corrosion resistance. From these properties most of the applications can be derived in which friction and wear play a role.

The first group of applications is determined by the requirement of a low coefficient of friction between sliding surfaces (Fig. 1). Other applications are for bearings and gears when lubrication with oil is not allowed or when the damping of vibrations is an important requirement. If durability plays a role, wear in addition to friction must be considered.

The other extreme, a high surface energy and a high adhesive force, can be obtained with other polymeric materials used as adhesive cements. If friction and wear of polymers are to be optimized in combination, we encounter a number of problems which may be demonstrated by the characteristic properties of two materials, both of which are used for bearings, PTFE (polytetrafluoroethylene) and PA (polyamide). PTFE is the material which shows minimum friction but a relatively large amount of wear; PA exhibits a small amount of wear in spite of its relatively high friction coefficient.[1-6] It is the purpose of this paper to discuss the reasons for this behavior. We begin with a brief description of molecular structure, morphology (microstructure), and bulk properties which are relevant to friction and wear. A discussion of adhesion, friction, and wear follows.

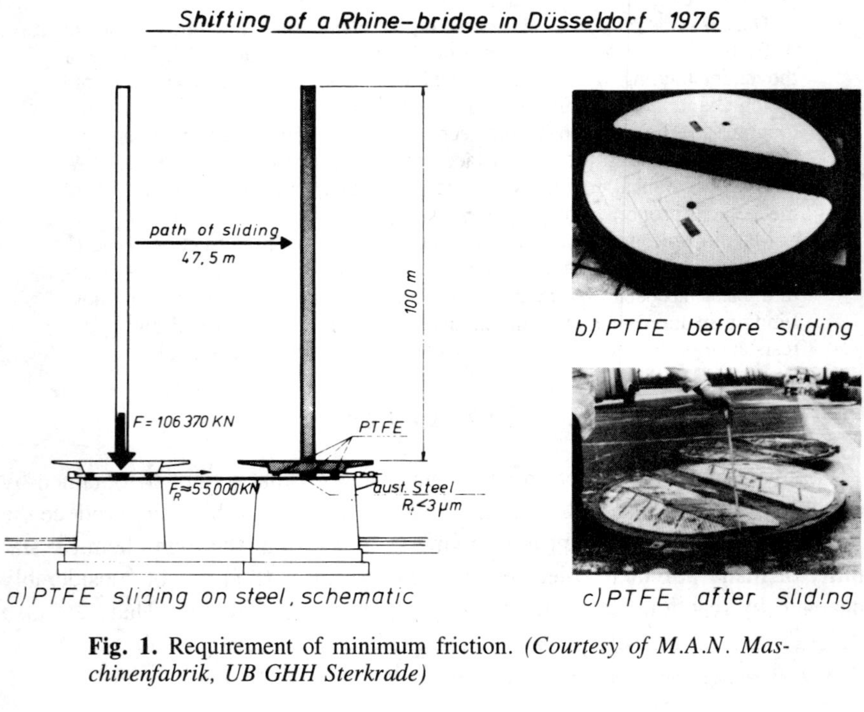

Fig. 1. Requirement of minimum friction. *(Courtesy of M.A.N. Maschinenfabrik, UB GHH Sterkrade)*

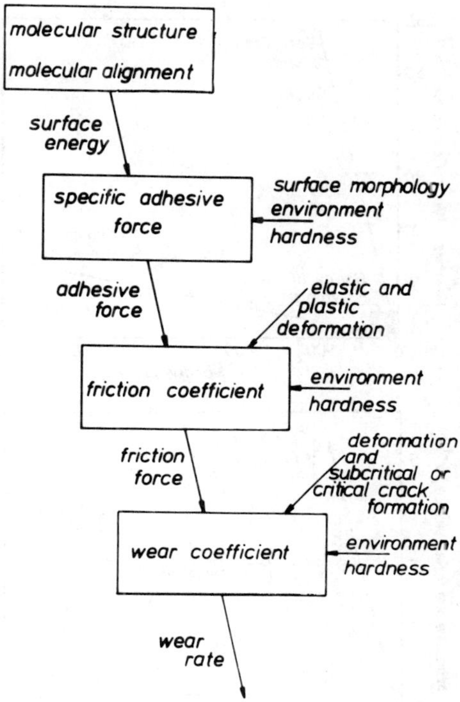

Fig. 2. Systematic survey of factors which determine friction and wear of polymers.

An outline of the paper is shown in Fig. 2. This logical sequence of the treatment follows from the physical background of the phenomena. For the understanding of adhesion, surface energies are required. For friction the mechanical properties, especially hardness and plastic deformation, come into play. For explaining wear, the mechanisms of separation of material from the surface are a factor. In addition, the melting temperature of the polymer and, much more so than in metals, the thermal conductivity are important bulk materials properties which we must consider. These materials properties will have to be distinguished from external conditions. These include temperature, pressure, velocity of sliding, initial surface roughness, the combination of two materials sliding on each other, chemical environment, and the special testing conditions for the friction and wear experiments.

Elastomers are not treated here, nor are the effects of lubrication. It is the purpose of our paper to emphasize the relations between the structural properties of selected polymeric materials and their friction and wear behavior.

In Fig. 3 the abrasive wear resistance of ceramics, metals, and polymers is compared.[7] It is quite evident that differences exist between these three groups. The most pronounced difference between polymers and other materials is simply due to the difference in hardness. Nevertheless, differences between various

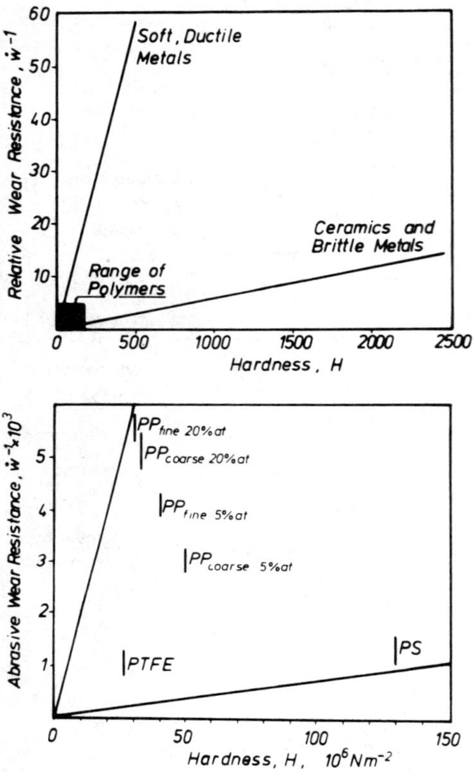

Fig. 3. Ranges of abrasive wear resistance of metallic and ceramic materials **(top graph)** *(after Ref 7)* and of thermoplastic polymers **(bottom graph)**.

species of polymers are important, which justifies the many research efforts designed to improve their properties. One should remember that "wear resistance" is not a simple physical property. It is rather a typical "system property." Contributing are a large number of physical properties and also the set of external conditions. The differences between polymers and other materials originate primarily from the fact that, because of their particular molecular structure, a different set of basic physical properties dominates this system. These differences cannot be understood without some knowledge of the microstructure of polymeric materials.

Structure and Properties of Polymeric Materials

Molecules

Many aspects of the frictional behavior of polymers are directly related to the molecular structure. Therefore, some characteristics of polymer molecules will be discussed here. First we must distinguish straight, stiff molecules from those which have a tendency to coil (Fig. 4). Straight molecules are able to form

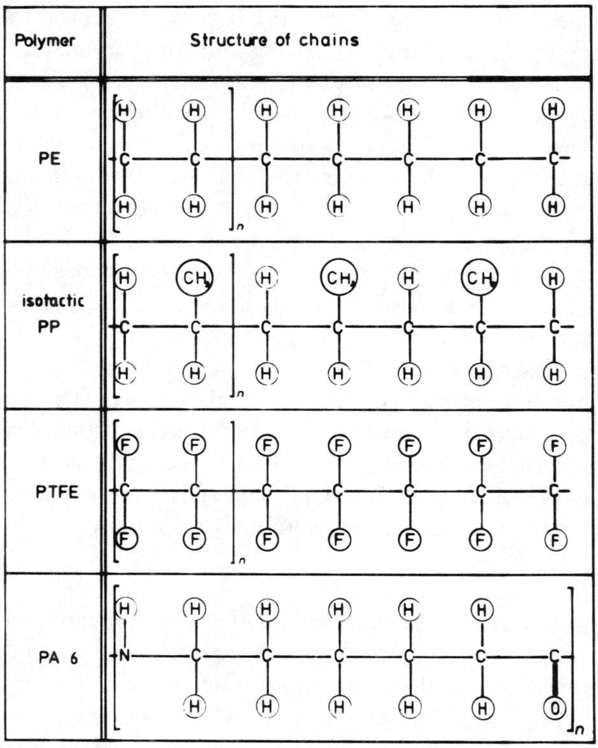

Fig. 4(a). Structure of the polymer molecules used for most of the experiments.

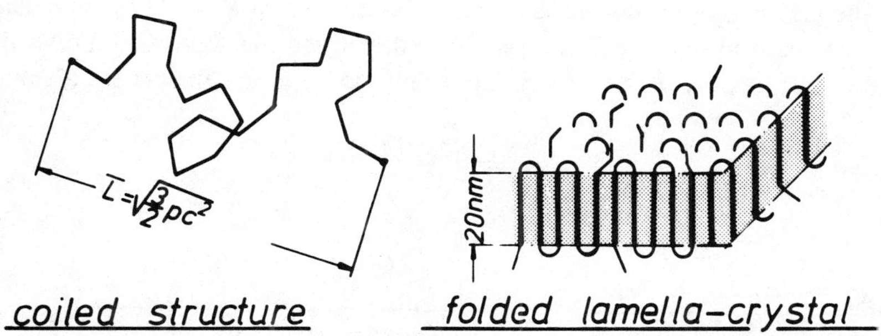

Fig. 4(b). Examples of molecular arrangement ($\bar{L}$ = average distance between chain ends).

crystals while coiled; branched molecules can only form glassy structures. Some chains form helices, as, for example, PTFE. Its large F-atoms cause great stiffness, which in turn leads to a high crystallinity, in spite of the weakness of the intermolecular bonds.

The molecule itself can be symmetric or asymmetric, depending on the position of side groups (radicals) (Fig. 5). A consequence of asymmetric shape of the molecule is the formation of net electric dipole moments, which in turn form the basis for strong intermolecular bonding. All high strength thermoplastic polymers are characterized by the existence of strong dipoles (polyvinyl chloride, PVC; polypropylene, PP) or of the still stronger hydrogen bonds (PA, polyamides). Side groups can be arranged in a disordered or ordered way (tacticity) (Fig. 6). High tacticity favors crystallization. In symmetric molecules the dipole moments of the individual bonds compensate each other so that the bonds between individual molecules become relatively weak. This weakness of the intermolecular bond coincides with a low surface energy of the material.

The strongest intermolecular bond is caused by cross-linking, which is effective in thermosetting polymers, elastomers, and cements. The existence of this type of bond excludes the possibility of plastic deformation. The strength of unsaturated bonds in the surface will determine the surface energy. High cohesion between molecules is favored by a high density of strong bonds—i.e., for a given type of bond, by high tacticity and high crystallinity.

Morphology

There is a wide variety of possibilities for the arrangement of molecules inside the bulk material, i.e., of morphologies of the polymer. The most important cases will be mentioned here. All thermosetting materials and elastomers and some thermoplastic materials are glasses. In the undeformed state they show a random arrangement of the molecular chains. The conformation of the molecules themselves varies between coiled and straight. Most thermoplastic materials are mixtures of crystalline and glassy regions.

The elementary crystalline element is the folded lamella. These lamellae in turn are stacked into packages, surrounded and tied together by non-crystalline portions of the microstructure. Some of the special arrangements are the micro-

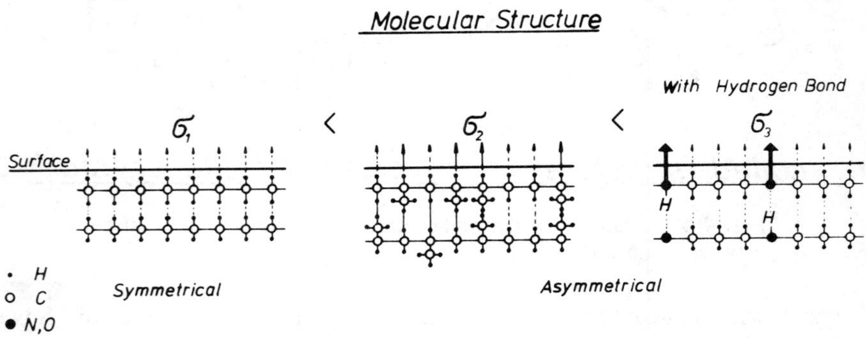

Fig. 5. Influence of different types of intermolecular bonding on surface energy and cohesion.

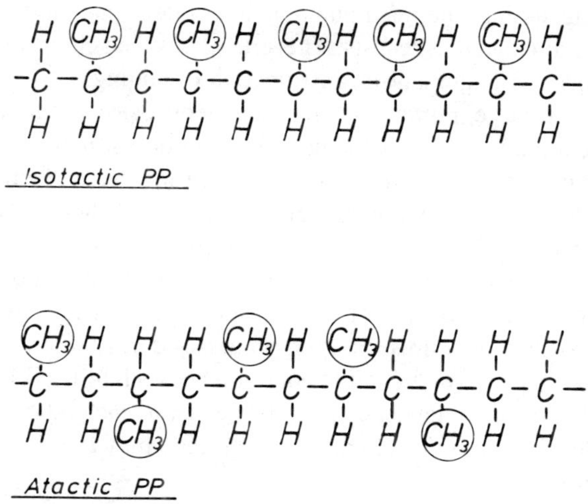

Fig. 6. Isotactic and atactic configuration of polypropylene molecules.

crystalline structure and the spherulitic structure. Cohesion between molecules increases with degree of crystallization. Cohesion of polyethylene (PE) and PTFE is predominantly due to their high crystallinity, while their specific intermolecular bonds are relatively weak. Intermolecular bonds in PP are stronger; however, a high tacticity is required to permit crystallization (Fig. 7).

In an intermediate range of temperature and strain rate all thermoplastic materials can deform plastically, and, as a consequence, the molecules become

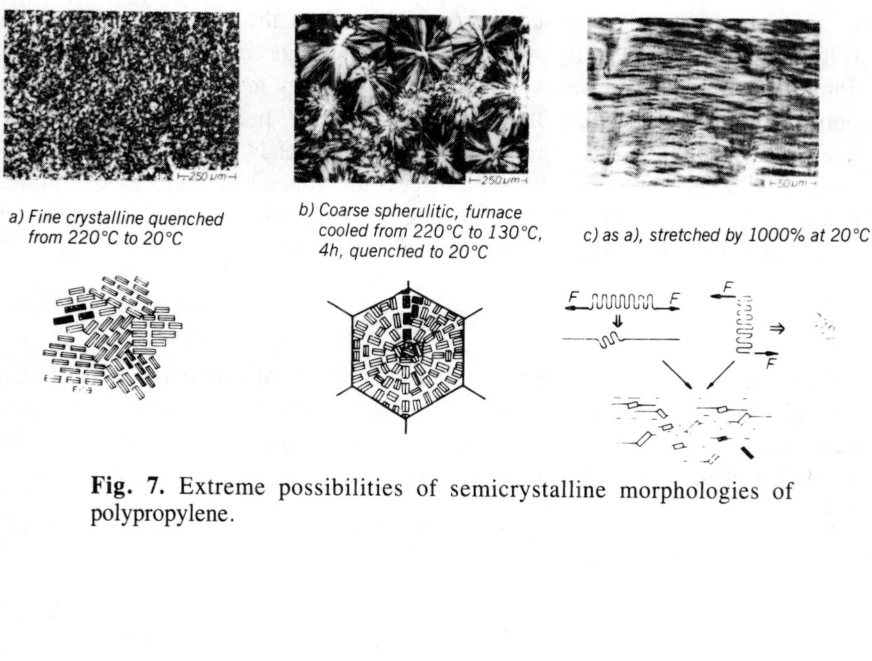

Fig. 7. Extreme possibilities of semicrystalline morphologies of polypropylene.

aligned. During sliding, the maximum amount of deformation in the surface can surpass that obtained in tensile tests. The structure is then characterized by a high degree of molecular alignment in the direction of sliding.

There exist a large number of heterogeneous structures in undeformed polymers. An important case is the coarse spherulitic structure in which small-molecular-weight portions have been rejected during crystallization, so that the boundary regions between the spherulites are amorphous. The amount of crystallinity and therefore cohesion is high inside the spherulites.

Often organic or inorganic admixtures are contained within polymeric materials. In these cases additional interfaces are produced which may be important, especially in connection with separation processes during wear. One function of such additives can be the reduction of adhesion between foils. For example, SiO_2 particles in PP reduce the tendency of foils to stick to each other. Such additives may induce decohesion by local cracking, and they may also form debris which can cause abrasion.

Bulk Physical Properties

The bulk properties of polymers are much different from those of metals in two respects. Mechanical properties vary over an extremely wide range, from high elastic modulus and brittle behavior at low temperatures through work hardening and relatively tough or rubber-elastic behavior at intermediate temperatures, to viscous behavior at still higher temperatures. The transition to the liquid state is continuous above the glass transition temperature of amorphous polymers. The existence of crystalline portions leads to a partially discontinuous transition to the liquid state at the melting temperature. Analogous transitions are found in the deformation behavior of rubbed surfaces. The surface temperature, however, can be raised considerably by frictional heat because of the low heat conductivity (Fig. 8). This is a property which polymers have in common with ceramics. Table 1 shows that their heat conductivities at ambient temperature are typically only one thousandth of those of metals and alloys. This property together with the low melting temperatures of most of the polymers leads to the particular sensitivity of all experimental results with respect to temperature, velocity of sliding, and load. This is one reason for a low degree of reproducibility of experimental results and the wide range of data found in the literature.

Adhesion

It is evident from Table 1 that the surface energy, σ, of polymeric materials can vary over a wide range because of differences in molecular structure. The phenomenon of adhesion is due principally to the reaction of surfaces and the formation of interfaces. The change in surface energy $\Delta\sigma$ is given by

$$\sigma_{AO} + \sigma_{BO} - \sigma_{AB} = \pm\Delta\sigma \approx f_{ad} \cdot d \quad [1a]$$

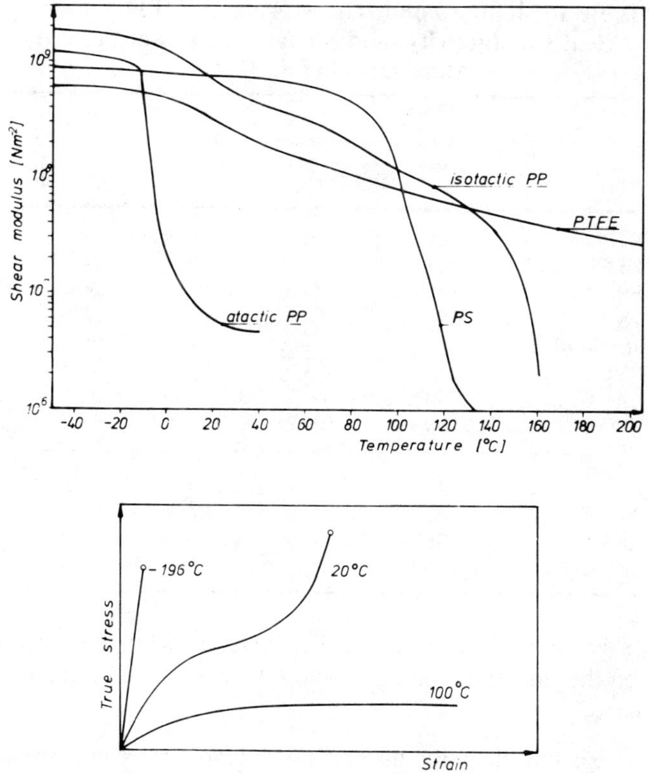

Fig. 8. Temperature dependence of mechanical properties: **(top)** shear modulus of several thermoplastics; **(bottom)** typical stress strain curves of PP.

where σ_{AO}, σ_{BO} are the surface energies when materials A, B have clean surfaces (as in a good vacuum), σ_{AB} is the surface energy of an AB interface, f_{ad} is the adhesion force per unit area and d is the intermolecular spacing. An attractive or a repulsive adhesion force can result:

$$f_{ad} \approx \frac{\Delta\sigma}{d} \quad [1b]$$

If two surfaces of like materials are brought into touch, an interface must form. It is identical with a kind of grain boundary containing only intermolecular bonds. If the two materials are different, an intermolecular phase-boundary forms. The energy depends on the specific energies of the original surfaces and on that of the newly formed interface. If the interface is a "grain-boundary," its energy, σ_{AA}, is usually much smaller than that of the original surfaces, σ_{AO}. For a given energy of the grain boundary, σ_{AA}, or interface, σ_{AB}, the resulting adhesive force is small, if the original surface energies were small.

Table 1. Melting Point, Glass Transition Temperature, Heat Conductivity, and Surface Energy of Several Materials (Ref 4, 37-41)

Material	Melting point (T_m), °C	Glass transition temperature (T_g), °C	Heat conductivity (λ), J/ms °C	Surface energy (σ), 10^{-3} J/m²
PTFE	327	126	0.244	22
HDPE	137	−120	0.337	24
PP	165	−18	0.221	26
PS	240	100	0.174	28.9
atactic PP	—	—	0.163	36
PVC	212	87	0.163	36.5
PA6.6	250	—	0.244	36.9
PMMA	160	105	0.186	38.3
SiO_2	2427	—	2.3	320
Ag	961	—	400	920
Cu	1083	—	380	1100
γ-Fe-18Cr-8Ni	1400	—	16	1400

From the balance of specific energies several useful conclusions can be obtained for the adhesive forces which act between different combinations of materials:

1. f_{ad} is maximum, if similar materials with high surface energies are brought into contact or react. This is the case for high melting temperature materials, if their surfaces are not modified by reactions with their environment.
2. f_{ad} is minimum if two materials with low surface energies react and if these two materials produce an interface with high energy. Very low surface energies are found in polymeric materials with a symmetrical molecular structure such as PTFE and PE (see Table 1).
3. The interfacial energy σ_{AB} increases with increasing difference in the nature of the bonding between the two components A and B. Consequently, adhesion decreases in the following sequence: materials which form strong bonds → equal materials → different materials with mutual miscibility → materials which are not miscible but of similar type of bonding → materials of different type of bonding which are immiscible. Examples of the last-mentioned case are many combinations of metals and polymers in which little possibility for chemical bonding exists during sliding. Therefore combinations of PE or PTFE with steel, titanium and other alloys are frequently found. An example of a metal-polymer system which does not fulfill this condition is PTFE with alloys containing alkaline metals such as sodium as alloying elements. In this case F atoms from the polymer may react with the alkaline metal atoms in the interface.

Friction

The friction force, F_R, is due to the energy which is dissipated during sliding.[8] If friction is caused exclusively by adhesion and decohesion in the interface, the friction coefficient μ is defined as follows:

$$F_{R_{ad}} = \mu_{ad} \cdot F = f_{ad} \cdot A = \frac{f_{ad}}{H} \cdot F. \qquad [2a]$$

However, decohesion of adhesive bonds is rarely the only contribution to friction. Only for very low surface energies and very low compressive force, F, is the frictional force determined by the adhesive tension, f_{ad}, and the effective asperity area, A. The area in turn depends on the surface hardness and the surface roughness of the sliding partners. The effective surface area, A, can be calculated, if a compressive force, F, is acting in addition to adhesion between the sliding materials. In this case A is reciprocally proportional to hardness for a certain range of forces (Eq 2).

From this discussion it is evident that the coefficient of friction is determined by at least two materials properties—namely, the surface energy and the hardness. This statement applies to thermosetting materials which are not able to deform plastically, and for which energy dissipated by elastic deformation can be neglected.

When both adhesion and plastic deformation contribute to friction, the friction force is given by

$$F_R = \mu \cdot F = (\mu_{ad} + \mu_{def})\, F. \qquad [2b]$$

In the case of thermoplastics, the frictional shear force which acts in the surface is usually higher than the critical shear stress needed to induce plastic deformation in the surface material. In this case deformation energy is dissipated and the surface is modified by molecular rearrangement.[9,10] The molecules are aligned in the direction of sliding and considerable amounts of work hardening occur in this direction. Cohesion in the other two principal directions is reduced. In many cases the energy dissipated by plastic deformation will surpass by far the energy used for decohesion of adhesive bonds in the surface. Consequently plastic deformation of the surface zone dominates the frictional force. This is always the case for abrasive friction, i.e. for hard particles plowing and cutting a surface.[11]

The change of the surface structure, which causes work hardening, can lead to a decreasing coefficient of friction due to the increase of hardness and a decreasing effective area A (Eq 2).[12] The effect of anisotropy of the surface energy may also contribute to the change of the friction coefficient with surface deformation.[13] The surface energy should decrease slightly as a result of molecular alignment parallel to the surface. This effect is under investigation and is not yet completely clarified (Fig. 9 and 10).

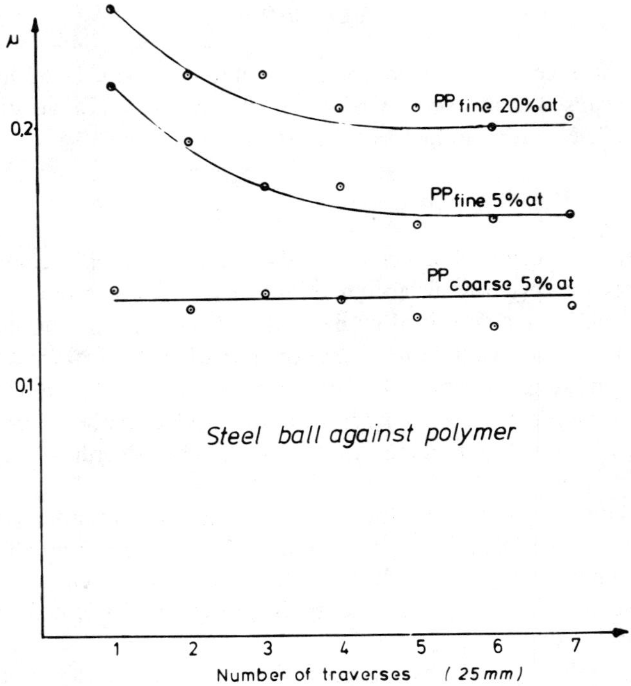

Fig. 9. Coefficients of friction of different types of PP as a function of the path of sliding. (Morphologies are given in Fig. 7; % at. = admixture atactic molecules, $f = 6 \cdot 10^4 \frac{N}{m^2}$, $v = 3.3 \cdot 10^{-3} \frac{m}{s}$.)

Figure 11 shows that the contribution of surface energy of different types of polymers usually cannot be neglected, if the origin of the coefficient of friction is analyzed. The data given in Table 2 indicate that the coefficient of friction increases with increasing surface energy of the polymeric material. Therefore it can be stated that, in contrast to metals in the atmosphere, the coefficient of friction of the polymer is partly determined by its surface energy. As in metals it is also affected by the plastic deformation behavior of the surface zone. The surface energy of metals in the atmosphere becomes ineffective because of shielding of the surface, e.g., by oxide layers. In polymers oxidation of the surface plays a role in unsaturated molecular structures like those found in some elastomers. Large dipole moments will favor the tendency for formation of adhesive layers of water on the surface. Thus, shielding of the polymer surface can occur when high surface energy polymers are exposed to a humid environment. This effect is well known for PA and its relatively low friction coefficient is explained by the lubricating effect of its water layer.

The nature of deformation in the surface of polymers is determined by the strong temperature dependence of their mechanical properties. If the temperature

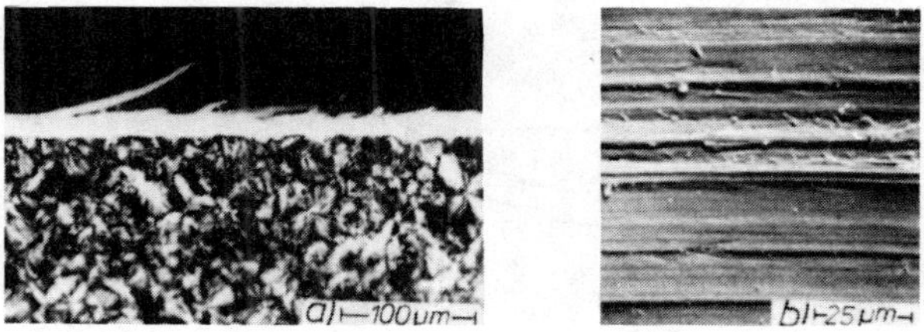

a) Fine crystalline; delamination of aligned molecules

b) Coarse spherulitic; less alignment of molecules in the surface, initiation of interspherulitic cracks

Fig. 10. Modification of surface structure of PP by frictional force: austenitic steel; surface roughness $R_t = 3$ µm, $f = 12 \cdot 10^4 \frac{N}{m^2}$, $v = 3.3 \cdot 10^{-3} \frac{m}{s}$.

dependence of the friction coefficient is analyzed, three ranges can be distinguished. In addition to adhesive friction at asperities and mechanically activated plastic deformation with work hardening of the surface, a transition to viscous flow or melting occurs at higher temperatures. Before the transition to viscous flow a maximum coefficient of friction, which can lead to stick-slip behavior, is sometimes found. This is due to surface softening, desorption of the water layer, and subsequent sticking of larger surface areas. At surface temperatures above the glass transition, T_g, the viscosity, η, decreases rapidly and the coefficient of friction becomes controlled by viscous flow in a layer of thickness Δx. In such cases the surface layer of the polymer behaves as a lubricant.[14,15]

$$F_R = \eta \frac{v}{\Delta x} \cdot A_o \qquad [3]$$

It can be assumed that under the conditions for validity of Eq 3, A_o is about equal to A. It may be noted that the maximum of μ is most pronounced in materials of high surface energy (PA), while for low surface energy materials (PTFE) the maximum, and therefore the tendency for stick-slip, is less pronounced or is absent.

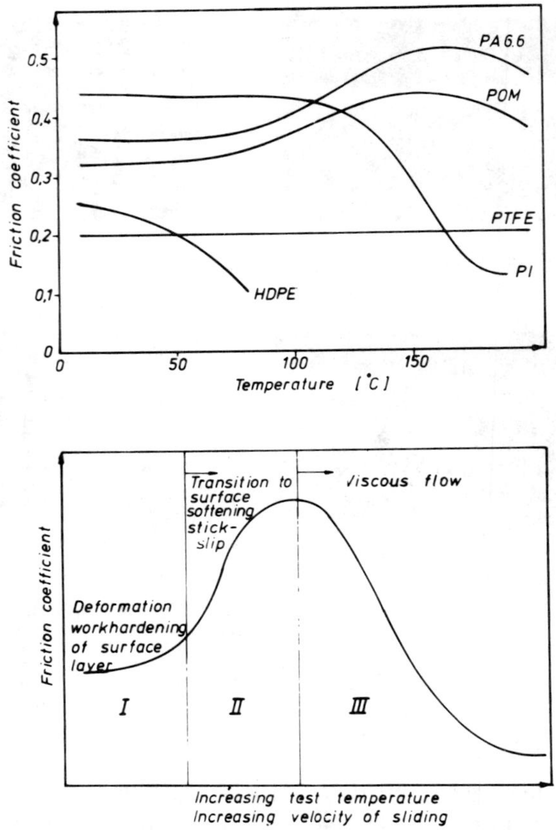

Fig. 11. (Top) Temperature dependence of the coefficient of friction for different polymers *(after Ref 3 and 4)*. **(Bottom)** Schematic survey of temperature dependence of the friction coefficient as dominated by different mechanisms.

Figure 11 (top) represents qualitatively the dependence of the coefficient of friction on the velocity of sliding. Because of the low heat conductivity an increased velocity is qualitatively equivalent to an increasing testing temperature.[16-18] The typical shape of this function (Fig. 11, bottom) shows three ranges: (*i*) at low temperatures a work-hardened surface layer forms as a result of molecular alignment; (*ii*) softening of the surface with increasing temperature favors adhesion over large surface areas and also the phenomenon of stick-slip; (*iii*) at still higher temperatures the frictional force becomes controlled by the decreasing coefficient of viscosity of a surface layer (Eq 3).

After the effect of surface energy and deformation behavior on the coefficient of friction has been demonstrated, the question should be raised whether the

Table 2. Friction Coefficients Against Austenitic Steel, Surface Energy, and Hardness of Several Polymers (Ref 37 and 38)

Polymer	Friction coefficient (μ)	Surface energy (σ), 10^{-3} Jm^{-2}	Hardness (H) at 20°C, 10^{6} Nm^{-2}
PTFE	0.11	22	32
HDPE	0.25	24	57
$PP_{fine\ 5\%\ at}$	0.3	26	40
$PP_{fine\ 20\%\ at}$	0.35	28	28
PS	0.46	28.9	140
$PP_{coarse\ 5\%\ at}$	0.27	30	50
$PP_{coarse\ 20\%\ at}$	0.4	30	30
PVC	0.6	36.5	115
atactic PP	—	36.8	$\ll$20
PA 6.6	0.35	36.9	100
PMMA	0.54	38.3	200

original morphology of the polymer is important for friction. Mixtures of PP molecules of different tacticity may serve as the first example of such effects (Fig. 12). Admixture of atactic molecules is associated with a reduced tendency for crystallization and therefore reduced cohesion. The coefficient of friction increases with the portion of atactic molecules. This effect can be explained by the decreasing hardness of material containing higher portions of atactic PP (Eq 2). In addition an increase in surface energy will further increase the coefficient of friction μ (Eq. 1).

A different morphological effect is found as a function of the angle, α, between the direction of chain alignment and the direction of sliding. Results are shown for PTFE and PP (Fig. 13). For these materials the coefficient of friction is minimum if the direction of sliding coincides with the direction of pre-aligned molecules ($\alpha = 0°$). This can be explained by the smaller chance for energy dissipation by plastic reorientation of aligned molecules which are strained in the direction of the molecular axis compared with the case of molecules strained perpendicular to their original direction.[19-21]

Figure 14 shows results of experiments with PP for which the morphology had been changed in a wide range between a fine crystal-glass mixture and a coarse spherulitic structure. These morphological conditions were produced by different heat treatments. They may be compared with data from plastically deformed material containing aligned molecules (Fig. 13).

The dependence of the coefficient of friction on morphology is qualitatively different if the polymer is rubbed against a material like steel which has a

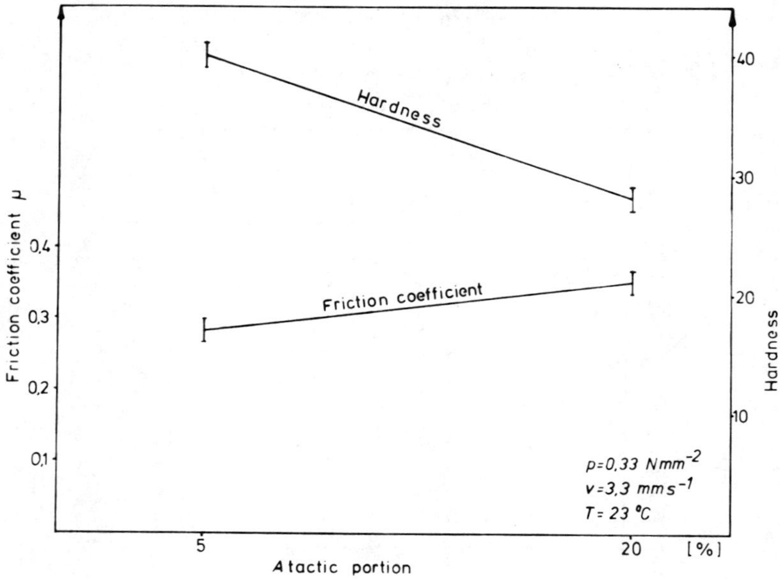

Fig. 12. Effect of molecular structure and morphology on the coefficient of friction. The atactic admixture increases surface energy and decreases crystallinity and therefore hardness. ($f = 12 \cdot 10^4 \frac{N}{m^2}$, $v = 3.3 \cdot 10^{-3} \frac{m}{s}$.)

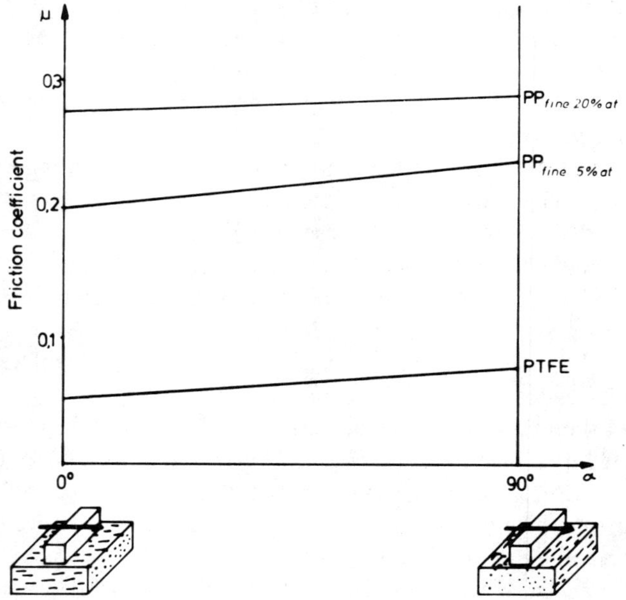

Fig. 13. Effect of the direction of sliding on the coefficient of friction in polymers with aligned molecules. ε(PTFE) = 600%; ε(PP) = 800%; $\varepsilon \equiv$ nominal amount of stretching ($f = 12 \cdot 10^4 \frac{N}{m^2}$, $v = 3.3 \cdot 10^{-3} \frac{m}{s}$).

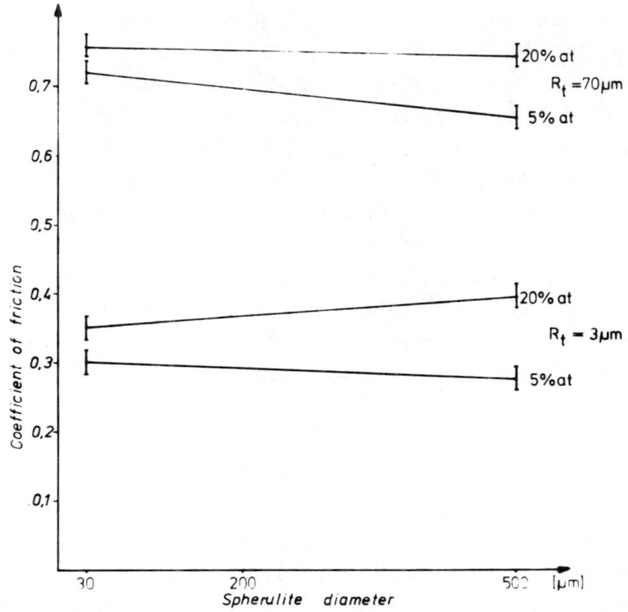

Fig. 14. Coefficient of friction of PP as a function of molecular structure (tacticity), morphology (see Fig. 7), and surface roughness (R_t). Abrasive friction with Al_2O_3 particles: diameter, 70 μm, R_t = 3 μm polished steel plate, $f = 3.5 \cdot 10^4 \frac{N}{m^2}$, $v = 3.3 \cdot 10^{-3} \frac{m}{s}$.

different surface roughness.* In the case of low surface roughness, a thin, highly deformed work-hardened layer can form in the surface of the polymer. The amount of work hardening is more pronounced when the original hardness of the material is low. For the case of higher surface roughness this layer is removed by abrasive wear more rapidly than it is re-formed. Therefore the original bulk mechanical properties (without surface work hardening) of the material become decisive for the coefficient of friction. This is an example of conditions for which the wear mechanism has a strong effect on friction.[22,23]

In PTFE intermolecular cohesion is extremely weak. Even for sliding on polished surfaces the deformed surface layer is removed by shear of the intermolecular bonds. This occurs much more easily in PTFE than in PP or even PA. The mechanism can be deduced from microstructural observations and measurements of the thickness of the deformed surface zone. The structure of this zone depends on the deformation behavior of the material under the frictional load and on the rate of removal of the deformed layer.

---

*The surface roughness, R_t, is defined as the maximum difference between top and bottom of the asperities.

The maximum amount of deformation is limited by surface fracture. In the case of brittle polymers the frictional force can induce cracking after small or negligible amounts of plastic deformation. In this case, only a limited amount of energy can be absorbed by plastic deformation before crack formation occurs. Under these conditions a transition to a lower friction coefficient is expected which is partially determined by crack formation and crack extension energy. For polymers this situation can be achieved either at very low test temperatures ($T << T_g$) or for mechanically heterogeneous materials. The latter case is demonstrated by the localized rupture of spherulite boundaries after negligible amounts of deformation in the interior of the spherulites themselves (Fig. 15). The friction coefficient for this morphology is low, especially if hard spherulites are built from isotactic molecules which provide a high amount of crystallinity (Fig. 14).

Wear

At the end of the previous section, some effects of wear and microcracking on friction have been discussed. On the other hand, the frictional force is generally the origin of wear. Dealing with wear implies that the mechanisms and the rate of separation from the surface are of predominant interest. Wear of polymers, as of other materials, can be subdivided into the mechanisms causing decohesion by (*a*) sliding of plane surfaces, (*b*) abrasion, and (*c*) erosion.

For (*a*) and (*b*) the wear coefficient, k, is often characterized by Archard's law,

$$\dot{w} = k\frac{f}{H} = k\frac{A}{A_o}, \qquad [4]$$

which states that the wear rate (dimensionless) is proportional to the effective area of the asperities A (Fig. 3); f is the compressive stress. The wear coefficient is determined by the mechanisms of separation of wear particles. It is constant, if the separation mechanisms are the same, and independent of hardness (see Fig. 3 for soft pure metals and for very brittle ceramics).

In polymers, however, the mechanisms of separation and consequently k can vary over a wide range.[24-31] A comparison of metals and polymers indicates that because of their very low hardness the wear rate of polymers is always higher than that of common metallic materials. If different polymers are compared, it should be expected that for low friction forces acting in the surface, a low wear rate is the consequence. This is contradicted by the fact that the material with the lowest surface energy and lowest friction force, PTFE, shows the highest wear rate. The wear rate of PA, with its higher friction force, is much smaller, at least for temperatures not too close to its melting temperature, where stick-slip can occur (Fig. 16). These observations indicate that for materials with low surface energy the strength of intermolecular bonds is so weak that separation is easy

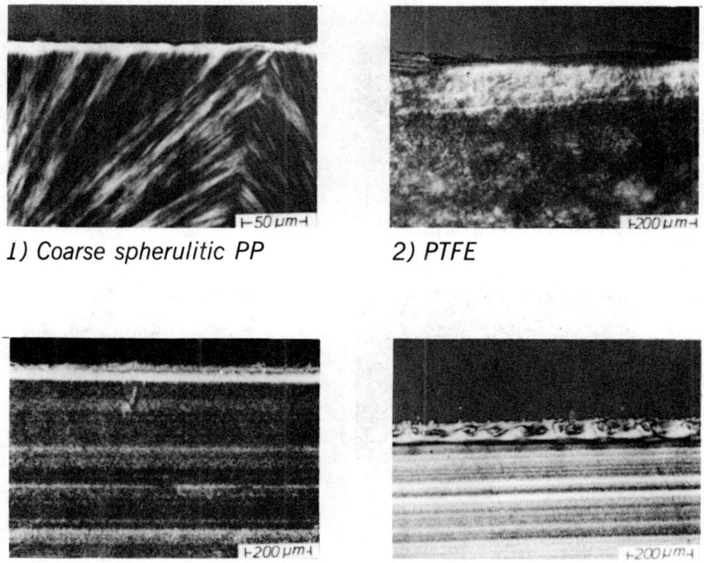

1) Coarse spherulitic PP
2) PTFE
3) PA
4) PE

1) PP coarse spherulitic: thin deformed surface layer (compare Fig. 10). **2)** PTFE, thin surface layer, delamination caused by weak intermolecular bonding. **3)** PA, thick work-hardened surface layer, strong intermolecular bonding. **4)** PE, thick work-hardened surface layer, intermolecular bonding stronger than PTFE, weaker than PA.

Fig. 15(a). Modification of surface structure of several polymers after sliding on austenitic steel (surface roughness, 3 μm, $f = 12 \cdot 10^4 \frac{N}{m^2}$, $v = 3.3 \cdot 10^{-3} \frac{m}{s}$).

even by moderate friction forces. This antagonistic characteristic of friction and wear of polymers limits their application. It is a challenge to search for molecular structures and morphological effects by which both wear and friction can be minimized.

For abrasive wear of semicrystalline PP the coefficient of friction increases with increased admixture of atactic molecules (Fig. 14). Nevertheless the wear rate decreases under these conditions, because the probability for decohesion is decreased because of better plastic deformability (lower hardness) of the non-crystalline portions of the structure. A high wear resistance is favored by lowering of the bulk hardness and increasing atactic admixtures (Fig. 17).

The effect of spherulite structure on the wear rate can be explained following the same reasoning (Fig. 18). A coarse spherulitic structure with a small portion of atactic molecules favors interspherulitic separation with little preceding plastic deformation. As a consequence this combination of molecular and morphological structure shows the highest wear rate, in spite of the high hardness and minimum friction.

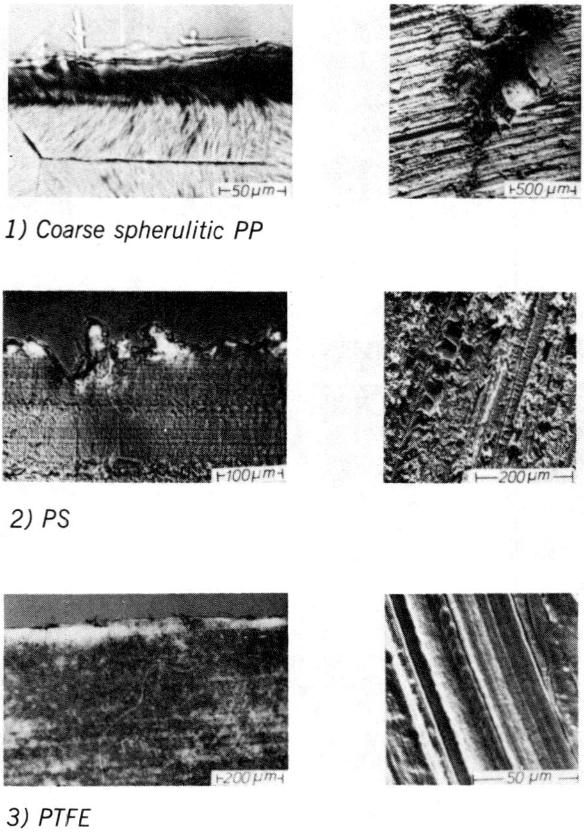

1) PP coarse spherulitic, thin surface layer with localized quasi-brittle rupture of spherulitic boundaries. **2)** PS, glassy, $T < T_g$, brittle cracking in the surface zone. **3)** PTFE, thin surface layer associated with plastic cutting of weakly bonded material.

Fig. 15(b). Modification of surface structure of several polymers by abrasive friction with SiO_2 particles (diameter, 70 µm, $f = 3.5 \cdot 10^4 \frac{N}{m^2}$, $v = 3.3 \cdot 10^{-3} \frac{m}{s}$).

In order to explain these results it is reasonable to consider mechanical properties which describe crack propagation (Fig. 19). For very brittle materials the wear rate will be controlled by fracture toughness, while for ductile materials it is determined by the deformation processes preceding separation.[32-36] There will be a transition for materials with an intermediate fracture toughness, for which maximum wear resistance can be expected. The transition point depends strongly on the wear system.

The effect of fracture energy is mostly negligible for sliding on polished surfaces, while it will become important for abrasion and for impact during erosion of polymers with low and even intermediate fracture toughness. For the

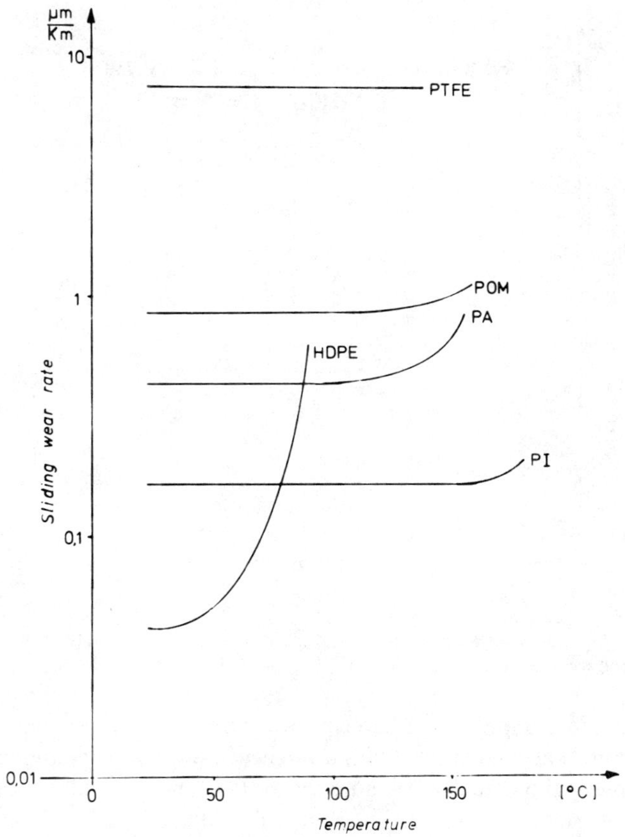

Fig. 16. Sliding wear rate of various polymers as a function of temperature: steel, $HR_c = 54$; $f = 10^7 \frac{N}{m^2}$. *(After Ref 3)*

interpretation of the different modes of decohesion, schematic drawings of the mechanisms are shown in Fig. 20. This has been supplemented by microscopic observations of these processes by light or scanning electron microscopy.

The high wear rate of PTFE can be explained from its molecular structure and deformation behavior. In the intermediate temperature range, the stiff molecules can be aligned and can shear in the direction of sliding. Subsequently the very weak intermolecular bonding forces allow easy separation by a delamination process. The behavior of PA indicates that a higher friction coefficient provides a larger shear force to form a deformed surface layer. However, delamination of this layer is much more difficult because of the stronger intermolecular forces. Adhesion of H_2O molecules and their penetration into the surface are additional factors which affect friction, surface deformation, and wear of PA in a complicated manner.

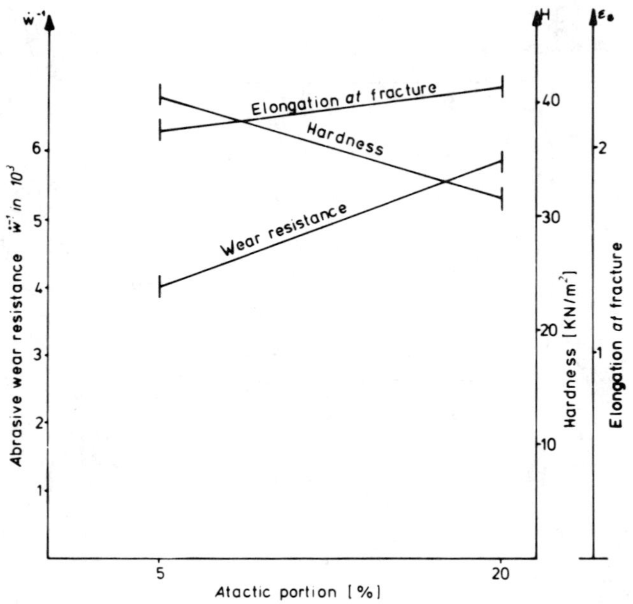

Fig. 17. Abrasive wear resistance of PP as a function of atactic admixture (compare Fig. 12): $f = 3.5 \cdot 10^4 \frac{N}{m^2}$, $v = 3.3 \cdot 10^{-3} \frac{m}{s}$.

The optimum compromise between the contradictory consequences of bond energy is at present achieved with high-molecular-weight polyethylene. Because of its symmetrical molecular structure the surface energy, and consequently the coefficient of friction, is only a little higher than that of PTFE. However, a high crystallinity and high density of the somewhat stronger intermolecular bonds produce better intermolecular cohesion. This effect can be supplemented further by using molecules of extremely large size (high-molecular-weight PE).

In addition to its low surface energy PTFE shows a second favorable physical property: a high melting temperature (Table 1). While PA, PE, and similar polymers show a high increase of friction and wear if temperatures of about 100°C are approached, this is not observed for PTFE. Its melting temperature is much higher than that of most other polymers. In addition, the rapid increase of wear caused by sticking is reduced for PTFE in the whole temperature range because of its low surface energy. A similar behavior is observed for polyimides (PI) (Fig. 16).[3] The friction characteristics of PTFE will provide for many applications of this material in spite of its moderate wear resistance, which seems to be unavoidable.

Combination of Friction and Wear

Three basic requirements are commonly encountered for the application of poly-

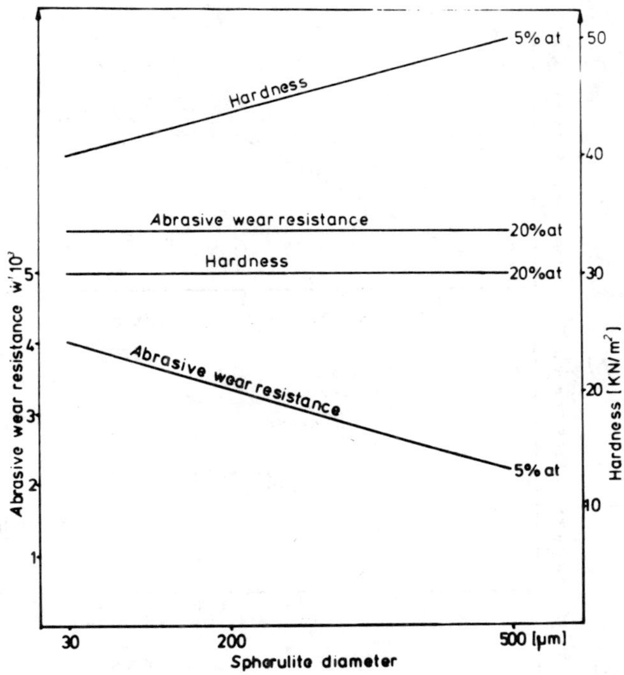

Fig. 18. Abrasive wear resistance of PP (different tacticity) as a function of morphology (see Fig. 7 and 14); $f = 3.5 \cdot 10^4 \frac{N}{m^2}$, $v = 3.3 \cdot 10^{-3} \frac{m}{s}$.

mers in sliding systems:

(*a*) $\mu \Rightarrow$ min., : PTFE
(*b*) $\dot{w} \Rightarrow$ min., : PA
(*c*) $\mu \cdot \dot{w} \Rightarrow$ min., : HDPE (high density polyethylene)

It could be shown that there is not a unique polymeric material to fulfill all these demands simultaneously. If a low frictional force is required for a wide range of temperatures, and if wear can be ignored (*a*), PTFE provides the optimum, mainly because of its low surface energy and high melting temperature. If a low wear rate at ambient temperature is most important, and if the magnitude of frictional forces is of secondary importance (*b*), PA will be suitable. Its most important feature is the high intermolecular cohesive forces which act against separation of wear particles, produce a high yield strength, and favor the formation of a lubricating layer in the surface by adhesion of water.

An optimum combination of the two properties is desirable for many applications [(*c*); Fig. 21]. This is, however, difficult to achieve, because weak intermolecular bonds are the prerequisite for a low surface energy, and therefore low friction. Strong bonds are required for high cohesion in the interior and high wear resistance. At the present stage of development one uses a species of molecules

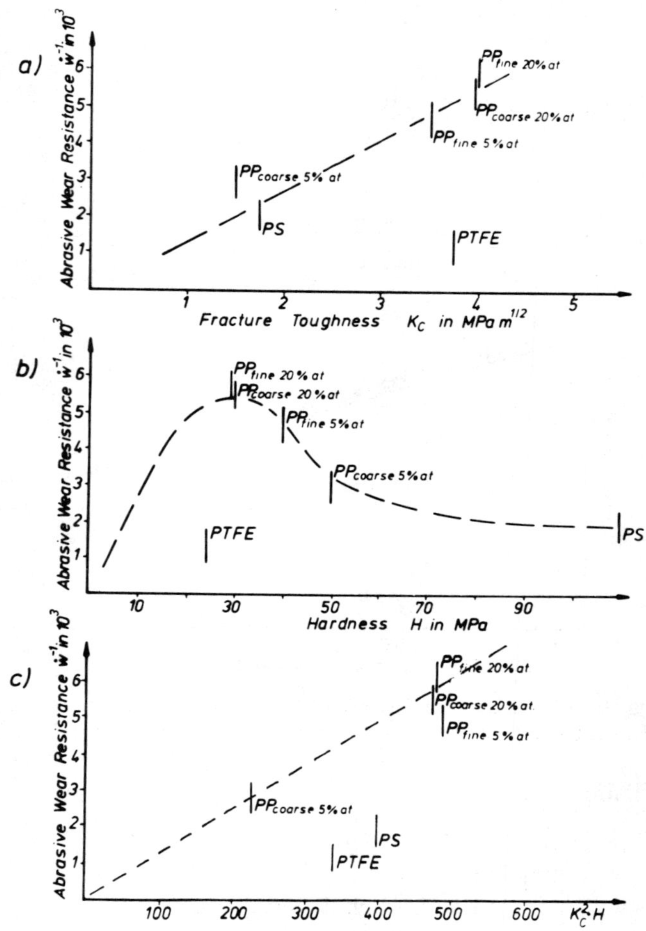

Fig. 19. Correlation of wear resistance with **(a)** fracture toughness, K_c; **(b)** hardness, H; **(c)** $K_c^2 \cdot H$.

which forms bonds that are relatively weak, but not too weak (PE). In HDPE a maximum bond density is achieved by crystallization, and entanglement is achieved by ultra-high molecular weights.

An ideal material should have a structure with weak bonds acting through the surface and with strong ones in the interior (Fig. 22). A material that preserves this structure while it is worn is not at present in sight. One possible way would be to form an adhesive layer by reactions with the environment. Such a layer should shield the strong bonds in the surface from the sliding material, and it should re-form continuously as soon as it has been removed. If the properties have to be independent of the environment (e.g., humidity), micromolecules

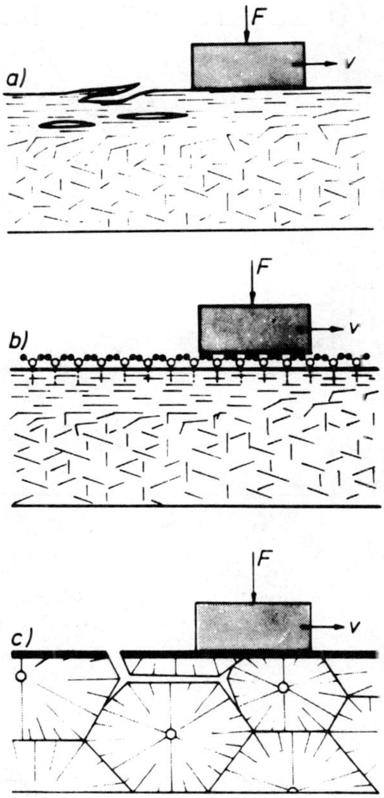

Fig. 20. Schematic representation of wear mechanisms in different polymeric structures and morphologies: **(a)** molecular alignment and delamination, fine crystalline structure, PTFE, PE, PP; **(b)** reduction of adhesive shear stress by adhesive layer of H_2O molecules, PA; **(c)** interspherulitic cracking in coarse spherulitic structure, PP.

that re-form the shielding layer by diffusion to the surface could be added tc the polymer.

Science and technology for the development of polymers with good friction and wear properties is not yet well developed. An improved understanding of the microstructural basis of these properties will surely provide us with new and improved materials. The potential for further lowering of the friction coefficient is minute because this depends only moderately on morphology. There is, however, a wide scope for improvements of the wear resistance by modifying molecules and molecular arrangements.

Summary

The origin of unlubricated friction and wear of polymers has been discussed, with emphasis on thermoplastic materials. Most of the experimental results which are reported were obtained with PTFE, PE, PP, PA, and PS (polystyrene).

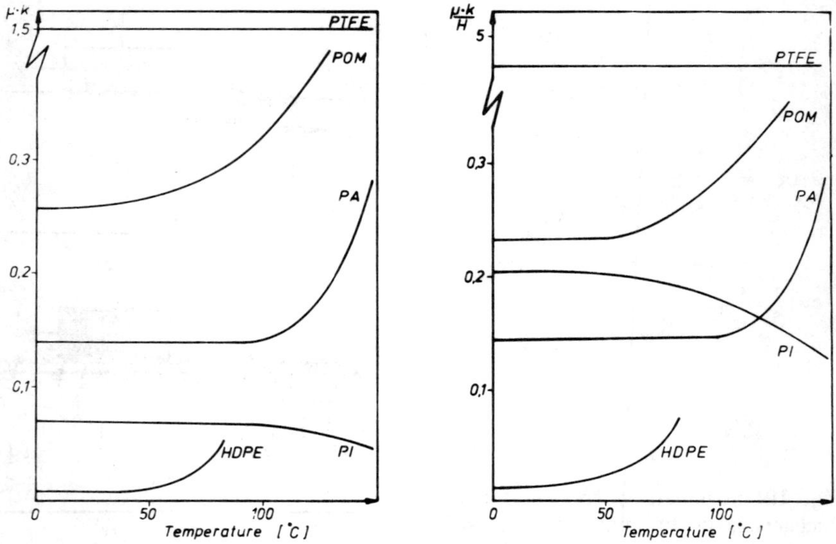

Fig. 21. Comparison of different polymers with respect to the combination of friction and wear as a function of temperature: **(left)** $\mu \bullet k = f(T)$; **(right)** $\mu \bullet \frac{k}{H} = f(T)$ (H = hardness at 20°C).

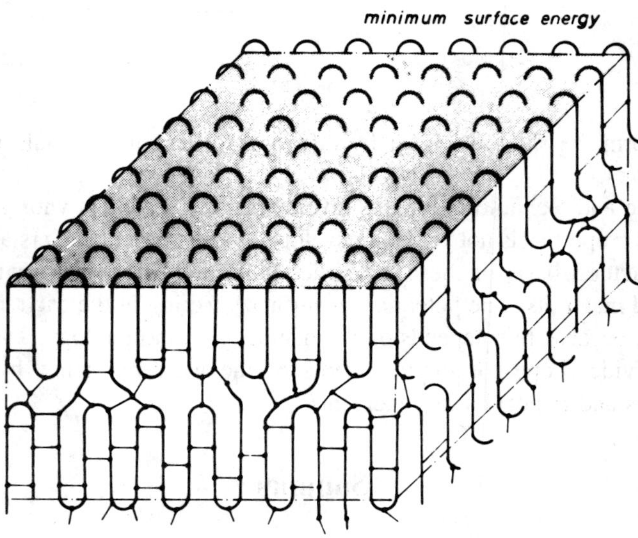

Fig. 22. Hypothetical structure of a polymer for minimum friction and wear.

The systematic discussion is based on the factors which affect the coefficient of friction: surface energy and elastic and plastic deformation, including hardness. For the understanding of wear, additional mechanical properties are required which characterize separation processes: crack initiation, and subcritical and critical crack growth. The differences between the friction and wear of metals and polymers are due to the low surface energy, high oxidation resistance, low hardness, low heat conductivity, and strong temperature dependence of all mechanical properties of polymers. Macroscopic properties are explained from molecular structure (type of polymers) and morphological structure (heat treatment). A structural basis has been used to discuss the limits of friction and wear properties. Finding materials which combine a low friction coefficient and high wear resistance is the most important concern for research in this field.

Acknowledgment

Thanks are due to DFG (Deutsche Forschungs Gemeinschaft, Schwerpunktprogramm: Physikalische Grundlagen des Fließ- und Deformationsverhaltens von Polymeren, Ho 325/13) for supporting this work. We gratefully acknowledge that the material used for our investigations was provided by BASF, D-6700 Ludwigshafen, and the information and pictures contained in Fig. 1 by M.A.N. Maschinenfabrik, Augsburg-Nürnberg, Aktiengesellschaft Unternehmensbereich GHH Sterkrade.

References

1. H. Uetz, V. Hakenjos, "Gleitreibungs- und Gleitverschleißversuche an Kunststoffen," Kunststoffe 59 (1961) 161.
2. K. Tanaka, Y. Uchiyama, S. Toyooka, "The Mechanism of Wear of PTFE," Wear 23 (1973) 153.
3. G. Erhard, E. Strickle, "Gleitelemente aus thermoplastischen Kunststoffen," Teil 1-3, Kunststoffe 62 (1972) 2, 232, 282.
4. R. P. Steijn, "Friction and Wear of Plastics," Metals Engineering Quarterly, ASM, May (1967) 9.
5. G. Halach, "Gleitreibungs- und Gleitverschleißverhalten von Kunststoffen in Abhängigkeit von verschiedenen Einflußgrößen," VDI–Gesellschaft Kunststofftechnik Jahrbuch, 1975.
6. J. J. Bikerman, "Sliding Friction of Polymers," J. Macromol. Sci.–Revs. Macromol. Chem., C 11 (1) (1974) 1.
7. M. M. Khrushchov, "Principles of Abrasive Wear," Wear 28 (1974) 69.
8. G. Fleischer, "Energetische Methode der Bestimmung des Verschleißes," Schmierungstechnik 4 (1973) 9.
9. H. Schonhorn, H. J. Frisch, G. L. Gains, Jr., "Surface Modification of Polymers and Practical Adhesion," Polymer Engineering and Science, July (1977) Vol. 17. No. 7, 440.
10. J. K. Krzeminski, E. Kocoalska, "Struktur und Eigenschaften von durch Reibung modifizierten Kunststoff–Oberflächen," Kunststoff 69 (1979) 482.

11. J. P. Giltrow, "A Relationship Between Abrasive Wear and the Cohesive Energy of Materials," Wear 15 (1970), 71.
12. H. Uetz, H. Breckel, "Reibungs- und Verschleiβversuche mit PTFE," Wear 10 (1967) 185.
13. K. Tanaka, T. Miyata, "Studies on the Friction and Transfer of Semi-Crystalline Polymers," Wear 41 (1977) 383.
14. K. Tanaka, Y. Uchiyama, "Friction, Wear and Surface Melting of Crystalline Polymers," Advances in Polymer Friction and Wear, ed. L. H. Lee, Plenum Press (1974).
15. K. Tanaka, S. Ueda, "The Mechanism of Wear of Polytetrafluoroethylene above the Melting Point," Wear 39 (1976) 323.
16. K. C. Ludema, D. Tabor, "The Friction and Visco-Elastic Properties of Polymeric Solids," Wear 9 (1966) 329.
17. S. Bahadur, K. C. Ludema, "The Viscoelastic Nature of the Sliding Friction of Polyethylene, Polypropylene and Copolymers," Wear 18 (1971) 109.
18. C. I. Chung, W. J. Hennessey, M. H. Tusim, "Frictional Behavior of Solid Polymers on a Metal Surface at Processing Conditions," Polymer Engineering and Science, January (1977), Vol. 17, No. 1, 9.
19. D. Tabor, D. E. Wynne Williams, "The Effect of Orientation on the Friction of PTFE," Wear 4 (1961) 391.
20. V. A. Beyi, V. V. Nevzorov, V. G. Savkin, A. I. Sviridenok, "Influence of Technological Factors on the Supramolecular Structure and Frictional Properties of the Surface Layers of Thermoplastics," Polymer Mechanics 12 (1976) 173.
21. G. V. Vinogradov, G. M. Bartenev, A. I. El' Kin, Yu. G. Yanowsky, V. N. Nikolayev, E. I. Frenkin, "Friction Properties of Crystalline Polymers," Brit. J. Appl. Phys., Ser. 2, Vol. 2, (1969) 1687.
22. V. A. Bely, V. G. Savkin, A. I. Sviridyonok, "Effect of Structure on Polymer Friction," Wear 18 (1971) 11.
23. G. V. Vinogradov, G. M. Bortenev, A. I. El 'Kin, V. K. Mikhaylov, "Effect of Temperature on Friction and Adhesion of Crystalline Polymers," Wear 16 (1970) 213.
24. R. Cortellucci, C. J. Heim, T. D. Koshy, P. J. Phillips, "Abrasion of Plastics," Wear 47 (1978) 397.
25. V. A. Belyi, A. I. Sviridyonok, V. A. Smurugov, V. V. Nevzorov, "Adhesive Wear of Polymers," Journal of Lubrication Technology, October 1977, 396.
26. D. C. Evans, J. K. Lancaster, "The Wear of Polymers," Treatise on Materials Science and Technology, Vol. 13, Academic Press, 1979.
27. M. Clerico, V. Patienno, "Sliding Wear of Polymeric Composites," Wear 53 (1979) 279.
28. M. K. Kar, S. Bahadur, "Micromechanism of Wear at Polymer-Metal Sliding Interface," International Conference on Wear of Materials, St. Louis, 1977.
29. J. K. Lancaster, "Friction and Wear," Polymer Science, ed. A. D. Jenkins, North-Holland Publishing Company, 1972.
30. H. Hosenthien, "Versuch der Berechnung des durch Werkstoffdämpfung bedingten Anteils an der Reibungszahl bei technisch trockener Reibung von Kunststoff gegen Stahl," Dissertation Universität Stuttgart, 1971.
31. B. J. Briscoe, D. Tabor, "Friction and Wear of Polymers," Polymer Surfaces, ed. D. T. Clark, W. J. Feast, John Wiley & Sons, 1978.
32. E. Hornbogen, "The Role of Fracture Toughness in the Wear of Metals," Wear 33 (1875) 251.

33. E. Hornbogen, "Metallurgical Aspects of Wear," VDI Fortschrittsberichte Reihe 5, No. 24, 1976.
34. K. H. Zum Gahr, D. V. Doane, "Optimizing Fracture Toughness and Abrasion Resistance in White Cast Irons," Metallurgical Transactions A 11A, April (1980) 613.
35. K. H. Zum Gahr, "How Microstructure Affects Abrasive Wear Resistance," Metal Progress, September 1979, 47.
36. A. Rosenfield, "Fracture Mechanics Applied to Sliding Wear," ASM Materials Science Seminar on Fundamentals of Friction and Wear of Materials, Pittsburgh, October 4-5, 1980.
37. H. Potente, R. Krüger, "Bedeutung polarer und disperser Oberflächenanteile von Plastomeren und Beschichtungsstoffen für die Haftfestigkeit von Verbundsystemen," Farbe und Lack 84 (1978) 72.
38. BASF-Datenblätter, BASF-Aweta KT/MEL, D-6700 Ludwigshafen.
39. H. Scholze, "Glas/ Natur, Struktur und Eigenschaften," Friedrich Vieweg & Sohn, Braunschweig, 1965.
40. E. Rabinowicz, "Friction and Wear of Materials," John Wiley & Sons, New York, 1965.
41. E. Hornbogen, "Werkstoffe," Springer-Verlag, 1979.

Appendix: List of Symbols and Abbreviations

A	Effective asperity area	m^2
A_o	Nominal area	m^2
d	Intermolecular spacing	m
$\pm f$	Specific compression force	Nm^{-2}
F	External compression force	N
$\pm f_{ad}$	Specific adhesion force	Nm^{-2}
F_R	Frictional force	N
H	Indentation hardness (Brinell, Vickers)	Nm^{-2}
HDPE	High density polyethylene	
k	Wear coefficient	–
PA	Polyamide	
PE	Polyethylene	
PI	Polyimide	
PP	Polypropylene	
PS	Polystyrene	
PTFE	Polytetrafluoroethylene	
R_t	Surface roughness	m
v	Velocity of sliding	ms^{-1}

$\dot{w}$	Wear rate; abraded mass of polymer per density of the polymer, sliding distance, and nominal contact area	–
$\dot{w}^{-1}$	Wear resistance	–
x	Variable distance of molecules	m
Δx	Thickness of viscous layers	m
μ	Coefficient of friction	–
μ_{ad}	Contribution of adhesion	–
μ_{def}	Contribution of deformation	–
σ_{AO}, σ_{BO}	Surface energy of A and B	Jm^{-2}
σ_{AB}	Interfacial energy between AB	Jm^{-2}
$\Delta\sigma$	Specific energy obtained by the reaction of two surfaces	Jm^{-2}
η	Effective viscosity in surface layer	$Pa \cdot s$

Wear Mechanisms in Ceramics

A. G. EVANS and D. B. MARSHALL
Department of Materials Science and Mineral Engineering
University of California
Berkeley, California 94720

Abstract

The mechanisms of material removal during wear are examined. Primary emphasis is placed on the lateral fracture mechanism, in an attempt to elucidate both its realm of importance and the concomitant material removal rates. Correlations with existent data are shown to be encouraging. The enhancement of material removal by heat generation in the plastic zone is examined and shown to be generally of minor significance. Some considerations of material removal below the fracture threshold are also presented.

Introduction

The wear of ceramics has been the subject of extensive empirical investigation.[1] For example, material removal rates have been measured on various materials as functions of the normal force, the horizontal (frictional) force, the wheel speed, etc. However, the detailed mechanisms of material removal have been only cursorily explored. Yet some understanding of mechanisms is an essential prerequisite to the development of optimum wear characteristics, especially for complex processes such as abrasive wear. The intent of the present paper is to examine possible mechanisms of material removal.

Several investigators[2-4] have recently recognized the important role of lateral fracture (Fig. 1a) in material removal. This mechanism of wear is explored in some detail, to determine both its realm of importance and the parameters that might influence the removal rate.

Lateral cracking occurs above a threshold force.[5,6] This threshold coincides with a transition from a relatively severe lateral fracture dominated material removal process at higher force levels to other less damaging wear mechanisms. The fracture threshold is thus an important concept in the study of wear, and the

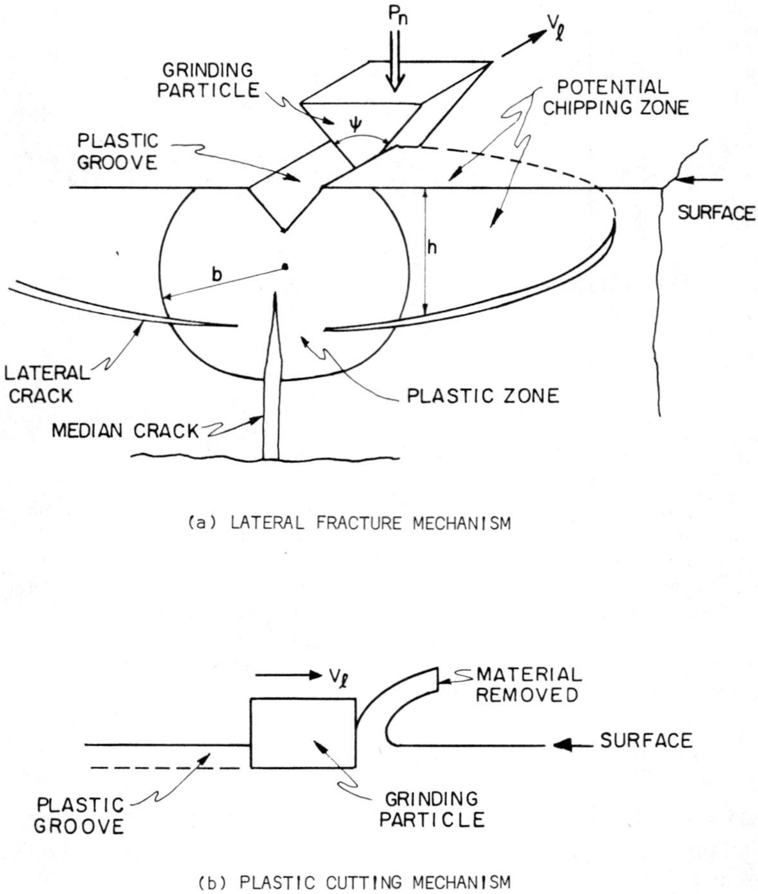

Fig. 1. Schematic of the mechanisms of material removal by lateral fracture and plastic cutting.

nature of the threshold should be given emphasis. The material removal mechanism below the fracture threshold is at present uncertain. An operative (although not necessarily exclusive) mechanism involves the plastic cutting that may occur during the grooving process (Fig. 1b).

The material removal rate above the fracture threshold evidently depends on the extent of the lateral fracture. An analysis of lateral crack extension will thus be used to predict wear rates as a function of the dominant material variables. Data for a range of materials will be correlated with the predictions.

The plastic flow associated with surface penetration must lead to a local temperature increase. This temperature rise inevitably results in thermal stresses that enhance the extension of the lateral cracks. The importance of thermal effects vis-à-vis residual stress motivated lateral fracture is examined in the third section of the paper. Then, some considerations of material removal below the fracture threshold are presented. Finally, some general implications for the development of wear resistant materials are discussed from the perspective provided by the present analyses.

The Lateral Fracture Model

The Threshold

The fracture threshold is a phenomenon with major significance for the control of wear, because it signifies a change in wear mechanism and a concomitant order-of-magnitude increment in wear rate. The general characteristics of fracture thresholds, at sites of plastic penetration, are reasonably well understood,[5-7] although detailed quantitative predictions are subject to some uncertainty. Of particular interest to the wear mechanism transition is the minimum threshold load, P*. This threshold load pertains in the presence of an adequate spatial density of crack precursors in the vicinity of the plastic zone,[6] a situation that is expected to prevail near surfaces subject to wear damage. The minimum threshold is dictated by the hardness, H, toughness, K_c, and elastic modulus, E, of the material, in accord with the relation

$$P^* = \zeta \left(\frac{K_c^4}{H^3}\right) f(E/H) \qquad [1]$$

where the function f(E/H) and the dimensionless constant ζ depend upon the type of crack. For lateral cracks, recent analysis[6] indicates that f is a slowly varying function of E/H and that $\zeta f(E/H) \approx 2 \times 10^5$. Threshold predictions for lateral fracture in several important material systems are summarized in Table 1. The threshold load levels are relatively low, but loads of this magnitude can exist at individual wear particles, depending upon the nature of the wear medium, the contact points, and the constraints imposed on the wear system. There have been few experimental studies of lateral fracture initiation, and validation of the predicted threshold characteristics is not possible at this time. Such studies are now in progress. However, it is noted that predictions of the radial fracture threshold using the same analytic procedures are in good accord with the more comprehensive experimental observations that exist for this system.[6,7]

The Lateral Crack Extension

Lateral cracks develop exclusively from the residual stress created by indentation.[8] The maximum extension of the crack is thus realized when the penetrating particle is removed. The cracking behavior can be analyzed by regarding the plastic zone as a precompressed spring which drives crack extension.[9,10] The crack advance is then dictated by the compliance of the material above the crack,

Table 1. Lateral Crack Nucleation Estimates(a)

Material	Toughness, $MPa\sqrt{m}$	Hardness, GPa	Threshold, N
Al_2O_3	2.2	20	0.4
Glass	0.7	6	0.2
MgO	1.1	8	0.9
Si_3N_4	5 (polycrystal)	16	31
Si_3N_4 (estimate)	~2 (single crystal)	16	~0.8

(a) The threshold in most brittle materials occurs at load levels wherein the scale of the cracks is on the order of (or smaller than) the grain size.[6] Single crystal values of toughness are thus pertinent unless the material is particularly fine grained. The thresholds are thus based on single crystal toughness levels unless otherwise stated.

treated as a thin elastic plate clamped at its outer edge to a rigid substrate. The specific crack extension depends, in part, upon the morphology of the lateral crack. For axisymmetric indentation (most pertinent to intermittent particle contact) the lateral crack length, c, is given by[10]

$$c = c^L \left[1 - (P_o^*/\hat{P})^{1/4}\right]^{1/2} \qquad [2]$$

where $\hat{P}$ is the peak load during particle penetration and P_o^* is an apparent threshold load for fully developed cracking. For peak loads well in excess of the threshold, the lateral crack length is given by

$$c \approx c^L = \alpha_1 \left(\frac{(E/H)^{2/5}}{K_c^{1/2} H^{1/8}}\right) \hat{P}^{5/8} \qquad [3a]$$

where α_1 is a material-independent constant that depends on the particle shape and can be determined by calibration on a material with well-characterized fracture properties (e.g., glass). For a linear plastic groove (as might pertain for a sliding particle contact) the equivalent result (Appendix, Eq A-9) is

$$c = \alpha_2 \left[\frac{(E/H)^{3/5}}{K_c^{1/2} H^{1/8}}\right] \hat{P}^{5/8} \qquad [3b]$$

where α_2 is another material-independent constant ($\alpha_2 \approx \alpha_1$). The only significant difference between the two solutions for c occurs in the (E/H) exponent. But, since the range of (E/H) values for most ceramics is less than an order of magnitude, the influence of the grinding morphology upon crack extension is considered to be of minor importance.

For present purposes, the indentation forces are considered to be sufficiently in excess of the apparent threshold to permit the direct use of Eq 3a and 3b for the prediction of crack extension. The wear behavior under these conditions will exhibit relatively simple functional dependencies on the dominant material variables (as described in the subsequent section). The predictions can thereupon be straightforwardly correlated with experimental results. The cracking behavior at intermediate loads is less extensive than that predicted from Eq 3a and 3b, and hence the wear predictions based on these equations represent an upper bound wear rate. A more complete analysis of crack extension at intermediate loads can be conducted if experimental correlations indicate that some important trends are inadequately predicted by neglect of the apparent threshold.

The above solutions for the lateral crack length pertain to conditions wherein all of the plastically deformed material is retained within the plastic zone, and thus they represent an upper bound crack driving force. When some of the plastically deformed material is removed by plastic cutting, the residual forces will be reduced and hence the lateral crack extension diminished. It is not known at this time whether the relative alleviation of the residual forces by cutting exhibits important material dependencies. Such effects would emerge from inadequate correlations of experimental results (between materials) with models based on Eq 3a and 3b.

The Material Removal Rate

The material removal analysis considers the extension of the lateral cracks normal to the motion of the abrading or sliding particle (Fig. 1a), since this distance defines the width of the potential material removal zone. The situation to be modelled is defined in Fig. 1(a); it consists of a particle or an asperity subject to a normal force P_n, moving with a prescribed lateral velocity v_ℓ.

The volume V_i of material removed by the passage of each particle that exceeds the fracture threshold force P^* is determined by the extent c_i and the depth h_i of the lateral fracture. The upper bound volume removal is

$$\hat{V}_i = 2h_i c_i \ell_i \qquad [4]$$

where ℓ_i is the distance of motion. The lateral fracture depth typically scales with the plastic zone radius b,[6,10] which, in turn, relates to the plastic indentation width, a, by[6]

$$b \sim a(E/H)^{2/5}\ , \qquad [5]$$

and the indentation width is related to the hardness and the peak normal penetration force by

$$a^2 \sim (\hat{P}_n/H) \qquad [6]$$

Substituting Eq 3a, 5, and 6 into Eq 4, the volume removed becomes

$$\hat{V}_i = \alpha_3 \frac{\hat{P}_n^{9/8}}{K_c^{1/2} H^{5/8}} (E/H)^{4/5} \ell_i \qquad [7]$$

where α_3 is a material-independent constant.

Implicit in the preceding analysis is the dominant influence of the normal force P_n upon the material removal process. The lateral force $P_\ell = \mu P_n$ (presumed to be dictated by the resistance of the material to plastic grooving) is considered incidental to material removal by lateral fracture. Some substantiation of this premise will emerge later from experimental correlations. At this time it is noted that, by analogy with the equivalent elastic problem,[11] the lateral force may tend to rotate the stress field in the direction of P_ℓ, in the sense that the tension is enhanced behind the contact zone and suppressed ahead of the contact. However, even if this occurs, there should be no appreciable influence on the stresses contained in the *plane* normal to P_n, stresses which have the primary effect on lateral crack extension. Hence, under quasi-static conditions, the width of the lateral fracture zone should be relatively insensitive to the magnitude of the lateral force.

At relatively high lateral velocities, local instabilities tend to develop. These instabilities presumably occur following a fracturing event and result in an abrupt increase in the local particle velocity, as manifest in the lateral force oscillations typically observed in studies of the motion of abrading particles. The impact events that result from the instabilities are likely to have impulse components both normal and parallel to the surface. The normal component affects the penetration, and thus has the more significant influence on the material removal process. The normal velocity, and the consequent magnitude of the lateral fracture, may depend upon the amplitude of the instability and hence indirectly upon the friction coefficient. These complexities may be of secondary significance, and the existence of dynamic effects should be sought in experimental results before embarking on an analytic study of the role of friction in the wear process.

Correlations With Experiment

Experimental results suitable for correlation with the present wear predictions include measurements of the influence of the normal force on the wear rate (observed for a specific material) and, more significantly, studies of the trends in the wear rate among different materials.* The influence of material properties on wear can be compared with the lateral fracture model either from measurements of the wear forces at a specified (constant) volume rate of removal or from removal rates at a constant level of applied force.

Direct measurements of the lateral crack width have been conducted on MgZn ferrite as a function of the normal force.[3] These results are consistent with the force exponent of 5/8 predicted for the lateral crack extension (Eq 3 and Fig. 2). The influence of the normal force on the volume removal rate has been studied for several materials by Wilshaw and Hartley.[12] These results (Fig. 3) indicate a

*Although all experimental results have been obtained on ceramic materials, the application of the model to hard metallic materials cannot be discounted.

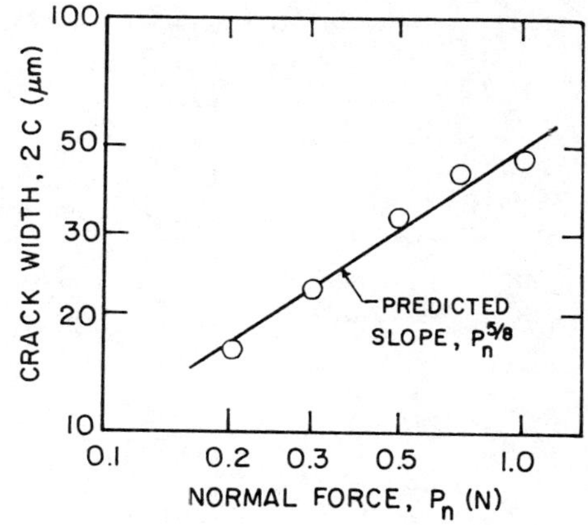

Fig. 2. Logarithmic plot showing the influence of the normal force on the extent of lateral fracture. *(Ref 3)*

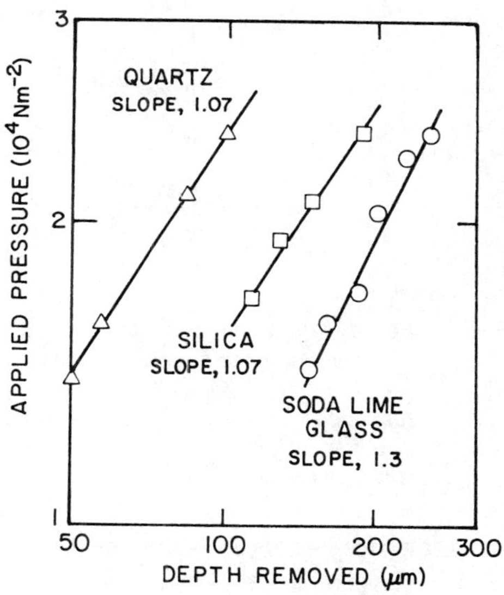

Fig. 3. Effect of the applied force on the material removal. *(Ref 12)*

force exponent that exceeds unity, as anticipated by the present analysis (Eq 7). The experimentally observed exponent ranges between 1.07 and 1.3.

The most comprehensive material removal data suitable for comparison with the present predictions are grinding data obtained by Koepke and Stokes.[13] These data comprise force measurements determined on different materials for constant volume removal conditions. For up-grinding conditions, which best simulate the behavior analyzed in the preceding section (because fracture chips are directly ejected from the system), the force data for a wide range of materials (Fig. 4) are in remarkable accord with the predicted material dependence, $(K_c^{1/2}H^{5/8})^{8/9}$. The superior correlation exists for the dependence on the normal force, while a more

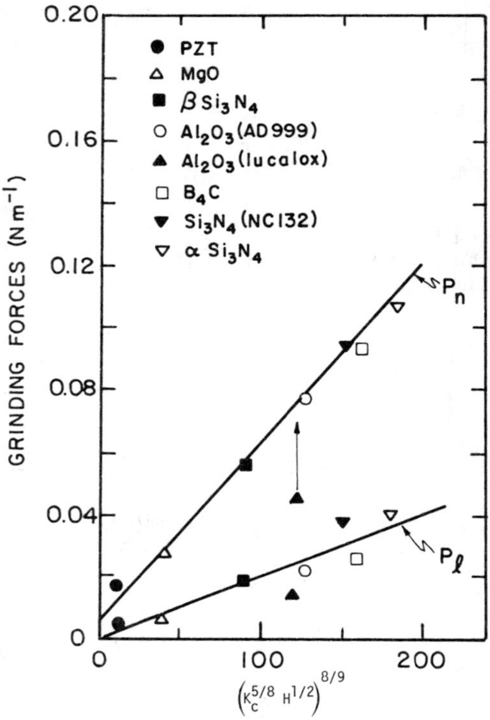

Fig. 4. Correlation of grinding forces at constant removal rates *(Ref 13)* with the toughness/hardness parameter predicted by the present analysis. The toughness values are in $MPa\sqrt{m}$ and the hardness values are for Knoop hardness. Because the data were taken with a multi-point grinding wheel, the measured grinding forces are normalized by the width of the cut.

modest correspondence obtains for the lateral force. These relative correlations tend to substantiate that the development of a lateral force is incidental to the material removal process. The one material that deviates from the correlation is Lucalox Al_2O_3. This is a relatively coarse grained Al_2O_3 that also exhibits unusual indentation fracture characteristics.[14] Under indentation conditions extensive grain boundary separation occurs, rather than the formation of discrete crack systems. Grinding in this (and similar) materials could thus proceed by grain ejection, and hence the material removal would be expected to exhibit a different dependence on material properties than anticipated by the lateral fracture model.

Machining data at constant force are also available for a range of materials.[15] However, the data exhibit an appreciable grain size dependence. This behavior presumably arises because the lateral fracture in the coarser grained materials extends over a relatively small number of grains, and hence the K_c and H values required for the volume removal predictions are smaller than the isotropic polycrystalline values. It is expedient, therefore, to conduct the comparison at the small extreme of grain size for which available polycrystalline material parameters are most likely to pertain. The resultant correlation is summarized in Fig. 5. The correct trends are predicted by the material parameter, $K_c^{-1/2}H^{-5/8}$, although the correlation is not as satisfying as that evident for the force measurements.

The relatively consistent trends in both the material removal rate and the grinding forces with the material parameters suggested by the simple lateral crack analysis indicate that most of the proposed complexities, if they occur, do not impose additional material-dependent influences upon the wear process.* Dynamic effects and residual force relaxation by plastic removal can presumably, therefore, be considered to introduce only small perturbations upon the wear characteristics anticipated from the quasi-static lateral fracture analysis, at least for the test conditions used and for materials within the range of properties encompassed by the existent experimental results.

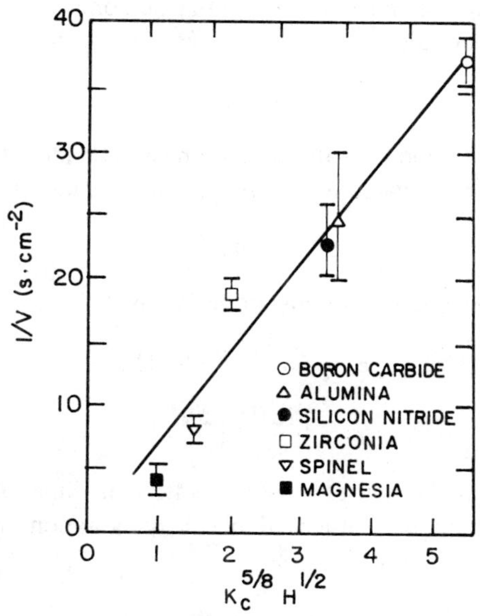

Fig. 5. Correlation of material removal rates at constant force *(Ref 15)* with the toughness/hardness parameter predicted by the present analysis.

*Recent abrasive wear studies reported by Moore and King[16] indicate a poor experimental correlation based on $K_c^{5/8}H^{1/2}$. However, the toughness values determined in the analysis are substantially inaccurate (e.g., 3.8 MPa$\sqrt{m}$ for soda lime glass compared with a generally accepted value of 0.7 MPa$\sqrt{m}$, 9.1 MPa$\sqrt{m}$ for Si_3N_4 compared with 4-5 MPa$\sqrt{m}$, etc.). Re-analysis of their results is thus needed before embarking on an assessment of the inadequacies of the simple lateral fracture model.

Thermal Effects

Heat is invariably generated during a grinding operation. This heat derives from the plastic work expended by the penetration of the abrading particle. The heat is generated within the plastic zone and leads to an expansion of this zone vis-à-vis the surrounding elastic material. The zone expansion is directly analogous to the effective expansion induced by plastic penetration.[9,10] Hence, heat generation can be considered to yield a lateral crack driving force that is simply superimposed on that associated with the residual indentation field. The ratio of crack driving forces attributed to these two sources of zone expansion will determine their relative influence upon the wear rate, by lateral fracture. In addition, it is noted that the zone expansion induced by heating creates shear stresses outside the isothermal plastic zone. These stresses tend to enlarge the plastic zone, increasing its volume from V to V*. This zone enlargement also contributes to the crack driving force ratio.

The work done in creating an indentation of volume ΔV is, simply, $H\Delta V$.[6] If this work is dissipated as heat within the plastic zone, the average temperature rise θ for adiabatic conditions is

$$\theta = H\Delta V/\rho c_p V^* \qquad [8]$$

where ρ is the density and c_p is the specific heat. The local temperature rise ΔT at a radial location r is related to the average temperature θ by[17]

$$\Delta T = 3\theta \ell n(b/r) \qquad [9]$$

The adiabatic unconstrained volume strain is thus†

$$\begin{aligned} e_T &= 3\alpha \int_{r=b}^{r=0} (r/b)^3 d\Delta T \\ &= \left(\frac{3\alpha H}{\rho c_p}\right)\left(\frac{\Delta V}{V^*}\right) \qquad [10] \end{aligned}$$

where α is the linear thermal expansion coefficient. Since $\Delta V/V$ is the unconstrained volume strain of isothermal plastic penetration, the adiabatic crack driving force ratio becomes

$$R = e_T/(\Delta V/V) = (3\alpha H/\rho c_p)\,(V/V^*) \qquad [11]$$

Inserting some typical values for the material parameters ($H \approx 10$ GPa, $\alpha \approx 5 \times 10^{-6} K^{-1}$, $\rho \approx 3 \times 10^3$ kgm^{-3}, $c_p \approx 600$ $Jkg^{-1}K^{-1}$) indicates that R is ~0.1 when $V^* \approx V$. Usually, therefore, thermal effects would be expected to exert a minor influence on the wear rate. A significant thermal effect may

†It is noted that the assumption of a uniform temperature rise θ within the plastic zone yields an identical result.

emerge, however, in hard materials that exhibit a large thermal expansion coefficient and a low specific heat.

Sub-Threshold Wear Rates

Below the lateral fracture threshold, material is removed by a plastic cutting process (Fig. 1b). The material removal rate is presumably dictated by the plastic penetration and should thus be uniquely related to the material hardness, the normal force, and the particle geometry:

$$V_i \sim \frac{\hat{P}_n \ell_i}{H} \cot(\psi/2) \qquad [12]$$

where ψ is the included angle of the penetrating particle. The volume removal rate induced by plastic cutting is about an order of magnitude smaller than that associated with the lateral fracture mechanism.[16] Appreciable wear resistance is thus experienced with this regime. The choice of wear conditions that exclude lateral fracture is evidently a highly desirable material/design objective.

Implications and Conclusions

The analysis of fracture-dominated wear processes in brittle solids indicates a reasonably good correlation with available experimental data. This correlation substantiates the strong importance of both the fracture toughness and the hardness upon the wear rate, high toughness and high hardness both being desirable for optimum wear resistance. A small influence of the thermal parameters of the material is also anticipated, but only for hard materials with a large thermal expansion coefficient or a small specific heat.

A major diminution in wear rate is associated with force levels below the fracture threshold. The force at the fracture threshold is strongly influenced by both the hardness and the toughness, such that fracture can be suppressed by a high toughness or a low hardness. Below this threshold, the material removal rate is substantially diminished and depends principally upon the hardness.

In general, materials with an optimum combination of high toughness and high hardness will be the most wear resistant. However, there is some ambiguity associated with the role of the hardness. Although a high hardness minimizes the wear rate both below and above the fracture threshold, the threshold force is diminished by an increase in hardness. Situations can arise, therefore, when a hardness increase in the vicinity of the threshold leads to an enhanced wear rate, as depicted in Fig. 6.

The influence of the fracture toughness also requires some qualification. The toughness needed to resist lateral fracture must be developed on the appropriate scale. Specifically, the microstructural scale must be appreciably smaller than the

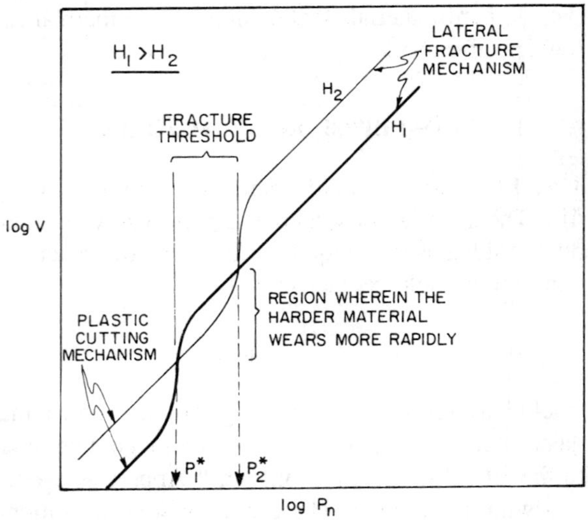

Fig. 6. Schematic indicating the wear rate as a function of normal force in two materials with a different hardness, H, but comparable levels of toughness, K_c, and of E/H.

length of the lateral fracture in order to experience the macroscopic fracture toughness. Some important effects of the dominant microstructural size (e.g., the grain size) are thus to be anticipated, especially at force levels in the vicinity of the threshold. These effects will require more detailed consideration on an individual basis.

Acknowledgment

The authors express their appreciation for the financial support of this research provided by the Office of Naval Research under Contract No. N0014-79-C-0159.

Appendix:

Lateral Crack Extension

Consider a two-dimensional lateral crack configuration created by a linear groove of length ℓ. The strain energy release rate is

$$\mathcal{G} = (P_r^2/2\ell)\, d\lambda/dc \qquad \text{[A-1]}$$

where P_r is the residual normal force exerted by the plastic zone on the surrounding elastic material, λ is the compliance of the elastic layer above the crack, and c is lateral crack length. Linear elasticity requires that[10]

$$P_r = P_r^o\,[1 - u_r/u_r^o] \quad \text{[A-2]}$$

where P_r^o is the initial residual force (at zero crack length), u_r is the opening displacement at the crack center, and u_r^0 is the opening displacement when the driving force relaxes to zero. For well-developed cracks (c>>h where h is the depth of the lateral fracture) the compliance may be approximated by*

$$\lambda = 4c^3/E\ell h^3 \quad \text{[A-3]}$$

Combining Eq A-1, A-2, and A-3, the stress intensity factor becomes

$$K \equiv \left(\frac{E\mathcal{G}}{1 - \nu^2}\right)^{1/2} = \left[\frac{6}{1 - \nu^2}\right]^{1/2} \frac{cP_r^o}{\ell h^3(1 + \beta c^3)} \quad \text{[A-4]}$$

where $\beta = (P_r^o/u_r^o)\,(4/E\ell h^3)$. Equating K to the critical value K_c, the lateral crack length becomes

$$1 + \beta c^3 = [6/(1 - \nu^2)]^{1/2}\, P_r^o\, c/\ell K_c h^{3/2} \quad \text{[A-5]}$$

which, to the same level of approximation used in Eq 3a, is

$$c \approx [6/(1 - \nu^2)]^{1/4}\,(Eh^{3/2}u_r^o/4K_c)^{1/2}. \quad \text{[A-6]}$$

Noting that (from Eq 5 and 6)

$$h \sim b \sim (E/H)^{2/5}(\hat{P}/H)^{1/2} \quad \text{[A-7]}$$

and that[9,10]

$$\begin{aligned} u_r^o &\sim \Delta V/\ell b \sim a^2/\ell b \\ &\sim (H/E)^{2/5}\,(\hat{P}/H)^{1/2} \end{aligned} \quad \text{[A-8]}$$

the lateral crack length becomes

$$c \sim \frac{(E/H)^{3/5}\,\hat{P}^{5/8}}{H^{1/8}\,K_c^{\,1/2}} \quad \text{[A-9]}$$

as in Eq 3b.

References

1. P. J. Gielisse and J. Stanisloa, The science of ceramic machining and surface finishing, NBS Special Technical Publication, *348*, 5 (1972).
2. M. V. Swain, Fracture mechanics of ceramics (ed. R. C. Bradt, D. P. H. Hasselman and F. F. Lange), Plenum, New York, *3*, 257 (1978).

---

*Eq A-3 is the standard solution for a thin elastic cantilever. The cantilever in the present model is disproportionately wide and may not extend to the edges of the material. The major influences of these deviations from ideal geometry are expected to be reflected in the numerical factor in Eq A-3 and in the introduction of a Poisson's ratio dependence.

3. J. D. B. Veldkamp, N. Hattu and V. A. C. Snyders, Fracture mechanics of ceramics, (ed. R. C. Bradt, D. P. H. Hasselman and F. F. Lange), Plenum, New York, *3*, 273 (1978).
4. A. G. Evans and T. R. Wilshaw, Acta Met., *24*, 939 (1976).
5. B. R. Lawn and A. G. Evans, J. Mater. Sci., *12*, 2195 (1977).
6. S. S. Chiang, D. B. Marshall and A. G. Evans, to be published.
7. J. Lankford and D. L. Davidson, J. Mater. Sci., *14*, 1662 (1979).
8. B. R. Lawn and M. V. Swain, J. Mater. Sci., *10*, 113 (1975).
9. B. R. Lawn, A. G. Evans and D. B. Marshall, J. Amer. Ceram. Soc. *63*, 574 (1980).
10. D. B. Marshall, B. R. Lawn and A. G. Evans, to be published.
11. G. M. Hamilton and L. E. Goodman, J. Appl. Mech., *33*, 371 (1966).
12. T. R. Wilshaw and N. E. W. Hartley, Proc. Third European Symposium on Comminution, Chemia GMBH, Weinheim/Bergestrasse (1972).
13. B. G. Koepke and R. J. Stokes, Science of Ceramic Machining and Surface Finishing, NBS Special Technical Publication, *562*, 75 (1979).
14. A. G. Evans, ASTM STP, *678* (1978) 112.
15. R. W. Rice and B. K. Speronello, J. Amer. Ceram. Soc., *59*, 330 (1976).
16. M. A. Moore and F. S. King, Wear, *60*, 123 (1980).
17. B. R. Lawn, B. J. Hockey and S. M. Wiederhorn, J. Amer. Ceram. Soc., *63*, 356 (1980).

Index

Primary emphasis in the index is on the mechanisms and theories of friction and wear. The many interconnecting paths of ideas and terminology expressed by the individual authors are recognized in an abundance of cross-references. Materials used in experiments or identified in accounts of experiences are mentioned when they are given special attention in text or illustrations. The symbol (F) following an entry signifies that information is presented in a figure; the symbol (T), in a table. —M.R.H.